Lecture Notes in Computer Science

Lecture Notes in Artificial Intelligence   13935

Founding Editor

Jörg Siekmann

Series Editors

Randy Goebel, *University of Alberta, Edmonton, Canada*
Wolfgang Wahlster, *DFKI, Berlin, Germany*
Zhi-Hua Zhou, *Nanjing University, Nanjing, China*

The series Lecture Notes in Artificial Intelligence (LNAI) was established in 1988 as a topical subseries of LNCS devoted to artificial intelligence.

The series publishes state-of-the-art research results at a high level. As with the LNCS mother series, the mission of the series is to serve the international R & D community by providing an invaluable service, mainly focused on the publication of conference and workshop proceedings and postproceedings.

Hisashi Kashima · Tsuyoshi Ide · Wen-Chih Peng
Editors

Advances in Knowledge Discovery and Data Mining

27th Pacific-Asia Conference
on Knowledge Discovery and Data Mining, PAKDD 2023
Osaka, Japan, May 25–28, 2023
Proceedings, Part I

 Springer

Editors
Hisashi Kashima ⓘ
Kyoto University
Kyoto, Japan

Wen-Chih Peng ⓘ
National Chiao Tung University
Hsinchu, Taiwan

Tsuyoshi Ide ⓘ
IBM Research, Thomas J. Watson Research
Center
Yorktown Heights, NY, USA

ISSN 0302-9743   ISSN 1611-3349 (electronic)
Lecture Notes in Artificial Intelligence
ISBN 978-3-031-33373-6   ISBN 978-3-031-33374-3 (eBook)
https://doi.org/10.1007/978-3-031-33374-3

LNCS Sublibrary: SL7 – Artificial Intelligence

General Chairs' Preface

On behalf of the Organizing Committee, we were delighted to welcome attendees to the 27th Pacific-Asia Conference on Knowledge Discovery and Data Mining (PAKDD 2023), held in Osaka, Japan, on May 25–28, 2023. Since its inception in 1997, PAKDD has long established itself as one of the leading international conferences on data mining and knowledge discovery. PAKDD provides an international forum for researchers and industry practitioners to share their new ideas, original research results, and practical development experiences across all areas of Knowledge Discovery and Data Mining (KDD). PAKDD 2023 was held as a hybrid conference for both online and on-site attendees.

We extend our sincere gratitude to the researchers who submitted their work to the PAKDD 2023 main conference, high-quality tutorials, and workshops on cutting-edge topics. We would like to deliver our sincere thanks for their efforts in research, as well as in preparing high-quality presentations. We also express our appreciation to all the collaborators and sponsors for their trust and cooperation.

We were honored to have three distinguished keynote speakers joining the conference: Edward Y. Chang (Ailly Corp), Takashi Washio (Osaka University), and Wei Wang (University of California, Los Angeles, USA), each with high reputations in their respective areas. We enjoyed their participation and talks, which made the conference one of the best academic platforms for knowledge discovery and data mining. We would like to express our sincere gratitude for the contributions of the Steering Committee members, Organizing Committee members, Program Committee members, and anonymous reviewers, led by Program Committee Co-chairs: Hisashi Kashima (Kyoto University), Wen-Chih Peng (National Chiao Tung University), and Tsuyoshi Ide (IBM Thomas J. Watson Research Center, USA). We feel beholden to the PAKDD Steering Committees for their constant guidance and sponsorship of manuscripts.

Finally, our sincere thanks go to all the participants and volunteers. We hope all of you enjoyed PAKDD 2023 and your time in Osaka, Japan.

April 2023   Naonori Ueda
                                                                           Yasushi Sakurai

PC Chairs' Preface

It is our great pleasure to present the 27th Pacific-Asia Conference on Knowledge Discovery and Data Mining (PAKDD 2023) as the Program Committee Chairs. PAKDD is one of the longest-established and leading international conferences in the areas of data mining and knowledge discovery. It provides an international forum for researchers and industry practitioners to share their new ideas, original research results, and practical development experiences from all KDD-related areas, including data mining, data warehousing, machine learning, artificial intelligence, databases, statistics, knowledge engineering, big data technologies, and foundations.

This year, PAKDD received a record number of 869 submissions, among which 56 submissions were rejected at a preliminary stage due to policy violations. There were 318 Program Committee members and 42 Senior Program Committee members involved in the reviewing process. More than 90% of the submissions were reviewed by at least three different reviewers. As a result of the highly competitive selection process, 143 submissions were accepted and recommended to be published, resulting in an acceptance rate of 16.5%. Out of these, 85 papers were primarily about methods and algorithms and 58 were about applications. We would like to thank all PC members and reviewers, whose diligence produced a high-quality program for PAKDD 2023. The conference program featured keynote speeches from distinguished researchers in the community, most influential paper talks, cutting-edge workshops, and comprehensive tutorials.

We wish to sincerely thank all PC members and reviewers for their invaluable efforts in ensuring a timely, fair, and highly effective PAKDD 2023 program.

April 2023

Hisashi Kashima
Wen-Chih Peng
Tsuyoshi Ide

Organization

General Co-chairs

Naonori Ueda	NTT and RIKEN Center for AIP, Japan
Yasushi Sakurai	Osaka University, Japan

Program Committee Co-chairs

Hisashi Kashima	Kyoto University, Japan
Wen-Chih Peng	National Chiao Tung University, Taiwan
Tsuyoshi Ide	IBM Thomas J. Watson Research Center, USA

Workshop Co-chairs

Yukino Baba	University of Tokyo, Japan
Jill-Jênn Vie	Inria, France

Tutorial Co-chairs

Koji Maruhashi	Fujitsu, Japan
Bin Cui	Peking University, China

Local Arrangement Co-chairs

Yasue Kishino	NTT, Japan
Koh Takeuchi	Kyoto University, Japan
Tasuku Kimura	Osaka University, Japan

Publicity Co-chairs

Hiromi Arai	RIKEN Center for AIP, Japan
Miao Xu	University of Queensland, Australia
Ulrich Aivodji	ÉTS Montréal, Canada

Proceedings Co-chairs

Yasuo Tabei   RIKEN Center for AIP, Japan
Rossano Venturini   University of Pisa, Italy

Web and Content Chair

Marie Katsurai   Doshisha University, Japan

Registration Co-chairs

Machiko Toyoda   NTT, Japan
Yasutoshi Ida   NTT, Japan

Treasury Committee

Akihiro Tanabe   Osaka University, Japan
Aya Imura   Osaka University, Japan

Steering Committee

Vincent S. Tseng   National Yang Ming Chiao Tung University,
                               Taiwan
Longbing Cao   University of Technology Sydney, Australia
Ramesh Agrawal   Jawaharlal Nehru University, India
Ming-Syan Chen   National Taiwan University, Taiwan
David Cheung   University of Hong Kong, China
Gill Dobbie   University of Auckland, New Zealand
Joao Gama   University of Porto, Portugal
Zhiguo Gong   University of Macau, Macau
Tu Bao Ho   Japan Advanced Institute of Science and
                               Technology, Japan
Joshua Z. Huang   Shenzhen Institutes of Advanced Technology,
                               Chinese Academy of Sciences, China
Masaru Kitsuregawa   University of Tokyo, Japan
Rao Kotagiri   University of Melbourne, Australia
Jae-Gil Lee   Korea Advanced Institute of Science &
                               Technology, Korea

Tianrui Li	Southwest Jiaotong University, China
Ee-Peng Lim	Singapore Management University, Singapore
Huan Liu	Arizona State University, USA
Hady W. Lauw	Singapore Management University, Singapore
Hiroshi Motoda	AFOSR/AOARD and Osaka University, Japan
Jian Pei	Duke University, USA
Dinh Phung	Monash University, Australia
P. Krishna Reddy	International Institute of Information Technology, Hyderabad (IIIT-H), India
Kyuseok Shim	Seoul National University, Korea
Jaideep Srivastava	University of Minnesota, USA
Thanaruk Theeramunkong	Thammasat University, Thailand
Takashi Washio	Osaka University, Japan
Geoff Webb	Monash University, Australia
Kyu-Young Whang	Korea Advanced Institute of Science & Technology, Korea
Graham Williams	Australian National University, Australia
Raymond Chi-Wing Wong	Hong Kong University of Science and Technology, Hong Kong
Min-Ling Zhang	Southeast University, China
Chengqi Zhang	University of Technology Sydney, Australia
Ning Zhong	Maebashi Institute of Technology, Japan
Zhi-Hua Zhou	Nanjing University, China

Contents – Part I

Classification

Clustering

Anomaly and Outlier Detection

BAARD: Blocking Adversarial Examples by Testing for Applicability, Reliability and Decidability

Xinglong Chang[1(✉)], Katharina Dost[1], Kaiqi Zhao[1], Ambra Demontis[2], Fabio Roli[3], Gillian Dobbie[1], and Jörg Wicker[1]

[1] The University of Auckland, Auckland, New Zealand
xcha011@aucklanduni.ac.nz,
{katharina.dost,kaiqi.zhao,g.dobbie,j.wicker}@auckland.ac.nz
[2] University of Cagliari, Cagliari, Italy
ambra.demontis@unica.it
[3] University of Genoa, Genoa, Italy
fabio.roli@unige.it

Abstract. Adversarial defenses protect machine learning models from adversarial attacks, but are often tailored to one type of model or attack. The lack of information on unknown potential attacks makes detecting adversarial examples challenging. Additionally, attackers do not need to follow the rules made by the defender. To address this problem, we take inspiration from the concept of Applicability Domain in cheminformatics. Cheminformatics models struggle to make accurate predictions because only a limited number of compounds are known and available for training. Applicability Domain defines a domain based on the known compounds and rejects any unknown compound that falls outside the domain. Similarly, adversarial examples start as harmless inputs, but can be manipulated to evade reliable classification by moving outside the domain of the classifier. We are the first to identify the similarity between Applicability Domain and adversarial detection. Instead of focusing on unknown attacks, we focus on what is known, the training data. We propose a simple yet robust triple-stage data-driven framework that checks the input globally and locally, and confirms that they are coherent with the model's output. This framework can be applied to any classification model and is not limited to specific attacks. We demonstrate these three stages work as one unit, effectively detecting various attacks, even for a white-box scenario.

Keywords: Adversarial Defense · Anomaly Detection · Applicability Domain · Evasion Attacks · White-box Adaptive Attacks

1 Introduction

Machine learning algorithms have shown promising results in many mission-critical fields, such as virtual drug screening [1] and autonomous driving [12].

H. Kashima et al. (Eds.): PAKDD 2023, LNAI 13935, pp. 3–14, 2023.
https://doi.org/10.1007/978-3-031-33374-3_1

Unfortunately, despite their high accuracy on benign examples, they are vulnerable to adversarial attacks, where malicious users exploit the classifiers' weakness by manipulating the input data [7]. Starting from a benign data point, attackers craft a small perturbation that allows them to achieve the desired outcome: misclassification of the input example. For example, by adding a small artifact to a stop sign, a self-driving vehicle can be fooled into misclassifying the stop sign as a speed limit sign, with the risk of causing a car crash [12].

Adversarial detectors extract features from unlabeled examples and use them to identify adversarial examples based on certain thresholds [21]. Existing detectors often suffer from the following issues: First, many detectors focus on detecting adversarial examples with only minimal perturbations [20] and tend to fail to detect stronger ones. Second, many defenses are built on a single assumption or one attack, i.e., adversarial examples lead to overly confident predictions from the classifier [11]. However, attackers are not constrained by such assumptions, as they can easily bypass such a detector by altering their strategy. Third, most defenses are tailored to a specific machine learning architecture and do not generalize to other models [23]. There is a lack of flexible detectors that can detect unseen attacks on various classifiers.

In cheminformatics, models are trained on a finite number of compounds because the data-collecting process is expensive and time-consuming. However, the chemical space is vast and diverse in its properties, so models trained on one part of the space may not work on others. Hence, models typically struggle to generalize unseen compounds. To avoid false predictions, *Applicability Domain* (AD) is a concept that defines a domain in which a model can perform reliably. Compounds that are outside this domain are rejected, as the model cannot make reliable predictions on them [1]. Similar to cheminformatics, adversarial detectors only have the information for known attacks. However, new attacks come out so frequently that it is impossible to cover all attacks. In this paper, instead of defending against previously unseen attacks, we focus on what the classifier can reliably predict, the training data. Inspired by the idea of a triple-stage AD originally introduced by Hanser *et al.* [9], we propose the BAARD framework, **B**locking **A**dversarial examples by testing for **A**pplicability, **R**eliability and **D**ecidability.

To identify unknown attacks, BAARD investigates the example from three different perspectives, utilizing the training data in the following ways: 1. *Applicability Stage* uses the training data to validate the input globally; 2. *Reliability Stage* confirms that the example can be backed up by training data locally; and 3. *Decidability Stage* checks the model's output to ensure it is coherent with the input. These three stages work as one unit to inspect the model's interpretation of an unlabeled example. As shown in Fig. 1, BAARD rejects the example if there is an inconsistency between the input and the model's prediction.

We summarize our contributions as follows:

- We are the first to demonstrate the effectiveness of linking two previously unlinked fields: the Applicability Domain in cheminformatics and adversarial detection in machine learning.

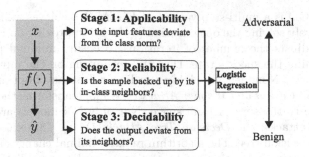

Fig. 1. An overview of BAARD. BAARD analyzes an example **x**, the classifier $f(\cdot)$, and its prediction $\hat{y}$ together by checking the Applicability, Reliability, and Decidability. Each stage outputs a score. The scores are used to train a logistic regression model to predict whether **x** is benign or adversarial.

- Inspired by the Applicability Domain, we propose the BAARD framework (Blocking Adversarial examples by testing Applicability, Reliability, and Decidability), which utilizes training data to systematically detect adversarial examples from three different perspectives.
- By designing an adaptive white-box attack targeting BAARD, we show that it is difficult to penetrate all three stages, even under the worst scenario.
- We demonstrate BAARD is highly portable. This simple yet effective framework can detect adversarial examples with various constraints on a wide range of classifiers, including classifiers that have been neglected previously despite being vulnerable to attacks, such as support vector machines and decision trees.

We introduce the adversarial threat model, attacks and detectors relevant to this paper in Sect. 2. Sect. 3 and 4 present the BAARD framework and demonstrate its effectiveness, respectively. Sect. 5 concludes this paper.

2   Background

This paper focuses on detecting *evasion attacks*, where the attacker crafts malicious inputs by adding perturbations to existing examples which can deceive the classifier to make unexpected predictions [7]. Evasion attacks are the most common adversarial attacks since it is easier for a malicious user to interact with the model at inference time.

Evasion Attacks. One of the earliest attacks on *neural network* (NN) models is the *Fast Gradient Sign Method* (FGSM) [8], a single-step attack that forms the adversarial example as: $\mathbf{x}' = \mathbf{x} - \epsilon \cdot \text{sign}(\bigtriangledown_{\mathbf{x}} \ell(\mathbf{x}, y))$ where **x** is a benign input, y is the targeted label, $\ell(\mathbf{x}, y)$ is the loss function used by the classifier, and hyperparameter ϵ controls the amount of perturbation. *Auto Projected Gradient Descent* (APGD) [6] is the latest improved version of *Projected Gradient Descent*

(PGD) [15]. PGD is a multi-step variant of FGSM. It achieves a higher success rate by iteratively solving the optimization problem. Improving on PGD, APGD dynamically adjusts the number of iterations to ensure minimal perturbation while maintaining the success rate. Directly optimizing on the input space can be difficult, since NN models are highly non-linear. Instead of optimizing on the input space, the *Carlini and Wagner Attack* (CW) [5] transforms the image from the pixel space to the simpler tanh space. Not only NN models are vulnerable to adversarial attacks, the *Decision Tree Attack* (DTA) [18] exploits the data structure of a decision tree. The algorithm makes minimal changes at each node and keeps traversing from the leaf to the root until the prediction from the classifier deviates from the legitimate class.

Detection. Detecting adversarial examples with indistinguishable perturbations (hard to recognize by human) has been studied extensively [20]. One common assumption is that if the adversarial perturbation is small enough, the legitimate class can be restored by adding or removing noise. Detectors, such as *Feature Squeezing* (FS) [22] and the *Positive and Negative representation* (PN) detector [13] are motivated by image reconstruction techniques. FS is a defense motivated by using image filters to restore adversarial examples. He *et al.* [10] pointed out that strong adversaries can easily bypass FS. The PN detector assumes an adversary cannot simultaneously deceive a classifier trained on both the original and color-negative images. Such techniques have clear limitations, a detector that uses images' properties cannot be generalized to other data types.

Another direction is to combine neighborhood relationship and noise generation. *Region-based Classification* (RC) [3] replaces the classifier with a region-based classifier by generating noisy samples centered at the example, and a decision is made via majority voting. Similar to RC, the *Odds are odd* (Odds) [19] detector assumes that adversarial examples are less robust to noise than benign examples. The assumption is that latent outputs significantly change when adding noise to an adversarial example. *Local Intrinsic Dimensionality* (LID) [14] is another neighbor-based algorithm that uses the intrinsic dimension metric by combining latent outputs from all hidden layers of a NN. The statistics are learned by comparing benign, noisy, and adversarial examples. ML-LOO [23] computes Leave-One-Out feature attribution maps on multiple hidden layers of a NN, and uses them to distinguish between benign and adversarial examples. Many detectors are based on certain assumptions of one type of attack. If the attacker's goal is to bypass the system, such a constraint may not apply [4]. A detection that is tailored to one attack is not robust against white-box attacks, where the attacker knows a particular defense is placed [21].

3   BAARD: Blocking Adversarial Examples

This paper connects cheminformatics' Applicability Domain with adversarial detection in machine learning. The goal of AD is to reject chemical compounds that the classifier cannot reliably predict. Therefore, AD analyzes the feature space and the classifier together to define a tight region around the training instances but omits the rest of the space [17]. Adversarial examples are perturbations of legitimate example, and remain similar to the original example. However, adversarial examples are designed to cause misclassifications leading to inconsistencies between the predicted labels of the adversarial example and its legitimate neighbors. This observation leads us to believe that the idea used in AD can effectively detect adversarial examples. BAARD consists of three stages as shown in Fig. 1. The rest of this section explains the working of each stage and their effectiveness when combined together.

Applicability Stage. In chemistry, this stage checks the compound to confirm it is appropriate for the model to make a prediction [9]. Here, we know the model is trained on the training data, so we check the input feature space by comparing it with the training data globally. We conduct a Z-test by computing mean and standard deviation of input features for each class from the training data. Given an example $\mathbf{x}$, the Z-score is defined by $\mathbf{z}_{\mathbf{x},\hat{y}} := (x - \mu_{X_{\text{train}},\hat{y}})/\sigma_{X_{\text{train}},\hat{y}}$, where $\mu_{X_{\text{train}},\hat{y}}$ and $\sigma_{X_{\text{train}},\hat{y}}$ are the mean and standard deviation for examples in the training data that have the same label as the model's prediction $\hat{y}$, and $\mathbf{z}_{\mathbf{x},\hat{y}}$ has the same dimension as $\mathbf{x}$. Because we are only interested in the extrema and Z-test is two-tailed, we define the Applicability Score as: S1 score $:= \max(|\mathbf{z}_{\mathbf{x},\hat{y}}|)$. The Applicability Stage inspects each feature of the new, unlabeled example, individually. It outputs a high score if any feature is significantly different from the training samples that match the classifier's predicted label.

Reliability Stage. Given a compound, this stage quantifies the relevance of information available to the model in chemistry. We implement this stage by examining the input locally using the compound's neighbors in the training set. Unlike the previous stage, which considered each input feature independently, this stage accounts for all features together using the neighborhood relationship.

Adversarial examples aim to minimize the perturbation while forcing the model to make classification errors [8]. This moves the legitimate input closer to the decision boundary, causing the predicted label to change and potentially placing the example far away from its new in-class neighbors. The reliability test is based on the distances between adversarial examples and their neighbors. These distances are often higher than the distances between legitimate examples and their neighbors.

Choosing an appropriate distance metric is essential when measuring nearest neighbors. The Euclidean distance (L_2-norm) is well suited for low-dimensional space, but Cosine similarity has shown more robust results in high-dimensional sparse features [13]. Cosine similarity between two feature vectors A and B is defined as: $S_C(A, B) := \sum_{i=1}^{n} A_i B_i / [(\sum_{i=1}^{n} A_i^2)^{\frac{1}{2}} (\sum_{i=1}^{n} B_i^2)^{\frac{1}{2}}]$, where n is the dimension of the feature vector, and $S_C \in [-1, 1]$. If S_C is close to 1, A and B are positive co-linear vectors. If $S_C = 0$, they are independent vectors, and if $S_C \approx -1$, they are strong opposite vectors. This means neither minimal nor maximal indicates A and B are close. To properly present the distance between

Algorithm 1. BAARD Stage 2 – Reliability Stage

Input: $\mathbf{x}$: unlabeled example, $\hat{y}$: its prediction, (X, Y): training set, k_{S2}: number of neighbors, and m_{S2}: sample size.
Output: $\texttt{S2\_score} \in [0, 2\pi]$
1: $X_{\hat{y}} \leftarrow$ Random sampling $\{(\mathbf{x}_1, y_1), \ldots, (\mathbf{x}_{m_{S2}}, y_{m_{S2}})\}$, where $\mathbf{x}_i \in X$, $y_i \in Y$, and $y_i = \hat{y}$
2: $D(\mathbf{x}, X_{\hat{y}}) \leftarrow$ Compute angular distances between example $\mathbf{x}$ and subset $X_{\hat{y}}$
3: $\texttt{S2\_score} \leftarrow \texttt{mean}(\texttt{top\_k}(D(\mathbf{x}, X_{\hat{y}}), k_{S2}))$ ▷ Compute the mean of top k_{S2} distances
4: **return** S2_score

---

two features using cosine similarity, we compute the angular distance D, which is defined as: $D(A, B) := \texttt{arccos}(S_C(A, B))/\pi$.

Algorithm 1 provides the pseudocode for this stage. It takes two hyperparameters: the number of nearest neighbors $k \in \mathbb{N}$, and the sample size m that limits the computational expense. For an unlabeled example $\mathbf{x}$, the S2 score is the mean distance of the k-nearest neighbors of $\mathbf{x}$ within a subset of training data where examples have the same label as the prediction $\hat{y}$. Because the angular distance is within $[0, 2\pi]$, the S2 score shares a similar scale as the S1 score. To reduce the computational cost, we randomly sample m instances from the training examples where the legitimate labels are the same as $\hat{y}$.

Decidability Stage. This stage confirms whether the model's output is coherent with the evidence from previous stages. Machine learning models operate under the assumption that similar examples have similar labels. Hence, a trained model can generalize to new and previously unseen examples. However, this is often violated when the model tries to predict maliciously crafted adversarial examples. The prediction of an adversarial example often conflicts with the predictions of its neighbors. As shown in Fig. 1, we use the local neighborhood relationship to check adversarial examples based on this property.

Algorithm 2 uses the same distance metric as in previous stages. The critical difference is that the entire training data are used regardless of their labels. We apply the Softmax function so the model outputs probability estimates. Given an example, we run a Z-test on its probability estimates based on its k neighbors.

Combining All Stages. A single stage may be effective on a certain type of attack, but no stage alone can cover all attacks. The Applicability and Reliability Stages both check the feature space but from different perspectives. Once we collect enough evidence from the input space, the Decidability Stage checks the output to ensure the model's output is coherent with the evidence. We fit a Logistic Regression model using the scores from BAARD on a hold-out training set to distinguish adversarial examples from legitimate inputs.

While being fast and memory-efficient, this approach has two issues when dealing with image data. 1. When the feature space is sparse, the S1 score becomes noise-sensitive. 2. The score varies under transformations, such as trans-

Algorithm 2. BAARD Stage 3 – Decidability Stage

Input: x: unlabeled example, $\hat{y}$: its prediction, (X, Y): training set, k_{S3}: number of
  neighbors, and m_{S3}: sample size.
Output: S3_score
 1: $S \leftarrow$ Random sampling $\{x_1, \ldots, x_{m_{S3}}\}$ where $x_i \in X$
 2: $D(x, S) \leftarrow$ Compute angular distances between example x and subset S
 3: $X' \leftarrow$ top_k$(D(x, S), k_{S3})$ ▷ Find top k-nearest neighbors.
 4: $P' \leftarrow$ Softmax$(f(X'))$ ▷ Compute probability estimates for neighbors.
 5: $\mu_{P'}, \sigma_{P'} \leftarrow$ mean$(P'),$ std(P') ▷ Compute mean and standard deviation vectors.
 6: $z \leftarrow \left| \frac{\text{Softmax}(f(x)) - \mu_{P'}}{\sigma_{P'}} \right|$
 7: **return** $z_{\hat{y}}$ ▷ $z_{\hat{y}}$ is the value of z index at $\hat{y}$.

lation and rotation. Images are commonly modeled by convolutional neural networks, because the convolutional layers can learn internal representation in a two-dimensional space. Hence, these latent outputs represent the extracted feature space learned by the model. We overcome the above issues by using the latent outputs after the convolutional layers but before the fully connected layer. Note that tabular data does not suffer from the same issues. Moreover, anything related to the training data can be calculated beforehand to speed up the algorithm at inference time.

4 Experiments

We evaluate BAARD by analyzing its parameters, deconstructing it, and testing it against attacks in both white-box and gray-box settings. We repeat the experiments five times to ensure robustness. To ensure reproducibility, all data, pre-trained classifiers, hyperparameters, additional results, and code are available at https://github.com/changx03/baard.

4.1 Experimental Setup

Data and Classifiers. We test BAARD on both image and tabular data. We acquire MNIST and CIFAR10 with default train-test split from PyTorch for image datasets. We use the model from Carlini and Wagner [5] for MNIST and ResNet18 from PyTorch for CIFAR10. The pre-trained models are available in our repository. We remove the misclassified examples and sample 1000 images for generating adversarial examples and another 1000 for validating the detectors from the test set. We acquire all tabular data from the UCI ML repository[1]. All tabular data use a 60-20-20 split. The SVM and *Decision Trees* (DT) models for tabular data use the default parameters. Additional datasets are tested and included in our repository.

---

[1] Source: https://archive.ics.uci.edu/ml.

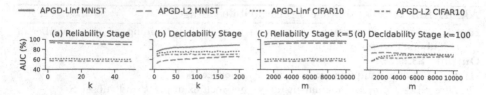

Fig. 2. Tuning hyperparameters for BAARD at the minimal adversarial perturbation. We first search for the optimal k, then tune the sample size m.

Attack Algorithms. We evaluated BAARD and other detectors under various attacks that are covered in Sect. 2, including PGD [15], APGD [6], CW-L_2 [5], and DTA [18]. We additionally include the results for FGSM [8], Boundary Attack [2], and DeepFool [16] in our repository. We define the adversary's goal to have examples misclassified as any class except the true one so all attacks are untargeted. To test attack strengths, when there are multiple L-norm constraints, we test both L_∞ and L_2 norm constraints. For each attack, we have considered a wide range of attack strengths. For instance, the parameter ϵ in APGD controls the amount of perturbation allowed [6]. We set the minimal value to where the attack has at least 95% success rate. The minimal ϵ for APGD is set to 0.22 and 4.0 for L_∞ and L_2 on MNIST, 0.01 and 0.3 on CIFAR10, respectively. In Table 1, these values are used as the "Low" ϵ, and the "High" is set to at least double the "Low" where there is a visible artifact on the example, but the legitimate label is still recognizable.

Evaluation Metrics. We report the *Area Under the Curve* (AUC) of the *Receiver Operating Characteristic* (ROC) curve as the performance metric. In practice, a single threshold may be selected based on the *False Positive Rate* (FPR). Hence, we also report TPRs when thresholds are chosen based on 5% FPRs (TPR@5FPR) when comparing different detectors.

4.2 Detection Results

Parameter Analysis. We treat each stage as an individual detector when tuning the hyperparameters. Since each stage's performance directly links to k and the sample size m is for speeding up the algorithm, we first find the optimal k while using the entire training set, then use the optimal k to tune m.

The values of k_{S2} and k_{S3} are different. As shown in Fig. 2, k_{S2} in the Reliability Stage becomes stable after the initial fluctuation. Reliability prefers a smaller k_{S2} value, as it checks the closest representation of $\mathbf{x}$ in the training samples with the same label as $\hat{y}$. Because Decidability finds neighbors from all training samples, a greater value of k_{S3} is preferred. Once k_{S2} and k_{S3} are chosen, the optimal m_{S2} and m_{S3} should be the minimum value while maintaining the detector's performance. Because the Reliability Stage uses the in-class training subset, the possible sample size is smaller than the Decidability Stage. Our results show that

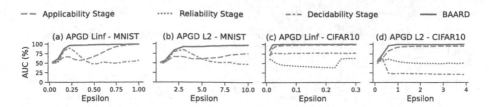

Fig. 3. BAARD's performance under decomposition against adversarial attacks with a full range of perturbations.

the detector's performance is sturdy after initial turbulence, suggesting that the sub-sampling has minimal impact on the overall performance. The experiment concludes that BAARD requires minimal tuning. We set k to 5 and 100 and m to 1000 and 5000 for the Reliability and Decidability Stages respectively for all image datasets.

Ablation Study. We decompose BAARD to investigate how each stage contributes to the overall performance. Figure 3 shows AUCs at various adversarial perturbations. Since attacks under a L_∞ constraint result in a significant deviation on the feature space [23], we find neither the S1 nor S2 score alone can detect such attacks. In Fig. 3d, the Decidability Stage's AUC (orange dotted line) goes lower than 50% when $\epsilon \geq 0.6$, indicating that the correlation between the S3 score and the detector's performance flip when ϵ increases. It means the classifier becomes more confident with the misclassified predictions when ϵ increases, leading to smaller S3 scores. Meanwhile, the S1 score becomes larger since the attack makes significant changes to the input. A low AUC on one stage indicates that stage alone is insufficient as a detector. However, by combining all stages, the results show BAARD is effective on a wide range of adversarial perturbations.

White-Box Evaluation. We address the robustness of BAARD against adaptive white-box attacks. To simultaneously attack the classifier and BAARD, the attacks' loss function is $\mathcal{L}^* := \mathcal{L} + \mathcal{L}_{S1} + \mathcal{L}_{S2} + \mathcal{L}_{S3}$, where $\mathcal{L}$ is the term for the evasion attack: $\mathcal{L} := -\text{CrossEntropy}(f(\mathbf{x}'), y_{\text{target}})$, and the rest of the terms are the losses for each stage. Because none of the stages are differentiable, a common approach is to apply gradient approximation [11]. Tramer *et al.* [21] pointed out

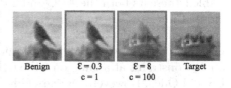

Benign $\varepsilon = 0.3$ $\varepsilon = 8$ Target
          $c = 1$ $c = 100$

Fig. 4. Apply our Adaptive White-box Targeted L_2 Attack to CIFAR10; When extreme parameters are used, it transforms a benign example into the target.

that gradient approximation tends to fail when the loss function includes multiple indifferentiable terms, a more robust approach is to find a target $\mathbf{x}_{\text{target}}$ that can pass the detector and use it as a reference. Hence, we propose an *Adaptive*

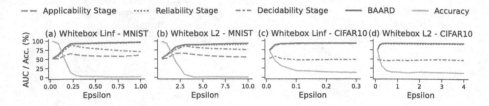

Fig. 5. BAARD's performance against Adaptive White-box Targeted attacks. The accuracy indicates the classifier's performance under such attacks.

White-box Targeted (AWT) attack as follows: we find the nearest neighbor from the training data based on the same feature space BAARD uses, as $\mathbf{x}_{target}$, and then minimize the difference between $\mathbf{x}$ and $\mathbf{x}_{target}$ to bypass S1 and S2. To avoid $f(\cdot)$ making over confident predictions, we use $f(\mathbf{x}_{target})$ as a reference to bypass S3. The new loss function becomes $\mathcal{L}^* := -\ell(f(\mathbf{x}), f(\mathbf{x}_{target})) - c\ell(\mathbf{x}, \mathbf{x}_{target})$, where both terms use the *Mean Squared Error* (MSE) loss and the hyperparameter c controls the ratio on how much $\mathbf{x}$ moves toward to $\mathbf{x}_{target}$. As shown in Fig. 4, if we relax the perturbation constraint ϵ and dial c to an extreme, the adversarial example becomes indistinguishable from the target.

We present the evaluation of BAARD against our AWT attacks in Fig. 5 with c set to 1. The attack can successfully deceive the classifier and bypass S1 or S3, but not all stages. Previous works show similar algorithms are effective on detectors with multiple loss functions, such as the Odds detector [21]. We find such attacks are ineffective on BAARD, as three stages work together, which are robust against AWT attacks under both L_2 and L_∞ constraints.

Gray-Box Benchmark. To benchmark the performance of BAARD against other detectors in Sect. 2, we use the same hold-out set to train logistic regression models for each dataset based on the features extracted from the detector. Table 1 presents both the AUC and TPR values obtained by varying the threshold of the regressors' outputs. BAARD performs consistently well across different classifiers under attacks with various strengths, showing outstanding performance on attacks with high perturbations. One outlier is the APGD attack with an L_2 constraint at a low ϵ on CIFAR10, where most detectors are weak, except FS and Odds. However, FS and Odds are tuned explicitly for low perturbations and completely fail to detect attacks with high perturbations. RC can apply to any classifier in theory, but it only performs well in CW2. Meanwhile, the detectors tailored to images and neural networks cannot apply to SVM and DT classifiers. No detector performs reliably on the PGD attacks on the Breast Cancer dataset. However, BAARD is substantially faster than detectors with similar performance, such as LID, Odds, and ML-LOO. In conclusion, BAARD is the most versatile detector tested that can reliably detect adversarial examples with various constraints on a wide range of classifiers.

Table 1. Performance of detectors. The AUC scores (%) on the left are computed from logistic regression. The right side shows the corresponding TPR at 5% FPR. "Low" and "High" indicate perturbations allowed for the attack.

Image Data	Attack / Perturbation / Detector	AUC-ROC (%) APGDinf Low	High	APGD2 Low	High	CW2	TPR@FPR5 (%) APGDinf Low	High	APGD2 Low	High	CW2
MNIST (CNN)	RC	72.9	51.6	50.8	51.5	99.9	49.0	0.0	0.0	0.0	100.0
	FS	99.7	74.0	73.7	68.1	100.0	99.3	1.7	26.1	3.3	100.0
	LID	60.6	98.1	43.4	80.8	62.8	17.5	91.8	6.6	43.3	14.3
	Odds	98.9	99.7	96.5	96.4	95.7	97.8	100.0	81.4	79.5	76.5
	ML-LOO	99.8	100.0	93.2	100.0	60.0	99.2	100.0	70.9	100.0	10.8
	PN	89.7	62.1	55.3	54.3	97.1	64.9	7.0	9.7	4.9	89.0
	BAARD	97.0	98.4	92.8	96.8	96.0	84.4	92.8	61.2	82.6	77.0
CIFAR10 (ResNet18)	RC	49.6	54.8	55.9	54.7	99.4	4.7	0.0	12.1	0.0	98.5
	FS	95.7	70.7	95.1	82.5	90.7	75.7	24.7	78.8	42.5	6.9
	LID	82.2	99.2	63.4	98.7	40.7	44.0	96.5	22.3	94.2	13.0
	Odds	98.0	67.4	97.2	80.9	96.1	95.2	0.2	95.5	2.6	83.1
	ML-LOO	67.0	99.6	58.6	99.2	66.4	28.0	98.9	16.8	97.5	10.8
	PN	76.6	54.2	75.4	58.1	66.8	18.1	7.5	17.0	9.7	10.8
	BAARD	81.6	100.0	70.2	99.2	89.0	35.4	100.0	16.4	96.1	85.2
Tabular Data	Attack (Model)	PGDinf (SVM)				DTA (DT)	PGDinf (SVM)				DTA (DT)
Banknote	RC	79.7	99.0			86.4	46.0	100.0			63.4
	FS, LID, etc.	-	-			-	-	-			-
	BAARD	96.5	100.0			95.9	87.0	100.0			89.9
Breast Cancer	RC	65.0	75.2			97.2	0.0	0.0			82.6
	FS, LID, etc.	-	-			-	-	-			-
	BAARD	77.7	52.0			96.8	21.8	7.6			85.8

5   Conclusion and Future Work

In this paper, we connected two previously unlinked domains: the Applicability Domain (AD) in cheminformatics and adversarial detection in machine learning. By sharing solutions to similar problems, both areas can benefit. We proposed BAARD, a novel adversarial detection framework inspired by AD. Our experiments showed its robustness against various adversarial evasion attacks, including those with strong perturbations. BAARD is portable and versatile enough to work with any classifier, removing the need for redesigning a defense. Our framework overcomes challenging issues in the field while maintaining comparable performance. In future research, we will explore how the insights we have gained from adversarial detection can be transferred into cheminformatics.

Acknowledgements. The authors wish to acknowledge the use of New Zealand eScience Infrastructure (NeSI) national facilities - https://www.nesi.org.nz.

References

1. Alvarsson, J., McShane, S.A., Norinder, U., Spjuth, O.: Predicting with confidence: using conformal prediction in drug discovery. J. Pharm. Sci. **110**(1), 42–49 (2021)
2. Brendel, W., Rauber, J., Bethge, M.: Decision-based adversarial attacks: reliable attacks against black-box machine learning models. In: ICLR (2018)
3. Cao, X., Gong, N.Z.: Mitigating evasion attacks to deep neural networks via region-based classification. In: ACSAC, pp. 278–287 (2017)

4. Carlini, N., Wagner, D.: Adversarial examples are not easily detected: bypassing ten detection methods. In: AISec, pp. 3–14 (2017)
5. Carlini, N., Wagner, D.: Towards evaluating the robustness of neural networks. In: IEEE SSP, pp. 39–57 (2017)
6. Croce, F., Hein, M.: Reliable evaluation of adversarial robustness with an ensemble of diverse parameter-free attacks. In: ICML, pp. 2206–2216. PMLR (2020)
7. Demontis, A., et al.: Why do adversarial attacks transfer? explaining transferability of evasion and poisoning attacks. In: USENIX Security, pp. 321–338 (2019)
8. Goodfellow, I.J., Shlens, J., Szegedy, C.: Explaining and harnessing adversarial examples. In: ICLR (2015)
9. Hanser, T., Barber, C., Marchaland, J., Werner, S.: Applicability domain: towards a more formal definition. SAR QSAR Environ. Res. **27**(11), 865–881 (2016)
10. He, W., Wei, J., Chen, X., Carlini, N., Song, D.: Adversarial example defenses: ensembles of weak defenses are not strong. In: USENIX WOOT, pp. 15–15 (2017)
11. Hu, S., Yu, T., Guo, C., Chao, W.L., Weinberger, K.Q.: A new defense against adversarial images: turning a weakness into a strength. In: NIPS 32 (2019)
12. Kloukiniotis, A., Papandreou, A., Lalos, A., Kapsalas, P., Nguyen, D.V., Moustakas, K.: Countering adversarial attacks on autonomous vehicles using denoising techniques: a review. In: IEEE OJ-ITS (2022)
13. Luo, W., Wu, C., Ni, L., Zhou, N., Zhang, Z.: Detecting adversarial examples by positive and negative representations. ASC **117**, 108383 (2022)
14. Ma, X., et al.: Characterizing adversarial subspaces using local intrinsic dimensionality. In: ICLR (2018)
15. Madry, A., Makelov, A., Schmidt, L., Tsipras, D., Vladu, A.: Towards deep learning models resistant to adversarial attacks. In: ICLR (2018)
16. Moosavi-Dezfooli, S.M., Fawzi, A., Frossard, P.: DeepFool: a simple and accurate method to fool deep neural networks. In: CVPR, pp. 2574–2582. IEEE (2016)
17. Netzeva, T.I., et al.: Current status of methods for defining the applicability domain of (quantitative) structure-activity relationships: the report and recommendations of ECVAM workshop 52. ATLA **33**(2), 155–173 (2005)
18. Papernot, N., McDaniel, P., Goodfellow, I.: Transferability in machine learning: from phenomena to black-box attacks using adversarial samples. arXiv preprint arXiv:1605.07277 (2016)
19. Roth, K., Kilcher, Y., Hofmann, T.: The odds are odd: a statistical test for detecting adversarial examples. In: ICML, pp. 5498–5507. PMLR (2019)
20. Tramer, F.: Detecting adversarial examples is (nearly) as hard as classifying them. In: ICML, pp. 21692–21702. PMLR (2022)
21. Tramer, F., Carlini, N., Brendel, W., Madry, A.: On adaptive attacks to adversarial example defenses. NIPS **33**, 1633–1645 (2020)
22. Xu, W., Evans, D., Qi, Y.: Feature squeezing: detecting adversarial examples in deep neural networks. arXiv preprint arXiv:1704.01155 (2017)
23. Yang, P., Chen, J., Hsieh, C.J., Wang, J.L., Jordan, M.: ML-LOO: detecting adversarial examples with feature attribution. In: AAAI, vol. 34, pp. 6639–6647 (2020)

Fast and Attributed Change Detection on Dynamic Graphs with Density of States

Shenyang Huang[1,2]($\boxtimes$)(iD), Jacob Danovitch[1,2](iD), Guillaume Rabusseau[2,3,4](iD), and Reihaneh Rabbany[1,2,4](iD)

[1] McGill University, Montreal, Canada
{shenyang.huang,jacob.danovitch}@mail.mcgill.ca
[2] Mila - Quebec AI Institute, Montreal, Canada
reihaneh.rabbany@mila.quebec
[3] DIRO, Université de Montréal, Montreal, Canada
guillaume.rabusseau@umontreal.ca
[4] Canadian Institute for Advanced Research (CIFAR) AI chair, Montreal, Canada

Abstract. How can we detect traffic disturbances from international flight transportation logs, or changes to collaboration dynamics in academic networks? These problems can be formulated as detecting anomalous change points in a dynamic graph. Current solutions do not scale well to large real world graphs, lack robustness to large amount of node additions / deletions and overlook changes in node attributes. To address these limitations, we propose a novel spectral method: Scalable Change Point Detection (SCPD). SCPD generates an embedding for each graph snapshot by efficiently approximating the distribution of the Laplacian spectrum at each step. SCPD can also capture shifts in node attributes by tracking correlations between attributes and eigenvectors. Through extensive experiments using synthetic and real world data, we show that SCPD (a) achieves state-of-the-art performance, (b) is significantly faster than the state-of-the-art methods and can easily process millions of edges in a few CPU minutes, (c) can effectively tackle a large quantity of node attributes, additions or deletions and (d) discovers interesting events in large real world graphs. Code is publicly available at https://github.com/shenyangHuang/SCPD.git.

Keywords: Anomaly Detection · Dynamic Graphs · Spectral Methods

1 Introduction

Anomaly detection is one of the fundamental tasks in analyzing dynamic graphs [5,16,17], with applications ranging from detecting disruptions in traffic networks, analyzing shifts in political environments and identifying abnormal events in communication networks. In this work we focus on identifying anomalous time points where the graph structure deviates significantly from the normal

H. Kashima et al. (Eds.): PAKDD 2023, LNAI 13935, pp. 15–26, 2023.
https://doi.org/10.1007/978-3-031-33374-3_2

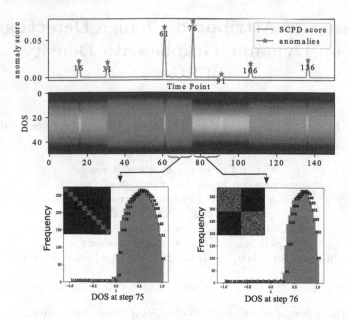

Fig. 1. SCPD utilizes the spectral density (approximated by Density of States (DOS)) to summarize the graph at each time point, change in DOS often indicate a change in graph distribution. The DOS becomes skewed after the number of communities decreases from ten to two in the SBM [10] hybrid experiment (see Sect. 5). The DOS is plotted for step 75 and 76 while the inset plots show the adjacency matrix of the graph.

behavior, also known as change point detection [12,13]. Detecting anomalies in dynamic graphs offers several challenges: real world graphs are often very large, their size can drastically evolve over time (e.g. nodes appearing and disappearing in social network graphs where nodes represent users) and complex information is associated with nodes in the graph (e.g., profile of users summarized as a set of attributes for each node).

Prior work on change point detection are limited by one or more of the following issues. *1). Lack of scalability*: modern networks often contains millions of edges and nodes, thus computationally intensive algorithms [12,14] can be difficult to apply on graphs with more than hundreds of nodes. *2). Overlooking attributes*: many networks also contain a diverse set of node attributes which evolve over time. No prior work has considered the evolution of node attributes and its relation with the graph structure. *3). Difficulty with evolving sizes*: real networks grow over time with new nodes often forming a large portion of the network. Methods such as [6,23] track a fixed set of nodes sampled from the initial time step, and are thus limited to detect changes happening within the initial set of nodes. Other approaches such as [12,13] summarize each snapshot with a vector dependent on the size of the snapshot. Therefore, as the graph grows, truncation on the summary vector is required to ensure a uniform vector size for all snapshots.

To address the above limitations, we propose Scalable Change Point Detection (SCPD), a novel change point detection method which detects both structural and node attribute anomalies in dynamic graphs. SCPD utilizes the distribution of eigenvalues (also known as the spectral density) of the Laplacian matrix as a low dimensional embedding of each graph snapshot. As change points induce a shift in graph distribution, they also cause changes in the spectral density. We leverage the Density of States (DOS) [4] framework to efficiently approximate the spectral density, allowing SCPD to scale to dynamic graphs with millions of nodes. Figure 1 illustrates the key idea of SCPD: to discretize the spectral density, the range of eigenvalues is divided into k bins and the number of eigenvalues within each bin is computed. As such, the number of bins k is not dependent on the size of the network. Therefore, SCPD can easily adapt to the evolving size of a dynamic graph. The main characteristics of SCPD are:

- **Accurate:** We show that SCPD achieves state-of-the-art performance in extensive synthetic experiments and can identify a number of major wars from the co-authorship network MAG-History of the History research community (while existing methods fail to adapt to the evolving size of this network).
- **Scalable:** SCPD has a linear time complexity with respect to the number of edges and is highly scalable. For example, on the MAG-History dataset with 2 million edges, SCPD runs in 29 s on a stock laptop with CPU.
- **Attributed:** To the best of our knowledge, SCPD is the first method to incorporate node attributes into change point detection for dynamic graphs. On our original COVID-flight dataset, SCPD leverages the country code of airports (nodes) to identify traffic disturbances due to flight restrictions specific to countries such as China and US.

2 Related Work

In this section, we review methods for change point and event detection. We compare compared SCPD to other approaches in Table 1. Note that current methods focuses on graph structural anomalies while SCPD is the first method to incorporate node attributes and satisfies all the desired properties.

Event Detection Idé and Kashima [13] uses the principal eigenvector of the adjacency matrix to represent the graph at each snapshot (called *activity vector*). Koutra et al. [14] formulated dynamic graphs as high order tensors and proposed to use the PARAFAC decomposition [3,9] to obtain vector representations for anomaly scoring. SPOTLIGHT [6] was proposed to spot anomalous graphs containing the sudden appearance or disappearance of large dense subgraphs.

Change Point Detection Wang et al. [23] modeled network evolution as a first order Markov process and use MCMC sampling to design the EdgeMonitoring method. Recently, Huang et al. [12] proposed Laplacian Anomaly Detection (LAD) which uses the exact singular values of the Laplacian matrix of each snapshot as the signature vector. SCPD employs a similar anomaly detection pipeline to LAD and also utilizes spectral information from the Laplacian

Table 1. SCPD is the only scalable method that detects both events and change points and also being the only method that accounts for attributes.

Method \ Property	Event	Change Point	Scalable	Evolving Size	Weights	Attributes
Activity vector [13]	✔		✔	✔	✔	
TENSORSPLAT [14]	✔				✔	
EdgeMonitoring [23]		✔	✔			
SPOTLIGHT [6]	✔		✔		✔	
LAD [12]	✔	✔		✔	✔	
SCPD [this paper]	✔	✔	✔	✔	✔	✔

matrix. However, computing Singular Value Decomposition (SVD) limits LAD to small graphs while SCPD is scalable to millions of nodes and edges.

Network Density of States Dong et al. [4] borrowed tools from condensed matter physics and added adaptation such as motif filtering to design an efficient approximation method for spectral density in large networks. Huang et al. [11] proposed a graph kernel which combines local and global density of states of the normalized adjacency matrix for the graph classification task. ADOGE [18] is an embedding method for exploratory graph analysis and graph classification on static graphs. To the best of our knowledge, our proposed SCPD is the first method to model spectral density for dynamic graphs.

3 Problem Formulation and Notations

We consider an undirected, weighted, dynamic graph $\mathbf{G}$ with node attributes (optional), as a sequence of graph snapshots, $\{\mathcal{G}_t\}_{t=1}^T$. Each $\mathcal{G}_t = (\mathcal{V}_t, \mathcal{E}_t, \mathbf{X}_t)$ represents the graph at time $t \in [1 \ldots T]$, where $\mathcal{V}_t$, $\mathcal{E}_t$ are the set of nodes and edges respectively, and $\mathbf{X}_t \in \mathbb{R}^{|\mathcal{V}_t| \times N_a}$ is the attribute matrix, where N_a is the number of attributes. An edge $(i, j, w) \in \mathcal{E}_t$ connects node i and node j at time t with weight w. We use $\mathbf{A}_t \in \mathbb{R}^{|\mathcal{V}_t| \times |\mathcal{V}_t|}$ to denote the adjacency matrix of $\mathcal{G}_t$.

Attribute Change Point Detection The goal of change point detection is to identify anomalous time steps in a dynamic graph, i.e. snapshots with *graph structures* that significantly deviate from the normal behavior. This often requires an anomaly score function measuring the graph structural difference between the current snapshot and the average behavior observed previously. In this work, we examine both *events*, one time change to the graph structure and *change points*, permanent alterations on the graph generative process. To the best of our knowledge, we are also the first work to incorporate node attributes in change point detection. In addition to detecting change points in the graph structure, the goal of *attribute* change point detection is to also identify time steps in which *the alignment* between node attributes and graph structure deviates significantly from the norm. For example, in a network with communities, if the distribution of an attribute conditioned on the community drastically change, we say that an attribute change point has happened.

4 Scalable Change Point Detection

To detect anomalous snapshots, we embed each graph snapshot into a low dimensional embedding called the signature vector based on the spectral density. Then, the normal behavior of the graph in the past is summarized into a vector. Lastly, we compare the signature from the current step with that of the past behavior and derive an anomaly score.

Designing Signature Vector Identifying change points require the comparison between multiple graph snapshot. In general, it is difficult to compare graphs directly as shown in the graph isomorphism problem [24]. Therefore, we want to embed each graph snapshot into a low dimensional vector, called the *signature vector* and facilitate comparisons between vectors rather than graphs. In this work, we choose the (global) density of states (DOS) of the Laplacian matrix as the signature vector as it has the following desirable property: 1.)*scalable*, DOS can scale to graphs with millions of nodes and edges, 2.)*independent of graph size*, DOS produces a fixed sized embedding independent of the number of nodes or edges in the graph, 3.) *incorporates attributes*, the local DOS can be used to model the alignment between node attributes and eigenvectors thus can be used to model attribute change points.

We use DOS to approximate the distribution of the Laplacian eigenvalues. The Laplacian eigenvalues captures many graph structures and properties [21] and have shown strong empirical performance for anomaly detection [12]. For example, the number of zero eigenvalues of the Laplacian matrix is equal to the number of connected components of the graph [22] and the eigenvectors of the Laplacian matrix provide an effective way to represent a graph in a 2D plane [8]. In addition, the eigenvalues of the Laplacian and their multiplicity reflect the geometry of many fundamental graphs such as complete graphs, star graphs and path graphs. However, computing all Laplacian eigenvalues of a graph requires $O(|\mathcal{V}| \cdot |\mathcal{E}|)$ which is only practical for small graphs, while computing DOS is scalable to large graphs. Later in this section, we show how to compute DOS efficiently and in Sect. 5 and 6, we demonstrate that DOS has state-of-the-art performance in change point detection.

Computing Anomaly Score After computing the signature vectors for each timestamp, now we explain how to detect anomalous snapshots. We assume that when an anomaly arrives, it would be significantly different from recent snapshots. Therefore, we extract the "expected" or "normal" behavior of the dynamic graph from a context window of size w from the past w signature vectors. To obtain unit vectors, $L2$ normalization is performed on the set of the signature vectors $\sigma_{t-w-1}, \ldots, \sigma_{t-1}$. Then, we stack the normalized vectors to form the context matrix $\mathbf{C}_t^w \in \mathbb{R}^{k \times w}$ of time t, where k is length of the signature vector. We compute the left singular vector of $\mathbf{C}_t^w$ to be the summarized normal behavior vector $\tilde{\sigma}_t^w$ (which can be seen as a weighted average over the context window). Smaller context window can detect more sudden or abrupt changes while a longer window can model gradual and continuous changes. Therefore, we use a short window with size w_s and a long window with size w_l to detect both events and change points.

Now we can compute the anomaly score at time t as $Z_t = 1 - \frac{\sigma_t^\top \tilde{\sigma}_t^w}{\|\sigma_t\|_2 \|\tilde{\sigma}_t^w\|_2} = 1 - \sigma_t^\top \tilde{\sigma}_t^w = 1 - \cos\theta$ where $\cos\theta$ is the cosine similarity between the current signature vector σ_t and the normal behavior vector $\tilde{\sigma}_t^w$. In this way, $Z \in [0,1]$ and when Z is closer to 1, the current snapshot significantly different from the normal behavior thus more likely to be an anomaly. The Z scores from windows of size w_s and w_l are then aggregated by the max operation. To emphasize the increase in anomaly score, we compute the difference in anomaly score with the previous step with $Z_t^* = \min(Z_t - Z_{t-1}, 0)$. Finally, the points with the largest Z^* are selected as anomalies. We show the Z^* score in all figures in this work.

Approximating Spectral Density For clarity, we drop the t subscript in this Section. The Laplacian matrix $\mathbf{L} \in \mathbb{R}^{|\mathcal{V}| \times |\mathcal{V}|}$ is defined as $\mathbf{L} = \mathbf{D} - \mathbf{A}$ where $\mathbf{D} \in \mathbb{R}^{|\mathcal{V}| \times |\mathcal{V}|}$, $\mathbf{A} \in \mathbb{R}^{|\mathcal{V}| \times |\mathcal{V}|}$ are the diagonal degree matrix and the adjacency matrix. In this work, we use the symmetric normalized Laplacian $\mathbf{L}_{sym} = \mathbf{D}^{-\frac{1}{2}} \mathbf{L} \mathbf{D}^{-\frac{1}{2}} = \mathbf{I} - \mathbf{D}^{-\frac{1}{2}} \mathbf{A} \mathbf{D}^{-\frac{1}{2}}$ to present the graph at each snapshot. Consider the eigendecomposition of $\mathbf{L}_{sym} = \mathbf{Q} \mathbf{\Lambda} \mathbf{Q}^T$ where $\mathbf{\Lambda} = diag(\lambda_1, \ldots, \lambda_{|\mathcal{V}|})$ and $\mathbf{Q} = [\mathbf{q}_1, \ldots, \mathbf{q}_{|\mathcal{V}|}]$ is an orthogonal matrix. We can now define Density of States or the spectral density as,

Definition 1 (Density of States (DOS)). *the global density of states or spectral density induced by* $\mathbf{L}_{sym}$ *is:*

$$\mu(\lambda) = \frac{1}{|\mathcal{V}|} \sum_{i=1}^{|\mathcal{V}|} \delta(\lambda - \lambda_i) \tag{1}$$

where δ is the Dirac delta function and λ_i is the i-th eigenvalue.

Intuitively, $\mu(\lambda)$ measures the portion of eigenvalues that are equal to λ. In practice, we discretize the range of λ into equal sized intervals and approximate how many λ falls within each interval. Therefore, across all intervals, the shape of the distribution of eigenvalues are approximated. We use the Kernel Polynomial Method (KPM) [4] to approximate the density function through an finite number polynomial expansion, in the dual basis of the Chebyshev basis and the spectrum is adjusted to be in $[-1, 1]$ for numerical stability. To incorporate attributes, we also consider Local Density of States:

Definition 2 (Local Density of States (LDOS)). *For any given input vector* $\mathbf{v} \in \mathbb{R}^N$, *the local density of states is:*

$$\mu(\lambda; \mathbf{v}) = \sum_{i=1}^{|\mathcal{V}|} |\mathbf{v}^T \mathbf{q}_i|^2 \delta(\lambda - \lambda_i) \tag{2}$$

where **v** *is an input vector,* λ_i, $\mathbf{q}_i$ *are the i-th eigenvalue and eigenvector.*

The term $\mathbf{v}^T \mathbf{q}_i$ acts as a weight on the ith bin of the spectral histogram. To incorporate node attributes, we set $\mathbf{v} = \mathbf{x}$ where $\mathbf{x} \in \mathbb{R}^{|\mathcal{V}|}$ is an attribute vector. This can be interpreted as the *alignment* between the node attribute vector and the graph structure of the group of nodes with such attribute. As the alignment is measured in each eigenvalue interval (similarly to DOS), we obtain a LDOS embedding of size k for each attribute and each possible category. By tracking this embedding over time, one can capture anomalous evolution specific to the given attribute. Here, categorical attributes are one-hot encoded, and numerical attributes are normalized by the sum. We use the Gauss Quadrature and Lanczos (GQL) [4] method to approximate the LDOS with attribute vectors.

Computational Complexity For unattributed dynamic graphs, SCPD has the complexity of $O(N_z \cdot N_m \cdot |\mathcal{E}|)$ for a given snapshot with $|\mathcal{E}|$ edges. N_z, N_m are hyperparameters in the KPM computation representing the number of probe vectors and Chebychev moments respectively. For all experiments, we set $N_z = 100$, $N_m = 20$. We also use $k = 50$ equal sized bins in the range of eigenvalues.

For attributed dynamic graphs, we use the GQL method to compute LDOS for attribute change point detection. GQL method performs the eigendecomposition of a tridiagonal matrix with $O(|\mathcal{V}|^2)$ worst case complexity. Note that in practice, such computation is very fast [18]. Therefore, for a given attribute on a dynamic graph, SCPD's time complexity is $O(\eta \cdot |\mathcal{E}| + |\mathcal{V}|^2)$ for a given snapshot. In practice, SCPD is very fast only costing 5 seconds to run on the COVID flight network with close to 1 million edges and 5 node attributes with an AMD Ryzen 5 1600 Processor and 16GB memory.

5   Synthetic Experiments

In this section, we conduct experiments with the Stochastic Block Model (SBM) [10] and the Barabási-Albert (BA) model [1] as synthetic graph generators and plant 7 ground truth anomalies for all experiments. We report the Hits@n metric same as in [12] and the execution time over 5 trials.

SBM Hybrid Experiment We follow the Hybrid setting in [12]. SBM [10] is used to generate equal sized communities with p_{in} being the intra-community connectivity and p_{out} being the cross-community connectivity. Change points are the merging or splitting of communities in the dynamic graph and events are one-time boosts in cross-community connectivity p_{out}. Figure 1 shows that SCPD perfectly identifies all the events and change points on a dynamic SBM graph. We also visualize the signature vectors (the computed DOS or distribution of eigenvalues) as a heatmap. The events (time point 16,61,91,136) corresponds to an energetic burst in the signature vector. And the change points correspond to the shifts in the distribution of Laplacian eigenvalues. Interestingly, the width of the distribution seem to correlate with the number of communities N_c.

BA Experiment We evaluate SCPD performance in a different graph distribution, the BA model. In this experiment, the change points correspond to the densification of the network (parameter m, increased number of edges attached

Table 2. SCPD can efficiently operate on large graphs while achieving the state-of-the-art performance. Each dynamic graph has 151 time steps. The results are Hits@7 averaged over 5 trials and the mean and standard deviations are reported. We consider a method not applicable (N/A) if the computation takes longer than 5 d.

Generator	SBM			BA	
Experiment	Hybrid		Evolving Size	Change Point	
Total Edges (millions)	0.8 m	56.9 m	1.0 m	0.6 m	5.5 m
SCPD (ours)	1.00 ± 0.00	1.00 ± 0.00	1.00 ± 0.00	1.00 ± 0.00	1.00 ± 0.00
LAD [12]	1.00 ± 0.00	N/A	1.00 ± 0.00	1.00 ± 0.00	N/A
SPOTLIGHT [6]	0.31 ± 0.06	0.57 ± 0.00	0.20 ± 0.07	0.06 ± 0.07	0.11 ± 0.11
SPOTLIGHTs	0.71 ± 0.00	0.71 ± 0.00	0.31 ± 0.06	1.00 ± 0.00	1.00 ± 0.00
EdgeMonitoring [23]	0.06 ± 0.11	0.00 ± 0.00	0.14 ± 0.00	0.06 ± 0.07	0.17 ± 0.11

from a new node to an existing node). SCPD is able to detect all change points in the BA model and the most drastic change in DOS happens when m changes from one to two and the graph becomes connected. This is because the number of zero eigenvalues in the Laplacian matrix corresponds to the number of connected components in the graph thus when the graph is connected, the smallest eigenvalue intervals become less energetic.

SBM Attribute Experiment We want to demonstrate SCPD's ability to detect anomalous evolution of the node attributes in a dynamic graph. A SBM model is used to construct communities for nodes while each node has a binary attribute. The attributes within a community can be either *homogeneous* or *heterogeneous*. In a homogeneous community, all nodes have the same attribute while half of all communities have label one while the other half have label two. In a heterogeneous community, each node has 0.5 probability being either one or two and the node attribute is no longer dependent on community structure. The change points are time points where the node attributes change to *homogeneous* or *heterogeneous*. SCPD is able to recover all change points (16,61,91 and 136) related to node attributes and detect both the change from homogeneous communities to heterogeneous ones as well as the reverse.

SBM Evolving Size Experiment We examine SCPD's ability to adapt to the evolving size of a dynamic graph (with a SBM as the graph generator). Initially, there are two communities with 300 nodes each. Later on, additional nodes are added and forming a total of 4 communities. Some change points involves only nodes from the initial step while some involves only newly added nodes. Only SCPD and LAD is able to correctly detect all anomalies while SPOTLIGHT and EdgeMonitoring can only detect changes local to the initial set of nodes. This shows that SCPD can effectively adapt to the evolving size of dynamic graph.

Summary of Results Table 2 compares the performance of SCPD with state-of-the-art methods on synthetic experiments. The SBM attribute experiment is not included as only SCPD can incorporate node attributes. The considered baselines include LAD [12], SPOTLIGHT [6] and EdgeMonitoring [23]. The original

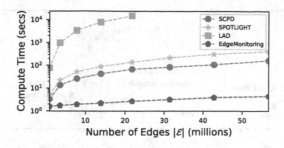

Fig. 2. Compute time comparison between different methods on the SBM hybrid experiment with varying number of edges.

SPOTLIGHT (with RRCF [7] detector) and our own variant, SPOTLIGHT with sum predictor, called SPOTLIGHTs are both included. Across all experiments, SCPD has the best overall performance. With the default RRCF anomaly detection pipeline, SPOTLIGHT [6] performs poorly on the BA model and middling performance on the SBM hybrid experiment. With the simple sum predictor introduced by us, SPOTLIGHTs is a much closer competitor with strong performance on the BA model and improved performance on the SBM hybrid experiment. However, SPOTLIGHTs is still not able to detect changes in the evolving size experiment and overall outperformed by SCPD. EdgeMonitoring [23] has low performance in the synthetic experiments due to its dependency on node ordering as well as the assumption that only a small percentage of edges would be resampled in a dynamic graph. The closest competitor to SCPD is LAD [12]. However, computing all the eigenvalues in LAD is prohibitively expensive on large graphs thus reported as not applicable.

In Fig. 2, we compare the computational time across different methods in the SBM hybrid experiment. The most expensive is LAD as it has worst case complexity cubic to number of nodes thus having poor trade-off between performance and efficiency. In contrast, both SCPD and SPOTLIGHT has complexity linear to the number of edges. However, SCPD outperforms SPOTLIGHT across all experiments shown in Table 2. Therefore, SCPD has the best trade-off between compute time and performance. Lastly, EdgeMonitoring has sublinear complexity to the number of edges however its performance is not ideal.

6   Real World Experiments

We empirically evaluate SCPD on two real world dynamic networks and cross reference anomalies detected by SCPD with significant events.

MAG History Co-authorship Network MAG-History is a co-authorship dynamic network extracted from the Microsoft Academic Graph (MAG) [2, 20] by identifying publications which are marked with the"History" tag. The processed dataset is an undirected dynamic graph from 1837 to 2018. There are 2.8 million projected edges across all time steps and 0.7 million nodes in total. To compute the DOS embedding for this dataset, SCPD only takes 30 s.

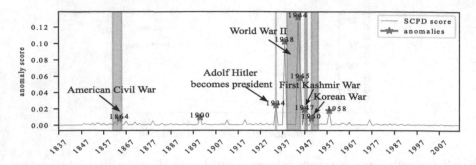

Fig. 3. SCPD detects significant historical events from the MAG-History dataset.

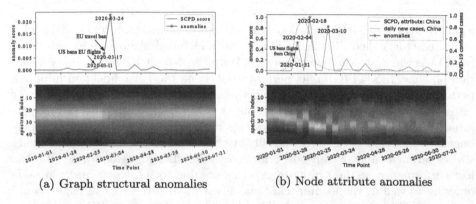

(a) Graph structural anomalies   (b) Node attribute anomalies

Fig. 4. a). SCPD detects the week of 03.17 and 03.24 as structural anomalies in the global flight network in 2020. On 03.17, the European Union closed its borders to travellers thus causing wide spread disruption in international flights. b). SCPD detects closure of flight routes to China due to COVID interventions at beginning of Feb 2020. The anomaly score and case numbers are normalized.

Figure 3 shows the anomalies detected by SCPD. Interestingly, many of the anomalies correspond to important historical events such as the American Civil War (1861–1865), Adolf Hitler's rise to power (1934), Second World War (1939–1945), First Kashmir War (1947–1948) and Korean War (1950–1953). The relation between the change in co-authorship graph structure and these historical events can be an interesting direction for future work. In Comparison, both variants of SPOTLIGHT miss the second world war as a top anomaly while EdgeMonitoring's output is noisy and many data points sharing high anomaly scores.

COVID Flight Network the COVID flight network[1] [15,19] is a dynamic air traffic network during the COVID-19 pandemic. The nodes are airports and each edge is an undirected timestamped tracked flight with the frequency as edge weight. We examine the period from 01-01-2020 to 07-27-2020. We use a full week as the duration of each snapshot to reduce the noise and variability

---

[1] https://zenodo.org/record/3974209/#.Yf62HepKguU.

from daily flights. Figure 4a shows the graph structural anomalies detected by SCPD using the DOS embeddings as signature vectors. The two weeks with the highest anomaly scores are 03-17-2020 and 03-24-2020. On 03–17, the European Union adopted a 30-day ban on non-essential travel to at least 26 European countries from the rest of the world (see here). On 03–11, the US President banned travel from 26 European countries. SCPD detects the disruptions by travel bans in the flight network. In comparison, SPOTLIGHT detects the week of 02–11 corresponding to flight restrictions on China while EdgeMonitoring also detects mid March as anomalies.

Figure 4b shows SCPD's anomaly scores when the node attribute is set to be an indicator vector for which nodes are Chinese airports. The detected anomalies lie mainly in February and early March because the COVID outbreak was first detected in China in January 2020. On 01-31-2020, the Trump administration suspended entry into the United States by any foreign nationals who had traveled to China in the past 14 d (see here). Therefore, the anomaly observed by SCPD on the week of 02-04-2020 is likely the directed result of the imposed travel restriction. Note that Fig. 4b shows that the peak of new daily cases in China[2] corresponds to peak in anomaly score, likely because of reduced domestic and international flights at that time. SCPD captures both the structural and attribute anomalies in the flight network.

7   Conclusion

In this work, we proposed a novel change point detection method: SCPD, to detect anomalous changes in the graph structure as well as node attributes in a dynamic graph. SCPD approximates the distribution of Laplacian eigenvalues as an embedding for the graph structure and Local DOS embeddings to measure the alignment between node attributes and the eigenvectors of the Laplacian at different frequency intervals. On synthetic experiments, SCPD achieves state-of-the-art performance while running efficiently on graphs with millions of edges. On two real world datasets, SCPD is able to capture structural and attribute change points corresponding to significant real world events.

Acknowledgement. This research was supported by the CIFAR AI chair program, NSERC PGS Doctoral Award and FRQNT Doctoral Award.

References

1. Barabási, A.L.: Philosophical transactions of the royal society a: mathematical, physical and engineering sciences. Netw. Sci. **371**(1987), 20120375 (2013)
2. Benson, A.R., Abebe, R., Schaub, M.T., Jadbabaie, A., Kleinberg, J.: Simplicial closure and higher-order link prediction. PNAS (2018)
3. Bro, R.: Parafac. tutorial and applications. Chemometrics and intelligent laboratory systems **38**(2), 149–171 (1997)

---

[2] https://www.worldometers.info/coronavirus/country/china/.

4. Dong, K., Benson, A.R., Bindel, D.: Network density of states. In: Proceedings of the 25th ACM SIGKDD International Conference on Knowledge Discovery & Data Mining. pp. 1152–1161 (2019)
5. Eswaran, D., Faloutsos, C.: Sedanspot: Detecting anomalies in edge streams. In: 2018 IEEE International Conference on Data Mining (ICDM). IEEE (2018)
6. Eswaran, D., Faloutsos, C., Guha, S., Mishra, N.: Spotlight: Detecting anomalies in streaming graphs. In: Proceedings of the 24th ACM SIGKDD International Conference on Knowledge Discovery & Data Mining (2018)
7. Guha, S., Mishra, N., Roy, G., Schrijvers, O.: Robust random cut forest based anomaly detection on streams. In: International conference on machine learning. pp. 2712–2721. PMLR (2016)
8. Hall, K.M.: An r-dimensional quadratic placement algorithm. Manag. Sci. 17(3), 219–229 (1970)
9. Harshman, R.A., et al.: Foundations of the PARAFAC procedure: models and conditions for an explanatory multimodal factor analysis (1970)
10. Holland, P.W., Laskey, K.B., Leinhardt, S.: Stochastic blockmodels: first steps. Social Netw. 5(2), 109–137 (1983)
11. Huang, L., Graven, A.J., Bindel, D.: Density of states graph kernels. In: Proceedings of the 2021 SIAM International Conference on Data Mining (SDM) (2021)
12. Huang, S., Hitti, Y., Rabusseau, G., Rabbany, R.: Laplacian change point detection for dynamic graphs. In: Proceedings of the ACM SIGKDD International Conference on Knowledge Discovery & Data Mining (2020)
13. Idé, T., Kashima, H.: Eigenspace-based anomaly detection in computer systems. In: Proceedings of the tenth ACM SIGKDD international conference on Knowledge discovery and data mining. pp. 440–449. ACM (2004)
14. Koutra, D., Papalexakis, E.E., Faloutsos, C.: Tensorsplat: Spotting latent anomalies in time. In: 2012 16th Panhellenic Conference on Informatics. IEEE
15. Olive, X.: Traffic, a toolbox for processing and analysing air traffic data. J. Open Source Softw. 4(39), 1518 (2019)
16. Peel, L., Clauset, A.: Detecting change points in the large-scale structure of evolving networks. In: Twenty-Ninth AAAI Conference on Artificial Intelligence (2015)
17. Ranshous, S., Harenberg, S., Sharma, K., Samatova, N.F.: A scalable approach for outlier detection in edge streams using sketch-based approximations. In: 2016 SIAM International Conference on Data Mining. SIAM (2016)
18. Sawlani, S., Zhao, L., Akoglu, L.: Fast attributed graph embedding via density of states. In: 2021 IEEE International Conference on Data Mining (ICDM) (2021)
19. Schäfer, M., Strohmeier, M., Lenders, V., Martinovic, I., Wilhelm, M.: Bringing up opensky: A large-scale ads-b sensor network for research. In: IPSN-14 Proceedings of the 13th International Symposium on Information Processing in Sensor Networks. pp. 83–94. IEEE (2014)
20. Sinha, A., Shen, Z., Song, Y., Ma, H., Eide, D., Hsu, B.J.P., Wang, K.: An overview of microsoft academic service (MAS) and applications. In: the 24th International Conference on World Wide Web. ACM Press (2015)
21. Spielman, D.A.: Spectral graph theory and its applications. In: 48th Annual IEEE Symposium on Foundations of Computer Science (FOCS'07). IEEE (2007)
22. Von Luxburg, U.: A tutorial on spectral clustering. Statist. comput. 17, 395–416 (2007). https://doi.org/10.1007/s11222-007-9033-z
23. Wang, Y., Chakrabarti, A., Sivakoff, D., Parthasarathy, S.: Fast change point detection on dynamic social networks. arXiv preprint arXiv:1705.07325 (2017)
24. Weisfeiler, B., Leman, A.: The reduction of a graph to canonical form and the algebra which appears therein. NTI Series 2(9), 12–16 (1968)

Outlying Aspect Mining via Sum-Product Networks

Stefan Lüdtke[1]([✉]), Christian Bartelt[2], and Heiner Stuckenschmidt[3]

[1] Center for Scalable Data Analytics and Artificial Intelligence (ScaDS.AI),
University of Leipzig, Leipzig, Germany
`stefan.luedtke@uni-leipzig.de`
[2] Institute for Enterprise Systems, University of Mannheim, Mannheim, Germany
[3] Data and Web Science Group, University of Mannheim, Mannheim, Germany

Abstract. Outlying Aspect Mining (OAM) is the task of identifying a subset of features that distinguish an outlier from normal data, which is important for downstream (human) decision-making. Existing methods are based on beam search in the space of feature subsets. They need to compute outlier scores for all examined subsets, and thus rely on simple outlier scoring algorithms.

In this paper, we propose SOAM, a novel OAM algorithm based on Sum-Product Networks (SPNs), a class of probabilistic circuits that can accurately model high-dimensional distributions. Our approach needs to fit an SPN only once, and leverages the tractability of marginal inference in SPNs to compute outlier scores in feature subsets. This way, computing outlier scores in subsets is fast, while being based on a flexible and accurate density estimator. We empirically show that SOAM clearly outperform the state-of-the-art method in search-based OAM, and even outperforms recent deep learning-based methods in the majority of the investigated cases. (Available at github.com/stefanluedtke/Sum-Product-Network-OAM).

Keywords: Outlying Aspect Mining · Outlier Interpretation · Sum-Product Network

1 Introduction

The identification of uncommon or anomalous samples (outliers) in a dataset is an important task in data science. Many outlier detection methods have been proposed, including classical methods based on notions of distance or density [8,18] as well as deep learning-based methods (see [12] for a review).

A less well investigated, but natural question is that of *outlying aspect mining* [3,17,22]: Given a sample classified as an outlier, which properties ("aspects") of the sample are the cause for this classification, i.e., which properties are specifically anomalous? This task has also been called *outlier interpretation* [9,24]

Supplementary Information The online version contains supplementary material available at https://doi.org/10.1007/978-3-031-33374-3_3.

H. Kashima et al. (Eds.): PAKDD 2023, LNAI 13935, pp. 27–38, 2023.
https://doi.org/10.1007/978-3-031-33374-3_3

or *outlying subspace detection* [25]. Figure 1 shows an example of OAM for the Wisconsin Breast Cancer dataset: Here, the green outlier (malignant case) has unusual texture and the red outlier has unusual shape, which can be relevant for downstream tasks like therapy decisions.

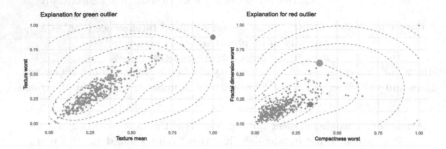

Fig. 1. Two explanations generated for outliers in the WBC dataset. Contours illustrate the (marginal) density of the SPN. Left: Features that best explain outlyingness of the green sample. Right: Feature that best explain outlyingness of the red sample. (Color figure online)

Most existing OAM methods are based on beam search to identify feature subsets in which the outlier score of a sample is maximal [3,17,22,23]. They require to compute outlier scores for each feature subset that is visited during beam search, which can be computationally very costly. Therefore, they use simple outlier scores, e.g. based on Kernel Density Estimators [3] or isolation scores in Nearest Neighbor Ensembles (iNNE) [17].

In this paper, we show that *Sum-Product Networks* (SPNs) [16] can be used for OAM in a natural way: SPNs are probabilistic models which can accurately model high-dimensional, mixed discrete-continuous distributions, while providing tractable marginal inference. For OAM, we only need to train an SPN once on the dataset, and can then compute outlier scores in feature subsets via marginal inference in the SPN. This way, computing outlier scores in subsets is fast, but based on a flexible and accurate density estimator. We call the resulting OAM method *SOAM* (SPN-based Outlying Aspect Mining).

Additionally, runtime of outlier score computation in SPNs does not depend on the feature subset size. Thus, in addition to the usual forward beam search, it becomes possible to use a backward selection search that starts with a large feature subset and prunes it iteratively. We find that both search strategies are complementary: Forward search achieves best results for low-dimensional data, and backward elimination is best for high-dimensional data.

We evaluate SOAM on a number of synthetic and real-world OAM tasks, and compare it with two types of methods: Search-based OAM which works fully unsupervised (like SOAM), and deep learning-based methods, which require outlier labels as additional inputs during training. SOAM outperforms the state-of-the-art search-based OAM method [17], and even outperforms the deep learning-based explanation methods [9,24] in the majority of the investigated cases.

2 Preliminaries and Related Work

2.1 Outlier Detection and Outlying Aspect Mining

Outlier Detection is the following unsupervised learning task: Given a dataset $\{\mathbf{x}^{(1)}, \ldots, \mathbf{x}^{(n)}\}$, classify each sample as either normal or outlier. Outlier detection algorithms usually compute a *scoring function* $f : \mathcal{X} \to \mathbb{R}$, which can be used for outlier classification by classifying all samples with $f(\mathbf{x}) > t$ as outliers, for a fixed threshold t.

There are different outlier detection methods that explicitly or implicitly define f. Classical methods include, for example, isolation forests [8], local outlier factor [2] or one-class support vector machines [18]. More recently, deep neural networks have been used for this task [12]. Most relevant to this paper are *probabilistic* methods, which assume that *normal* data was generated from a distribution $p(\mathbf{x}; \theta)$ with parameters θ. An outlier is a sample which is unlikely to be drawn from $p(\mathbf{x}; \theta)$. That is, they use the scoring function $f(\mathbf{x}) = -p(\mathbf{x}; \theta)$. Parametric as well as non-parametric density estimators have been considered for $p(\mathbf{x}; \theta)$, e.g., Gaussian mixtures [15] or Kernel Density Estimation [19].

Outlying Aspect Mining. (OAM) is the task of retrieving a subset of features in which a sample is specifically anomalous [3,17,23,25]. More formally, let $D \subseteq \{1, \ldots, n\}$ be a set of indices, and let $\mathbf{x}_D$ denote the projection of $\mathbf{x}$ onto the subspace indicated by D. OAM is the task of identifying D, such that $f(\mathbf{x}_D)$ is maximized for a given sample $\mathbf{x}$.

The naive approach of computing $f(\mathbf{x}_D)$ individually for each subspace $D \subseteq \{1, \ldots, n\}$ quickly becomes infeasible due to the combinatorial explosion in $|D|$. Therefore, existing OAM methods [3,17,23,25] usually perform a greedy beam search that iteratively adds dimensions to D, and focus on simple scoring functions f. For example, [3] use a kernel density estimator (KDE) to compute outlier scores for each visited subspace. [23] build on this work, replacing the KDE with a faster, grid-based density estimator.

To select the best subspace, simply returning the subspace D where $f(\mathbf{x}_D)$ is maximized is not usually not appropriate, because scoring functions for different dimensionalities are usually not directly comparable [22]. Thus, *dimensionality-unbiased* scores like *z-score* normalization

$$z(\mathbf{x}, D, \mathbf{X}) = \frac{f(\mathbf{x}_D) - \mu(\mathbf{X}_D)}{\sigma(\mathbf{X}_D)} \tag{1}$$

w.r.t. the training dataset $\mathbf{X}$ or a rank transformation have been proposed [3]. They allow to compare scores between different dimensionalities, thus allowing to identify a subspace D in which $\mathbf{x}_D$ is most outlying, relative to other samples.

OAM via Explainability Methods. In contrast to search-based OAM methods, feature subsets that best explain the outlyingness of sample can also be obtained via algorithm-agnostic, local explainability methods. Several methods tailored towards the outlier detection task have been proposed [9,24]: COIN [9] learns a classifier ensemble that separates outliers from clusters of nearby normal data and use the classifier weights as feature importance. ATON [24] trains a neural network, consisting of an embedding layer and a subsequent self-attention layer. From the attention weights, feature importance weights can be obtained, which can be converted to an explaining feature subset by thresholding.

In both methods, the training data needs to contain outlier labels. When these labels are computed by an outlier detection algorithm, they explain the decisions of that outlier detector.

To emulate the OAM task, it is customary to apply the explanation methods to the *ground truth* outlier labels. This way, explanations are not negatively influenced by incorrect outlier labels, posing the most fair comparison to OAM algorithms. This procedure has, for example, been used by the authors of COIN and ATON to compare their methods to OAM methods [9,24].

2.2 Sum-Product Networks

Representation. A Sum-Product Network (SPN) [16] is a rooted directed acyclic graph representing a probability distribution over a sequence of random variables (RVs) $\mathbf{X} = X_1, \ldots, X_n$. Each node represents a distribution p_N over a subset $\mathbf{X}_{\phi(N)} \subseteq \mathbf{X}$, where $\phi(N) \subseteq \{1, \ldots, n\}$ is called the scope of the node N. In the following, ch(N) denotes the children of node N. An SPN contains tree types of nodes: Leaf nodes, product nodes and sum nodes. A product node represents a factorized distribution $p(\mathbf{X}_{\phi(N)}) = \prod_{C \in \mathrm{ch}(N)} p_C(\mathbf{X}_{\phi(C)})$. A sum node represents a mixture distribution $p_N(\mathbf{X}_{\phi(N)}) = \sum_{C \in \mathrm{ch}(N)} w_C \, p_C(\mathbf{X}_{\phi(C)})$ with mixture weights w_C. A leaf node directly represents a (tractable) univariate or multivariate distribution. *Decomposability* (children of product nodes have pairwise disjoint scopes) and *completeness* (children of sum nodes have identical scope) ensure that an SPN actually represents a valid probability distribution. By definition, the distribution represented by an SPN is the distribution defined by its root node. Early research on SPNs focused on categorical distributions or simple parametric leaf distributions like Gaussians [16]. More recently, SPNs with piecewise polynomial leaf distributions have been used to model continuous and mixed data [10].

Inference. The appealing property of SPNs is that any marginal distribution $p(\mathbf{X}'=\mathbf{x}')$ for a subset $\mathbf{X}' \subset \mathbf{X}$ can be computed efficiently. Intuitively, this is possible because summation over the marginalized RVs can be "pushed down" into the leaf nodes of the SPN [14]. Thus, marginal inference reduces to marginalization of the leaves and evaluating the internal nodes of the SPN once. As leaves are usually chosen such that marginal inference in leaf distributions is possible in constant time, marginal inference is linear in the number of nodes of the SPN. Specifically, when the leaf distributions are univariate, the value of marginalized leaves can simply be set to 1.

Learning. Early learning algorithms focused on structure learning [4,10,21]. Most prominently, LearnSPN [4] is a greedy structure learning algorithm, which creates a tree-structured SPN in a top-down fashion. It recursively tests for independence of RVs (in which case it creates a product node and recurses), and otherwise clusters the data into subsets, creates a corresponding sum node and recurses. [10] proposed an extension of LearnSPN which also works for continuous and mixed domains. Recently, [13] proposed a learning algorithm which first initializes a random SPN structure and then learns parameters via EM. This way, parameter learning can leverage fast, parallel GPU computations.

3   Outlying Aspect Mining via SPNs

The central challenge in OAM is to efficiently compute $f(\mathbf{x}_D)$ for subspaces D. For probabilistic outlier detection methods, this task is equivalent to computing a marginal distribution $f(\mathbf{x}_D) = -p(\mathbf{x}_D; \theta)$. Such a marginal is obtained by integrating over all RVs $\mathbf{X} \setminus \mathbf{X}_D$. More formally, let $\bar{D} = \{1, \ldots, n\} \setminus D$, and denote $\bar{D} = \{\bar{D}_1, \ldots, \bar{D}_k\}$. The outlier score in subspace D is given by

$$f(\mathbf{x}_D) = -p(\mathbf{x}_D; \theta) = \int_{x_{\bar{D}_1}} \cdots \int_{x_{\bar{D}_k}} p(\mathbf{x}; \theta) \, \mathrm{d}x_{\bar{D}_1} \ldots \mathrm{d}x_{\bar{D}_k} \qquad (2)$$

Explicitly computing such marginals is intractable for many expressive density estimators. Instead, the strategy taken by existing OAM methods [3,23] is to project the training samples to the subspace D, and estimate the parameters of the model $p(\mathbf{x}_D; \theta)$ from those samples.

We propose to use an SPN to represent the joint density $p(\mathbf{x}; \theta)$. Time complexity of evaluating a marginal probability in Eq. 2 is linear in the number of nodes of the SPN [16], independently of the number of RVs that are marginalized—and irrespective of the number of original training samples, in contrast to approaches that perform parameter estimation for each subspace.

In addition, SPNs are flexible and powerful density estimators, reaching state-of-the-art performance in several density estimation tasks. Thus, they should be able to accurately model $p(\mathbf{x}; \theta)$, opening up the potential for increased OAM performance, compared to conventional OAM methods that have to rely on simple density estimators.

We call the resulting method *SOAM* (Sum Product Network-based Outlying Aspect Mining). In the following, we discuss search strategies as well as dimensionality selection strategies utilized in SOAM in more detail.

Search Strategies. As computing outlier scores for all $2^n - 1$ feature subspaces quickly becomes infeasible with increasing number of features n, a search strategy that only explores promising subspaces is required. **Forward beam search**, which greedily adds features has been used for this task before [22]. More specifically, the beam search keeps a set of B hypotheses (subspaces). In each step and for each hypothesis, it greedily adds that feature to the hypothesis that maximizes the outlier score of the sample in the extended feature set. Search is carried out until a maximum depth S. The search algorithm is shown in Algorithm 1.

Algorithm 1. forwardBeamSearch($\mathbf{x}$,S,B,θ)

Input: Outlier $\mathbf{x}$ of dimensionality n, maximum explanation size S, beam width B, distribution parameters θ (e.g., as an SPN)

Output: For each $k \in \{1, \ldots, n\}$, a subspace D of size k in which x_D is most anomalous

$D_1 \leftarrow$ ARGLOWEST$(B, D, \theta, \mathbf{x})$ ▷ Store B most outlying dimensions

$D_1^{(\text{best})} \leftarrow$ ARGLOWESTDENSITIES$(1, D, \theta, \mathbf{x})$ ▷ Overall most outlying dimension, needed as return value later

for $k \in \{2, \ldots, S\}$ **do**

    $D_k \leftarrow \{\}$

    for $D_{k-1}^{(i)} \in D_{k-1}$ **do** ▷ For all hypotheses, get all candidate subspaces of size k

        $D_k \leftarrow D_k \cup \{D_{k-1}^{(i)} \cup d \,|\, d \in D\}$

    end for

    $D_k \leftarrow$ ARGLOWEST$(B, D_k, \theta, \mathbf{x})$

    $D_k^{(\text{best})} \leftarrow$ ARGLOWESTDENSITIES$(1, D_k, \theta, \mathbf{x})$

end for

return $D_1^{(\text{best})}, \ldots, D_S^{(\text{best})}$

function ARGLOWEST$(B, D_k, \theta, \mathbf{x})$ ▷ B subspaces from D_k where $\mathbf{x}$ is least likely

    $L \leftarrow \{p(\mathbf{x}_d; \theta) \,|\, d \in D_k\}$

    return $\{d \,|\, d \in D_k, \text{rank}(p(\mathbf{x}_d; \theta), L) \leq B\}$

end function

At depth k, each hypothesis consists of k features, and $n - k$ features need to be explored (where n is the overall number of features). Thus, up to depth k, $\sum_{i=1}^{k} n - k < n\,k$ feature subspaces are explored per hypothesis. In SPNs, computing an outlier score (a marginal probability density) amounts to evaluating the SPN (with N nodes) once, resulting in an overall time complexity of beam search-based explanation of $\mathcal{O}(N\,n\,S)$, where S is the maximum search depth.

Intuitively, beam search works well when a sample that has high outlier score in a feature set of size k also has a high outlier score in one of the subsets of size $k - 1$. When this is not the case, beam search can fail to find reasonable explanations, as pointed out by [24].

To alleviate this problem, we propose a top-down, **backward elimination** search strategy to identify outlying subspaces. Instead of greedily adding dimensions, the search algorithm starts with the full feature set, and then greedily removes one feature at a time, so that in resulting feature subspace, the outlyingness of the sample is maximal (compared to all other subspaces of that size). The algorithm is shown in Algorithm 2. Intuitively, when a sample is an outlier in k-dimensional subspace, it cannot be a complete inlier in any $(k+1)$-dimensional subspace. Thus, starting from high dimensionality and only removing features can lead to more accurate results than bottom-up beam search.

At iteration k of backward elimination, the feature subset consists of $n - k$ features. For each of the $n - k$ subsets of size $n - k - 1$, an outlier score needs to be computed. The algorithm runs for n iterations, resulting in $\sum_{k=0}^{n}(n-k) < n^2$

Algorithm 2. backwardElimination($\mathbf{x}, \theta$)

Input: Outlier $\mathbf{x}$ of dimensionality n, distribution parameters θ (e.g., as an SPN)
Output: For each $k \in \{1, \ldots, n\}$, a subspace D of size k in which x_D is most anomalous
$D_n \leftarrow \{1, \ldots, n\}$
for $k \in \{n-1, \ldots, 1\}$ **do**
    $d_k \leftarrow \underset{d \in D}{\operatorname{argmax}} \; p(\mathbf{x}_{D_k \setminus d}; \theta)$
    $D_{k-1} \leftarrow D_k \setminus \{d_k\}$
end for
return $D_1, \ldots, D_{n-1}$

---

explored subsets. Thus, overall runtime complexity of backward elimination is $\mathcal{O}(n^2 N)$, where N is the number of nodes of the SPN.

Dimensionality Selection. Both beam search and backward elimination result in an outlier score for each visited feature subset. As a last step, one of the subsets needs to be selected as most outlying. Simply selecting the subset with lowest outlier score might not be optimal, because scores for different dimensionalities are usually not directly comparable. Specifically, the densities $p(\mathbf{x}_D; \theta)$ will typically be smaller for larger dimensionality of $\mathbf{x}_D$. [22] introduce *dimensionality-unbiasedness* as a desideratum for outlier scores to allow for such comparison. Dimensionality-unbiasedness can be achieved, for example, by z-score transformation (see Eq. 1). However, such transformations are computationally inefficient as outlier scores need to be computed for all samples instead of only the query sample.

Instead, we propose to use the *elbow* method to select the optimal feature subset size (which has been, for example, used for determining the optimal number of clusters in k-means clustering [1]): In real datasets, we often observe a large difference between the minimal log density of all examined feature subsets of size k and $k+1$ for a given sample. In this case, we assume the subspace of size $k+1$ as most outlying for that sample. More concretely, we compute differences between subsequent lowest log density, and then return the lowest-dimensional subspace where the difference is larger than a threshold κ. When a difference of at least κ never occurs, we return the single feature with lowest univariate density.

4   Experimental Evaluation

Goal of the experiments was to evaluate the performance of SOAM, compared to state-of-the-art OAM methods. Specifically, we evaluated the F1 score of retrieved outlying subspaces on a number of synthetic and real-word datasets. Additionally, we compared the forward beam search (SOAM$_f$) and backward elimination (SOAM$_b$) search strategies for SOAM. Experiments regarding the dimensionality selection strategies (elbow method, z-score transformation) and runtime are shown in the supplementary material.

4.1 Data Sets

Synthetic Datasets. We used 21 synthetic datasets[1] created by [6]. Each dataset consists of 10, 20, 30, 40, 50, 75 or 100 features (3 datasets per number of features) and contains 1000 samples, 19 to 136 of which are outliers. The datasets were created in such a way that each outlier is easily detectable in a pre-defined, 2- to 5-dimensional feature subset (which varies between outliers), but is an inlier in any lower-dimensional projection of the data.

Real-World Datasets. Additionally, we evaluated OAM performance on nine real-world datasets[2] provided by [24]. To cope with the lack of ground-truth outlying subspaces, they created explanation labels as follows: First, each dataset was reduced to its ten first principal components. Then, for each dataset and each feature subset of that dataset, three outlier detection algorithms (Isolation Forests [8], COPOD [7] and HBOS [5]) were applied to the subspace. The explanation label of an outlier was defined to be the feature subset where the outlier score is maximal (w.r.t. the algorithm). From the available twelve datasets, we selected those nine datasets where at least one of the three outlier detection algorithms could achieve more than 0.5 ROC AUC, to ensure that the notion of outliers was sensible.

4.2 Experiments

We compared SOAM to the following state-of-the-art OAM algorithms:

- **SiNNE** [17] is the latest contribution in a line of search-based OAM algorithms including [22,23]. Instead of its predecessors, the approach uses a *dimensionality-unbiased* outlier score function that does not require post-hoc normalization. It has been empirically shown to outperform other search-based OAM methods [17].
- **COIN** [9] is an explainability method which fits a set of classifiers to a labeled dataset to separate outliers from clusters of nearby normal data, and uses the weights in the classifiers as feature importance values.
- **ATON** [24] is a state-of-the-art neural network model for outlier explantion based on attention.

Note that COIN and ATON are *explanation* methods, i.e., in contrast to the OAM methods SOAM and SiNNE, they require outlier labels as additional input. Here, we supply COIN and ATON with the ground truth outlier labels during training (which are available for these benchmark datasets), similar to [9,24], to allow for the most fair comparison to OAM methods.

We used implementations of SiNNE, COIN and ATON provided by [24][3]. We used the SPFlow library [11] for fitting and inference in SPNs, and the LearnSPN algorithm [4] for SPN structure learning.

---

[1] Available at www.ipd.kit.edu/mitarbeiter/muellere/HiCS.
[2] Available at github.com/xuhongzuo/outlier-interpretation.
[3] github.com/xuhongzuo/outlier-interpretation.

Table 1. OAM performance (F1 score of retrieved relevant subspaces and F1 score rank) for synthetic datasets. SiNNE did not finish in less than 5,000 s for $D \geq 30$. Note that COIN and ATON were additionally supplied with outlier ground truth labels during training, which was not required by SOAM and SiNNE.

D	SOAM$_f$	SOAM$_b$	SiNNE	ATON	COIN
10	0.867 (2)	0.799 (5)	0.86 (3)	0.806 (4)	**0.933 (1)**
20	**0.668 (1)**	0.646 (4)	0.65 (3)	0.589 (5)	0.667 (2)
30	0.562 (2)	**0.676 (1)**	0.54 (3)	0.497 (4)	0.427 (5)
40	0.399 (2)	**0.634 (1)**	–	0.348 (3)	0.261 (4)
50	0.351 (2)	**0.682 (1)**	–	0.3 (3)	0.227 (4)
75	0.355 (2)	**0.698 (1)**	–	0.205 (3)	0.158 (4)
100	0.267 (2)	**0.611 (1)**	–	0.154 (3)	0.118 (4)
Mean	0.496 (2)	**0.678 (1)**	–	0.414 (3)	0.399 (4)

All SPN learning hyperparameters were set to fixed values across all experiments and datasets: We used Gaussian leaf distributions for real features and categorical leaf distributions for categorical features. During row splits, the data was partitioned via EM for Gaussian Mixture Models, using 2 mixture components. The Randomized Dependence Coefficient was used as independence test, setting $\alpha = 0.6$. For beam search, we used a fixed beam width of 10, and set the elbow threshold to $\kappa = \exp(1)$. This choice of SPN hyperparameters was based on [21]. Optimization of these hyperparameters on a validation set could improve SOAM performance further, but was not attempted here as these fixed parameters already achieved good performance.

In all cases, the entire dataset was used for fitting the models. SOAM and SiNNE models were trained with fully unsupervised data, while COIN and ATON additionally required the ground truth outlier labels. Outlier *explanation* labels (i.e., for each outlier, the outlying subspaces) were only used for evaluation.

5   Results

Synthetic Data. We first evaluated the quality of the explanations (in terms of F1 score of retrieved dimensions) on the synthetic datasets. We evaluated both forward beam search and backward elimination search.

Table 1 shows F1 scores of the different OAM methods. For each data dimensionality D, mean F1 scores of the three datasets of that dimensionality are reported. Both SOAM variants outperformed the state-of-the-art methods (except for $D = 10$), with an increasingly large difference in F1 for increasing D. With regards to the two search strategies, it can be seen that backward elimination outperformed beam search for higher-dimensional cases. SOAM$_b$ (SOAM with backward elimination search) is the only method where F1 score did not decrease substantially for larger data dimensionality, achieving good OAM performance even for $D = 100$. We suspect that this is due to the fact that beam

Table 2. OAM performance (F1 score of retrieved relevant subspaces and F1 score rank) for real-world datasets. The three rows for each dataset correspond to the three ground truth explanation labels. Note that COIN and ATON were additionally supplied with outlier ground truth labels during training, which was not required by SOAM and SiNNE.

dataset	SOAM$_f$	SOAM$_b$	SiNNE	ATON	COIN
arrhythmia	**0.742 (1)**	0.726 (2)	0.564 (4)	0.676 (3)	0.367 (5)
	0.635 (1)	0.577 (3)	0.499 (4)	0.596 (2)	0.398 (5)
	0.695 (2)	**0.751 (1)**	0.473 (4)	0.557 (3)	0.273 (5)
ionosphere	**0.644 (1)**	0.488 (4)	0.482 (5)	0.622 (3)	0.629 (2)
	0.59 (2)	0.452 (5)	0.454 (4)	**0.671 (1)**	0.573 (3)
	0.658 (1)	0.564 (4)	0.433 (5)	0.618 (3)	0.647 (2)
letter	**0.701 (1)**	0.519 (5)	0.668 (2)	0.665 (3)	0.562 (4)
	0.641 (2)	0.388 (5)	0.614 (3)	**0.664 (1)**	0.554 (4)
	0.778 (1)	0.752 (2)	0.616 (3)	0.545 (4)	0.403 (5)
optdigits	**0.754 (1)**	0.45 (5)	0.654 (3)	0.671 (2)	0.607 (4)
	0.725 (1)	0.472 (5)	0.622 (3)	0.672 (2)	0.593 (4)
	0.887 (1)	0.871 (2)	0.58 (3)	0.557 (4)	0.298 (5)
pima	0.589 (2)	0.538 (5)	0.588 (3)	**0.673 (1)**	0.553 (4)
	0.632 (2)	0.515 (5)	0.557 (4)	**0.65 (1)**	0.586 (3)
	0.747 (1)	0.656 (2)	0.441 (4)	0.531 (3)	0.415 (5)
satimage	0.604 (2)	**0.612 (1)**	0.429 (5)	0.585 (3)	0.429 (4)
	0.661 (2)	0.59 (3)	0.41 (5)	**0.664 (1)**	0.539 (4)
	0.746 (2)	**0.823 (1)**	0.442 (4)	0.541 (3)	0.247 (5)
wbc	**0.718 (1)**	0.63 (2)	0.57 (4)	0.604 (3)	0.56 (5)
	0.552 (2)	0.447 (5)	0.499 (3)	**0.601 (1)**	0.461 (4)
	0.679 (1)	0.659 (2)	0.502 (5)	0.579 (4)	0.639 (3)
wineRed	0.436 (3)	0.366 (5)	0.505 (2)	**0.661 (1)**	0.429 (4)
	0.432 (4)	0.367 (5)	0.493 (2)	**0.652 (1)**	0.45 (3)
	0.491 (1)	0.407 (4)	0.361 (5)	0.481 (2)	0.408 (3)
wineWhite	0.526 (3)	0.454 (4)	0.531 (2)	**0.619 (1)**	0.436 (5)
	0.469 (4)	0.428 (5)	0.528 (2)	**0.605 (1)**	0.497 (3)
	0.569 (1)	0.529 (2)	0.388 (4)	0.479 (3)	0.38 (5)
Mean	**0.641 (1)**	0.557 (3)	0.515 (4)	0.609 (2)	0.479 (5)

search is susceptible to missing relevant dimensions when their number increases (and the beam width stays constant), whereas backward elimination is more stable w.r.t. dimensionality.

Real Data. Next, we evaluated OAM performance on the real-world datasets processed by [24]. The results for ATON, COIN and SiNNE were taken directly from the paper introducing ATON [24]. Table 2 shows the empirical results. For these datasets, SOAM$_f$ (with forward search) outperformed the state-of-the-art in 17 out of 27 cases (63 %). Here, forward beam search generally outperformed backward elimination, which is consistent with results for the synthetic data: As these datasets were preprocessed to be at most 10-dimensional, forward beam search with a beam width of 10 was still able to identify explanations correctly.

Overall, the empirical results are encouraging: For the high-dimensional synthetic data, SOAM achieved a new state-of-the-art, and for the low-dimensional real-world data, our approach still outperformed state-of-the-art methods in 63% of the cases (and did not require outlier labels, in contrast to COIN and ATON).

6   Discussion and Conclusion

In this paper, we proposed an OAM method that utilizes an SPN as density estimator. SPNs are accurate, flexible density estimators, in contrast to density estimators previously used for OAM. Due to the tractability of marginal inference of SPNs, OAM is still efficient. We empirically showed that our approach can retrieve subspaces where samples are most outlying more accurate than existing methods, clearly outperforming state-of-the-art OAM methods, and even outperforming deep learning-based methods (which require outlier labels as additional inputs) in the majority of the cases. Specifically, in contrast to existing methods, the proposed backward elimination search, enabled by used of SPNs, can maintain a high accuracy when the data dimensionality increases.

Here, we only investigated OAM for tabular data. Applying SPNs to the closely related task of *image anomaly localization* [20] is a possible next step. For this task, efficient SPN training algorithms and implementations suitable for image data, like the recently proposed Einsum Networks [13], are an attractive option.

Acknowledgements. Stefan Lüdtke acknowledges the financial support by the Federal Ministry of Education and Research of Germany and by the Sächsische Staatsministerium für Wissenschaft Kultur und Tourismus in the program Center of Excellence for AI-research "Center for Scalable Data Analytics and Artificial Intelligence Dresden/Leipzig", project identification number: ScaDS.AI

References

1. Aggarwal, C.C.: Data Mining. Springer, Cham (2015). https://doi.org/10.1007/978-3-319-14142-8
2. Breunig, M.M., Kriegel, H.P., Ng, R.T., Sander, J.: LOF: identifying density-based local outliers. In: Proceedings of the 2000 ACM SIGMOD International Conference on Management of Data, pp. 93–104 (2000)
3. Duan, L., Tang, G., Pei, J., Bailey, J., Campbell, A., Tang, C.: Mining outlying aspects on numeric data. Data Min. Knowl. Discov. **29**(5), 1116–1151 (2015). https://doi.org/10.1007/s10618-014-0398-2
4. Gens, R., Domingos, P.: Learning the structure of sum-product networks. In: International Conference on Machine Learning, pp. 873–880. PMLR (2013)
5. Goldstein, M., Dengel, A.: Histogram-based outlier score (HBOS): a fast unsupervised anomaly detection algorithm. In: KI-2012: Poster and Demo Track 9 (2012)
6. Keller, F., Muller, E., Bohm, K.: HICS: high contrast subspaces for density-based outlier ranking. In: 2012 IEEE 28th International Conference on Data Engineering, pp. 1037–1048. IEEE (2012)
7. Li, Z., Zhao, Y., Botta, N., Ionescu, C., Hu, X.: COPOD: copula-based outlier detection. In: 2020 IEEE International Conference on Data Mining (ICDM), pp. 1118–1123. IEEE (2020)

8. Liu, F.T., Ting, K.M., Zhou, Z.H.: Isolation forest. In: 2008 Eighth IEEE International Conference on Data Mining, pp. 413–422. IEEE (2008)
9. Liu, N., Shin, D., Hu, X.: Contextual outlier interpretation. In: Proceedings of the 27th International Joint Conference on Artificial Intelligence, pp. 2461–2467 (2018)
10. Molina, A., Vergari, A., Di Mauro, N., Natarajan, S., Esposito, F., Kersting, K.: Mixed sum-product networks: a deep architecture for hybrid domains. In: Thirty-Second AAAI Conference on Artificial Intelligence (2018)
11. Molina, A., et al.: SPFlow: an easy and extensible library for deep probabilistic learning using sum-product networks (2019)
12. Pang, G., Shen, C., Cao, L., Hengel, A.V.D.: Deep learning for anomaly detection: a review. ACM Comput. Surv. (CSUR) **54**(2), 1–38 (2021)
13. Peharz, R., et al.: Einsum networks: fast and scalable learning of tractable probabilistic circuits. In: International Conference on Machine Learning, pp. 7563–7574. PMLR (2020)
14. Peharz, R., Tschiatschek, S., Pernkopf, F., Domingos, P.: On theoretical properties of sum-product networks. In: Artificial Intelligence and Statistics, pp. 744–752. PMLR (2015)
15. Pimentel, M.A., Clifton, D.A., Clifton, L., Tarassenko, L.: A review of novelty detection. Signal Process. **99**, 215–249 (2014)
16. Poon, H., Domingos, P.: Sum-product networks: a new deep architecture. In: Proceeding of the UAI (2011)
17. Samariya, D., Aryal, S., Ting, K.M., Ma, J.: A new effective and efficient measure for outlying aspect mining. In: Huang, Z., Beek, W., Wang, H., Zhou, R., Zhang, Y. (eds.) WISE 2020. LNCS, vol. 12343, pp. 463–474. Springer, Cham (2020). https://doi.org/10.1007/978-3-030-62008-0_32
18. Schölkopf, B., Platt, J.C., Shawe-Taylor, J., Smola, A.J., Williamson, R.C.: Estimating the support of a high-dimensional distribution. Neural Comput. **13**(7), 1443–1471 (2001)
19. Schubert, E., Zimek, A., Kriegel, H.P.: Generalized outlier detection with flexible kernel density estimates. In: Proceedings of the 2014 SIAM International Conference on Data Mining, pp. 542–550. SIAM (2014)
20. Venkataramanan, S., Peng, K.-C., Singh, R.V., Mahalanobis, A.: Attention guided anomaly localization in images. In: Vedaldi, A., Bischof, H., Brox, T., Frahm, J.-M. (eds.) ECCV 2020. LNCS, vol. 12362, pp. 485–503. Springer, Cham (2020). https://doi.org/10.1007/978-3-030-58520-4_29
21. Vergari, A., Di Mauro, N., Esposito, F.: Simplifying, regularizing and strengthening sum-product network structure learning. In: Appice, A., Rodrigues, P.P., Santos Costa, V., Gama, J., Jorge, A., Soares, C. (eds.) ECML PKDD 2015. LNCS (LNAI), vol. 9285, pp. 343–358. Springer, Cham (2015). https://doi.org/10.1007/978-3-319-23525-7_21
22. Vinh, N.X., et al.: Discovering outlying aspects in large datasets. Data Min. Knowl. Discov. **30**(6), 1520–1555 (2016). https://doi.org/10.1007/s10618-016-0453-2
23. Wells, J.R., Ting, K.M.: A new simple and efficient density estimator that enables fast systematic search. Pattern Recogn. Lett. **122**, 92–98 (2019)
24. Xu, H., et al.: Beyond outlier detection: outlier interpretation by attention-guided triplet deviation network. In: Proceedings of the Web Conference 2021, pp. 1328–1339 (2021)
25. Zhang, J., Lou, M., Ling, T.W., Wang, H.: Hos-miner: a system for detecting outlying subspaces of high-dimensional data. In: Proceedings of the 30th International Conference on Very Large Data Bases (VLDB'04), pp. 1265–1268. Morgan Kaufmann Publishers Inc. (2004)

TSI-GAN: Unsupervised Time Series Anomaly Detection Using Convolutional Cycle-Consistent Generative Adversarial Networks

Shyam Sundar Saravanan[1], Tie Luo[1(✉)] [ID], and Mao Van Ngo[2]

[1] Missouri University of Science and Technology, Rolla, MO 65401, USA
{ssdmw,tluo}@mst.edu
[2] Singapore University of Technology and Design, Singapore 487372, Singapore
vanmao_ngo@sutd.edu.sg

Abstract. Anomaly detection is widely used in network intrusion detection, autonomous driving, medical diagnosis, credit card frauds, etc. However, several key challenges remain open, such as lack of ground truth labels, presence of complex temporal patterns, and generalizing over different datasets. This paper proposes TSI-GAN, an unsupervised anomaly detection model for time-series that can learn complex temporal patterns automatically and generalize well, i.e., no need for choosing dataset-specific parameters, making statistical assumptions about underlying data, or changing model architectures. To achieve these goals, we convert each input time-series into a sequence of 2D images using two encoding techniques with the intent of capturing temporal patterns and various types of deviance. Moreover, we design a reconstructive GAN that uses convolutional layers in an encoder-decoder network and employs *cycle-consistency loss* during training to ensure that inverse mappings are accurate as well. In addition, we also instrument a *Hodrick-Prescott filter* in post-processing to mitigate false positives. We evaluate TSI-GAN using 250 well-curated and harder-than-usual datasets and compare with 8 state-of-the-art baseline methods. The results demonstrate the superiority of TSI-GAN to all the baselines, offering an overall performance improvement of 13% and 31% over the second-best performer MERLIN and the third-best performer LSTM-AE, respectively.

Keywords: Anomaly detection · time series · unsupervised learning · generative adversarial networks

1 Introduction

Anomaly detection aims to identify sub-sequences of various lengths that are considered abnormal within a context represented by data. Accurate and automated anomaly detection is crucial to a wide range of applications including network security, smart manufacturing, autonomous driving, and digital healthcare. Time-series data is ubiquitous in almost all application domains; hence,

H. Kashima et al. (Eds.): PAKDD 2023, LNAI 13935, pp. 39–54, 2023.
https://doi.org/10.1007/978-3-031-33374-3_4

time-series anomaly detection has been actively studied for years, especially recently using machine learning. However, it remains a very challenging task for three key reasons: (i) labels for abnormal data are often rare, preventing proper training of supervised learning models; (ii) real-world time-series data is often subject to noise and characterized by complex temporal patterns that are difficult to identify; (iii) different datasets have different properties and thus often require a specific choice of parameters (e.g., using domain knowledge) for anomaly detectors to work well, making them hard to generalize.

To address these challenges, we propose a novel generative adversarial network (GAN) architecture called TSI-GAN for *unsupervised* time series anomaly detection. First, we encode the input time series to images to capture the temporal correlation and various types of deviance present in the time series, which explains part of our approach, TSI, which stands for *Time Series to Images*. This encoding also allows us to leverage GAN's outstanding performance on tasks of image generation [6] and image-to-image translation [9]. Second, we design a GAN with two critics and two generators that consist of convolutional layers in order to reconstruct the encoded images and obtain effective reconstruction errors. The purpose of the GAN is to learn a generalized distribution of normal samples such that it produces reconstruction errors that are (i) large on anomalous inputs and (ii) small on normal data even in the presence of noise and time non-stationarity. We also take a fully nonparametric approach throughout our design pipeline and as a result our model does not make any assumptions about the underlying data and does not require choosing parameters for each dataset, or altering model architectures like [25].

In addition, GAN-based methods typically sample a random latent and optimize it using gradient descent as a separate step during *inference* to find the latent representation that would yield an accurate inverse mapping for *each sample* [11,18]. This is highly inefficient on large datasets and impractical for real-time applications as proven by [25]. In contrast, we train an encoder-decoder network in our GAN with *cycle consistency loss* to obtain the latent representation of the inverse mapping automatically and immediately, making our inference almost instantaneous.

Third, as a further enhancement we address false positives (alarms), which are often a pain point in existing anomaly detection methods. To this end, we post-process the reconstruction errors using the *Hodrick-Prescott filter* [7] and then combine the errors from two encoding channels using a weighted sum. This way, we obtain a reliable anomaly score vector which leads to reduced false positives.

In summary, this paper makes the following contributions:

- We introduce TSI-GAN, a novel convolutional cycle-consistent GAN architecture that learns to reconstruct 2D-encoded complex 1D time-series data and produces reliable reconstruction errors for detecting non-trivial time series anomalies without any labels, and in real-time.
- We address the challenge of model generalization by taking a fully nonparametric approach throughout our design pipeline. As a result, our method

makes no assumptions about underlying data and requires no manual parameter choice, or changing model architectures.

- We mitigate false alarms as a common issue in anomaly detection, by post-processing the reconstruction error using a filtering technique and a weighting strategy.
- We benchmark TSI-GAN against eight state-of-the-art baseline methods on 250 well-curated and harder-than-usual datasets. The results validate our approach as the best performer overall, with a large winning margin over other methods.

Our results are fully reproducible, with code open-sourced at https://github.com/LabSAINT/TSI-GAN.

2 Related Work

Due to the importance of anomaly detection in many applications, research in this field has been active for years. While statistical methods are traditionally applied, machine learning and especially deep learning-based approaches have recently received increasingly more attention due to their attractive performance. These methods can generally be classified into:

Proximity-based methods classify a data point as a point anomaly or a sub-sequence as a collective anomaly when its locality is sparsely populated. These methods can be further classified into *cluster-based methods* such as k-means clustering [3], *distance-based methods* such as k-nearest neighbors [1], and *density-based methods* such as DBSCAN [2]. The main drawback of these methods when applied to time series anomaly detection is that they require the number of anomalies to be known a priori and are unable to capture temporal patterns. Time-series discord discovery [24] is a recently proposed distance-based method that identifies very unusual subsequences in a time series. Under this category, Nakamura et al. introduced MERLIN [13], which is considered to be the state-of-the-art for anomaly detection in univariate time series and is included as a baseline in our experiments.

Prediction-based methods try to predict future values of a time series and classify a data point as an anomaly if the predicted value differs from the real data by more than a specified threshold. Time series forecasting methods such as ARIMA [15] can be used, but they often require extensive examination and preprocessing of data and are sensitive to parameters. Several deep-learning approaches have been proposed to overcome these limitations. For example, Hundman et al. [8] proposed an LSTM model with dynamic thresholding (LSTM-DT) to make predictions and reduce false positives.

Reconstruction-based methods learn a latent low-dimensional representation of the input time-series data and try to reconstruct the input based on the representation. The assumption is that anomalies will lose information when mapped to the latent space and thus will not be reconstructed accurately, producing a larger reconstruction error. Hence, reconstruction error is measured at each time step and thresholding techniques are applied to detect the anomalies.

Several deep learning approaches have been proposed including LSTM-based Autoencoder (LSTM-AE) [14], Dense Autoencoder (DENSE-AE) [17], DONUT [22] which uses a Variational Autoencoder (VAE), and GAN-based methods [5,11,18]. TadGAN [5] presents a recent study using GAN to perform this task and it is considered to be state-of-the-art in terms of GAN-based methods.

However, TadGAN uses 1-D representation and requires the sampling interval of input data to be known for data preprocessing; otherwise, anomalies that do not have extreme amplitude (either high or low relative to other points) will not be detected. This is a notable limitation because most anomalies in the real world are complicated rather than just simple amplitude spikes or dips. Another related work is T2IVAE [23], which transforms time series to images and uses VAE to reconstruct the input time series. However, VAEs are prone to overfitting and often reconstruct anomalous samples quite accurately, resulting in unreliable reconstruction errors. Even though T2IVAE attempts to reduce this risk by employing an adversarial training strategy in the last five training epochs, the overfitting effect remains rather prominent.

We take a GAN-based approach instead of VAE because we find that GAN is strongly averse to the overfitting phenomenon when it comes to infrequent anomalous samples and unlike TadGAN we 2D encode the input time series and use CNN layers in our GAN to learn feature maps as if learning from images. This way, we are able to encode temporal information/correlation and capture various types of deviance and thus obtain more accurate and reliable anomaly scores based on reconstruction errors.

There are also **commercial tools** including Microsoft Azure Anomaly Detector [16] and LinkedIn Luminol [12]. Azure uses spectral residual (SR) from the saliency detection domain [16] and CNN to learn a discriminating threshold. The output is a sequence of labels indicating if a particular timestamp is anomalous. Luminol uses the Bitmap detector algorithm [20] which divides input time series into chunks and calculates the frequency of similar chunks to calculate anomaly scores. These commercial tools are included as baselines in our experiments as well.

3   Encoding Time-series to Images

The core idea behind encoding the time-series to images is that if any time step is anomalous, then the row and column corresponding to that time step in the encoded image will be significantly different from other normal pixels (see Fig. 1) and thus could be easily detected by a reconstruction-based model. Consider an input time series $\mathring{X} = \{x_1, x_2, ..., x_T\}$, where T is the time series length. We use a sliding window with window size W and step size S to divide $\mathring{X}$ into N overlapping sub-sequences, $\mathring{X}_k = \{x_{k+1}, x_{k+2}, \ldots, x_{k+W}\}$, where $k = 0, \ldots, N-1$ and $N = \lfloor \frac{T-W}{S} \rfloor$. We set $W = 64$ and $S = 1$ and convert each window of size 64 into a two-channel image of size $64 \times 64 \times 2$, using two time-series encoding techniques: Gramian Angular Field (GAF) [19] and Recurrence Plot (RP) [4].

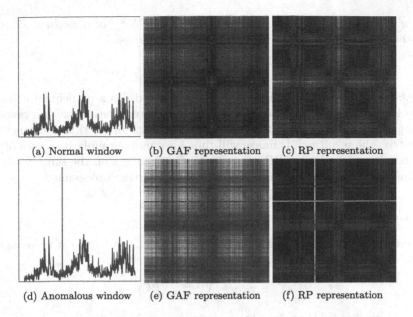

(a) Normal window   (b) GAF representation   (c) RP representation

(d) Anomalous window   (e) GAF representation   (f) RP representation

Fig. 1. Illustration of normal (top row) and anomalous (bottom row) windows encoded by GAF and RP, for an example time-series.

3.1 Gramian Angular Field (GAF)

Given a sub-sequence $\mathring{X}_k = \{x_{k+i}\}_{i=1}^{W}$ at time step k, GAF rescales all the observations into the interval $[-1, 1]$ and calculates $\bar{X}_k = \{\bar{x}_{k+i}\}_{i=1}^{W}$, where

$$\bar{x}_{k+i} = \frac{(x_{k+i} - \max(\mathring{X}_k)) + (x_{k+i} - \min(\mathring{X}_k))}{\max(\mathring{X}_k) - \min(\mathring{X}_k)}$$

Next, we represent each rescaled $\bar{X}_k$ using polar coordinates, as radius $r = t_{k+i}/W$ where $t_{k+i} \in \mathbb{N}$ is the timestamp, and angular $\phi = \arccos(\bar{x}_{k+i}) \in [0, \pi]$. This polar conversion produces a one-to-one mapping with a unique inverse function and preserves absolute temporal relation (as opposed to Cartesian coordinates). Thus, we can identify the temporal correlation at different time intervals by calculating the trigonometric sum between each point within the sub-sequence:

$$X_k^{GAF} = (\bar{X}_k)^T \otimes \bar{X}_k - \left(\sqrt{I - (\bar{X}_k)^2}\right)^T \otimes \sqrt{I - (\bar{X}_k)^2}$$

where X_k^{GAF} is a $W \times W$ matrix, I is the unit row vector $[1, 1, ..., 1]$ ($\bar{X}_k$ is a row vector too), and $\otimes$ represents outer product.

3.2 Recurrence Plot (RP)

A recurrence plot (RP) [4] is an image that represents the distance between observations extracted from a sub-sequence time series. Given a sub-sequence window $\mathring{X}_k = \{x_{k+i}\}_{i=1}^{W}$, we calculate a RP matrix X_k^{RP} of dimension $W \times W$ where each element at row a and column b is defined as

$$x_{k,(a,b)}^{RP} = \Theta(\epsilon - \|x_{k+a} - x_{k+b}\|), \ \forall a, b \in \{1, \cdots, W\} \tag{1}$$

where $\Theta(\cdot) : \mathbb{R} \to \{0, 1\}$ is a Heaviside function, and ϵ is a predefined distance threshold. In this work, instead of using binary representation, we use raw distances $\|x_{k+a} - x_{k+b}\|$ (without the need for choosing ϵ or $\Theta(\cdot)$) to construct the RP matrix; the resulting 2D image will thus have more granularity scales of the distances. In order to align RP images with GAF images on the same scale, we scale the RP matrix into the range $[-1, 1]$ before further processing.

3.3   Combining Two Channels

After encoding the series using GAF and RP, respectively, we treat them as two channels and stack them along the channel axis to obtain:

$$X_k = Stack(X_k^{GAF}, X_k^{RP})$$

Since we have divided the original time series $\mathring{X}$ into N overlapping sub-sequence windows of size W, and encoded each window as a 2-channel image of shape $[W \times W \times 2]$, we thus finally obtain a sequence of images $\mathbb{X} = \{X_k\}_{k=1}^{N}$.

4   The TSI-GAN Model

4.1   Model Architecture

Reconstruction-based anomaly detection methods learn a model that maps input data (in our case, an image with two channels) to the latent low-dimensional space and then reconstructs the input using the latent representation. The objective is to train a model that captures a generalized latent representation of the *normal* patterns, such that anomalies will *not* be reconstructed accurately and hence result in a larger reconstruction error. In our proposed method, we learn two mapping functions, $\mathcal{E} : \mathcal{X} \to \mathcal{Z}$ and $\mathcal{G} : \mathcal{Z} \to \mathcal{X}$, where $\mathcal{X}$ represents the input domain, $\mathcal{Z}$ represents the latent domain for which Gaussian distribution $\mathcal{N}(0, 1)$ is used. For any given input image at time step k, denoted by X_k, the model tries to reconstruct it as $X_k \to \mathcal{E}(X_k) \to \mathcal{G}(\mathcal{E}(X_k)) \approx \hat{X}_k$.

The entire model architecture is presented in Fig. 2. We model the above mapping functions as Generators, where $\mathcal{E}$ acts as an encoder which maps the input image to the latent space using convolution layers, and $\mathcal{G}$ acts as a decoder which transforms the latent representation to a reconstructed input image using transposed convolution. We use two Critics $\mathcal{C}_x$ and $\mathcal{C}_z$: $\mathcal{C}_x$ regulates the decoder

$\mathcal{G}$ by trying to distinguish real images X from the reconstructed images $\mathcal{G}(\mathcal{E}(x))$; $\mathcal{C}_z$ regulates the encoder $\mathcal{E}$ by trying to liken the latent representation $\mathcal{E}(x)$ to the Gaussian noise z. The L_2-norm will be used in our cycle consistency loss which we describe later in Sect. 4.2.

4.2   Loss Function and Training Strategy

We use two loss functions: (1) Wasserstein loss, to match the distribution of generated images with the distribution of input images, and (2) cycle consistency loss, to ensure the desired mapping route $X_k \to Z_k \to \hat{X}_k$.

Wasserstein Loss: We train the generator $\mathcal{G}$ and its critic $\mathcal{C}_x$ with Wasserstein loss:

$$\min_{\mathcal{G}} \max_{\mathcal{C}_x \in \mathbf{C}_x} L_X(\mathcal{C}_x, \mathcal{G}) \triangleq \mathbb{E}_{x \sim \mathbb{P}_X}[\mathcal{C}_x(x)] - \mathbb{E}_{z \sim \mathbb{P}_Z}[\mathcal{C}_x(\mathcal{G}(z))] \tag{2}$$

Similarly, for Encoder $\mathcal{E}$ and its Critic $\mathcal{C}_z$, the loss function is defined as:

$$\min_{\mathcal{E}} \max_{\mathcal{C}_z \in \mathbf{C}_z} L_Z(\mathcal{C}_z, \mathcal{E}) \triangleq \mathbb{E}_{z \sim \mathbb{P}_Z}[\mathcal{C}_z(z)] - \mathbb{E}_{x \sim \mathbb{P}_X}[\mathcal{C}_z(\mathcal{E}(x))] \tag{3}$$

where $\mathbf{C}_x$ and $\mathbf{C}_z$ are the set of all the 1-Lipschitz functions.

We also add a gradient penalty regularization term to both (2) and (3) to ensure a 1-Lipschitz continuous Critic so that Wasserstein Loss validly approximates the *Earth Mover's Distance* [10]. The complete architecture is presented in Table 1 and Fig. 2.

Cycle Consistency Loss:
The GAN model described above is able to map X_k to a desired Z_k. However, the inverse mapping of Z_k back to $\hat{X}_k$ is not guaranteed by training with just Wasserstein losses alone. This is because those losses only ensures distribution similarity but not instance similarity. To this end, Schlegl et al. [18] proposed an iterative approach where they

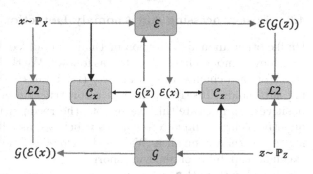

Fig. 2. TSI-GAN model architecture: $\mathcal{E}$ is the encoder and $\mathcal{G}$ the decoder; $\mathcal{C}_x$ and $\mathcal{C}_z$ are the critics.

sample a random latent and optimize it using gradient descent as a separate
step during *inference* to find the best Z_k that would generate $\mathcal{G}(\mathcal{E}(X_k))$ that is
most similar to the input image X_k. However, this method suffers from large
search space and is inefficient for large datasets and real-time applications, as
shown by Zenati et al. [25]. Hence, we use *cycle-consistency loss* [26] to train the
generators $\mathcal{E}$ and $\mathcal{G}$:

$$\min_{\mathcal{E}} L_{CL}(\mathcal{E}) \triangleq \mathbb{E}_{x \sim \mathbb{P}_X} \|x - \mathcal{G}(\mathcal{E}(x))\|_2 + \mathbb{E}_{z \sim \mathbb{P}_Z} \|z - \mathcal{E}(\mathcal{G}(z))\|_2 \qquad (4)$$

For $\mathcal{G}$ we only use the forward consistency loss, as the backward consistency loss
(i.e., $\mathbb{E}_{z \sim \mathbb{P}_Z} \|z - \mathcal{E}(\mathcal{G}(z))\|_2$) has been integrated into (4) and thus is not necessary
for $\mathcal{G}$:

$$\min_{\mathcal{G}} L_{CL}(\mathcal{G}) \triangleq \mathbb{E}_{x \sim \mathbb{P}_X} \|x - \mathcal{G}(\mathcal{E}(x))\|_2. \qquad (5)$$

Final Objective. Combining the objectives (2), (3), (4), (5) we arrive at the
final objective:

$$\min_{\{\mathcal{E},\mathcal{G}\}} \max_{\{C_x \in \mathbf{C}_x, C_z \in \mathbf{C}_z\}} L_X(C_x, \mathcal{G}) + L_Z(C_z, \mathcal{E}) + L_{CL}(\mathcal{E}) + L_{CL}(\mathcal{G}) \qquad (6)$$

4.3   Post-processing and Anomaly Detection

Unlike other anomaly detection methods, we added a post-processing procedure
to achieve a more reliable detector, as follows. We then extract the reconstruction
error for each channel as ϵ_{gaf} and ϵ_{rp} (refer lines 1–5 in Algorithm 1). Calculating thresholds directly on the raw reconstruction error will lead to many false
positives. To mitigate this, we smooth the reconstruction error to suppress frequently occurring minor error peaks which are usually caused by normal behavior rather than anomalies. We use the *Hodrick-Prescott filter* [7] because of its
excellent capability of removing short-term fluctuations in data since we are only
concerned with peaks that persist for a sustained period of time. It extracts a
smooth trend r from a given sequence ϵ of length N by solving:

$$\min_{r} \left(\sum_{k=1}^{N} (\epsilon_k - r_k)^2 + \lambda \sum_{k=2}^{N-1} [(r_{k+1} - r_k) - (r_k - r_{k-1})]^2 \right)$$

Algorithm 1: Anomaly Detection using TSI-GAN

// Compute reconstruction errors

1 for $k = 1, ..., N$ do
2 $\quad \hat{X}_k \leftarrow \mathcal{G}(\mathcal{E}(X_k))$
3 $\quad \hat{X}_k^{GAF}, \hat{X}_k^{RP} \leftarrow$ extract GAF & RP channels from $\hat{X}_k$
4 $\quad \epsilon_{gaf_k} \leftarrow \sum_{i=1}^{W} \sum_{j=1}^{W} (X_{k,i,j}^{GAF} - \hat{X}_{k,i,j}^{GAF})^2$
5 $\quad \epsilon_{rp_k} \leftarrow \sum_{i=1}^{W} \sum_{j=1}^{W} (X_{k,i,j}^{RP} - \hat{X}_{k,i,j}^{RP})^2$

// Post-processing to obtain anomaly scores

6 for $ch \in \{gaf, rp\}$ do
7 $\quad \epsilon_{ch} \leftarrow$ HP(ϵ_{ch}) // Hodrick-Prescott filter
8 $\quad peaks_{ch} \leftarrow$ find_peaks(ϵ_{ch})
9 $\quad \sigma_{ch} \leftarrow \frac{(peaks_{ch}[0] - peaks_{ch}[1])}{peaks_{ch}[0]} + 1$

10 $score\_vec \leftarrow \sigma_{gaf} \times \epsilon_{gaf} + \sigma_{rp} \times \epsilon_{rp}$

// Detect anomalies

11 $mean \leftarrow$ mean($score\_vec$)
12 for $k = 1, ..., N$ do
13 $\quad$ if $score\_vec_k > mean$ then
14 $\quad\quad pred_k = true$
15 $\quad$ else
16 $\quad\quad pred_k = false$

17 Group consecutive $pred_k$'s into $\{seq_i\}_{i=1}^{L}$

// Pruning to reduce false alarms

18 $\{m_i\} \leftarrow$ max($\{seq_i\}$)
19 $\{m_i\} \leftarrow$ sort($\{m_i\}, descending = true$)
20 sort $\{seq_i\}$ in the same order of $\{m_i\}$
21 for $i = 1, ..., L$ do
22 $\quad p_i \leftarrow (m_{i-1} - m_i)/m_{i-1}$
23 $\quad$ if $p_i < \theta$ then
24 $\quad\quad$ reclassify $\{seq_j\}_{j=i}^{L}$ as normal
25 $\quad\quad$ break;

After smoothing ϵ_{gaf} and ϵ_{rp}, we find the local (neighborhood) peaks in each channel and sort them in descending order to calculate a confidence level $\sigma \in [1, 2]$ for each channel:

$$\sigma = \frac{peaks[0] - peaks[1]}{peaks[0]} + 1,$$

where $peaks[0]$ and $peaks[1]$ are the first and the second highest peaks in the smoothed reconstruction errors, respectively. The idea is that when the difference between these two peaks is large, that channel is assumed to be more confident about its detection of the anomaly and hence weighed higher in the final anomaly score. This score is defined by combining the two reconstruction errors ϵ_{gaf} and ϵ_{rp} using their respective confidence level (see line 10 in Algorithm 1). Here it is defined as a vector (of length N) because each of the N windows will have an anomaly score. For the weight σ, if there are multiple anomalies, the difference between $peaks[0]$ and $peaks[1]$ will be small and thus σ will be smaller than the other channel if the other channel detects a single outlier, which is desired since outliers are rare by definition and thus single outliers are more likely than multiple. Otherwise, if both channels detect multiple, they will be weighted by similar σ's.

After obtaining the anomaly score for each window, we calculate the mean anomaly score over all the windows and any window that exceeds this threshold is flagged as an anomaly. Following that, consecutive anomalous windows will be grouped together to form a sequence (i.e., collective anomaly). Finally, we use an anomaly pruning approach (lines 17–25 in Algorithm 1) introduced by

Hundman et al. [8], to further mitigate false positives. The above post-processing and detection procedures are formulated in Algorithm 1.

5 Performance Evaluation

5.1 Datasets

We use the UCR 2021 anomaly detection dataset[1] which contains 250 sub-datasets collected from a variety of sources. Unlike commonly used datasets such as Yahoo, Numenta, and NASA which are found to have numerous flaws [21] including incorrect ground truth labels, triviality of the anomalies, and unrealistic anomaly density, this UCR dataset is carefully curated, harder to detect and is much more reliable. Moreover, this dataset contains a combination of point, collective, and contextual anomalies as well as amplitude, seasonal, and trend anomalies, which offers a good variety for evaluation.

We choose a total of 6 categories from this dataset and each category contains 4–13 original sub-datasets; each original sub-dataset comes with a distorted duplicate by adding artificial fluctuations. Therefore, the number of sub-datasets is doubled. The only exception is the `Noise` category in which the sub-datasets are chosen from multiple other categories with Gaussian noise added. A brief description of each category is as follows: `AirTemperature` consists of hourly air temperature between 03/01 and 03/31 from 2009 to 2019, collected from CIMIS station 44 in Riverside, CA. `PowerDemand` consists of Italian power demand data between1/1/1995 and 5/31/1998.

Table 1. Architecture of our proposed TSI-GAN. Transp. Conv: Transposed Convolution; BN: Batch Normalization; LN: Layer Normalization respectively; LReLU: LeakyReLU; Lrn. rate: learning rate.

Operation	Kernel	Strides	Units	BN?	Activation
Encoder					
Convolution	7×7	3×3	48	✓	ReLU
Convolution	5×5	3×3	96	✓	ReLU
Convolution	4×4	2×2	192	✓	ReLU
Convolution	2×2	1×1	z_dim	✗	−
Decoder					
Transp. Conv	2×2	1×1	192	✓	LReLU
Transp. Conv	4×4	2×2	96	✓	LReLU
Transp. Conv	5×5	3×3	48	✓	LReLU
Transp. Conv	7×7	3×3	2	✗	Tanh
Critic X					
Convolution	7×7	3×3	48	LN	LReLU
Convolution	5×5	3×3	96	LN	LReLU
Convolution	4×4	2×2	192	LN	LReLU
Convolution	2×2	1×1	1	✗	−
Critic Z					
Fully Conn.			50	LN	ReLU
Fully Conn.			25	LN	ReLU
Fully Conn.			1	✗	−
Hyperparams.					
z_dim	100	Lrn. rate (α)	1e-4	Iterations	5000
Optimizer	RMSProp	Wt decay (λ_{wd})	1e-4	BatchSize	128

---

[1] UCR 2021 anomaly detection dataset: https://bit.ly/3V2n6FY.

Table 2. Statistics of Datasets used in our experiments.

Property	Dataset						
	AirTemperature	PowerDemand	InternalBleeding	EPG	NASA T-1	Noise	All datasets
# Sub-datasets	14	8	26	12	10	16	250
# Data Points	98208	239448	194992	359304	113488	629494	19353766
# Anomalous Points	398	1688	3018	1292	644	3134	49363
# (% tot.)	0.004%	0.007%	0.015%	0.003%	0.005%	0.004%	0.002%

`InternalBleeding` consists of the arterial blood pressure measurements of pigs. EPG is collected from an insect known as Asian Citrus Psyllid, recorded using an Electropalatography (EPG) apparatus. `NASA T-1` is collected from NASA Mars Science Laboratory (MSL) dataset that consists of spacecraft telemetry signals. Detailed statistics of each category and all the datasets is presented in Table 2.

5.2 Performance Metrics

In real-world application scenarios, most anomalies happen in the form of collective anomalies and hence we use the window-based rules introduced by Hundman et al. [8]: (1) If an anomalous window overlaps any predicted window, a true positive (TP) is recorded; (2) If a predicted window does not overlap with any anomalous window, a false positive (FP) is recorded; (3) If an anomalous window does not overlap with any predicted window, a false negative (FN) is recorded. Based on this set of rules, we calculate *Precision* and *F1-Score* as the performance metrics.

5.3 Experimental Results

Table 3 reports the average F1-Score on the original and distorted datasets for each category, and in the last column, the F1-Score and Precision averaged over all the 250 datasets.

Table 3. Average F1-Score on original and distorted datasets for each category, as well as F1-Score and Precision averaged over all the 250 datasets.

Model	AirTemperature		PowerDemand		InternalBleeding		EPG		NASA T-1		Noise	All 250 datasets	
	Orig.	Distor.	Orig.	Distor.	Orig.	Distor.	Orig.	Distor.	Orig.	Distor.		F1	Precision
TSI-GAN	**1.0**	**0.833**	**0.667**	**0.667**	0.846	0.474	**0.5**	**0.556**	**0.933**	0.267	0.479	**0.468**	**0.445**
MERLIN	0.054	0.18	0.04	0.071	**0.926**	**0.721**	0.354	0.191	0.613	**0.6**	**0.49**	0.414	0.402
LSTM-AE	0.389	0.611	0.375	0.583	0.654	0.308	0.222	0.444	0.533	0.333	0.208	0.355	0.301
DONUT	0.611	0.444	0.083	0.1	0.59	0.564	0.278	0.167	0.333	0.533	0.458	0.351	0.325
LSTM-DT	0.778	**0.833**	0.25	0.5	0.615	0.449	0.222	0.222	0.6	**0.6**	0.271	0.32	0.289
DENSE-AE	0.194	0.111	0.0	0.0	0.231	0.077	0.222	0.222	0.2	0.0	0.271	0.159	0.136
TadGAN	0.0	0.133	0.0	0.0	0.282	0.24	0.233	0.189	0.267	0.2	0.171	0.131	0.092
Azure	0.181	0.199	0.083	0.196	0.099	0.176	0.167	0.167	0.007	0.017	0.084	0.05	0.037
Luminol	0.022	0.021	0.078	0.089	0.118	0.046	0.037	0.088	0.009	0.014	0.019	0.049	0.021

Overall, it is observed
that TSI-GAN achieves
an F1-Score of 0.468 and
Precision of 0.445, out-
performing all the base-
line methods. More specifically, TSI-GAN offers an
improvement of 13% and
31% on F1-score over
the second and the third
best methods, MERLIN
(0.414) and LSTM-AE
(0.355), respectively. We
note that 85–95% of the
improvement was attri-
buted to GAN, while 5-
15% was attributed to
post-processing. When the
individual categories are

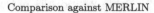
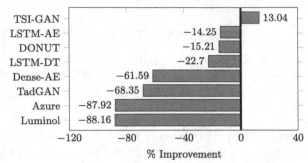

Fig. 3. Comparing all anomaly detection methods against MERLIN in terms of F1 score averaged across all 250 datasets, expressed as a percentage of improvement.

considered, TSI-GAN performs the best on AirTemperature, PowerDemand, EPG for both original and distorted datasets and wins over other methods by a significant margin; it also offers competitive performance on other categories (InternalBleeding, NASA-T1 Distorted, and Noise) as well.

Using MERLIN as a benchmark, we measure the performance difference between each method and MERLIN in Fig. 3. It indicates that TSI-GAN is the only one that offers a positive performance improvement while all the other methods underperform MERLIN.

Among all the deep learning-based methods, LSTM-AE performs the best, with an average F1-Score of 0.355; DONUT comes in second with a slightly lower score of 0.351. We examine the possible reasons for their shortfall as compared to TSI-GAN and how our approach overcomes them. As we mentioned earlier, autoencoder-based methods carry the risk of overfitting anomalies during train-ing, by reconstructing anomalous samples just as accurately as normal samples. DONUT which employs VAE has this tendency as can be seen in Fig. 4. This is also a plausible reason for the underperformance of other autoencoder-based models such as LSTM-AE.

In contrast, TSI-GAN uses an adversarial training strategy which makes our model largely immune to this behavior. However, while TadGAN and many others alike also use adversarial learning, they are unable to capture anomalies that are not amplitude spikes or dips unless dataset-specific parameters such as sampling interval are known. The reason is that they do not instrument feature engineering to capture anomalies that deviate in seasonality, trend, etc., and therefore tend to only detect extremely high or low amplitude points in the

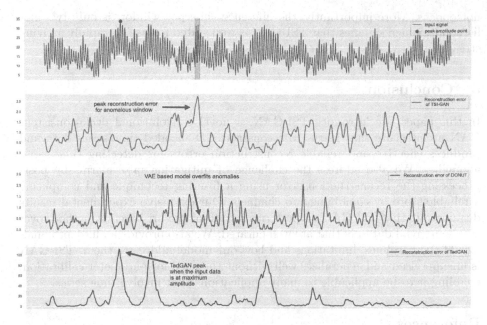

Fig. 4. Illustration of TSI-GAN vs. DONUT vs. TadGAN when applied to an example time series. The translucent red interval depicts the ground-truth anomaly. (Color figure online)

Table 4. Training and Inference time of TSI-GAN

	AirTemperature			PowerDemand		
	# of Samples	Total time	Per-window time	# of Samples	Total time	Per-window time
Training	7996	251 s	0.06 s	29772	246 s	0.01 s
Inference	4083	10 s	0.002 s	11862	27 s	0.002 s

input as can be observed in Fig. 4. It is the main reason why TadGAN only performs well on datasets in which all anomalies are either amplitude spikes or dips; such anomalies, however, are *trivial* to detect as pointed out by [21]. On the other hand, TSI-GAN uses GAF and RP encoding which substantially enhances its ability to detect various types of non-trivial deviance, as can be observed in Fig. 4.

Time Efficiency. We report the training and inference time of TSI-GAN in Table 4. The times are measured on a NVIDIA RTX 3070 GPU with 8GB of VRAM along with AMD Ryzen 7 5800H @ 3.20 GHz CPU.

We can see that the training time remains almost constant irrespective of the number of training samples, the reason is that we train for iterations and not epochs. More importantly, the inference time per window is only two milliseconds, which signifies that TSI-GAN is well suited for use on rapidly arriving streaming data.

6 Conclusion

In this paper, we introduce TSI-GAN, a novel convolutional cycle-consistent GAN architecture that learns to reconstruct 2D-encoded time-series data and produces effective and reliable reconstruction errors for detecting time series anomalies. We also address the challenge of mitigating false alarms by post-processing the reconstruction error using a filtering technique and computing a reliable score by combining two channels. Our extensive experimental results demonstrate that TSI-GAN outperforms 8 state-of-the-art baseline methods over 250 non-trivial datasets that are well-curated. We also provide an in-depth analysis of the baselines' limitations and how our model addresses them. TSI-GAN is unsupervised and generalizes well without the need for parameter calibration, enabling it to be applicable to many applications that involve time series.

References

1. Angiulli, F., Pizzuti, C.: Fast outlier detection in high dimensional spaces. In: Elomaa, T., Mannila, H., Toivonen, H. (eds.) PKDD 2002. LNCS, vol. 2431, pp. 15–27. Springer, Heidelberg (2002). https://doi.org/10.1007/3-540-45681-3_2
2. Çelik, M., Dadaşer-Çelik, F., Dokuz, A.Ş.: Anomaly detection in temperature data using DBSCAN algorithm. In: 2011 International Symposium on Innovations in Intelligent Systems and Applications, pp. 91–95. IEEE (2011)
3. Chawla, S., Gionis, A.: k-means-: a unified approach to clustering and outlier detection. In: Proceedings of the 2013 SIAM International Conference on Data Mining, pp. 189–197. SIAM (2013)
4. Eckmann, J.P., Kamphorst, S.O., Ruelle, D., et al.: Recurrence plots of dynamical systems. World Sci. Seri. Nonlinear Sci. Ser. A **16**, 441–446 (1995)
5. Geiger, A., Liu, D., Alnegheimish, S., Cuesta-Infante, A., Veeramachaneni, K.: Tadgan: time series anomaly detection using generative adversarial networks. In: 2020 IEEE International Conference on Big Data (Big Data), pp. 33–43. IEEE (2020)
6. Gulrajani, I., Ahmed, F., Arjovsky, M., Dumoulin, V., Courville, A.C.: Improved training of wasserstein gans. In: Guyon, I., et al. (eds.) Advances in Neural Information Processing Systems, vol. 30. Curran Associates, Inc. (2017)
7. Hodrick, R.J., Prescott, E.C.: Postwar us business cycles: an empirical investigation. J. Money Credit Banking 1–16 (1997)
8. Hundman, K., Constantinou, V., Laporte, C., Colwell, I., Soderstrom, T.: Detecting spacecraft anomalies using LSTMS and nonparametric dynamic thresholding. In: Proceedings of the 24th ACM SIGKDD International Conference on Knowledge Discovery & Data Mining, pp. 387–395 (2018)

9. Isola, P., Zhu, J.Y., Zhou, T., Efros, A.A.: Image-to-image translation with conditional adversarial networks. In: Proceedings of the IEEE Conference on Computer Vision and Pattern Recognition, pp. 1125–1134 (2017)

10. Levina, E., Bickel, P.: The earth mover's distance is the mallows distance: some insights from statistics. In: Proceedings Eighth IEEE International Conference on Computer Vision (ICCV), vol. 2, pp. 251–256. IEEE (2001)

11. Li, D., Chen, D., Jin, B., Shi, L., Goh, J., Ng, S.-K.: MAD-GAN: multivariate anomaly detection for time series data with generative adversarial networks. In: Tetko, I.V., Kůrková, V., Karpov, P., Theis, F. (eds.) ICANN 2019. LNCS, vol. 11730, pp. 703–716. Springer, Cham (2019). https://doi.org/10.1007/978-3-030-30490-4_56

12. Linkedin: Luminol (2015). https://github.com/linkedin/luminol. Access 15 May 2022

13. Nakamura, T., Imamura, M., Mercer, R., Keogh, E.: Merlin: parameter-free discovery of arbitrary length anomalies in massive time series archives. In: 2020 IEEE International Conference on Data Mining (ICDM), pp. 1190–1195. IEEE (2020)

14. Nguyen, H., Tran, K.P., Thomassey, S., Hamad, M.: Forecasting and anomaly detection approaches using LSTM and LSTM autoencoder techniques with the applications in supply chain management. Int. J. Inf. Manage. **57**, 102282 (2021)

15. Pena, E.H., de Assis, M.V., Proença, M.L.: Anomaly detection using forecasting methods Arima and HWDs. In: 2013 32nd international conference of the Chilean computer science society (SCCC), pp. 63–66. IEEE (2013)

16. Ren, H., et al.: Time-series anomaly detection service at Microsoft. In: Proceedings of the 25th ACM SIGKDD International Conference on Knowledge Discovery & Data Mining, pp. 3009–3017 (2019)

17. Sakurada, M., Yairi, T.: Anomaly detection using autoencoders with nonlinear dimensionality reduction. In: Proceedings of the MLSDA 2014 2nd Workshop on Machine Learning for Sensory Data Analysis, pp. 4–11 (2014)

18. Schlegl, T., Seeböck, P., Waldstein, S.M., Schmidt-Erfurth, U., Langs, G.: Unsupervised anomaly detection with generative adversarial networks to guide marker discovery. In: Niethammer, M., et al. (eds.) IPMI 2017. LNCS, vol. 10265, pp. 146–157. Springer, Cham (2017). https://doi.org/10.1007/978-3-319-59050-9_12

19. Wang, Z., Oates, T.: Encoding time series as images for visual inspection and classification using tiled convolutional neural networks. In: Workshops at the Twenty-Ninth AAAI Conference on Artificial Intelligence (2015)

20. Wei, L., Kumar, N., Lolla, V.N., Keogh, E.J., Lonardi, S., Ratanamahatana, C.A.: Assumption-free anomaly detection in time series. In: SSDBM, vol. 5, pp. 237–242 (2005)

21. Wu, R., Keogh, E.: Current time series anomaly detection benchmarks are flawed and are creating the illusion of progress. IEEE Trans. Knowl. Data Eng. (2021)

22. Xu, H., et al.: Unsupervised anomaly detection via variational auto-encoder for seasonal KPIS in web applications. In: Proceedings of the 2018 World Wide Web Conference on World Wide Web, pp. 187–196. International World Wide Web Conferences Steering Committee (2018)

23. Xu, L., et al.: NVAE-GAN based approach for unsupervised time series anomaly detection. arXiv preprint arXiv:2101.02908 (2021), not peer-reviewed or published, code not available

24. Yankov, D., Keogh, E., Rebbapragada, U.: Disk aware discord discovery: finding unusual time series in terabyte sized datasets. Knowl. Inf. Syst. **17**(2), 241–262 (2008)

25. Zenati, H., Romain, M., Foo, C.S., Lecouat, B., Chandrasekhar, V.: Adversarially learned anomaly detection. In: 2018 IEEE International conference on data mining (ICDM), pp. 727–736. IEEE (2018)
26. Zhu, J.Y., Park, T., Isola, P., Efros, A.A.: Unpaired image-to-image translation using cycle-consistent adversarial networks. In: Proceedings of the IEEE International Conference on Computer Vision, pp. 2223–2232 (2017)

Achieving Counterfactual Fairness
for Anomaly Detection

Xiao Han[1], Lu Zhang[2], Yongkai Wu[3], and Shuhan Yuan[1(✉)]

[1] Utah State University, Logan, UT 84322, USA
{xiao.han,shuhan.yuan}@usu.edu
[2] University of Arkansas, Fayetteville, AR 72701, USA
lz006@uark.edu
[3] Clemson University, Clemson, SC 29634, USA
yongkaw@clemson.edu

Abstract. Ensuring fairness in anomaly detection models has received much attention recently as many anomaly detection applications involve human beings. However, existing fair anomaly detection approaches mainly focus on association-based fairness notions. In this work, we target counterfactual fairness, which is a prevalent causation-based fairness notion. The goal of counterfactually fair anomaly detection is to ensure that the detection outcome of an individual in the factual world is the same as that in the counterfactual world where the individual had belonged to a different group. To this end, we propose a counterfactually fair anomaly detection (CFAD) framework which consists of two phases, counterfactual data generation and fair anomaly detection. Experimental results on a synthetic dataset and two real datasets show that CFAD can effectively detect anomalies as well as ensure counterfactual fairness.

Keywords: Anomaly Detection · Counterfactual Fairness

1 Introduction

Anomaly detection, which aims to detect samples that are deviated from the normal ones, has a wide spectrum of applications. Recently, deep anomaly detection models, powered by complex deep neural nets, have made promising progress in effectively detecting anomalies. Besides effectiveness, researchers recently notice the importance of taking the societal impact of anomaly detection into consideration as many anomaly detection tasks involve human individuals. Fairness as one fundamental component to build trustworthy AI has received much attention. Recent studies have shown that anomaly detection models can incur discrimination against certain groups. For example, a deep anomaly detection model could overly flag black males as anomalies [16]. In the scenarios of credit risk analysis, anomaly detection models predict more females as anomalies [15].

Several fair anomaly detection models have been proposed, which ensure no discrimination against a particular group based on the sensitive feature [1,3,14–16]. However, these approaches mainly focus on achieving association-based fairness notions like demographic parity. Recent studies have demonstrated

H. Kashima et al. (Eds.): PAKDD 2023, LNAI 13935, pp. 55–66, 2023.
https://doi.org/10.1007/978-3-031-33374-3_5

the importance of treating fairness as causation-based notions that concern the causal effect of the sensitive feature on the model outcomes [2, 8, 11]. Counter-factual fairness is one important causation-based fairness notion [9]. It considers that a model is fair if, for a particular individual, the model outcome in the factual world is the same as that in the counterfactual world where the individual had belonged to a different group. To the best of our knowledge, no studies have been conducted to ensure counterfactual fairness in anomaly detection.

In this work, we focus on counterfactual fairness for anomaly detection with the goal to ensure that the detection outcomes remain consistent in both the factual and counterfactual worlds. Achieving counterfactual fairness for anomaly detection is challenging. First, we can only observe the factual data. The counterfactual data are unobservable and cannot be obtained by simply changing the sensitive feature of the factual data. This is because the data generation is governed by an underlying causal mechanism where any intervention on one feature will subsequently affect the values of other features. Second, in anomaly detection, we can only observe factual normal data. Building a detection model which ensures the detection results be unchanged for individuals across the factual and counterfactual worlds while also preserving high anomaly detection performance imposes additional challenges.

To tackle the above challenges, we propose a Counterfactually Fair Anomaly Detection (CFAD) framework. We do not require the knowledge of the causal graph and structural equations but only assume that the data generation follows a generalized linear Structural Causal Model (SCM). We use an autoencoder as the base anomaly detection model where the anomaly score of a sample is derived based on the reconstruction error of the autoencoder. Then, we propose a two-phase approach. In the first phase, motivated by [12] which leverages the graph autoencoder for causal structure learning from observed data, we develop an approach to generate counterfactual data based on a graph autoencoder. In the second phase, we apply adversarial training [6, 10] on a vanilla autoencoder to achieve counterfactual fairness for anomaly detection. The idea is to ensure that the hidden representations of factual and counterfactual data derived from the encoder cannot be distinguished by a discriminator. As a result, the reconstruction error, i.e., anomaly score, will not differ much between the factual and counterfactual data, leading to similar detection results for both factual and counterfactual data.

2   Preliminary

Structural Causal Model (SCM). Our work adopts Pearl's Structural Causal Model (SCM) [13] as the prime methodology for defining and measuring counterfactual fairness. Throughout this paper, we use the upper/lower case alphabet to represent variables/values.

Definition 1. *An SCM is a triple $\mathcal{M} = \{U, V, F\}$ where*

1) U is a set of exogenous variables that are determined by factors outside the model. A joint probability distribution $P(u)$ is defined over the variables in U.

2) V is a set of endogenous variables that are determined by variables in $U \cup V$.

3) F is a set of deterministic functions $\{f_1, \ldots, f_n\}$; for each $X_i \in V$, a corresponding function f_i is a mapping from $U \cup (V \setminus \{X_i\})$ to X_i, i.e., $X_i = f_i(X_{pa(i)}, U_i)$, where $X_{pa(i)} \subseteq V \setminus \{X_i\}$ called the parents of X_i, and $U_i \subseteq U$.

An SCM is often illustrated by a causal graph $\mathcal{G}$ where each observed variable is represented by a node, and the causal relationships are represented by directed edges $\rightarrow$. In this graphical representation, the definition of parents is consistent with that in the SCM.

Inferring causal effects in the SCM is facilitated by the do-operator which simulates the physical interventions that force some variable $X \in V$ to take a certain value x. For an SCM $\mathcal{M}$, intervention $do(X = x)$ is equivalent to replacing original function in F with $X = x$. After the replacement, the distributions of all variables that are the descendants of X may be changed. We call the SCM after the intervention the submodel, denoted by $\mathcal{M}[x]$. For any variable $Y \in V$ which is affected by the intervention, its interventional variant in submodel $M[x]$ is denoted by $Y[x]$.

Counterfactuals. Counterfactuals are about answering questions such as for two variables $X, Y \in V$, whether Y would be y had X been x in unit (or situation) $U = u$. Such question involves two worlds, the factual world represented by $\mathcal{M}$ and the counterfactual world represented by $\mathcal{M}[x]$, and hence cannot be answered directly by the do-operator. When the complete knowledge of the SCM is known, the counterfactual quantity can be computed by the three-step process:

1) Abduction: Update $P(u)$ by evidence e to obtain $P(u|e)$.
2) Action: Modify $\mathcal{M}$ by performing intervention $do(x)$ to obtain the submodel $\mathcal{M}[x]$.
3) Prediction: Use modified submodel $\mathcal{M}[x]$ with updated probability $P(u|e)$ to compute the probability of $Y = y$.

3 Counterfactually Fair Anomaly Detection

3.1 Counterfactual Fairness

We start by defining counterfactual fairness in the context of anomaly detection. Following the typical anomaly detection setting, we assume a training set $\mathcal{D} = \{d^{(n)}\}_{n=1}^{N}$ which consists of N normal samples/individuals and a test set that consists of both normal samples and anomalies. Each sample is given by $d^{(n)} = \{s^{(n)}, x^{(n)}\}$ where S denotes a binary sensitive variable and $X = \{X_i \mid i = 1 : m\}$ denotes all other variables (i.e., profile attributes). We then use Y to denote the anomaly label. For representation, we use $S = \{s^+, s^-\}$ to denote advantage and disadvantage groups respectively, and use $Y = \{0, 1\}$ to denote normal samples and anomalies respectively. The goal is to learn a detection model for computing an anomaly score $g(x^{(n)})$ based on the profile attributes for each individual n, which can be used to judge whether it is an anomaly.

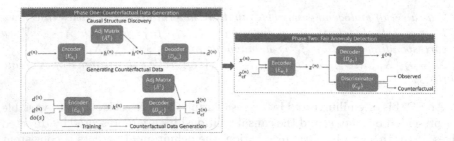

Fig. 1. Framework of CFAD

To define counterfactual fairness, similar to [9], for each individual $d^{(n)}$ we consider its instance in the counterfactual world $\mathcal{M}_s$ by flipping the value of its sensitive variable to the opposite s (i.e., s^+ becomes s^- and vice versa), denoted by $d_{\text{cf}}^{(n)} = \{s, x_{\text{cf}}^{(n)}\}$ where $x_{\text{cf}}^{(n)}$ represents the profile attributes in the counterfactual world. Note that $x_{\text{cf}}^{(n)}$ may not be the same as $x^{(n)}$ due to the causal relation between S and X in the underlying data generation mechanism. Then, counterfactual fairness is defined as:

Definition 2. *An anomaly detection model is counterfactually fair if for each individual n we have* $g(x^{(n)}) = g(x_{cf}^{(n)})$.

3.2 Overview of Counterfactually Fair Anomaly Detection (CFAD)

The goal of CFAD is to train an anomaly detection model on $\mathcal{D}$ that can: (1) effectively detect anomalies, and (2) ensure counterfactual fairness. To achieve this goal, CFAD consists of two phases, counterfactual data generation and fair anomaly detection. Counterfactual data generation is to generate a counterfactual dataset $\mathcal{D}_{\text{cf}} = \{d_{\text{cf}}^{(n)}\}_{n=1}^{N}$ of $\mathcal{D}$ in which each counterfactual sample is generated by the submodel which flips the value of the sensitive variable to its counterpart. To this end, we assume a generalized linear SCM and develop a novel graph autoencoder for data generation. In the second phase, we make use of a standard autoencoder for anomaly detection where the anomaly score is derived based on the reconstruction error. To achieve fairness, we develop an adversarial training framework to train the autoencoder by taking the factual and counterfactual data as inputs. The idea is to make the hidden representations of the autoencoder not encode the information of the sensitive variable so that intervening the sensitive variable would not change the detection outcome. Figure 1 shows the framework of CFAD.

3.3 Phase One: Counterfactual Data Generation

We assume that the data generation follows a generalized linear SCM, which is a common assumption in gradient-based causal discovery. To ease representation, we also assume that S has no parents in the SCM. Our method can easily extend to cases where S has parents by keeping the values of S's parents unchanged in the counterfactual world since the intervention on S has no influence on its parents. Thus, W.L.O.G. the structural equation of each variable X_i in X can be written as follows.

$$X_i = A_{1,i} \cdot f(S) + \sum_{X_j \in X_{pa(i)} \backslash \{S\}} A_{j,i} \cdot f(X_j) + U_i, \qquad (1)$$

where $f(\cdot)$ can be any linear/nonlinear function and $A_{j,i}$ is an element in the adjacency matrix $A \in \mathbb{R}^{(m+1) \times (m+1)}$ which indicates the weights of the generalized linear SCM. Each sample $d^{(n)} = \{s^{(n)}, \{x_i^{(n)} \mid i = 1 : m\}\}$ satisfies Eq. (1). Following the Abduction-Action-Prediction process, from Eq. (1), we have

$$u_i^{(n)} = x_i^{(n)} - A_{1,i} \cdot f(s^{(n)}) - \sum_{X_j \in X_{pa(i)} \backslash \{S\}} A_{j,i} \cdot f(x_j^{(n)}).$$

Meanwhile, by performing intervention to flip $s^{(n)}$ to its counterpart s, the structural equation of counterfactual variable $X_i[s]$ in the submodel $\mathcal{M}[s]$ of Eq. (1) is given by

$$X_i[s] = A_{1,i} \cdot f(s) + \sum_{X_j \in X_{pa(i)} \backslash \{S\}} A_{j,i} \cdot f(X_j[s]) + U_i. \qquad (2)$$

Note that S is fixed to s by the intervention and U_i is not affected by the intervention. Denoting the counterfactual of $d^{(n)}$ by $d_{cf}^{(n)} = \{s, \{x_i^{(n)}[s] \mid i = 1 : m\}\}$, it should satisfy Eq. (2). Thus, we have

$$x_i^{(n)}[s] = A_{1,i} \cdot f(s) + \sum_{X_j \in X_{pa(i)} \backslash \{S\}} A_{j,i} \cdot f(x_j^{(n)}[s]) + u_i^{(n)},$$

which leads to

$$x_i^{(n)}[s] = A_{1,i} \cdot f(s) + \sum_{X_j \in X_{pa(i)} \backslash \{S\}} A_{j,i} \cdot f(x_j^{(n)}[s]) + x_i^{(n)} - A_{1,i} \cdot f(s^{(n)}) - \sum_{X_j \in X_{pa(i)} \backslash \{S\}} A_{j,i} \cdot f(x_j^{(n)}). \qquad (3)$$

Finally, we compute the value of $x_i^{(n)}[s]$ according to Eq. (3) following the topological order and derive $d_{cf}^{(n)}$ from the observational data.

The challenge in the above derivation is how to estimate function $f(\cdot)$ and adjacency matrix A of the SCM. Next, we develop a causal structure discovery approach based on the graph autoencoder as proposed in [12].

Causal Structure Discovery. We estimate the adjacency matrix of the SCM defined in Eq. (1) by a graph autoencoder model with parameters $\{\theta_1, \phi_1, \hat{A}\}$. Specifically, an encoder is first adopted to derive the hidden representation of a sample $d^{(n)}$, i.e., $h^{(n)} = E_{\theta_1}(d^{(n)})$, where $E_{\theta_1}(\cdot)$ is parameterized by a multilayer neural network. Then, the message passing operation is applied on the hidden representation, i.e., $h'^{(n)} = \hat{A}^T h^{(n)}$, where $\hat{A}$ is a parameter matrix. Finally, a decoder is used to reconstruct the original input from $h'^{(n)}$, i.e.,

$$\hat{d}^{(n)} = D_{\phi_1}(h'^{(n)}) = D_{\phi_1}(\hat{A}^T E_{\theta_1}(d^{(n)})),$$

where $D_{\phi_1}(\cdot)$ is parameterized by a different multilayer neural network. Note that both the encoder $E_{\theta_1}(\cdot)$ and the decoder $D_{\phi_1}(\cdot)$ work in a variable-wise manner in order to preserve the order of the message passing in the SCM. To train the graph autoencoder model, the objective function is defined as:

$$\mathcal{L}_{\text{GAE}}(A, \theta_1, \phi_1) = \frac{1}{2N} \sum_{n=1}^{N} \|d^{(n)} - \hat{d}^{(n)}\|_2^2 + \lambda \|\hat{A}\|_1 \text{ s.t. } tr(e^{\hat{A} \odot \hat{A}}) - m - 1 = 0,$$

where the constraint $tr(e^{\hat{A} \odot \hat{A}}) - m - 1 = 0$ is to ensure acyclicity in the graph. After training, matrix $\hat{A}$ will be a good estimation of the adjacency matrix A.

One challenge in applying the graph autoencoder to our work is that, although the graph autoencoder can accurately estimate the adjacency matrix $\hat{A}$, it does not produce a good reconstruction of the input sample, which implies that it does not accurately estimate the function $f(\cdot)$ in the SCM. In order to generate the counterfactual data, the reconstructed sample with high fidelity is critical. Hence, we improve the graph autoencoder by adding another decoder that focuses on data reconstruction, where the trained matrix $\hat{A}$ and the encoder $E_{\theta_1}(\cdot)$ are reused in this step.

In particular, we similarly feed each sample $d^{(n)}$ to trained encoder $E_{\theta_1}(\cdot)$ to obtain the corresponding hidden representation. Then, in order to be consistent with the structural equations Eq. (1), different from [12] where the message passing operation is applied in the representation space, we first use a new variable-wise decoder $D_{\phi_1'}$ to transform the hidden representation back to the original data space, and then aggregate the message from the neighbors based on matrix $\hat{A}$. As a result, the reconstruction process of each sample is given by the following equation.

$$\hat{d}^{(n)} = \hat{A}^T D_{\phi_1'}(E_{\theta_1}(d^{(n)})).$$

The objective function is to reconstruct the input with $\hat{A}$ and θ_1 fixed:

$$\mathcal{L}_{\text{D}}(\phi_1') = \frac{1}{2N} \sum_{n=1}^{N} \sum_{i=1}^{d} \|d_i^{(n)} - \hat{d}_i^{(n)}\|_2^2.$$

After training, we obtain the approximated mapping function $\hat{f} = D_{\phi_1'} \circ E_{\theta_1}$.

Generating Counterfactual Data. Given estimated adjacency matrix $\hat{A}$ and function $\hat{f}$, for each sample $d^{(n)}$, we generate its counterfactual $d_{\text{cf}}^{(n)}$ following the Abduction-Action-Prediction process. We first intervene $s^{(n)}$ to its counterpart s and compute $\hat{f}(s)$. Then, we sort all variables in X in a topological order and compute $\hat{x}_i^{(n)}[s]$ iteratively according to Eq. (3) where A and f are replaced by their estimators $\hat{A}$ and $\hat{f}$. Finally, we obtain $\hat{\mathcal{D}}_{cf} = \{\hat{d}_{cf}^{(n)}\}_{n=1}^N$, where $\hat{d}_{cf}^{(n)} = \{s, \{\hat{x}_i^{(n)}[s] \mid i = 1 : m\}\}$.

3.4 Phase Two: Fair Anomaly Detection

We use the autoencoder as the base model for anomaly detection, which is trained to minimize the reconstruction errors of normal samples. It is worth noting that a fully-connected autoencoder model is used here which is different from the variable-wise autoencoder used in the previous section for counterfactual data generation. Meanwhile, to achieve counterfactual fairness, we leverage the idea of adversarial training to make the hidden representations derived by the autoencoder not encode the information of the sensitive variable. To this end, we develop a pre-training and fine-tuning framework to ensure the effectiveness of anomaly detection as well as counterfactual fairness. The reason for adopting the pre-training and fine-tuning training approach instead of the end-to-end training is that some counterfactual samples in $\hat{\mathcal{D}}$ could be anomalies. If we include all samples in $\hat{\mathcal{D}}$ to train the autoencoder model, the performance of anomaly detection can be damaged. Hence, we use samples in $\mathcal{D}$ to pre-train the autoencoder model. Then, during fine-tuning, we slightly update the autoencoder so that the effectiveness of anomaly detection and counterfactual fairness can be balanced. Finally, we do not use the sensitive variable and only use the non-sensitive variables X to train the anomaly detection model.

To be more specific, in the pre-training phase, given the training set with normal samples $\mathcal{D}$, an encoder first maps each sample $x^{(n)}$ to a hidden representation $z^{(n)} = E_{\theta_2}(x^{(n)})$, and then a decoder aims to reconstruct the original input from the hidden representation $\hat{x}^{(n)} = D_{\phi_2}(z^{(n)})$. The objective function is to minimize the reconstruction error of normal samples:

$$\mathcal{L}_{AE}(\theta_2, \phi_2) = \frac{1}{2N}\sum_{n=1}^N \|d^{(n)} - D_{\phi_2} \circ E_{\theta_2}(x^{(n)})\|_2^2.$$

After pre-training the autoencoder model, in order to achieve counterfactual fairness, we further incorporate the adversarial training strategy to further fine-tune the autoencoder model so that the hidden representation $z^{(n)}$ derived by the encoder is free of the information of the sensitive variable. To this end, for each sample $d^{(n)} = \{s^{(n)}, x^{(n)}\}$ and its counterfactual sample $\hat{d}_{cf}^{(n)} = \{s, \hat{x}_{cf}^{(n)}\}$, we first derive the hidden representations, $z^{(n)}$ and $z_{cf}^{(n)}$, respectively, by feeding them to the encoder E_{θ_2}. Then, a discriminator C_ψ is applied on $z^{(n)}$ and $z_{cf}^{(n)}$ to predict whether the hidden representations are from observed or counterfactual samples, which is a binary classification task. We parameterize the discriminator C_ψ by a multilayer neural network with the sigmoid function as the output layer and use the negative of the standard cross-entropy loss for binary classification tasks as the objective function to train the discriminator:

$$\mathcal{L}_C(\theta_2, \psi) = \frac{1}{N}\sum_{n=1}^N [\log(C_\psi(z^{(n)})) + \log(1 - C_\psi(z_{cf}^{(n)}))].$$

The discriminator is trained to accurately separate the hidden representations of observed and counterfactual samples. Meanwhile, to make the hidden representation derived from the encoder invariant to the change of sensitive attribute, the adversarial game is to train the encoder E_{θ_2} to fool the discriminator C_ψ but still be good for reconstructing the original input. As a result, the objective function can be defined as a minimax problem:

$$\min_{\theta_2,\phi_2} \max_{\psi} \mathcal{L}_{AE}(\theta_2, \phi_2) + \lambda \mathcal{L}_C(\theta_2, \psi), \qquad (4)$$

where λ is a hyper-parameter to balance the reconstruction error and adversarial loss. Besides minimizing the reconstruction error $\mathcal{L}_{AE}$, the encoder also tries to maximize the cross-entropy loss for the discriminator $\mathcal{L}_C(\theta_2, \psi)$. Once the discriminator is unable to distinguish the hidden representations from factual or counterfactual data, we expect that both factual and counterfactual samples have similar reconstruction errors.

After training, the anomaly score for a new sample $d = \{s, x\}$ is computed based on the reconstruction error:

$$g(x) = \|x - D_{\phi_2} \circ E_{\theta_2}(x)\|_2^2.$$

If the anomaly score $g(x) > \tau$, where τ is a hyperparameter of the model, we label the sample as anomalous, i.e., $\hat{y} = 1$.

4   Experiments

4.1   Experimental Setup

Datasets. We conduct experiments on a synthetic dataset and two real-world datasets, Adult and COMPAS. Table 1 summarizes the statistics of three datasets.

Table 1. Statistics of datasets.

	Synthetic		Adult		COMPAS	
	Training	Test	Training	Test	Training	Test
Normal (Y=0)	12000	4000	12000	4000	2000	1283
Abnormal (Y=1)	N/A	400	N/A	800	N/A	384

Synthetic Dataset. We first build a synthetic dataset with 21 variables where we can obtain the ground truth of counterfactuals. We first randomly generate the adjacency matrix A of a causal graph using the Erdős-Rényi model [17] where one node is defined as a root node for representing the sensitive variable S.

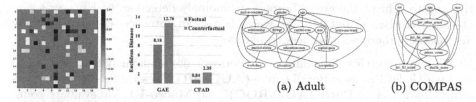

Fig. 2. Adjacency   **Fig. 3.** Results on   **Fig. 4.** Learned causal graphs.
matrix A data generation.

Figure 2 shows the generated adjacency matrix A. The value of S is randomly generated with binarized value $\{-1, 1\}$ to indicate sensitive and non-sensitive groups. Then, similar to [12], the rest 20 variables are generated based on the following data generating procedure: $X = 3A^T \cos(X + 1) + U$, where U is a standard Gaussian noise. Finally, one leaf node is selected as the decision attribute Y for determining anomalies. Specifically, for each sample, if the value of Y is greater than 0.85 quantile or smaller than 0.01 quantile, we label this sample as an anomaly, i.e., $Y = 1$. If the value of Y is between 0.3 and 0.7 quantiles, we label the sample as normal, i.e., $Y = 0$. Meanwhile, for both training and test sets, for 50% of the samples, their corresponding counterfactuals have labels that are different from the factual ones.

Adult Dataset. Adult is a real-world dataset with 14 features [5]. We treat *"gender"* as the sensitive attribute and samples with *"income > 50k"* as anomalies. We normalize all continuous features and binarize all categorical features. Figure 4a shows the causal graph on Adult learned in Phase One of our approach. Meanwhile, as we do not know the ground truth of counterfactuals, we use the generated counterfactual samples for measuring counterfactual fairness.

COMPAS Dataset. COMPAS is another real-world dataset [4], which consists of 8 features. We consider "race" as the sensitive attribute, where "African-American" and "Caucasian" are the disadvantage and advantage groups, respectively, and treat "recidivists" as anomalies. Similar to Adult, we normalize all continuous features and binarize all categorical features. Figure 4b shows the learned causal graph.

Baselines. We compare CFAD with the following baselines: 1) Principal Component Analysis (**PCA**), which is a dimensional reduction based anomaly detection approach; 2) One-class SVM (**OCSVM**), which is a one-class classification model that can detect outliers based on the observed normal samples; 3) Isolation Forest (**iForest**), which is a widely used tree-based anomaly detection model; 4) Autoencoder (**AE**), which is trained on normal data and widely-used for anomaly detection based on the deep autoencoder structure; 5) Deep Clustering based Fair Outlier Detection (**DCFOD**) [15], which adopts the adversarial training to achieve the group fairness in anomaly detection; 6) Fairness-aware Outlier Detection (**FairOD**) [14], which is also an autoencoder-based anomaly detection approach with fairness regularizers.

Evaluation Metrics. We evaluate the performance of anomaly detection based on Area Under Precision-Recall Curve (**AUC-PR**), Area Under Receiver Operating Characteristic Curve (**AUC-ROC**), and **Macro-F1**. We evaluate counterfactual fairness by computing the **changing ratio** of the samples whose detection outcomes are different from those for their corresponding counterfactuals, i.e., $changing\_ratio = \frac{\sum_{n=1}^{N} \mathbb{1}[\hat{y}^{(n)} \neq \hat{y}_{cf}^{(n)}]}{N}$, where $\mathbb{1}[\cdot]$ is the indicator function.

Implementation Details. Regarding baselines, we use Loglizer [7] to evaluate PCA, OC-SVM, and iForest. We implement FairOD and DCFOD based on public source code [15]. By default, the threshold τ for anomaly detection is set based on the 0.95 quantile of reconstruction errors (AE, FairOD, and CFAD) or distance to the normal center (DCFOD) in the training set. Our code on CFAD is available online[1].

4.2   Experimental Results

Counterfactual Data Generation. We first evaluate the performance of counterfactual data generation in the synthetic dataset by comparing CFAD with GAE [12] in terms of Euclidean distance between the generated and ground-truth samples. As shown in Fig. 3, on the factual data, CFAD achieves a much lower reconstruction error compared with GAE. More importantly, for counterfactual data generation, CFAD is much better compared with GAE. It indicates that by incorporating a variable-wise decoder $D_{\phi'_1}$ for data generation, CFAD can generate counterfactual samples with high fidelity.

Table 2. Anomaly detection on synthetic and real datasets with threshold $\tau = 0.95$. For AUC-PR, AUC-ROC, and Macro-F1, the higher the value the better the effectiveness; for Changing Ratio, the lower the value the better the fairness.

Method	Synthetic Dataset				Adult Dataset				COMPAS Dataset			
	AUC-PR	AUC-ROC	Macro-F1	Changing Ratio	AUC-PR	AUC-ROC	Macro-F1	Changing Ratio	AUC-PR	AUC-ROC	Macro-F1	Changing Ratio
PCA	0.992	0.999	0.908	0.478	0.238	0.582	0.476	0.261	0.365	0.642	0.595	0.268
OC-SVM	0.776	0.953	0.477	0.399	0.282	0.638	0.482	0.285	0.337	0.593	0.488	0.376
iForest	0.190	0.693	0.570	0.271	0.312	0.658	0.570	0.279	0.311	0.567	0.564	0.415
AE	0.957	0.996	0.883	0.461	0.349	0.640	0.608	0.590	0.344	0.616	0.581	0.407
DCFOD	0.383	0.832	0.721	0.212	0.249	0.623	0.533	0.071	0.260	0.569	0.466	0.067
FairOD	0.580	0.873	0.689	0.261	0.222	0.621	0.531	0.131	0.265	0.548	0.493	0.068
CFAD	0.947	0.996	0.930	0.199	0.319	0.589	0.576	0.057	0.314	0.596	0.539	0.049

Anomaly Detection. We further evaluate the performance of anomaly detection in terms of effectiveness as well as fairness. Table 2 shows the evaluation results. We report the mean value after five runs.

---

[1] https://github.com/hanxiao0607/CFAD.

Synthetic Dataset. CFAD can well balance the effectiveness and fairness in anomaly detection with high AUC-PR, AUC-ROC, and Macro-F1 and a low changing ratio. AE can achieve good performance on anomaly detection, but its changing ratio is high. DCFOD and FairOD, which achieve group fairness in anomaly detection, both have relatively low changing ratios, but their effectiveness in anomaly detection is not satisfactory.

Real Datasets. We have similar observations on the Adult and COMPAS datasets. CFAD achieves good performance in both effectiveness and fairness. For baselines that have no fairness component, their performance is good in terms of the effectiveness in anomaly detection, but they all have high changing ratios. Similarly, although DCFOD and FairOD have relatively low changing ratios, their effectiveness is much worse than other approaches.

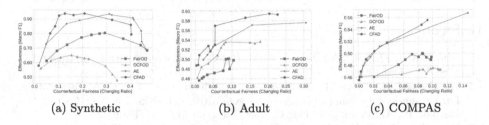

(a) Synthetic   (b) Adult   (c) COMPAS

Fig. 5. Trade-off between effectiveness and fairness.

Trade-off Between Effectiveness and Fairness. We further investigate the trade-off between effectiveness and fairness by varying the threshold as different quantiles of reconstruction errors or distances in the training set. We plot the effectiveness and fairness of each threshold setting of four approaches CFAD, AE, DCFOD, and FairOD in Fig. 5, where the x-axis is the changing ratio (counterfactual fairness), the y-axis indicates the Macro-F1 score (effectiveness), and each dot in the line indicates the result from one threshold. The dots from right to left indicate the performance based on quantiles including $\{0.8, 0.85, 0.9, 0.95, 0.97, 0.98, 0.99, 0.995, 0.999\}$. Ideally, we expect an anomaly detection model can achieve a high Marco-F1 score with a low changing ratio, which is the top left corner of the figure.

As shown in Fig. 5, CFAD performs best when the effectiveness trades off with fairness, as CFAD is closest to the top left corner of the figure. Specifically, on the Synthetic dataset, CFAD achieves much higher Macro-F1 values (effectiveness) with similar changing rates (fairness) compared with DCFOD and FairOD. Meanwhile, for most of the thresholds chosen based on quantiles, CFAD has higher Macro-F1 and lower changing ratios compared with AE. On the Adult and COMPAS datasets, CFAD can have higher Macro-F1 values and lower changing ratios compared with DCFOD and FairOD.

5   Conclusions

In this work, we have developed a counterfactually fair anomaly detection (CFAD) framework, which is able to effectively detect anomalies and also ensure counterfactual fairness. The core idea of CFAD is to generate counterfactual data governed by a learned causal structure based on the proposed graph autoencoder model. Then, by using a vanilla autoencoder as the anomaly detection model, an adversarial training strategy is adopted to ensure the representations derived by the autoencoder without the information of sensitive attributes. After that, counterfactual fairness is achieved by having similar reconstruction errors for both factual and counterfactual samples. The experimental results show that CFAD can achieve counterfactually fair anomaly detection while well-balancing the trade-off between effectiveness and fairness.

Acknowledgement. This work was supported in part by NSF 1910284 and 2103829.

References

1. Almanza, M., Epasto, A., Panconesi, A., Re, G.: k-clustering with fair outliers. In: WSDM. ACM (2022)
2. van Breugel, B., Kyono, T., Berrevoets, J., van der Schaar, M.: DECAF: generating fair synthetic data using causally-aware generative networks. In: NeurIPS (2021)
3. Deepak, P., Abraham, S.S.: Fair Outlier Detection. In: WISE (2020)
4. Dressel, J., Farid, H.: The accuracy, fairness, and limits of predicting recidivism. Sci. Adv. **4**(1), eaao5580 (2018)
5. Dua, D., Graff, C.: UCI machine learning repository (2017). http://archive.ics.uci.edu/ml
6. Edwards, H., Storkey, A.: Censoring representations with an adversary. arXiv:1511.05897 (2016)
7. He, S., Zhu, J., He, P., Lyu, M.R.: Experience report: System log analysis for anomaly detection. In: ISSRE. IEEE (2016)
8. Kilbertus, N., Rojas-Carulla, M., Parascandolo, G., Hardt, M., Janzing, D., Schölkopf, B.: Avoiding discrimination through causal reasoning. In: NIPS (2017)
9. Kusner, M.J., Loftus, J.R., Russell, C., Silva, R.: Counterfactual Fairness. In: NeurIPS (2018)
10. Madras, D., Creager, E., Pitassi, T., Zemel, R.: learning adversarially fair and transferable representations. arXiv:1802.06309 (2018)
11. Nabi, R., Shpitser, I.: Fair inference on outcomes. In: AAAI (2018)
12. Ng, I., Zhu, S., Chen, Z., Fang, Z.: A graph autoencoder approach to causal structure learning. In: NeurIPS Workshop on Machine Learning and Causal Inference for Improved Decision Making (2019)
13. Pearl, J.: Causality, 2nd edn. Cambridge University Press, Cambridge (2009)
14. Shekhar, S., Shah, N., Akoglu, L.: FairOD: fairness-aware outlier detection. In: AIES. AAAI/ACM (2021)
15. Song, H., Li, P., Liu, H.: Deep clustering based fair outlier detection. In: SIGKDD. ACM (2021)
16. Zhang, H., Davidson, I.: Towards fair deep anomaly detection. In: FAcct. ACM (2021)
17. Zheng, X., Aragam, B., Ravikumar, P.K., Xing, E.P.: DAGs with no tears: continuous optimization for structure learning. In: NeurIPS (2018)

The Common-Neighbors Metric Is Noise-Robust and Reveals Substructures of Real-World Networks

Sarel Cohen[1] , Philipp Fischbeck[2(✉)] , Tobias Friedrich[2] ,
and Martin Krejca[3]

[1] The Academic College of Tel Aviv-Yaffo, Tel Aviv, Israel
`sarelco@mta.ac.il`
[2] Hasso Plattner Institute, University of Potsdam, Potsdam, Germany
`{philipp.fischbeck,tobias.friedrich}@hpi.de`
[3] LIX, CNRS, Ecole Polytechnique, Institut Polytechnique de Paris,
Palaiseau, France
`martin.krejca@polytechnique.edu`

Abstract. Real-world networks typically display a complex structure
that is hard to explain by a single model. A common approach is to
partition the edges of the network into disjoint simpler structures. An
important property in this context is *locality*—incident vertices usually
have many common neighbors. This allows to classify edges into two
groups, based on the number of the common neighbors of their incident
vertices. Formally, this is captured by the *common-neighbors* (CN) metric, which forms the basis of many metrics for detecting *outlier* edges.
Such outliers can be interpreted as noise or as a substructure.

We aim to understand how useful the metric is, and empirically analyze several scenarios. We randomly insert outlier edges into real-world
and generated graphs with high locality, and measure the metric accuracy for partitioning the combined edges. In addition, we use the metric
to decompose real-world networks, and measure properties of the partitions. Our results show that the CN metric is a very good classifier that
can reliably detect noise up to extreme levels (83% noisy edges). We
also provide mathematically rigorous analyses on special random-graph
models. Last, we find the CN metric consistently decomposes real-world
networks into two graphs with very different structures.

Keywords: Noise · Clustering · Networks

1 Introduction

The structure of real-world processes across a large variety of scientific domains,
such as biology, ecology, sociology, or technology, typically results in highly complex networks [3,14]. These networks display many structural properties, such as
high *heterogeneity* (many different vertex degrees) and high *locality* (vertices that

share a large common neighborhood are likely to be connected), which seem to play a crucial role for reasoning about the networks [4]. Thus, it comes as no surprise that these properties are utilized in order to decompose complex networks into simpler ones. A prominent approach for this task is *graph clustering* [18].

Graph clustering aims to partition the vertices of a network into sets such that vertices from the same set have a similar value based on some metric, for example, the nearest neighbors of each vertex [22]. An important special case of clustering, also typically performed as a pre-processing step in clustering [6], is *outlier detection* [1], which aims to separate vertices with suspicious metric values from the rest. Algorithms for outlier detection vary in the amount of information they utilize. Some settings consider graphs annotated with features [10,13,15]. Other settings work exclusively with the structure of the network, that is, its vertices and edges [20]. Many approaches define a metric for *vertices* [9].

An alternative approach is to classify the *edges* of a network instead of its vertices [2]. In this setting, outlier detection is the opposite of link prediction [12]. Results for edge outlier detection are scarce, with the article by Zhang, Kiranyaz, and Gabbou [21] being the most extensive one. The authors consider different edge metrics based on the *common-neighbors* (CN) metric, which counts the number of shared vertices of the two vertices incident to a given edge. The authenticity of an edge is determined by how largely its metric score differs from the expected score of an edge, assuming the outlier-free graph follows a certain random-graph distribution. This approach is evaluated on real-world networks with randomly added edges. Although the real-world networks do not necessarily match the theoretical assumptions required for the authenticity of an edge, the authors show that their different metrics typically achieve an area-under-ROC-curve value of at least 0.85. This shows these metrics are rather robust to noise, making edge outlier detection a promising tool for *noise detection* in networks.

Contribution. Motivated by the good performance of the metrics by Zhang, Kiranyaz, and Gabbou [21], we focus on the usefulness of the *pure* CN metric for edge outlier detection, that is, we use the CN metric without *any* assumptions about the underlying graph model. Our intention is to use the CN metric in order to partition the edge set of a graph into two sets, each of which represents the connections of a different graph. Ideally, the two resulting graphs differ in locality, a very defining graph property, as we remarked above. If one of the resulting graphs is close to a (random) noise graph, then our setting resembles noise detection in graphs. However, it is more general than that, as we do not require any of the graphs to follow a noise model.

Setting. We consider *mixed graphs*, which are the superposition of two graphs defined over the same set of vertices but with different edge sets. One of the two graphs that make up the mixed graph is the *base graph*, which we consider to be the graph that consists of no outlier edges. The other graph is the *overlay graph*. We apply the CN metric to the mixed graph and evaluate how well it can separate the base graph from the overlay graph.

Methodology. We evaluate the performance of the CN metric empirically in three different settings (Sect. 4). In the first setting (Sect. 4.1), both the base and the overlay graph follow well-established random-graph distributions. As base graphs, we use graphs that place their vertices randomly with respect to a geometry: random geometric graphs [16] and hyperbolic random graphs [11]. These models have a high locality, the second one also high heterogeneity. For the overlay graph, we use the Erdős–Rényi model [8], that is, we add edges independently, each with the same probability. As this model does not make use of locality, the separation should work well.

In the second setting (Sect. 4.2), we exchange the base graph for real-world networks. The overlay graph still follows the Erdős–Rényi model. Thus, the edges of the overlay graph remain to not follow any locality. Here, we aim to see how sufficient the natural locality of real-world networks is for a good separation.

In the final setting (Sect. 4.3), the *mixed* graph is a real-world network, i.e., we have no ground truth information anymore. The aim is to see how well the CN metric separates a real-world network into two distinct graphs. To this end, we vary the threshold that determines when an edge is classified as an outlier, and we compare graph properties in the resulting base and overlay graphs.

Results. For all three settings, the CN metric performs very well. For the first setting (Sect. 4.1), the CN metric achieves an area-under-ROC-curve (AUC) value of at least 0.96—in many cases of at least 0.98. These results hold even for extreme scenarios where the amount of random/outlier edges is 5 times the amount of edges in the base graph. This shows that the CN metric is immensely robust with respect to non-local noise.

For the second setting (Sect. 4.2), the quality depends more on the base graph, with some settings having a (still rather high) AUC value of 0.80, whereas others have a value of over 0.90. This shows the locality of real-world networks is high enough such that non-local noise is well detected. However, our experiments indicate the quality also depends on other graph properties like graph density.

For the final setting (Sect. 4.3), the number of components as well as the global clustering coefficient (GCC) of the two resulting graphs indicate that the CN metric does indeed classify non-local edges as overlay edges, as the GCC of the base graph increases with the removal of overlay edges, and the number of connected components also quickly increases.

In addition to our empirical results, we prove mathematically rigorously what the expected CN score of an edge in mixed graphs is, with respect to whether the edge was present in the base graph (Theorem 1) or only in the overlay graph (Theorem 2). In these analyses, we assume that the base graph is a random geometric graph and the overlay graph an Erdős–Rényi graph, which we assume to be sparse. We find that the expected difference between the CN score of an edge in the mixed graph that is already present in the base graph versus the score if the edge is only present in the overlay graph is in the order of magnitude of the expected vertex degree of the base graph. Thus, a higher expected vertex degree of the base graph makes it easier to detect outliers.

Conclusion. Our results indicate that the CN metric is very well suited for classifying real-world networks into two distinct, simpler networks. Neither the CN metric nor the classification method require any problem-specific knowledge. Especially, the CN metric is highly robust to noise. This all suggests that the simple CN metric is a very good tool for handling the detection of outlier edges.

2   Preliminaries

Let $\mathbf{N}$ denote the set of all natural numbers (incl. 0). For all $m, n \in \mathbf{N}$, let $[m..n] := [m, n] \cap \mathbf{N}$, and let $[n] := [1..n]$. We consider undirected, simple graphs $G = (V, E)$, with *vertices* in V and *edges* in E. For all $v \in V$, we denote the *(exclusive) neighborhood* of v by $\Gamma_G(v) = \{u \in V \mid \{u, v\} \in E\}$. Further, let $\binom{V}{2} := \{\{u, v\} \mid u, v \in V \wedge u \neq v\}$ denote the set of all unordered pairs over V.

2.1   Setting

We consider *mixed graphs* $G = (V, E)$ that are the superposition a *base graph* $G_\mathrm{b} = (V, E_\mathrm{b})$ and an *overlay graph* $G_\mathrm{o} = (V, E_\mathrm{o})$, that is, $E = E_\mathrm{b} \cup E_\mathrm{o}$. We say that *G is composed of* G_b *and* G_o.

We consider the *common-neighbors (CN) metric*. For a graph $G = (V, E)$, the CN metric (over G) is the function $\mathrm{cn}_G \colon \binom{V}{2} \to [0..|V| - 2]$ that maps each pair of vertices to the size of their shared neighborhood. That is, for all $\{u, v\} \in \binom{V}{2}$, it holds that $\mathrm{cn}_G(\{u, v\}) = |\Gamma_G(u) \cap \Gamma_G(v)|$. Note that u and v are not accounted for, as $u \notin \Gamma_G(u)$ and $v \notin \Gamma_G(v)$. We call $\mathrm{cn}_G(\{u, v\})$ the *CN score* of $\{u, v\}$.

2.2   Random-Graph Models

We consider various formal random-graph models, which we introduce in the following. In addition to those, we also consider (deterministic) real-world networks, which we explain in Sect. 4.2. For all of the following models, when we introduce a graph, it actually represents a random element following a distribution over the set of all graphs that can be constructed as described. This distribution is defined implicitly via the random choices for how the vertices and/or edges are drawn. We do not introduce special notation for such a distribution.

Random Geometric Graphs. A *random geometric graph* (RGG) is a graph $G = (V, E)$ with $V \subset [0, 1]^2$ together with a *radius* $r \in [0, 1/\sqrt{2}]$. The vertices of an RGG lie in the unit torus, that is, for all $u, v \in V$, the distance between u and v is wrapping around the borders, formally, $\mathrm{dist}(u, v) := \sqrt{|u_1 - v_1|_\mathrm{o}^2 + |u_2 - v_2|_\mathrm{o}^2}$, where, for all $i \in [2]$, it holds that $|u_i - v_i|_\mathrm{o} := \min\{|u_i - v_i|, 1 - |u_i - v_i|\}$.

The vertices of an RGG are placed independently and uniformly at random into the unit torus, that is, the probability for a vertex to be placed in an area of size $A \in [0, 1]$ is A. After placing the vertices, the edges are determined deterministically by connecting two vertices if and only if their distance is at most r. That is, $E = \{\{u, v\} \in \binom{V}{2} \mid \mathrm{dist}(u, v) \leq r\}$. Since a vertex u is connected to another vertex v if and only if v is in a circle of radius r around u, the expected degree of u is $(|V| - 1)\pi r^2$.

Erdős–Rényi Graphs. An *Erdős–Rényi graph* (ER) is a graph $G = (V, E)$ together with an *edge probability* $p \in [0, 1]$. In contrast to an RGG, the vertices of an ER have no geometric interpretation and can be anything. The edges of G are all drawn independently, each with probability p. That is, for each $\{u, v\} \in \binom{V}{2}$, it holds that $\Pr[\{u, v\} \in E] = p$. Since a vertex u is connected to another vertex v with probability p, the expected degree of u is $(|V| - 1)p$.

Hyperbolic Random Graphs. A *hyperbolic random graph* (HRG) is a graph $G = (V, E)$ together with a power-law exponent $\beta \in (2, 3)$ and a radius R. All vertices are positioned in a disk of radius R in the hyperbolic plane according to a probability distribution based on β, and two vertices are connected by an edge if and only if their hyperbolic distance is at most R. The expected average degree can be controlled via R, while β determines the exponent of the power-law degree distribution. The resulting graphs have high heterogeneity and locality.

Randomness in Mixed Graphs. When we consider mixed graphs G composed of a base graph G_b and an overlay graph G_o, we make sure that at most one model determines how vertices are placed. This guarantees that no random choices conflict with each other, so G is well-defined. Since G_b and G_o have their own edges, the randomness in drawing the edges cannot conflict with each other.

3   Theoretical Results

We consider mixed graphs $G = (V, E)$ composed of an RGG $G_{\mathrm{rgg}} = (V, E_{\mathrm{rgg}})$ with radius $r \in [0, 1/4]$ as base graph and an ER $G_{\mathrm{er}} = (V, E_{\mathrm{er}})$ with edge probability $p \in [0, 1]$ as overlay graph. We mathematically analyze the CN score of an edge $e \in E$, depending on whether e is present in the base graph or not (Sect. 3.2). Our main results are Theorems 1 and 2, which show together that for $p = o(1)$ (with respect to $|V|$), that is, the overlay graph is not dense, the expected difference of the CN score of e with respect to whether it is present in the base graph or not is in the order of nr^2, which is the same order as the expected vertex degree in an RGG. Thus, the higher the expected vertex degree of the base graph, the further the CN scores in the mixed graph differ from edges present in the base graph and those only present in the overlay graph.

Before we introduce and discuss the results, we discuss important properties relevant to the results. These revolve around the probabilities for vertices to lie at a certain distance with respect to two given vertices u and v, whose CN score we are interested in. We omit proofs due to space restrictions.

3.1   Probabilities of Vertex Placements

Let u and v be vertices from a mixed graph $G = (V, E)$ based on an RGG $G_{\mathrm{rgg}} = (V, E_{\mathrm{rgg}})$ of radius $r \in [0, 1/4]$ and an ER $G_{\mathrm{er}} = (V, E_{\mathrm{er}})$ with edge probability $p \in [0, 1]$. In order to determine how much the CN score of u and v

changes from G_{rgg} to G, we calculate how likely it is for other vertices to have edges to u and v, both in E_{rgg} and in E_{er}. In the following, we first determine the probability of a vertex being connected to both u and v in G_{rgg}. Then, we determine the probability of a vertex that is not a common neighbor of u and v in G_{rgg} to be a common neighbor in G.

Common Neighbors in the Base Graph. In this setting, the shared area of the two circles of radius r around u and v is important. We call this area $\mu(u \cap v)$, and we remark it is the probability of a vertex to be in $\Gamma_{G_{\text{rgg}}}(u) \cap \Gamma_{G_{\text{rgg}}}(v)$, as they are drawn uniformly at random. Based on this, we derive the expectation of $\mu(u \cap v)$ with respect to whether u and v are themselves connected.

Lemma 1. *Let $G_{\text{rgg}} = (V, E_{\text{rgg}})$ be an RGG with radius $r \in [0, 1/4]$. Furthermore, let $\{u, v\} \in \binom{V}{2}$ and let $R := \{\{u, v\} \in E\}$. Then*

$$\mathrm{E}[\mu(u \cap v) \mid R] = \frac{4\pi - 3\sqrt{3}}{4}r^2 \ \text{and} \ \mathrm{E}[\mu(u \cap v) \mid \overline{R}] = \frac{3\sqrt{3}\pi r^2}{4(1 - \pi r^2)}r^2. \tag{1}$$

Common Neighbors in the Mixed Graph. We consider the probability of a vertex w to be a common neighbor of u and v in G, given that it is not a common neighbor in G_{rgg}. This happens because of one of the following reasons.

1. $w \in \Gamma_{G_{\text{rgg}}}(u)$: In this case, $w \notin \Gamma_{G_{\text{rgg}}}(v)$. Since $w \in \Gamma_G(u) \cap \Gamma_G(v)$, there is an edge in $E_{\text{er}} \setminus E_{\text{rgg}}$.
2. $w \in \Gamma_{G_{\text{rgg}}}(v)$: This case is symmetric to the previous one when exchanging u with v, as all vertices are handled symmetrically in RGGs.
3. $w \in \overline{\Gamma_{G_{\text{rgg}}}(u) \cup \Gamma_{G_{\text{rgg}}}(v)}$: In this case, there are two edges in $E_{\text{er}} \setminus E_{\text{rgg}}$.

The following lemma determines the probability of w falling into one of these three cases.

Lemma 2. *Let $G = (V, E)$ be a mixed graph composed of an RGG $G_{\text{rgg}} = (V, E_{\text{rgg}})$ with radius $r \in [0, 1/4]$ as base graph and an ER $G = (V, E_{\text{er}})$ with edge probability $p \in [0, 1]$ as overlay graph. Furthermore, let $\{u, v\} \in \binom{V}{2}$ and $w \in V \setminus \{u, v\}$. Last, let O denote the event $\{w \notin \Gamma_{G_{\text{rgg}}}(u) \cap \Gamma_{G_{\text{rgg}}}(v)\}$, and let R denote the event $\{\{u, v\} \in E_{\text{rgg}}\}$. Then, abbreviating $a := (3\sqrt{3})/4$,*

$$\Pr[w \in \Gamma_G(u) \cap \Gamma_G(v) \wedge O \mid R] = pr^2(2a - (\pi + a)p) + p^2 \ \text{and} \tag{2}$$

$$\Pr[w \in \Gamma_G(u) \cap \Gamma_G(v) \wedge O \mid \overline{R}] = p\pi r^2 \left(2(1 - p) - (2 - p)\frac{ar^2}{1 - \pi r^2}\right) + p^2. \tag{3}$$

3.2   The CN Score of Different Edges

Using the probabilities from Sect. 3.1, we derive the expected CN score of an edge in the mixed graph. The following theorem assumes that the edge is already present in the base graph. Afterward, we consider the case that the edge is only present in the overlay graph. At the end, we conclude.

Theorem 1. *Let $G = (V, E)$ be a mixed graph over $n \in \mathbf{N}_{\geq 2}$ vertices composed of an RGG $G_{\mathrm{rgg}} = (V, E_{\mathrm{rgg}})$ with radius $r \in [0, 1/4]$ as base graph and an ER $G = (V, E_{\mathrm{er}})$ with edge probability $p \in [0, 1]$ as overlay graph. Furthermore, let $\{u, v\} \in E$, let $N_G = \mathrm{cn}_G(\{u, v\})$, let $N_{G_{\mathrm{rgg}}} = \mathrm{cn}_{G_{\mathrm{rgg}}}(\{u, v\})$, let q denote the left expected value from Eq. (1), let s denote the probability from Eq. (2), and let R denote the event $\{\{u, v\} \in E_{\mathrm{rgg}}\}$. Then*

$$\mathrm{E}[N_G \mid R] = \mathrm{E}[N_{G_{\mathrm{rgg}}} \mid R] + (n-2)s \ \text{ and } \ \mathrm{E}[N_{G_{\mathrm{rgg}}} \mid R] = (n-2)q.$$

The following theorem shows how the CN score changes if the edge is only in the overlay graph. It looks similar to Theorem 1 but considers other probabilities.

Theorem 2. *Let $G = (V, E)$ be a mixed graph over $n \in \mathbf{N}_{\geq 2}$ vertices composed of an RGG $G_{\mathrm{rgg}} = (V, E_{\mathrm{rgg}})$ with radius $r \in [0, 1/4]$ as base graph and an ER $G = (V, E_{\mathrm{er}})$ with edge probability $p \in [0, 1]$ as overlay graph. Further, let $\{u, v\} \in E$, let $N_G = \mathrm{cn}_G(\{u, v\})$, let $N_{G_{\mathrm{rgg}}} = \mathrm{cn}_{G_{\mathrm{rgg}}}(\{u, v\})$, let q denote the right expected value from Eq. (1), let s be the probability from Eq. (3), let R denote the event $\{\{u, v\} \notin E_{\mathrm{rgg}}\}$, and let K denote the event $\{\{u, v\} \in E_{\mathrm{er}}\}$. Then*

$$\mathrm{E}[N_G \mid \overline{R}, K] = \mathrm{E}[N_{G_{\mathrm{rgg}}} \mid \overline{R}, K] + (n-2)s \ \text{ and } \ \mathrm{E}[N_{G_{\mathrm{rgg}}} \mid \overline{R}, K] = (n-2)q.$$

Let q_{rgg} and s_{rgg}, respectively, denote q and s from Theorem 1, and let q_{er} and s_{er} be defined analogously with respect to Theorem 2. If $(q_{\mathrm{rgg}} + s_{\mathrm{rgg}})$ and $(q_{\mathrm{er}} + s_{\mathrm{er}})$ are sufficiently separated, then so are the respective CN scores for edges in the mixed graph that are present in the base graph or only in the overlay graph, which makes separating these two edge types not difficult. By Lemma 1, it holds that $q_{\mathrm{rgg}} - q_{\mathrm{er}} = \left(\pi - 3\sqrt{3}/(4(1 - \pi r^2))\right)r^2 = \Theta(r^2)$, which is non-negative for all $r \in [0, 1/4]$. Similarly, by Lemma 2, we get that $s_{\mathrm{rgg}} - s_{\mathrm{er}} = -(2-p)pr^2(\pi - (3\sqrt{3})/4 - \pi^2 r^2)/(1 - \pi r^2) = -\Theta(pr^2)$, which is non-positive for all $r \in [0, 1/4]$ and all $p \in [0, 1]$. Due to the difference of the signs, a general comparison is difficult. However, assuming that the overlay graph is sparse, that is, $p = o(1)$, we see that $(q_{\mathrm{rgg}} + s_{\mathrm{rgg}}) - (q_{\mathrm{er}} + s_{\mathrm{er}}) = \Theta(r^2)$. Thus, the difference in the expected CN score of edges present in the base graph and those only present in the overlay graph is $\Theta(r^2 n)$, which is in the same order as the expected vertex degree of an RGG. Thus, an increased average degree in the base graph results in a larger expected difference in scores.

4   Empirical Results

We present empirical findings on the quality of the CN metric for different scenarios. We first consider scenarios where we know both the base graph and the overlay graph. As base graph, we consider two random graph models (Sect. 4.1) as well as real-world networks (Sect. 4.2). Last, we consider the case where the mixed graph is a real-world network, and we partition its edges according to the CN metric (Sect. 4.3). We briefly explain how we carry out our study.

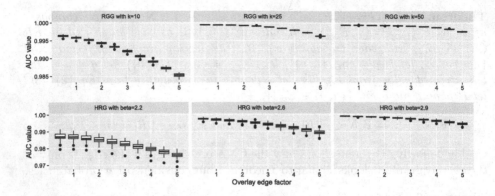

Fig. 1. (Top) The AUC score for an RGG as base graph and an ER as overlay graph. We fix the number of vertices to 5 000 and the expected average degree of the base graph $k \in \{10, 25, 50\}$. The overlay edge factor varies from 0.5 to 5, and we display 50 samples per configuration. (Bottom) The AUC score for an HRG as base graph and an ER as overlay graph. We fix the number of vertices to 5 000, expected average degree of the base graph $k = 25$, and vary the power-law degree exponent $\beta \in \{2.2, 2.6, 2.9\}$. The overlay edge factor varies from 0.5 to 5, and we display 50 samples per configuration.

AUC Metric. When evaluating the quality of the CN metric for separation of the two known edge sets, we measure the well-established *area-under-the-ROC-curve (AUC)* score. This measure is commonly used for classification models and provides an aggregate measure for the true-positive and false-positive rate of a binary classifier across all possible thresholds. We treat our scenario as a binary classification task, with base edges being positive. The AUC essentially is the probability that a random positive example has a higher score than a random negative example, i.e., that the CN score of a random base edge is higher than that of a random overlay edge. A random metric would yield an AUC score of 0.5, while a perfect metric would yield 1.0.

Experimental Setup. Our Python implementation uses the libraries Net-worKit [19] and igraph [7] for generating and analyzing graphs. They provide implementations for random graph models and graph properties. All experiments were run on a system with an Apple M1 chip and 16 GB RAM. However, note we do not consider run times, and all experiments were finished in minutes. All code and data is published at https://github.com/PFischbeck/cn-noise-experiments.

4.1 Graph Model as Base Graph

We consider two graph models as base graph, which are known to be highly clustered due to the use of an underlying geometry in the generation process. As base graph, we consider RGGs as well as HRGs (see Sect. 2.2 for details). As overlay graph, we consider ERs with an expected number of edges relative to the number of edges in the base graph. For example, an *overlay edge factor*

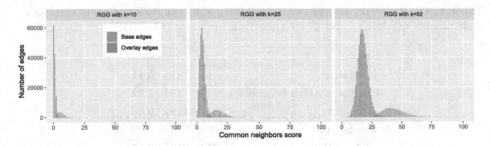

Fig. 2. The distribution of the CN scores for an RGG with 5 000 vertices and varying expected degree k as base graph, and an ER as overlay graph with overlay edge factor 5. The color shows whether the edges are from the base (red) or overlay graph (blue). (Color figure online)

of 2 means that there are twice as many overlay edges as there are base edges, in expectation. For a fixed model configuration and overlay edge factor, we take 50 samples and display them as box plots.

Random Geometric Graphs. For RGGs as base graph, we fix the number of vertices to 5 000 and vary the expected average degree to be 10, 25, and 50. Figure 1 (top) shows the resulting AUC scores for varying overlay edge factors.

One clearly sees that in all scenarios, the AUC score is very high, staying above 0.98. As one would expect, an increased overlay edge factor leads to lower scores, as the overlay edges make it harder to tell the two edge sets apart. The dependence on the average degree seems to consist of two parts. First, there is an increase of the AUC score for increased average degree, as predicted in Sect. 3. In addition, for higher average degree, the increase in overlay edges has a reduced effect on the AUC score. Recall that the number of overlay edges is relative to the number of base edges and thus also scales with increased average degree.

In order to understand this behavior better, we also provide a view on the distribution of scores for the two edge partitions. We fix an overlay edge factor of 5 and look at one sample for all three considered average degrees. Figure 2 shows the score distribution for these configurations.

As the average degree is increased, the CN scores increase for both base and overlay edges. However, they also increase their variance, and thus their overlap increases. Nonetheless, the high average degree still makes it easy to distinguish between the high number of edges outside of the overlap for $k = 50$.

Hyperbolic Random Graphs. For HRGs as base graph, we fix the number of vertices to 5 000, expected average degree $k = 25$, and vary the power-law degree exponent to be 2.2, 2.6, and 2.9. Figure 1 (bottom) shows the resulting AUC scores for varying overlay edge factors.

Across all three configurations, the AUC score is relatively high, although not as high as for the RGGs as base graph (Fig. 1 (top)). Recall that a lower power-law exponent corresponds to a more heterogeneous degree distribution, leading to many low-degree and few high-degree vertices. For base graphs with

Table 1. The real-world networks we use as base graph, with their number of vertices n, their number of edges m, and global clustering coefficient (GCC).

Graph	n	m	GCC
advogato	6 k	43 k	0.11
bio-WormNet-v3-benchmark	2 k	79 k	0.72
ca-HepPh	11 k	118 k	0.66
ia-digg-reply	30 k	86 k	0.02
soc-brightkite	57 k	213 k	0.11
web-indochina-2004	11 k	48 k	0.57

low power-law exponent, edges connected to low-degree vertices have low CN scores, making them harder to differentiate from the overlay edges. This leads to a lower AUC score. Further, a higher overlay edge factor yields a lower AUC score. This is because the CN score of overlay edges is increased by other overlay edges. As the power-law exponent increases, the variance of the AUC score decreases.

4.2   Real-World Network as Base Graph

We consider various real-world networks as base graph, with an ER as overlay graph. The real-world networks are shown in Table 1. They are part of the NetworkRepository collection [17], and we use them in a cleaned format [5]. The networks are from different contexts (including biological, social, and web networks) and vary both in graph size and in their locality. We measure locality via the *global clustering coefficient (GCC)*, which can be interpreted as the probability that a triplet of vertices with at least two edges also has the third edge. Thus, it is an indicator for how clustered or local a graph is.

For every real-world network, we add an ER overlay graph with the same number of vertices as the base graph, and we vary the overlay edge factor from 0.5 to 5. We take 50 samples per configuration (recall that the ER overlay graph is random), and we consider the resulting AUC score of the CN scores.

The AUC scores for almost all real-world networks are at a high level, even with 5 times as many overlay edges as base edges. In addition, for most graphs, the AUC score remains constant as the overlay edge factor varies. The exceptions are the graphs bio-WormNet-v3-benchmark and ca-HepPh. Both graphs have few vertices and high clustering, which might lead to higher CN scores for overlay edges, both via other overlay edges and base edges.

Overall, there is a strong relation between the graph clustering (via the global clustering coefficient) and the AUC score of the CN metric. The ia-digg-reply network has a very low GCC and low AUC scores. Based on our experiments, we think the metric quality depends on several graph properties, including clustering, degree heterogeneity, graph density, and number of low-degree vertices.

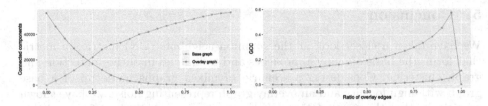

Fig. 3. The number of connected components and the global clustering coefficient of the base graph (red) and overlay graph (blue) when splitting the edges of the soc-brightkite network according to the CN metric. The ratio r_{split} defines that the $r_{\text{split}} \cdot |E|$ edges with lowest CN score are classified as overlay edges. (Color figure online)

The results are under the assumption that the real-world base graphs do not contain any overlay edges themselves, which cannot be known. In order to better understand this real-world edge set, we also consider real-world networks as the mixed graph in the following section.

4.3 Real-World Graph as Mixed Graph

In the experiments above, we had control over the base and overlay graph and thus were able to evaluate the quality of the CN score based on this ground truth. However, when partitioning a given graph without ground truth, we have to turn to other properties. In particular, if this metric does indeed help partition the given graph into a local, clustered structure and a global, random structure, this should be reflected in the properties of the two partition sets. We investigate this here. To this end, we take the real-world network soc-brightkite and treat it as a mixed graph. We measure the CN scores of its edges and sort the edges according to this score, with ties solved uniformly at random. For a fixed ratio r_{split}, the $r_{\text{split}} \cdot |E|$ edges with the lowest score are classified as overlay edges, while the remaining edges are classified as base edges. We build the base graph and overlay graph according to this edge partitioning, and we measure the global clustering coefficient as well as the number of connected components of the two parts. Figure 3 shows the resulting values for varying ratio r_{split}.

As the split ratio increases, the number of components of the base graph quickly rises, with an average of roughly two vertices per component for $r_{\text{split}} = 0.3$. On the other hand, the number of components of the overlay graph quickly decreases, which indicates that the edges classified as overlay edges are in fact global in the sense that they often connect previously disconnected components.

Also, as more edges are classified as overlay edges, the global clustering component of the base graph increases, indicating the overlay edges are indeed non-local, leaving local edges responsible for high clustering untouched. This is also seen in the very low clustering coefficient of the overlay graph even for $r_{\text{split}} = 0.9$.

5   Conclusion

We have taken a closer look at the common-neighbors (CN) metric—a metric that forms the basis of many approaches and techniques in outlier detection and graph clustering. Considering a scenario of mixed graphs made up of a base graph with high locality and an overlay graph representing noise, we have shown empirically that the simple CN metric is very accurate and robust for partitioning the edge set, even in the presence of much noise. In addition, the metric can handle real-world networks and partition them into two edge sets of differing properties, helping understand the underlying structures. Our theoretical analysis also gives indications to why the metric works for simple graph models.

A better understanding of this foundational metric is the basis for understanding and designing improved metrics in the fields of outlier detection and graph clustering. We have shown how the metric relates to locality and clustering, and our work indicates interesting related questions. In particular, it would be helpful to further analyze the metric for more complex graph models, including different noise models. In addition, it would be valuable to determine the other factors besides locality that influence the quality of the CN metric, including the degree distribution or density.

References

1. Aggarwal, C.C.: Outlier Detection in Graphs and Networks, pp. 369–397 (2017)
2. Aggarwal, C.C., He, G., Zhao, P.: Edge classification in networks. In: ICDE, pp. 1038–1049 (2016)
3. Albert, R., Barabási, A.L.: Statistical mechanics of complex networks. Rev. Mod. Phys. **74**, 47–97 (2002)
4. Bläsius, T., Fischbeck, P.: On the external validity of average-case analyses of graph algorithms. In: 30th Annual European Symposium on Algorithms (ESA 2022), vol. 244, pp. 21:1–21:14 (2022). https://doi.org/10.4230/LIPIcs.ESA.2022.21
5. Bläsius, T., Fischbeck, P.: On the External Validity of Average-Case Analyses of Graph Algorithms (Data, Docker, and Code), May 2022
6. Chakrabarti, D.: AutoPart: parameter-free graph partitioning and outlier detection. In: Boulicaut, J.-F., Esposito, F., Giannotti, F., Pedreschi, D. (eds.) PKDD 2004. LNCS (LNAI), vol. 3202, pp. 112–124. Springer, Heidelberg (2004). https://doi.org/10.1007/978-3-540-30116-5_13
7. Csardi, G., Nepusz, T.: The igraph software package for complex network research. InterJournal Complex Systems, 1695 (2006)
8. Erdős, P., Rényi, A.: On random graphs I. Publicationes Mathematicae **6**, 290–297 (1959)
9. Hautamaki, V., Karkkainen, I., Franti, P.: Outlier detection using k-nearest neighbour graph. In: ICPR, vol. 3, pp. 430–433 (2004)
10. Kou, Y., Lu, C.T., Dos Santos, R.F.: Spatial outlier detection: a graph-based approach. In: ICTAI, vol. 1, pp. 281–288 (2007)
11. Krioukov, D., Papadopoulos, F., Kitsak, M., Vahdat, A., Boguñá, M.: Hyperbolic geometry of complex networks. Phys. Rev. E **82**, 036106 (2010)
12. Lü, L., Zhou, T.: Link prediction in complex networks: a survey. Physica A **390**(6), 1150–1170 (2011)

13. Mansour, R.F., Abdel-Khalek, S., Hilali-Jaghdam, I., Nebhen, J., Cho, W., Joshi, G.P.: An intelligent outlier detection with machine learning empowered big data analytics for mobile edge computing. Clust. Comput. (2021)
14. Newman, M., Barabási, A., Watts, D.: The Structure and Dynamics of Networks. Princeton Studies in Complexity, Princeton University Press (2011)
15. Pandhre, S., Gupta, M., Balasubramanian, V.N.: Community-based outlier detection for edge-attributed graphs. CoRR abs/1612.09435 (2016)
16. Penrose, M.: Random Geometric Graphs, vol. 5. OUP Oxford (2003)
17. Rossi, R.A., Ahmed, N.K.: The network data repository with interactive graph analytics and visualization. In: AAAI (2015)
18. Schaeffer, S.E.: Graph clustering. Comput. Sci. Rev. 1(1), 27–64 (2007)
19. Staudt, C.L., Sazonovs, A., Meyerhenke, H.: NetworKit: a tool suite for large-scale complex network analysis (2015)
20. Suri, N.N.R.R., Murty, N.M., Athithan, G.: Outlier Detection: Techniques and Applications. Springer, Cham (2019). https://doi.org/10.1007/978-3-030-05127-3
21. Zhang, H., Kiranyaz, S., Gabbouj, M.: Outlier edge detection using random graph generation models and applications. J. Big Data 4(1), 1–25 (2017). https://doi.org/10.1186/s40537-017-0073-8
22. Zhang, H., Kiranyaz, S., Gabbouj, M.: Data clustering based on community structure in mutual k-nearest neighbor graph. In: TSP, pp. 1–7 (2018)

An Effective WGAN-Based Anomaly Detection Model for IoT Multivariate Time Series

Sibo Qi, Juan Chen, Peng Chen[✉], Peian Wen, Wenyu Shan, and Ling Xiong

School of Computer and Software Engineering, Xihua University,
Jinniu District, Chengdu 610039, China
chenpeng@mail.xhu.edu.cn

Abstract. This paper studies an effective unsupervised deep learning model for multivariate time series anomaly detection. Since multivariate time series usually have problems of insufficient labeling and highly-complex temporal correlation, effectively detecting anomalies in multivariate time series data is particularly challenging. To solve this problem, we propose a model named Wasserstein-GAN with gradient Penalty and effective Scoring (WPS). In this model, Wasserstein Distance with Gradient Penalty helps to capture the data regularities between generator output and real data, thus improving the training stability. Meanwhile, an effective scoring function that consists of reconstruction error, discrimination error, and prediction error is designed to evaluate the accuracy of the abnormal prediction and recall. The experimental results show that compared with the suboptimal baseline model, our proposed WPS obtains 17.68% and 10.41% improvement in prediction precision and F1 score, respectively.

Keywords: Multivariate Time Series · Anomaly detection · Deep Learning · Generative Adversarial Network · Internet of Things

1 Introduction

The evolution of the Big Data era and the emergence of the Internet of Things (IoT) created a large volume of time-series data. This information is derived via networked sensors and actuators in smart buildings, factories, power plants, and data centers. These continuous monitoring data are required to detect anomalies and confirm that the working environment remains normal. Anomalies are defined as time steps that reflect abnormal patterns of system behavior. Anomaly detection

This research was supported by the Science and Technology Program of Sichuan Province under Grant No. 2020JDRC0067, No. 2023JDRC0087, and No. 2020YFG0326, and the Talent Program of Xihua University under Grant No. Z202047 and No. Z222001.

S. Qi and J. Chen—Contributed to the work equally and should be regarded as co-first authors.

H. Kashima et al. (Eds.): PAKDD 2023, LNAI 13935, pp. 80–91, 2023.
https://doi.org/10.1007/978-3-031-33374-3_7

is often used for mission-critical purposes because it allows staff to take action to investigate and resolve the underlying problem before it causes a disaster.

As better effective in multivariate time series analysis, deep learning-based anomaly detection methods have attracted more research interest increasingly [17]. Among them, Generative Adversarial Networks (GANs) [9], which have achieved excellent results in areas such as graphics, have also led to a steady focus on GAN-based time series anomaly detection algorithms [3,10,12]. The main idea behind GANs is to let the generator network constantly correct the generated synthetic data, which is more similar to the training data under the supervision of the discriminator network, to fit the real data as accurately as possible. Through the adversarial training process between the generator and discriminator networks, the generator can achieve strong generation performance by learning to produce the distribution of data samples that closely resemble real data. However, the GANs usually suffer from training instability caused by vanishing gradient and mode collapse since the insufficient expression of distribution distance based on the Jensen-Shannon (JS) divergence [3,10,12]. In addition, due to the anomalies' sparsity and temporal data's noise, the anomalies' pattern may be similar to the normal one, which becomes hard to distinguish.

Subtle anomalies are still challenging in anomaly detection [17]. This phenomenon is more prominent in some complex systems with solid connections between different data sources, latent noise, and the sparsity of anomalies, making the abnormal pattern more difficult to distinguish and detect. Therefore, the existing detection techniques still suffer from a high false alarm rate. This case is more common for data with multiple dimensions closely related, such as cloud centers, which will impose a considerable burden on staff in production environments [4]. It illustrates the importance and urgency of reducing the false alarm rate while ensuring the detection recall of multivariate time series data.

In our model, we overcome the problem of the GAN-based method by Wasserstein Distance with Gradient Penalty (GP). It could correctly measure the distance between the data distribution of the real and generated samples. Additionally, we consider a new LSTM-based factor as a predictor of participation in anomaly scoring to enhance the robustness of the generative adversarial architecture.

In summary, our key contributions to this work are:

- Our proposed method accurately measures the distance of distributions between the data from the generator and the real sample according to the Wasserstein Distance versus GP, which greatly improves the training stability and mitigates the pattern collapse problem of the GAN-based time series anomaly detection model.
- We use an enhanced co-optimization strategy that combines three kinds of errors to recognize subtle anomalies and reduce the false alarm rate.
- Extensive experiments manifest that WPS outperforms the suboptimal model by 17.68% and 10.41% in precision and F1 on three public datasets and show more in addressing the problem of unsupervised multivariate time series anomaly detection in cloud systems.

2   Related Work

Anomaly detection challenges often employ unsupervised machine learning algorithms due to the rarity and difficulty of acquiring labels. Classic methods are mainly proximity-based [2,6,15,20,21], using distance measures to quantify the similarity between objects with the advantages of low computational cost and rapidity. However, these methods usually require a priori knowledge and the number of anomalies and cannot capture temporal correlations. Therefore, we mainly discuss the implementation of multivariate time series anomaly detection based on deep learning, which can be classified into prediction-based and reconstruction-based methods.

Prediction-Based and Reconstruction-Based Methods. Prediction-based methods [11,19] learn feature representations by predicting the current data instances using the representations of the previous instances within a temporal window as the context [17]. The residual between the predicted and real value of test data computes the anomaly score. When the difference between the predicted input and the original input of a data point exceeds a certain threshold, the data point is identified as an anomaly. Reconstruction-based methods [1,16,18,22,24] learn a model to capture the latent structure (low-dimensional representations) of the given time series data and then create a synthetic reconstruction of the data to compare the differences with the real data to calculate the reconstruction error score [17]. The heuristic for using this technique in anomaly detection is that the learned feature representations are enforced to learn essential data regularities to minimize reconstruction errors; anomalies are difficult to reconstruct from the resulting representations and thus have large reconstruction errors [17]. The methods [4,5,7,13,14,23] based on Generative Adversarial Networks (GANs) [9] are the latest development of deep learning-based unsupervised anomaly detection, which gained popularity due to their promising performance. However, the limitation of these methods is that the vanishing gradient and mode collapse caused by JS divergence still exists. It is caused by the distance between the data distribution generated by the generator and the real sample distribution cannot be calculated correctly. We address this issue by using Wasserstein distance with GP.

Co-Optimization Strategy. Some studies observed the limitation of single anomaly determination strategy and started using a co-optimization strategy to improve anomaly detection accuracy. For example, MAD-GAN [13] and TanoGAN [5] combine discrimination and reconstruction errors to find subtle anomalies. MTAD-GAT proposes a co-optimization strategy that combines the prediction and reconstruction errors output by RNN networks to achieve improved anomaly detection precision and reduce false positives. However, they both made only preliminary explorations in this area. To achieve this goal, we adopt an enhanced co-optimization strategy combining three error types: reconstruction, discrimination, and prediction.

3   Methodology

3.1   Unsupervised Time Series Anomaly Detection

Given a time series $\mathbf{x} = [x_1, x_2, \ldots, x_T]$ denote a multivariate time-series of length T, where $\mathbf{x}_t \in \mathbb{R}^M$ is its value at time t and M is the number of features in the input indicates the total number of sensors within the same entity in a realistic system. x_i denotes the entries with indices a point of multivariate time series. To prepare a training dataset $\mathbf{X} \in \mathbb{R}^{N \times S}$, $\mathbf{x}_{1:T}$ is split into N time series of length S. We write $\mathbf{x}_{i,1:S}$ to denote the ith time series in $\mathbf{X}$. Similarly, $\mathbf{z} = [z_1, z_2, \ldots, z_T]$ is a set of multivariate sub-sequences taken from a random space to represent white noise.

3.2   WPS Model

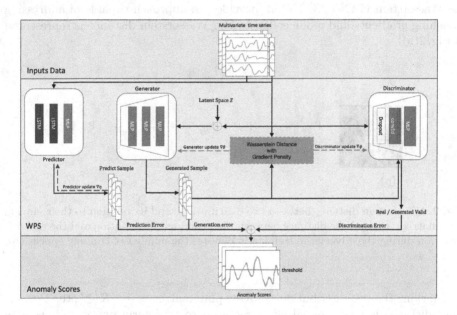

Fig. 1. The WPS architecture. The generator and discriminator get better fitting performance of data distribution by the Wasserstein distance with gradient penalty.

Overall Architecture. The WPS, which is the overall architecture illustrated in Fig. 1, consists of the generative adversarial architecture and the simultaneous training predictor. For the generative adversarial architecture, with the input data $\mathbf{x}_{i,1:S}, i = 1, \ldots, N$, the generator $G(\cdot; \boldsymbol{\theta})$ with parameter $\boldsymbol{\theta}$ reconstructs the values for time steps in $[t_0 + 1, t_0 + \tau]$ conditioning on the sum of $\mathbf{x}_{i,t_0+1:S}$, and $\mathbf{z}$, where $t_0 + \tau = S$. τ is the number of time steps $G(\cdot; \boldsymbol{\theta})$ is trained to reconstruct. That is, $\hat{\mathbf{x}}_{t_0+1:S} = G(\mathbf{x}_{t_0+1:S} + \mathbf{z}; \boldsymbol{\theta})$. The discriminator $D(\cdot; \boldsymbol{\omega})$ with parameter

ω then validates the closeness between the output value of $G(\cdot;\boldsymbol{\theta})$ and the true value of target range. For the predictor, given an input time series $\mathbf{x}_{i,1:S}, i = 1,\ldots,N$, our model $P(\cdot;\boldsymbol{\eta})$ with parameter $\boldsymbol{\eta}$ predicts the values for time steps in $[t_0 + 1, t_0 + \tau]$ conditioning on $\mathbf{x}_{i,1:t_0}$, where $t_0 + \tau = S$. τ is the number of time steps $P(\cdot;\boldsymbol{\eta})$ is trained to predict. That is, $\widehat{\mathbf{x}}'_{t_0+1:t_0+\tau} = P(\mathbf{x}_{i,1:t_0};\boldsymbol{\eta})$. The time ranges $[1,t_0]$ and $[t_0 + 1 : S]$ are referred to as the conditioning range and target range, respectively.

Wasserstein GAN. It is believed that the process of Generative Adversarial Networks(GAN) [9] involves the generator (G) and the discriminator (D) in order to force the distribution of generated data to be close to the actual distribution. This can be considered as two agents playing a mini-max game with a value function $V(D,G)$. In the classic GAN, the JS divergence can't correctly compute the proximity between the not intersecting real and generated distribution, which usually leads to weak performance. As an alternative to classical GAN, the Wasserstein GAN (WGAN) [3] provides an approach capable of addressing vanishing gradients and mode collapse. The Wasserstein distance is represented in Fig. 2(a).

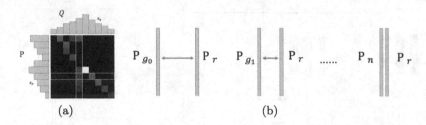

(a)   (b)

Fig. 2. Wasserstein distance between two distributions and its relation to their moving plan matrix v (a) and the distance between the generated distribution and the real distribution during the adversarial training, n denotes the number of training epochs (b).

Marking the two distribution's moving plan matrix as $v \in \mathcal{R}^{I \times J}$, the Wasserstein distance is the minimal effort required to transform one distribution to another. Matrix elements with the brightest colors reflect more motion from matrix element $\mathbb{P}$ to matrix element $\mathbb{Q}$. Despite having an equal area of 1, $\mathbb{P}$ and $\mathbb{Q}$ differ in their shape. A further illustration of the relationship can be found in Fig. 2(a), where x_q is a column-wise summation of v, while x_p is a row-wise summation. The values of all the elements in matrix v add up to 1. The Wasserstein distance, a method of finding an v that minimizes the transformation cost, is found by finding an optimal v. Plan v has an average distance $B(\cdot)$ as

$$B(v) = \sum_{x_p, x_q} v(x_p, x_q) \|x_p - x_q\| \tag{1}$$

The function of Wasserstein distance $W(\cdot)$ formulated as

$$W(\mathbb{P}, \mathbb{Q}) = \min_{v \in \varPi} B(v) \tag{2}$$

Figure 2(b) illustrates that the Wasserstein distance accurately measures the distance between $\mathbb{P}_r$ and $\mathbb{P}_g$ distributions at the beginning of the confrontation training. Once JS divergence intersects, this distance remains the same. That's why classic GAN training is erratic, with mode collapse occurring in such situations.

Gradient Penalty. Gulrajani et al. [10] find that WGAN may produce low-quality data samples by the discriminator's use of weight cropping. To overcome this disadvantage, the penalty term proposed forces the discriminator to keep the gradient norm of the discriminator to stay close to 1. The alternative approach further solves the vanishing gradient problems in WGAN, and the objective function is as follows.

$$L = \mathop{\mathbb{E}}_{\tilde{x} \sim \mathbb{P}_g} [D(\tilde{x})] - \mathop{\mathbb{E}}_{x \sim \mathbb{P}_r} [D(x)] + \lambda \mathop{\mathbb{E}}_{\bar{x} \sim \mathbb{P}_{\bar{x}}} \left[\left(\|\nabla_{\bar{x}} D(\bar{x})\|_2 - 1 \right)^2 \right] \tag{3}$$

where $\bar{x} = \epsilon x + (1 - \epsilon)\tilde{x}$ and $\epsilon \sim U[0, 1]$. The Gradient Penalty (GP) coefficient λ determines the stringency of the control constraint.

Adversarial Training. We use Wasserstein distance with GP instead of JS divergence in our method. In the adversarial training phase, our model first updates the discriminator and then updates the generator with the validates from the discriminator. The object of the generator is to generate high-quality dummy samples that are as similar as possible to the original time series. The generator is trained with the updated loss function as

$$L_G = -\frac{1}{N} \sum_{i=1}^{N} D(\hat{\mathbf{x}}_{i, t_0+1:S}) \tag{4}$$

The discriminator is trained with the updated loss function expressed as

$$L_D = \frac{1}{N} \sum_{i=1}^{N} D\left(\hat{\mathbf{x}}_{i, t_0+1:S}\right) - \frac{1}{N} \sum_{i=1}^{N} D\left(\mathbf{x}_{i, t_0+1:S}\right) + \lambda \frac{1}{N} \sum_{i=1}^{N} \left(\left(\|\nabla_{\bar{x}_i} D\left(\bar{x}_i\right)\|_2 - 1 \right)^2 \right) \tag{5}$$

where $\bar{x}_i = \epsilon \mathbf{x}_{i, t_0+1:S} + (1 - \epsilon)\hat{\mathbf{x}}_{i, t_0+1:S}$ and $\epsilon \sim U[0, 1]$.

Simultaneous Predictor Training. The predictor is trained in parallel with the generative adversarial architecture. In each epoch, the predictor gets the conditioning range data as input to predict the target range data same as the generator and discriminator get. Here we are using Mean Square Error (MSE) Loss, so the predictor objective function is

$$L_P = \frac{1}{N} \sum_{i=1}^{N} \parallel \mathbf{x}_{i,t_0+1:S} - \widehat{\mathbf{x}}'_{i,t_0+1:S} \parallel^2 \tag{6}$$

3.3   Anomaly Scoring

The anomaly scores are obtained by the model processing the test dataset. Anomalies are those with anomaly scores higher than the threshold. We integrate the reconstruction error (Rscore) of the generator, the discrimination error (Dscore) of the discriminator, and the prediction error (Pscore) of the predictor as the anomaly score (ADscore), and are formulated as follows.

$$Rscore = \sum_{i=1}^{N} \parallel \mathbf{x}_{i,t_0+1:S} - G\left(\mathbf{x}_{t_0+1:S} + \mathbf{z}\right) \parallel^2 \tag{7}$$

$$Dscore = \sum_{i=1}^{N} \left(-D(G\left(\mathbf{x}_{t_0+1:S} + \mathbf{z}\right)) + 1\right) \tag{8}$$

$$Pscore = \sum_{i=1}^{N} \parallel \mathbf{x}_{i,t_0+1:S} - P\left(\mathbf{x}_{i,1:t_0}\right) \parallel^2 \tag{9}$$

$$ADscore = \alpha Rscore + \beta Dscore + \gamma Pscore \tag{10}$$

In this equation, $\alpha + \beta + \gamma = 1$ is utilized to parameterize the empirically set trade-off between Rscore, Dscore, and Pscore. During anomaly detection, anomalies are identified by anomaly scores and specific thresholds.

4   Experiment

4.1   Experimental Setup

Datasets. We use three publicly available datasets to verify the effectiveness of our model, namely SMD (Server Machine Dataset) [22], SMAP (Soil Moisture Activate Passive satellite), and MSL (Mars Science Laboratory rover). SMD is real-time data from 28 cloud platform servers running, including 38 dimensions. SMAP and MSL are spacecraft datasets collected by NASA [11], including 55 and 25 dimensions, respectively. For our experiments, we select 50,000 data points from the training and test sets of SMD, and 20,000 data points from the training and test sets of MSL and SMAP.

Evaluation Metrics. Since our datasets are unbalanced, we adopt the standard evaluation metrics that contain Precision (Pre), Recall (Rec), and F1 Score (F1) in anomaly detection tasks, which are suitable for unbalanced data. We use the approach from [22] to search the threshold that calculates the best F1-score, which iterates all possible anomaly thresholds and picks the optimal one.

Baselines. We compare multiple unsupervised methods in multivariate time series anomaly Methods. Local Outlier Factor (LOF) [6] assigns to each object a degree of being an outlier; the degree depends on how isolated the object is concerning the surrounding neighborhood, clustering the high degree points as anomalies. MADGAN [13] is the first method to apply GAN on multivariate time series anomaly detection combined LSTM network. In our implementation, the RNN layer has been replaced by the MLP layer (Multilayer Perceptron) for simplicity and facility. USAD [4] is a fast and low-cost anomaly detection model based on encoder-decoder adverse architecture, which uses two-phase adversarial training to obtain a more stable training process and adjustable parameters to amplify stumble anomalies. OmniAnomaly [22] is a prior-driven stochastic model for multivariate time series anomaly detection that directly returns the posterior reconstruction probability of the multivariate time series input. The log of the probability serves as the channel-wise score, which is summed across channels to get the anomaly score [8].

Implementation. Our method was implemented by Pytorch. Considering the temporal information in the sequence data, we normalize the multivariate time series data by MinMaxScaler and then use sliding windows to divide it into subseries. We set the historical window size for the predictor to 6 to predict the value at the next timestamp. The kernel size of the 1-D convolution layer in the discriminator was set to 3 with a padding of 1. The dimensions of MLP for the generator and discriminator are half those of the input time series. Adam optimizer was used to train the generator, discriminator, and predictor models, and the learning rate was initialized to 1e−4 for the generator and discriminator and 2e-3 for the predictor. We set the hidden dimensions of the LSTM layer as 64. In the anomaly detection phase, we empirically set the ADscore of α, β as 0.35, 0.15, and γ as 0.5 after a series of explorations and tests. The training epochs of all models are set as 15. Our experiments were conducted on a server with an NVIDIA 2080Ti graphics card, an Intel Core i9-10900K CPU at 3.70GHz, and 32GB of RAM.

4.2 Baseline Comparisons

We compare multiple proximity-based and reconstruction-based methods as baselines. Based on precision, recall, and F1 scores, Table 1 presents the anomaly detection performance of WPS and baselines on three datasets. Among all methods, LOF and MADGAN perform the least. A multivariate time series anomaly detection requires interdependence modeling, which LOF lacks due to its design

Table 1. Experimental results of our model and the baseline methods. The best performance is presented as bold, and the second best as underlined.

Method	SMD (subset)			SMAP (subset)			MSL (subset)			overall		
	Pre	Rec	F1	Pre	Rec	F1	Pre	Rec	F1	mean(Pre)	mean(Rec)	mean(F1)
LOF	0.6678	0.2626	0.3770	_0.5795_	1.0000	_0.7338_	_0.8446_	0.8588	_0.8517_	_0.6973_	0.7071	0.6541
MADGAN	0.6305	0.8959	0.7401	0.2305	1.0000	0.3747	0.8078	_0.8634_	0.8347	0.5563	0.9198	0.6498
USAD	0.6795	0.9115	0.7786	0.2837	1.0000	0.4420	0.7677	**0.9282**	0.8404	0.5770	**0.9466**	0.6870
OmniAnomaly	_0.8740_	**0.9716**	_0.9202_	0.5006	1.0000	0.6672	0.6974	0.8028	0.7464	0.6907	_0.9248_	_0.7779_
WPS	**0.9920**	_0.9355_	**0.9639**	**0.6777**	1.0000	**0.8079**	**0.9516**	0.8084	**0.8742**	**0.8738**	0.9146	**0.8820**

for handling multivariate data without temporal information. Also, its lower recall is intolerable in practical application scenarios. The recall of MADGAN is at a decent level. However, its potential vanishing gradient and mode collapse problems typical of GAN-based models severely affect its precision and F1 score performance. Thanks to two-phase stability adversarial training and parameter settings that can be adjusted for anomaly detection sensitivity, USAD achieves the highest recall among all baselines. However, its precision and F1 score are still unsatisfactory. It is due to its use of only the basic autoencoder instead of a more efficient method that can capture temporal correlations as the core component of the model. OmniAnomaly is the suboptimal method. By learning resilient representations of multivariate time series data based on LSTM-VAE, it has the capacity to cope with explicit temporal dependency among stochastic variables. Part of its poor performance could be attributed partly to the use of static scoring functions [8]. Our model significantly achieves an overall improvement of 17.68% in precision and 10.41% in the best F1 scores on these three datasets compared to the suboptimal model. We also notice that our model's best F1 scores in the SMD (96.39%) dataset are more remarkable than in SMAP and MSL (80.79%, 87.42%). In datasets like SMD, the closer feature dependence leads the anomalies more challenging to detect. This points out that our approach may be more appropriate to be applied to datasets with multiple tightly related dimensions.

4.3 Convergence Analysis

To validate the effectiveness of Wasserstein distance and GP in mitigating the vanishing gradient and mode collapse problems of the GAN-based model, we compared the convergence of WPS, MADGAN, and WPS(w/o GP) on the SMD dataset. To visualize the fluctuation of the convergence, we normalized the average training Mean Square Error (MES) by MinMaxScaler. As shown in Fig. 3(a), the average training MSE of MADGAN converges to a constant value after a brief decrease on the SMD subset, and then the fluctuations become more dramatic as the epoch increases. It is a typical manifestation of the mode collapse and vanishing gradient problems in the GAN-based model, which uses JS divergence. In contrast, our model shows a continuous and steady decrease after the rapid drop in the average training MSE. We also validate the convergence in WPS and WPS(w/o GP), a variant of GAN based on JS divergence. We calculate the average training MSE from the SMD subset at each training epoch, as

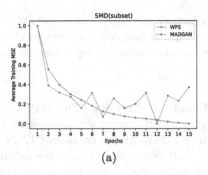
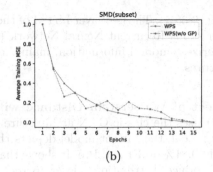

Fig. 3. Convergence analysis of WPS on SMD compared with MADGAN (a) and WPS(w/o GP) (b).

Fig. 3(b) shows. Similar to Fig. 3(a), WPS decreases steadily, without fluctuation, as the training phase progresses during the initial training epochs. Both of them strongly validate the effectiveness of Wasserstein distance and GP used in our model to alleviate the vanishing gradient and mode collapse problems, thus improving the convergence efficiency of the model.

4.4 Ablation Experiments

Table 2. Ablation experiments results. The best performance is presented as bold, and the second best as underlined.

Method	SMD (subset)			SMAP (subset)			MSL (subset)			overall		
	Pre	Rec	F1	Pre	Rec	F1	Pre	Rec	F1	mean(Pre)	mean(Rec)	mean(F1)
WPS	**0.9920**	0.9355	**0.9639**	0.6777	1.0000	**0.8079**	0.9516	0.8084	**0.8742**	**0.8738**	**0.9624**	**0.8820**
w/o DS	0.9186	**0.9365**	0.9274	0.5278	1.0000	0.6909	0.9392	0.8084	0.8689	0.7952	0.9586	0.8291
w/o PS	0.9059	0.8922	0.8990	0.5982	1.0000	0.7486	0.4731	**0.8930**	0.6185	0.6591	0.7884	0.7554
w/o GP	0.8010	0.9250	0.8590	0.4498	1.0000	0.6205	0.4826	**0.8930**	0.6266	0.5778	0.8025	0.7020

We undertake ablation analysis to evaluate the significance of various models in WPS and the results shown in Table 2. The necessity of the different components is analyzed as follows.

Effect of Discrimination Error. Without the discriminating error, the variants of WPS, i.e., WPS(w/o DS), reduced by 4.47% on F1 and 6.83% on Pre compared to our method. It indicates that discriminator validation can aid in recognizing anomalies during the anomaly scoring stage. It is also an essential factor that using the one-dimensional convolutional layer enhances the represent extractability of the discriminator.

Effect of Prediction Error. In this variant, i.e., WPS(w/o PS), we remove the prediction error, resulting in a significant decrease in F1(12.34%) and

Pre(19.38%). The two-layer LSTM of the predictor can explain this result. Moreover, t his Recurrent Neural Network (RNN) has a memory function to characterize temporal information, mainly to get a more accurate representation of features.

Effect of Wasserstein Distance with Gradient Penalty. The WPS(w/o GP) uses the classic GAN architecture employing JS divergence. Compared to our model, the variant model depicts the most substantial decreases of 12.78% and 20.81% on F1 and Pre. It shows that GP and Wasserstein distance combine to produce distributions closer to real data during the training process. This leads to significant improvements in performance.

5   Conclusion

In this work, we proposed our Predictive Wasserstein Generative Adversarial Network with GP (WPS) approach, which stably converges in the training stage in anomaly detection and can recognize subtle abnormal patterns with the assistance of a parallel training predictor. We use an enhanced co-optimization strategy that combines three errors to obtain more accurate anomaly scores, boosting the model's ability to find subtle anomalies. On datasets from different domains, WPS greatly outperforms the baseline in terms of accuracy, training stability, and false alarms. Future work can consider real-time prediction and the interpretability of anomalies to further improve the practicality of the approach.

References

1. Abdi, H., Williams, L.J.: Principal component analysis. Wiley Interdiscipl. Rev.: Comput. Stat. **2**(4), 433–459 (2010)
2. Angiulli, F., Pizzuti, C.: Fast outlier detection in high dimensional spaces. In: Elomaa, T., Mannila, H., Toivonen, H. (eds.) PKDD 2002. LNCS, vol. 2431, pp. 15–27. Springer, Heidelberg (2002). https://doi.org/10.1007/3-540-45681-3_2
3. Arjovsky, M., Chintala, S., Bottou, L.: Wasserstein generative adversarial networks. In: International Conference on Machine Learning, pp. 214–223. PMLR (2017)
4. Audibert, J., Michiardi, P., Guyard, F., Marti, S., Zuluaga, M.A.: USAD: unsupervised anomaly detection on multivariate time series. In: Proceedings of the 26th ACM SIGKDD International Conference on Knowledge Discovery & Data Mining, pp. 3395–3404 (2020)
5. Bashar, M.A., Nayak, R.: Tanogan: time series anomaly detection with generative adversarial networks. In: 2020 IEEE Symposium Series on Computational Intelligence (SSCI), pp. 1778–1785. IEEE (2020)
6. Breunig, M.M., Kriegel, H.P., Ng, R.T., Sander, J.: LOF: identifying density-based local outliers. In: Proceedings of the 2000 ACM SIGMOD International Conference on Management of Data, pp. 93–104 (2000)
7. Chen, P., et al.: Effectively detecting operational anomalies in large-scale IoT data infrastructures by using a GAN-based predictive model. Comput. J. (2022)

8. Garg, A., Zhang, W., Samaran, J., Savitha, R., Foo, C.S.: An evaluation of anomaly detection and diagnosis in multivariate time series. IEEE Trans. Neural Networks Learn. Syst. **33**(6), 2508–2517 (2021)
9. Goodfellow, I., et al.: Generative adversarial networks. Commun. ACM **63**(11), 139–144 (2020)
10. Gulrajani, I., Ahmed, F., Arjovsky, M., Dumoulin, V., Courville, A.C.: Improved training of wasserstein GANs. In: Advances in Neural Information Processing Systems, vol. 30 (2017)
11. Hundman, K., Constantinou, V., Laporte, C., Colwell, I., Soderstrom, T.: Detecting spacecraft anomalies using LSTMS and nonparametric dynamic thresholding. In: Proceedings of the 24th ACM SIGKDD International Conference on Knowledge Discovery & Data Mining, pp. 387–395 (2018)
12. Jabbar, A., Li, X., Omar, B.: A survey on generative adversarial networks: variants, applications, and training. ACM Comput. Surv. (CSUR) **54**(8), 1–49 (2021)
13. Li, D., Chen, D., Jin, B., Shi, L., Goh, J., Ng, S.-K.: MAD-GAN: multivariate anomaly detection for time series data with generative adversarial networks. In: Tetko, I.V., Kůrková, V., Karpov, P., Theis, F. (eds.) ICANN 2019. LNCS, vol. 11730, pp. 703–716. Springer, Cham (2019). https://doi.org/10.1007/978-3-030-30490-4_56
14. Li, Y., Peng, X., Zhang, J., Li, Z., Wen, M.: DCT-GAN: dilated convolutional transformer-based GAN for time series anomaly detection. IEEE Trans. Knowl. Data Eng. **35**, 3632–3644 (2021)
15. Liu, F.T., Ting, K.M., Zhou, Z.H.: Isolation forest. In: 2008 Eighth IEEE International Conference on Data Mining, pp. 413–422. IEEE (2008)
16. Malhotra, P., Vig, L., Shroff, G., Agarwal, P., et al.: Long short term memory networks for anomaly detection in time series. In: ESANN, vol. 2015, p. 89 (2015)
17. Pang, G., Shen, C., Cao, L., Hengel, A.V.D.: Deep learning for anomaly detection: a review. ACM Comput. Surv. (CSUR) **54**(2), 1–38 (2021)
18. Park, D., Hoshi, Y., Kemp, C.C.: A multimodal anomaly detector for robot-assisted feeding using an LSTM-based variational autoencoder. IEEE Robot. Autom. Lett. **3**(3), 1544–1551 (2018)
19. Pena, E.H., de Assis, M.V., Proença, M.L.: Anomaly detection using forecasting methods Arima and HWDS. In: 2013 32nd International Conference of the Chilean Computer Science Society (SCCC), pp. 63–66. IEEE (2013)
20. Schölkopf, B., Williamson, R.C., Smola, A., Shawe-Taylor, J., Platt, J.: Support vector method for novelty detection. In: Advances in Neural Information Processing Systems, vol. 12 (1999)
21. Schubert, E., Sander, J., Ester, M., Kriegel, H.P., Xu, X.: DBSCAN revisited, revisited: why and how you should (still) use DBSCAN. ACM Trans. Database Syst. (TODS) **42**(3), 1–21 (2017)
22. Su, Y., Zhao, Y., Niu, C., Liu, R., Sun, W., Pei, D.: Robust anomaly detection for multivariate time series through stochastic recurrent neural network. In: Proceedings of the 25th ACM SIGKDD International Conference on Knowledge Discovery & Data Mining, pp. 2828–2837 (2019)
23. Wen, P., Yang, Z., Wu, L., Qi, S., Chen, J., Chen, P.: A novel convolutional adversarial framework for multivariate time series anomaly detection and explanation in cloud environment. Appl. Sci. **12**(20), 10390 (2022)
24. Xu, H., et al.: Unsupervised anomaly detection via variational auto-encoder for seasonal KPIS in web applications. In: Proceedings of the 2018 World Wide Web Conference, pp. 187–196 (2018)

Association Rules

RL-Net: Interpretable Rule Learning
with Neural Networks

Lucile Dierckx[1,2]([✉]) [iD], Rosana Veroneze[2,3] [iD], and Siegfried Nijssen[1,2] [iD]

[1] TRAIL Institute, Louvain-la-Neuve, Belgium
[2] ICTEAM/INGI, UCLouvain, Louvain-la-Neuve, Belgium
`{lucile.dierckx,rosana.veroneze,siegfried.nijssen}@uclouvain.be`
[3] FEEC/DCA, Unicamp, Campinas-SP, Brazil

Abstract. As there is a need for interpretable classification models in
many application domains, symbolic, interpretable classification models
have been studied for many years in the literature. Rule-based models
are an important class of such models. However, most of the common
algorithms for learning rule-based models rely on heuristic search strategies developed for specific rule-learning settings. These search strategies
are very different from those used in neural forms of machine learning,
where gradient-based approaches are used. Attempting to combine neural and symbolic machine learning, recent studies have therefore explored
gradient-based rule learning using neural network architectures. These
new proposals make it possible to apply approaches for learning neural
networks to rule learning. However, these past studies focus on unordered
rule sets for classification tasks, while many common rule-learning algorithms learn rule sets with an order. In this work, we propose RL-Net,
an approach for learning ordered rule lists based on neural networks. We
demonstrate that the performance we obtain on classification tasks is
similar to the state-of-the-art algorithms for rule learning in binary and
multi-class classification settings. Moreover, we show that our model can
easily be adapted to multi-label learning tasks.

Keywords: Interpretability · Pattern Set Mining · Rule Learning ·
Binary Neural Networks

1 Introduction

Organizations are increasingly using Machine Learning models to help decision-
making. For many application domains (such as medicine, health care, criminal
justice, and education), interpretability is essential in addition to predictive performance. Therefore, white-box models are preferable to black-box models in
these scenarios.

Rule-based classification models are an important class of interpretable models, which provide symbolic white-box models that are expressed as simple IF-
THEN rules. A distinction can be made between rule *sets* and rule *lists*. In a

© The Author(s), under exclusive license to Springer Nature Switzerland AG 2023
H. Kashima et al. (Eds.): PAKDD 2023, LNAI 13935, pp. 95–107, 2023.
https://doi.org/10.1007/978-3-031-33374-3_8

rule list, the rules have an order, and the first rule of which all conditions in the IF-part are satisfied is used to perform a prediction. An advantage of this app-roach is that the resulting models are also interpretable on classification tasks with more than two classes: one rule is used to perform the prediction. Models based on rule sets typically rely on voting, where all rules vote for classes, or they only work for two classes. This makes these methods less interpretable. For this reason, we focus on rule lists in this work.

Roughly speaking, two classes of approaches for learning rule-based models can be distinguished. A common strategy for rule learning relies on *pattern mining*. Traditional pattern mining is formulated as the problem of computing $\text{Th}(\mathcal{L}, \varphi, \mathcal{D}) = \{\pi \in \mathcal{L}|\varphi(\pi, \mathcal{D}) \text{ is true}\}$, where $\mathcal{D}$ is the dataset, $\mathcal{L}$ is a language of patterns, and φ is a constraint, often based on support [9,10]. The size of the search space of this problem is exponential in the size of $\mathcal{L}$.

The number of all patterns satisfying the constraints is usually too large. Thus, patterns are often post-processed in a step-wise procedure to become useful [10]. In the first step, the patterns that meet the constraints are enumerated. In the second step, some patterns are selected and combined. Again, we have another search space of exponential size, in this case in the size of $\text{Th}(\mathcal{L}, \varphi, \mathcal{D})$.

Most methods adopt heuristics to select and combine the patterns, which is commonly the case for associative classification proposals, such as CBA [12] and CMAR [11]. These methods solve one particular instance of the *pattern set mining* problem, which consists in computing $\textbf{Th}(\mathcal{L}, \varphi, \psi, \mathcal{D}) = \{\Pi \subseteq \text{Th}(\mathcal{L}, \varphi, \mathcal{D})|\psi(\Pi, \mathcal{D}) \text{ is true}\}$, where ψ expresses constraints that have to be sat-isfied by the overall pattern set [9,10]. The major drawback of the step-wise procedure is that it does not scale well.

The second class of methods scales better. Instead of first mining patterns, these approaches learn the rules themselves also using heuristics; typically, they use a heuristic to iteratively add the most promising condition to a rule. A well-known representative of this class is the RIPPER [5] algorithm. While, as a consequence, these approaches find rules more quickly, the heuristics are often specific to one learning task and may have as effect that the algorithm overlooks good rules.

In many application domains of machine learning, recent advances in learning deep neural networks have led neural network techniques to become the state-of-the-art. Search strategies in the neural network literature are very different, and are often based on the use of gradient descent techniques. The success of neural methods has had as effect that many learning problems have now been phrased and solved using these gradient descent techniques. However, traditional neural networks are not interpretable models.

Therefore, the research community has been looking for strategies to com-bine symbolic forms of Artificial Intelligence with techniques based on neural net-works, leading to techniques for neuro-symbolic AI. In the case of rule learning, this could lead to a combination of the interpretability of rule-based models with the search strategies employed when learning neural networks. Recently, Qiao et al. [14] and Fischer & Vreeken [7] developed pattern set mining strategies that rely on binarized neural networks. Adopting a neural network trained with gradient descent methods has several advantages. Indeed, all advances in the area of neural networks have the potential to be leveraged for pattern set mining. This includes stochastic well-developed gradient descent algorithms, well-developed loss functions, well-developed regularization concepts, sophisticated development frameworks, and powerful computing platforms. However, none of these approaches studied how to learn *ordered* rule models for a wide range of learning tasks, including multi-class and multi-label classification tasks, and it is not clear how well neural network-based techniques would work on this task.

In this work, we focus on such problems by extending the Decision Rules Network (DR-Net) proposal of Qiao et al. [14]. We incorporated the possibilities of (a) using hierarchy among the rules, hence adding the possibility of learning classifiers based on *rule lists* in addition to *rule sets*, and (b) solving multi-class classification problems. Furthermore, the consequent part of the rules (i.e., the class labels) is fixed in DR-Net, so all learned rules have the same consequent. We instead learn the class label of each rule together with the condition. Lastly, our proposal can easily be tweaked to solve multi-label classification problems.

This paper is organized as follows: Sect. 2 gives an overview of the related work. Section 3 presents the architecture of our proposed interpretable Rule Learning neural Network (RL-Net). The datasets used, the models for comparison, the experimental protocol, as well as the obtained results, are presented in Sect. 4. Finally, Sect. 5 concludes this paper.

2 Related Work

The use of neural networks to learn rule-based classifiers is still in its early stages.

A first example is the work of Beck and Fürnkranz [2,3] which learns rule sets to perform binary classification using a network structure. However, the network weights are learned using a greedy heuristic instead of a differentiable approach.

Yang et al. [16] presented the Deep Neural Decision Tree (DNDT) method that mimics the structure of a decision tree using a neural network architecture. In this proposal, the weights are trained with a gradient descent algorithm. The splitting value for each attribute and the rule labels are learned during the training. The model is thus suitable for binary and multi-class classification. The key limitation of their proposal is that it is not scalable w.r.t the number of features. In their experiments, they could only find an accurate single tree for datasets with at most 12 features. Moreover, the limitation of a tree structure makes it impossible to learn arbitrary rule lists.

The Explainable Neural Rule Learning (ENRL) method [15] also learns rules in a differentiable manner coupled with a neural network structure. The Explainable Condition Module (ECM) is the building block of the method. It comprises a feature, an operator, and a value, learning atomic propositions such as $age \geq 18$. Based on the atomic propositions, ENRL adopts a complete binary tree topology to express multiple rules, and the problem of seeking appropriate rules is transformed into a neural architecture search. ENRL creates an ensemble of trees, and the final decision is made by a voting mechanism, which makes this method less interpretable. Also, it is limited to binary classification.

The Decision Rules Network (DR-Net) method [14] learns rule sets for binary classification. The model is composed of three layers: the input layer, the Rules layer, and the OR layer. The *Rules layer* learns the rules. The number of neurons in this layer is a user-defined parameter that sets the maximum number of rules. Its regularization term controls the length of the rules. The *OR layer* chooses which rules to use and which to ignore. Its regularization term controls the number of rules that will be used in the classifier. The network training is done in two alternating phases, one for the Rules layer and one for the OR layer. This method does not identify rule lists, but rule sets.

As can be seen, most existing methods focus on binary classification, and none of them can find arbitrary rule lists. In this work, we contribute to a neural network for learning rule-based classifiers that are fully interpretable, and suitable for binary and multi-class classification. Our proposal takes full advantage of the advances in neural network literature.

3 Approach

This section introduces the details of our contribution.

3.1 RL-Net

Our Rule Learning neural Network (RL-Net) was conceived to learn interpretable rule lists that can perform multi-class classification. It can also be easily tweaked to be used in multi-label experiments. RL-Net employs the structure of a neural network as well as its gradient optimization learning methods.

The network is composed of four layers, as depicted in Fig. 1. The first layer is the input layer that receives the dataset's features. It is connected to the rule layer, where the rule conditions are learned. The next layer expresses the hierarchy among the rules, which is necessary to learn a rule list instead of a rule set. Finally, the output layer assigns a specific class label to each rule.

Each layer is presented in more detail in what follows. The method implementation can be found in our GitHub repository on https://github.com/luciledierckx/RLNet.

Input Layer. We assume that the features that are fed to the network are binary. As discussed in the rule layer description, there is no need to duplicate the input dataset to express the negation of a feature because the network can express that by itself. The number of nodes in this layer is equal to the number of binarized features of the dataset.

Rule Layer. This layer mimics the behavior of logical ANDs. It is composed of r nodes, where r is a user-defined parameter that specifies the number of rules to be learned. This layer and the input layer are the same as in DR-Net [14],

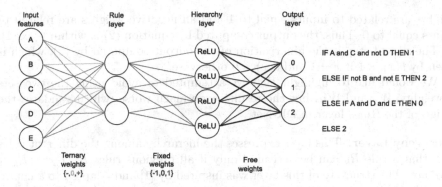

Fig. 1. Global architecture of RL-Net for learning a rule list composed of three rules and the default rule (*else*) for an input dataset consisting of five binary features and three class labels. Green (resp. red) weights represent weights with a positive (resp. negative) value. The edges in bold represent the edges with the highest weight for each node in the hierarchy layer. Connections between neurons that are not represented are non-trainable zero weights.

while the next ones are different. Each weight of this layer can either be negative, zero, or positive to represent the fact of using a feature in the rule (+), using the negation of that feature (−), or not considering that feature for the rule (0). As it is not simple to learn discrete weights with a gradient descent algorithm, the ternary weights W_T are obtained by an element-wise product of two other matrices:

$$W_T = W_S \circ W_H \tag{1}$$

The weights in the matrices W_S and W_H are floating-point numbers. The weights in W_S will converge to a positive or a negative value during the training, thus deciding if we use the positive form of a given attribute or its negation. The weights in W_H are referred to as hidden weights. They will decide whether an attribute is used in a rule or is ignored. We ensure that these weights converge to 1 or 0 thanks to the method discussed by Louizos et al. [13] to approximate binary random variables with a Bernoulli distribution. To these hidden weights, a sparsity-based regularization [13] is added to push the weights toward zero and, therefore, obtain shorter rules.

The neurons of the rule layer have to mimic the behavior of a logical AND, that outputs *true* (1) when all rule conditions are met or *false* (0) otherwise. This is done in two steps, as proposed in [14].

In the first step, a neuron from the rule layer performs the following operation:

$$y = \sum w_i \cdot x_i - \sum_{w_i > 0} w_i + 1, \tag{2}$$

where $w_i \in W_T$ and $x_i \in \{0, 1\}$ is the value of the binary feature i. In this formulation, the bias has a dynamic value that depends on the number of positive weights for the neuron. Note that $y = 1$ can only be obtained when all positive weights are related to inputs equal to 1 and all negative weights are related to inputs equal to 0. Thus, the output computed by equation (2) is within $(-\infty, 1]$.

The second step is the binarization of the output computed by (2), which is given by $b(x) = 1$ if $x = 1$ and $b(x) = 0$ otherwise.

We would like to highlight that we attempted to avoid using the two sets of weights, W_S and W_H, but the results were much worse, which validates the design of the Rules layer in DR-Net.

Hierarchy Layer. This layer expresses the hierarchy among the different rules such that a rule R_k can be activated only if all previous rules $R_1, R_2, ..., R_{k-1}$ were not. The structure of this layer was inspired by [1] and adapted to a neural network architecture. The weights of this layer are set in the initialization and are not trainable, as illustrated in Fig. 1. The activation of a neuron of this layer is given by a ReLU function. The neuron k in the rule layer represents the rule R_k in the rule list. It is connected to a neuron l of the hierarchy layer by an edge with weight w_{kl}, where w_{kl} is 1 when $l = k$, -1 when $l < k$, and 0 otherwise.

The last neuron of the hierarchy layer represents the default rule (*else*). It is the only neuron of this layer that uses a bias with a value equal to 1. This ensures that the default rule will be applied when none of the previous rules are applicable. Thus, the number of nodes in this layer is equal to $r+1$. As the input values of the hierarchy layer are binary (0 or 1), its output will also be binary.

Output Layer. The last layer of our network learns the label associated with every rule condition. The number of neurons in this layer is equal to the number of class labels in the input dataset. The weights are free, but we add L2-regularization. The activation function is the softmax. As only one rule at a time is active, only the label with the highest activation (i.e., with the highest weight for the active rule) is considered at prediction time. Thus, the learned rules are fully interpretable.

3.2   Training and Tuning of RL-Net

Standard neural network techniques are applied in the training of RL-Net. Indeed, we used the Adam optimization algorithm with the cross-entropy loss function. A callback on the validation loss is also applied.

As observed in standard neural networks, the performance of RL-Net for strongly imbalanced datasets can be improved by using a class-balanced version of the loss. Therefore, a balanced version of the loss is used in the first e_b epochs, where e_b is a user-defined parameter. This parameter can be set to zero for datasets for which there is no need for a balanced loss.

Other important parameters of RL-Net refer to the hidden weights W_H. These weights are initialized using a normal distribution with mean μ and variance σ^2, which directly impacts the probability of using (or not) an attribute in a rule. Thus, the choice of these parameters influences the length (specificity) of the rules in the initialization, making the search for a good local optimum more or less hard. Other hyper-parameters that must be tuned are the learning rate, the weight of the sparsity-regularization term of the rule layer, as well as the weight of the L2-regularization of the output layer.

3.3   RL-Net for Multi-label Classification

The RL-Net architecture was conceived for multi-class classification, but one of its advantages is that it can easily be transformed into a basic multi-label classifier with two minor changes. The first modification is that the activation of the output layer must be the sigmoid activation instead of the softmax. The second one concerns the loss function, which must be designed for multi-label classification, such as binary cross-entropy loss, focal loss, Huber loss, multi-label margin loss, MSE loss, and L1 loss. For our experiments, we choose the binary cross-entropy loss.

4   Experiments

Our experiments were designed to answer the following research questions: How does RL-Net compare against its basis, DR-Net, for binary classification? How does RL-Net compare against state-of-the-art rule-based classifiers, RIPPER and CART, for binary and multi-class classification? Are the simple changes in RL-Net for multi-label classification enough to achieve a satisfactory performance?

4.1   Datasets

We selected 7 binary and 6 multi-class datasets for our experiments, among which all binary datasets used in the DR-Net paper [14]. We also chose 2 multi-label datasets to perform a first evaluation on multi-label classification. The datasets all come from the UCI Repository [6] except heloc[1], house[2], yeast[3], and scene(See footnote 3).

From the DR-Net paper [14], we used adult census (adult), magic gamma telescope (magic), fico heloc (heloc), and home price prediction (house). To these, we added internet advertisements (ads), king-rook vs. king-pawn (chess), and mushroom (mushroom). For multi-class classification, we chose car evaluation (car), nursery (nursery), contraceptive method choice (contraceptivemc), page blocks classification (pageblocks), pen-based recognition of handwritten digits (pendigits), and sensorless drive diagnosis (drive). We kept the different classes of the multi-class datasets untouched except for nursery where we merged the class "very_recom" and "recommend" as they represented respectively 2.531% and 0.015% of the class distribution. The multi-label experiments were made with the yeast (yeast) and scene (scene) datasets.

---

[1] https://community.fico.com/s/explainable-machine-learning-challenge.
[2] https://www.openml.org/d/821.
[3] https://www.uco.es/kdis/mllresources/.

Table 1. Characteristics of the different datasets: The column #Attributes binarized presents the number of attributes after the different preprocessing steps while all the other columns are computed from the unprocessed dataset.

Binary and multi-class datasets	#Rows	#Attributes	#Attributes binarized	Proportion of each class			
Adult	48842	14	128	0.24, 0.76			
Magic	19020	10	90	0.35, 0.65			
House	22784	16	132	0.70, 0.30			
Heloc	10459	23	147	0.48, 0.52			
Mushroom	8124	22	111	0.52, 0.48			
Chess	3196	36	38	0.52, 0.48			
Ads	3279	1559	1577	0.86, 0.14			
Nursery	12960	8	26	0.33, 0.03, 0.33, 0.31			
Car	1728	7	21	0.70, 0.22, 0.04, 0.04			
Pageblocks	5473	10	88	0.90, 0.06, 0.01, 0.02, 0.02			
Pendigits	10992	16	135	10 classes with equal proportions			
Contraceptivemc	1473	9	34	0.43, 0.35, 0.23			
Drive	58509	48	432	11 classes with equal proportions			
Multi-label datasets	#Rows	#Attributes	#Attributes binarized	#Labels	Cardinality	Density	Distinct
Yeast	2417	103	927	14	4.237	0.303	198
Scene	2407	294	2646	6	1.074	0.179	15

4.2　Data Preprocessing

Regarding the data used for training, the first step was to remove the data samples for which the percentage of missing values was $\geq 40\%$. Next, we removed the features for which the percentage of missing values was $\geq 40\%$. The remaining missing values were replaced by the most frequent value in the case of categorical attributes, and by the mean value for numerical features. Lastly, we applied the same feature binarization as the one implemented for the DR-Net. This binarization applies one-hot encoding to the categorical attributes, and quantile discretization to the numerical ones followed by ordinal scaling [8].

The main properties of each dataset before and after the preprocessing are presented in Table 1. The preprocessed datasets are the input for all methods used in our experiments. In that way, we can be sure that any observed difference in performance comes from the model itself and not from the data preprocessing.

4.3　Algorithms

We compare RL-Net with two state-of-art rule-based classifiers, CART [4] and RIPPER [5], as well as with DR-Net, which is the basis for RL-Net. We used CART from the *scikit-learn* library and RIPPER from *Weka* (*JRip*). We used the authors' implementation of DR-Net.

4.4　Protocol

Some of the datasets (namely adult, pendigits, yeast, and scene) have a predefined test set. In this case, it was used as the test dataset. Otherwise, we created a test dataset by selecting 25% of the data samples in a stratified fashion.

For RL-Net and DR-Net, we tuned the hyper-parameters using stratified 10-fold cross-validation. The number of rules ranges from 2 to 20 in our experiments, but we fixed it to 10 in the hyper-parameter tuning for a matter of time. For both algorithms, we tuned the following hyper-parameters: the number of epochs, the learning rate, and the weight of the sparsity-regularization term. The batch size was set to 5% of the dataset size.

For RL-Net, we also tuned the weight of the L2-regularization of the output layer, the number of epochs for the balanced loss e_b, and the mean value μ of the normal distribution used for the initialization of the hidden weights W_H.

We compare the algorithms using the same number of rules. Therefore, we do not need to train the OR layer of DR-Net. Accordingly, we set the weight of the OR layer regularization term to zero, and the network training was focused on the Rules layer. For CART, the number of rules is controlled through the user-defined parameter *max_leaf_nodes*. There is no user-defined parameter to control the number of rules in RIPPER's implementation. However, the *minimal weights of instances within a split* parameter influences the number of rules in the classifier. So, we varied this parameter to find the desired number of rules.

For CART, we also tuned the *criterion* using stratified 10-fold cross-validation.

The remaining hyper-parameters of all algorithms were left to their default values. For ease of reproduction, all details about the hyper-parameter tuning are available in our GitHub repository.

4.5 Results

Figure 2 shows the performance of RL-Net and its competitors in terms of accuracy. We set the number of rules from 2 to 20 as our focus is on obtaining interpretable models. RL-Net and DR-Net were run 10 times for each dataset because these methods can get stuck in a poor local optimum depending on the random initialization. The results of these runs are exhibited in a box-plot format.

The binary classification performance is presented in Figs. 2(a)-(g). When we compare RL-Net to DR-Net, we see that it is not possible to say that one of them always performs better than the other. It actually depends on the dataset. For some datasets, such as adult, mushroom, and chess, RL-Net has higher accuracy, for others like ads, it is the other way around. In some other cases, the best-performing method depends on the number of rules considered, such as for magic, house, and heloc. RL-Net has a large performance variability on heloc and ads datasets (with a maximal variability of 10%). In contrast, DR-Net has a large performance variability on the chess dataset (with a maximal variability of 35%). This variability is a drawback of neural networks with such non-standard layers, but it does not stop both networks from achieving competitive performance when the learning does not get stuck in a poor local optimum. RIPPER's performance is generally better than the two neural network approaches, but the difference is not large. RL-Net outperforms CART in a wide range of cases. The performance of the CART decision tree is the most affected when the number of rules is low.

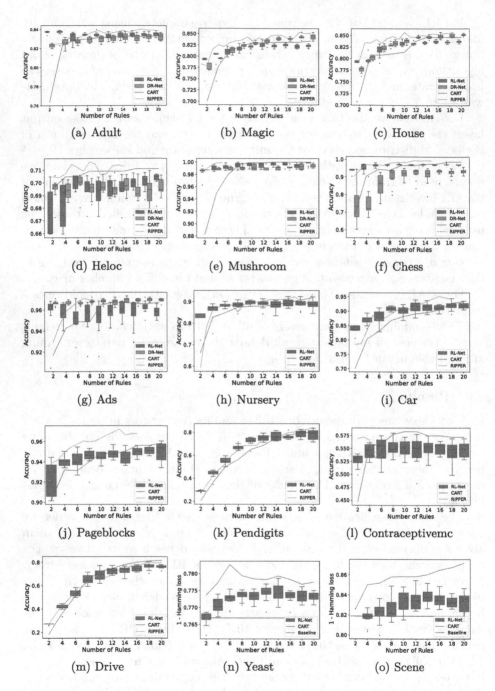

Fig. 2. Performance of our RL-Net (blue), the DR-Net (orange), Ripper (green), and CART (red) versus the number of rules, on binary datasets (a to g), multi-class datasets (h to m), and multi-label datasets (n to o). (Color figure online)

Indeed, the length of the rules in the tree is limited by the maximum number of rules, while it is not the case for the other algorithms.

Figures 2(h)-(m) present the results for multi-class classification. RL-Net can achieve performance as high as RIPPER for nursery, car, pendigits, and drive datasets. RIPPER outperforms RL-Net for the pageblocks dataset, and when using 2 to 4 rules for the contraceptivemc dataset. Concerning CART, RL-Net can achieve identical performances even though some runs get stuck in poor local optimums as for the binary case. From these results, we can thus clearly see that RL-Net works well on multi-class classification.

The results for multi-label classification are presented in Figs. 2(n)-(o). In addition to CART, we also compare our results with a baseline that, for each class label, predicts the most frequent value (true or false). For both datasets, RL-Net has a lower performance than CART. For the yeast dataset, RL-Net follows CART's performance but scores 1 to 1.5% lower. RL-Net's result for two rules is similar to the baseline one, but RL-Net's performance improves with the number of rules, being up to 1.2% better. For the scene dataset, RL-Net can obtain a lower performance than the baseline when it gets stuck in a poor local minimum, but it clearly outperforms the baseline. However, CART is considerably better than RL-Net for the scene dataset. From these results, we note that using RL-Net for multi-label classification has potential, but its performance is not state-of-the-art yet. This experiment is a proof of concept, indicating that exploring this direction could yield good results.

5   Conclusion

Building on the interest in combining neural and symbolic machine learning, in this work we explored gradient-based rule learning using neural network architectures. We implemented and presented our RL-Net method to learn binary and multi-class rule lists using a neural network approach. We showed that with minor adaptations RL-Net can be used to learn multi-label classifiers. We compared our proposal to some other state-of-the-art algorithms for binary and multi-class classification. We also evaluated the potential of RL-Net for learning multi-label tasks. From our results, we concluded that RL-Net is a proper method for learning fully interpretable binary and multi-class classifiers. It does not always achieve the highest performance, but it is never far from the best. Regarding multi-label classification, some additional work should be done to increase RL-Net's performance, but our network architecture is easily compatible with this task, indicating that RL-Net has potential in the integration of rule learning with other neural network-based approaches. Future works include making the method less susceptible to a bad initialization, improving the method for multi-label classification, and integrating RL-Net further with other research in the neural network literature, such as transfer learning, semi-supervised learning, or active learning.

Acknowledgments. This work was supported by Service Public de Wallonie Recherche under grant n°2010235 - ARIAC by DIGITALWALLONIA4.AI, and under grant n°2110107 - SERENITY2 by WIN2WAL. We would also like to thank FAPESP, Brazil (Grants No. 2017/21174-8 and 2020/00123-9) for the financial support.

Computational resources have been provided by the supercomputing facilities of the Université Catholique de Louvain (CISM/UCL) and the Consortium des Équipements de Calcul Intensif en Fédération Wallonie Bruxelles (CÉCI) funded by the Fond de la Recherche Scientifique de Belgique (F.R.S.-FNRS) under convention 2.5020.11 and by the Walloon Region.

References

1. Aoga, J.O.R., Nijssen, S., Schaus, P.: Modeling pattern set mining using Boolean circuits. In: Schiex, T., de Givry, S. (eds.) CP 2019. LNCS, vol. 11802, pp. 621–638. Springer, Cham (2019). https://doi.org/10.1007/978-3-030-30048-7_36
2. Beck, F., Fürnkranz, J.: An empirical investigation into deep and shallow rule learning. Front. Artif. Intell. 4 (2021)
3. Beck, F., Fürnkranz, J.: An investigation into mini-batch rule learning. arXiv preprint arXiv:2106.10202 (2021)
4. Breiman, L., Friedman, J.H., Olshen, R.A., Stone, C.J.: Classification and regression trees. Routledge (2017)
5. Cohen, W.W.: Fast effective rule induction. In: Twelfth International Conference on Machine Learning, pp. 115–123. Elsevier (1995)
6. Dua, D., Graff, C.: UCI machine learning repository (2017). http://archive.ics.uci.edu/ml
7. Fischer, J., Vreeken, J.: Differentiable pattern set mining. In: Proceedings of the 27th ACM SIGKDD Conference on Knowledge Discovery & Data Mining, pp. 383–392 (2021)
8. Ganter, B., Wille, R.: Formal Concept Analysis: Mathematical Foundations. Springer Science & Business Media (2012). https://doi.org/10.1007/978-3-642-59830-2
9. Guns, T., Nijssen, S., De Raedt, L.: Evaluating pattern set mining strategies in a constraint programming framework. In: Huang, J.Z., Cao, L., Srivastava, J. (eds.) PAKDD 2011. LNCS (LNAI), vol. 6635, pp. 382–394. Springer, Heidelberg (2011). https://doi.org/10.1007/978-3-642-20847-8_32
10. Guns, T., Nijssen, S., De Raedt, L.: K-pattern set mining under constraints. IEEE Trans. Knowl. Data Eng. **25**(2), 402–418 (2011)
11. Li, W., Han, J., Pei, J.: CMAR: accurate and efficient classification based on multiple class-association rules. In: Proceedings 2001 IEEE International Conference on Data Mining, pp. 369–376. IEEE (2001)
12. Liu, B., Hsu, W., Ma, Y., et al.: Integrating classification and association rule mining. In: KDD, vol. 98, pp. 80–86 (1998)
13. Louizos, C., Welling, M., Kingma, D.P.: Learning sparse neural networks through L0 regularization. In: 6th International Conference on Learning Representations, ICLR 2018 - Conference Track Proceedings (2018)
14. Qiao, L., Wang, W., Lin, B.: Learning accurate and interpretable decision rule sets from neural networks. In: Proceedings of the AAAI Conference on Artificial Intelligence, vol. 35, pp. 4303–4311 (2021)

15. Shi, S., Xie, Y., Wang, Z., Ding, B., Li, Y., Zhang, M.: Explainable Neural Rule Learning. In: WWW 2022 - Proceedings of the ACM Web Conference 2022, pp. 3031–3041. Association for Computing Machinery, Inc (2022)
16. Yang, Y., Morillo, I.G., Hospedales, T.M.: Deep neural decision trees. In: ICML Workshop on Human Interpretability in Machine Learning. arXiv preprint arXiv:1806.06988 (2018)

Classification

The Causal Strength Bank: A New Benchmark for Causal Strength Classification

Xiaosong Yuan[1,2], Renchu Guan[1,2], Wanli Zuo[1,2], and Yijia Zhang[3(✉)]

[1] College of Computer Science and Technology, Jilin University, Changchun, China
yuanxs19@mails.jlu.edu.cn, {rcguan,wlzuo}@jlu.edu.cn
[2] Key Laboratory of Symbolic Computation and Knowledge Engineering,
Ministry of Education, Changchun, China
[3] College of Electronic Countermeasures, National University of Defense Technology,
Hefei, China
yijia18@mails.jlu.edu.cn

Abstract. Causal relation extraction is essential in the causality discovery of natural language processing. The development of causal relation extraction from the model-driven is staggering, so we resort to the data-driven method. More causal information is necessary because most current datasets only label the locations of causal entities or events, which may restrict the learning capacity of models. In this paper, we introduce a novel benchmark causal strength classification and corresponding dataset, Causal Strength Bank (CSB), consisting of a Chinese dataset (C-CSB) and an English dataset (E-CSB) which merge causal strength, causal polarity, and causal entity. To ensure credibility, we select four canonical English datasets and clean Wikipedia passages for the Chinese corpus. The corpus is then annotated and cross-checked by professional annotators in two stages, ensuring the accuracy of CSB. We evaluate various baseline methods on CSB and show that causal strength information benefits causal relation extraction, demonstrating the value of the proposed dataset. Our dataset is available at https://github.com/yuanxs21/CSB-dataset.

Keywords: Causality extraction · Causal strength classification · Benchmark dataset

1 Introduction

Causal relation extraction is a branch of relation extraction task whose goal is to identify the causality in the texts and locate the position of causal objectives. It is critical for commonsense reasoning, question answering, and decision support. These years, copious datasets and models are proposed to improve the performance of general models on the task [1–7]. Regretfully, either well-defined datasets or model-based methods focus on extracting causal entities or events from texts, which can merely identify the existence of causality. They all ignore the strength and polarity between causal objectives, which can help models learn to describe causality precisely.

© The Author(s), under exclusive license to Springer Nature Switzerland AG 2023
H. Kashima et al. (Eds.): PAKDD 2023, LNAI 13935, pp. 111–122, 2023.
https://doi.org/10.1007/978-3-031-33374-3_9

Previous datasets such as SemEval-2010 Task 8 [3], Event Storyline [2], and CausalTime Bank [8] have focused on the general relationship between two entities, and have aggregated time cues or other order clues to enhance model performance. However, existing models show hesitant results around a fixed level or only a slight improvement. Additionally, the proportion of causality samples in these datasets is small, which limits their ability to support deep learning models that require large amounts of training data.

-3   The *earthquake* happened in Tokyo caused over 6, 020 people *death*.

3   The *economy* created *jobs* at a surprisingly robust pace in January, the government reported on Friday, evidence that America's economic stamina has withstood any disruptions caused so far by the financial tumult in Asia.

-1   The *slowdown* raises *questions* about the economy's strength because spending fueled much of the third-quarter GNP growth.

0   Michael Barry is one of the most respected riders in the peloton and the author of two books and many fascinating articles.

Fig. 1. Sentences annotated with both causal entities and strength labels from E-CSB, we highlight the causal strength of a sentence with blue style and causal entities with *red italic*. (Color figure online)

Many causal relation extraction models aim to solve problems in distinct datasets, such as the RHNB [5] divide causal connectives of SemEval-2010 Task 8 into other classes as a new category feature; similar to RHNB, Dual-CET [6] is a cause-effect network to discover co-occurrence patterns and evolution rules of causation. Still, the adaptability of such well-designed models to other datasets is weak. With time clue, [9] is a common inference framework to extract causal entities in CausalTime Bank. However, such a complex model cannot apply to time-independent contexts. Some models also focus on self-annotated data, [4] extracting cause-effect pairs in web texts annotated by a template; [7] devising a multi-head attention network to discover implicit causality from web texts. Various models are with data-caused problems, and we thus solve these problems by proposing a new dataset type in this work.

Extracting causal relationships is a meaningful exercise [10]. However, more causal information should be considered, and the causal strength is critical. In the first sentence of Fig 1, only "earthquake" and "death" are labeled; no more causal message. Among these cases, we can figure that "death" caused by "earthquake" and the "questions" raised by "slowdown" conveys different causal strengths. Furthermore, the "death" caused by "earthquake" states a severe negative effect, while the pair of "jobs" and "economy" apparently is positive. Those samples express different causal polarities. Although [11] computes the causal strength by the co-occurrence of words in a whole corpus, the statistical method considers no context. [12] extends it by increasing the expression number of a single word but is still limited by hard-statistic methods. Many works show that the strength information about causal relationships would advance the causal studies [11, 12].

There is an isolation between causality extraction and causal strength for now. However, we find that the two tasks may boost the performance of each other. Suppose we could utilize both causal strength and entity information. Albeit these two aspects in sentences are crucial for more detailed causality obtaining, no appropriate dataset is available in this research as far as we know. Considering the above, we introduce a novel benchmark dataset in this paper. We collect several datasets and define a set of rules to annotate each sentence's causal strength and polarity. Eventually, we get a brand new dataset Causal Strength Bank(CSB), consisting of the English Causal Strength Bank(E-CSB) and Chinese Causal Strength Bank(C-CSB) with 3 strength and 2 polarity categories, 7 classes in total.

We conduct several experiments to demonstrate the effectiveness of our benchmark and dataset. Because a multi-task model is more capable of aggregating relevant information for various tasks than a single-task model [14], we conduct a multi-task experiment on E-CSB to show the improvement of causal relation extraction coming from causal strength. Experiments based on multi-task learning architecture perform better than single causality extraction, highlighting our dataset's significance. Our main contributions can be summarized as follows:

- We develop a new task named causality strength classification, whose goal is to classify the causality strength expressed by sentences and design a set of annotation rules for a causal strength dataset.
- We build a novel causal strength classification dataset by collecting and annotating the raw corpus of Chinese and English from reliable sources, and we analyze the dataset from different perspectives.
- We design and conduct single-task and multi-task learning experiments with baseline models on the datasets; results show that causal strength classification is beneficial for causal relationship extraction.

2   Related Work

Although the popularity of causal strength is not on par with causality extraction, there are still some standard datasets and influential models in causal effect estimation from texts.

Datasets. [15] present a commonsense knowledge base, Weltmodell, including lots of causal knowledge, and it is automatically generated from over 3.5 million English language books. CausalNet [11] is a framework that automatically harvests a network of causal-effect terms from a prominent web corpus.

Models. Methods based on statistics compute a score to represent the causality strength for given causes and effects, like PMI [16], utilizing mutual information and lexicography. [17] first introduces the PMI method into commonsense causal strength estimating on a substantial amount of web texts. Beyond the CausalNet, [11] modify the PMI with sufficient and necessary causality. [12] created a list of multi-word to extend [11] method, which considered a multi-word expression as a unit of an event as well as a single word.

However, estimating the causal effect in texts by statistical methods is deficient for prior knowledge. We consider it a critical thing that introduces human judgment of a specific causality into the causal relation extraction task. Thus we present a new dataset for causal effect estimating research.

3 The Causality Strength Datasets

In this work, our initiative is to facilitate causality discovery research for more accuracy and detail. Therefore, we present a new benchmark named the causal strength classification task and a new fine-grained dataset containing more causal information, i.e., the causal strength and polarity. There are three challenges during the dataset development process: data scarcity, data dirty, and data labeling. We manage these problems by collecting corpus, data process, and manually annotating, which is quite different from the former works [11,12]. In addition, we also report analyses of the dataset from different perspectives.

3.1 Data Collection and Processing

Extending the existing relation extraction datasets is an ideal way to precisely measure causal relationships in texts. We notice that researchers worldwide prefer datasets of English for relation extraction. Comparatively, Chinese has not received enough attention in research as the world's most spoken language. As a result, Chinese datasets are substantially smaller in number and scope than English datasets. To enhance the study of causality in English while compensating for the lack of corpus in Chinese, we collect and process data from the Chinese and English corpora by distinct strategies.

-3 1979年1月，法国对博卡萨的支持开始减少——当月在首都班戈发生的骚乱导致了对平民的大屠杀。

-2 1960年代底特律的工会及退休金制度长期下来造成底特律庞大的财政负担，这些抚恤金支付到生命结束再加上人口逐渐的减少，使得市政府负债膨胀至无法负荷。

1 在语法学中词形变化指单词的变化，可以导致语法功能改变，进而使其代表的意义也有所改变。

0 相干逻辑的显著特征是它们是次协调逻辑：矛盾的存在不会导致逻辑爆炸。

Fig. 2. Examples with causal strength label from C-CSB, we highlight the causal strength before a sentence with blue style.

A few widely used relation extraction datasets with labeled causal entities/events are available. To take full advantage of them, we pick out some curated ones from them since the sentences have varying degrees of causality and a polarity message. Considering the quality and popularity, we integrate the

SemEval-2010 Task 8, CTB, ESC, and **SemEval-2020 Task 5** [13] as our raw English corpus. Then, all sentences with labeled causal relations are sent to invited annotators, who would deeply reannotate based on the original version. As for sentences labeled with other relationships, we choose another strategy to tackle.

There is no authoritative dataset for the Chinese corpus, so we build it from zero. First, we fetch the **Wikipedia official corpus**[1], since it has dedicated personnel to review and maintain each entry, and we believe the knowledge descriptions of Wikipedia are objective and reliable. Another reason is that Wikipedia covers topics broadly, which may stimulate information fusion among data in different areas. In the processing, we first employ the Wikipedia Extractor[2] to standardize the texts in the wiki-corpus and obtain traditional Chinese text files. After that, we convert its format into simplified Chinese and utilize the OpenCC[3]. To figure out sentences that deliver causal relations, possibly from the simplified Chinese corpus, we also build a causal trigger word vocabulary to filter them. As the Wikipedia corpus consists of numerous documents and the length is relatively long, we split the documents and confine each sentence within 120 Chinese characters, ensuring the sentences are complete. Then sentences that probably convey causal relation are delivered to annotators.

3.2   Data Labeling

We use a two-step annotation process to ensure accuracy and reliability. Causal strength is annotated with four categories: *none*, *normal*, *moderate*, and *strong*. We also add binary polarity labels to describe the effect of causal relationships, with _*pos* indicating a positive impact and _*neg* indicating a negative impact. Examples of negative impacts include unnatural death, pollution, and extinction caused by human activity.

We refer to international safety accident-level standards to distinguish different levels of negative causal strengths in our labeling procedure. For accidents resulting in fewer than three deaths, injuries to fewer than ten people, or economic damages of less than one million dollars, annotators label corresponding sentences as *normal_neg*. Accidents resulting in more than three but fewer than ten deaths, or causing economic damages of more than one million dollars but less than five million dollars, are labeled as *moderate_neg*. Severe accidents resulting in ten or more deaths, more than 50 serious injuries, or direct economic damages exceeding $50 million are labeled as *strong_neg*.

We also label positive causality strengths, which lack a clear distinction like safety accidents. To establish a reference point, we exploit the opposite of safety accidents and consider the impact range of benefits produced by causes that lead to positive outcomes. We categorize a generally favorable event as *normal_pos* if it benefits fewer than 10 people. Annotators label a cause as *moderate_pos* if

---

[1] https://dumps.wikimedia.org/.
[2] http://medialab.di.unipi.it/wiki/Wikipedia_Extractor.
[3] https://github.com/BYVoid/OpenCC.

it affects more than 10 but less than 50 people and leads to some improvements. Causes that have a significant positive impact on more than 50 people or promote human society's progress are labeled as *strong_pos*.

The data is divided into equal parts for each annotator based on their primary language (5 for Chinese and 7 for English). Annotators individually label their assigned parts and then label two additional parts completed by others. After this, each sentence has three labels. In the third phase, annotators make the final decision on each sentence, and if the three labels are within a variation of 1 and have the same polarity, we take their average as the final label. Otherwise, we have the other two annotators review the sentence, and if it remains heavily ambiguous, we discard it.

Non-causal sentences are labeled with the *none* category using a dedicated script. The individual causal strengths and polarities are then combined to produce final causal strength labels, such as *pos_strong* or *neg_moderate*, for each example in the dataset. The final English Causal Strength dataset (E-CSB) includes 8,293 sentences, while the Chinese Causal Strength dataset (C-CSB) comprises 6,240 sentences. Examples from both datasets are provided in Fig. 1 and Fig. 2 and can be used for tasks such as causal relation extraction and causal strength classification.

3.3   Dataset Analysis

To facilitate further research by interested researchers in this field, we comprehensively analyze the datasets from 3 perspectives: label distribution, lengths of sentences, and the number of conjunction in each classification.

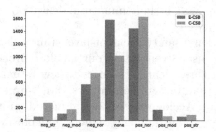

Fig. 3. Labels distribution of E-CSB and C-CSB.

Label Distribution. There are a total of 7 types of labels. The labels in E-CSB are extremely imbalanced at first. Therefore, we randomly select 4,000 sentences of other relations in SemEval-2010 Task 8, then annotate them with *none* and add them to E-CSB. This operation mitigates the imbalance in E-CSB; the final causal strength distribution of the dataset is in Fig 3. However, the overall data volume of *pos_* tags is still less than that of *neg_* tags, which may base on the fact that the mass media is good at attracting attention with negative news.

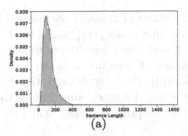 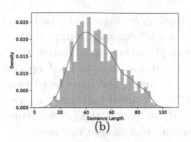

(a) (b)

Fig. 4. Lengths of sentences in (a) E-CSB and (b) C-CSB.

Sentence Length. Sentence length refers to the number of words in a sentence. Because of the perplexity in natural language understanding, the longer a sentence is, the more difficult it is to understand. In computational linguistics, it also affects computational complexity, influencing the performance of a model. In both E-CSB and C-CSB, we count information on sentence lengths, discovering that most sentences are between 100 and 200 characters long, and the sizes of only very few sentences are beyond 400 or 200. The sentence length information of E-CSB and C-CSB is in Fig. 4.

Table 1. The types of conjunctions in the E-CSB and their number show that most of them can express the logical progression of cause and effect in a given context.

Conjective	As	Because	For	However	If	Since	So	While
Number	884	341	1,038	107	471	156	658	204

Numbers of Conjunctions. Conjunctive play a crucial role in the causality expression of texts. Contexts of conjunctions carry entities/events with a causal relationship and can also indicate a sentence's complexity. We show the conjunctive information in Table 1. The types of conjunctions include transition, sequence, causality, juxtaposition, etc. For example, in an article, *for* and *as* frequently lead to the causes of previous events, and the causality represented by this narrative approach is implicit.

Aside from the above, we also look into the domains of the datasets. We discover that the contents of CSB mainly focus on medical, historical, and authoritative news reports, demonstrating the datasets' diversity and complexity.

4   Method

Single-task Learning The concept of single-task learning is presented in contrast to multi-task learning. Most famous machine learning models typically focus on optimizing for a particular metric, and as the name implies, it involves learning only one task at a time. Based on the single-task learning strategy, we implement several representative baseline models to conduct experiments on the above two tasks and observe the performance of these models on these two tasks.

Multi-task Learning. Multi-task learning is a branch of transfer learning closer to real-world scenarios. Suppose there are m learning tasks $\{\tau_i\}_{i=1}^{m}$. All the tasks or a subset of them are related; the purpose of multi-task learning is to learn m tasks simultaneously to level up the performance of learning a model for each task τ_i by using the knowledge in other tasks [14]. By integrating intrinsically linked tasks, multi-task learning can dig out internal connections among these tasks. If we design the architecture of these tasks properly, they will provide additional information mutually.

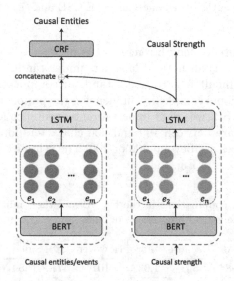

Fig. 5. Multi-task model for causal strength classification and relation extraction.

Joint Causal Strength and Entity Extraction Model. The model architecture is in Fig. 5. The left part of the joint model aims to extract cause and effect entities, while the right part intends to classify the causal strength of given sentences. The left consists of a pre-trained BERT [18], an LSTM [21] layer, and a Conditional Random Field (CRF) [22] layer, and the right part is the same as the left except for the CRF. For both the two parts, we employ BERT as the base model to obtain high-quality word representations and use LSTM to encode texts in low-dimension vector space. For the left, between the LSTM layer and the CRF layer, the outputs of the LSTM layer in the right part will be concatenated with the intermediate variables of the LSTM in the left, and the concatenation then will be sent to the CRF.

The observed sequence is the word sequence, and the output labels are the Causal BIO attributes corresponding to each word. The model will classify each term by the same number of feature functions and globally optimized values. The prediction process is to compute the score of the tags with each feature, and the label with the highest score is the predicted result.

5 Experiment

Our experiments include both single-task and multi-task settings. In the single-task experiments, we evaluate the performance of several classic models on causal relation extraction and causal strength classification tasks. In the multi-task experiments, we examine the potential of causal strength information in improving causality extraction. Specifically, we conduct experiments on E-CSB to demonstrate how causal strength can enhance the accuracy of causality discovery.

Table 2. Experimental results on causal relation extraction task, we take the precision, recall, and F1 (%) as metrics.

Model	Precision	Recall	F1
BiLSTM	73.93	59.05	65.66
BiLSTM+CRF	66.21	64.52	65.34
BERT	73.28	75.47	74.36
BERT+BiLSTM	74.06	74.62	74.34
BERT+CRF	77.64	73.50	75.51
BERT+BiLSTM+CRF	76.25	74.81	75.52

5.1 Datasets and Evaluation Metrics

As we present a new dataset, we must evaluate its effectiveness and positive influence on causality discovery research. In the experiments, we split the dataset into training and validation sets in the ratio of 9:1. For metrics, we adopt the accuracy, precision, recall, and F1 metrics of the causal relation extraction task. As for the causal strength classification task, accuracy is the only metric.

5.2 Baselines

Causal Strength Classification. We take the classic machine learning models as baselines in the causal strength classification experiments to observe the performance of our proposed task with different feature extraction methods.

SVM. Support vector machine (SVM) is a classic model in text classification and is effective in many classification tasks [19]. We modify the original SVM structure to fit the requirement of our experiment.

TextCNN. The model [20] assumes each word in a sentence can be denoted by a vector. Hence, a sentence is a matrix, allowing convolution operations to perform on the matrix-like image data.

BiLSTM. LSTM's success lies in adding an input gate, a forget gate, and an output gate based on the traditional RNN models. We adopt the BiLSTM model [21], integrating the sequence information of front and back directions.

Table 3. Experimental results on Chinese and English causal strength classification task, we take the accuracy (%) as metric.

Models	C-CSB	E-CSB
SVM	74.32	78.11
TextCNN	80.41	83.39
BiLSTM	83.24	86.65
BERT	87.52	91.54

BERT. It abandons the RNN architecture, effectively solving the long-term dependency problem in NLP. We feed the hidden state of the CLS token to a linear classifier to predict the causal strength in the experiments.

Causal Relation Extraction. Since we present BERT and BiLSTM in the baselines of causality strength classification, here we introduce the joint models in causal relation extraction. To be noticed, we use CRF in the relation extraction models because of its stable performance on the sequence labeling task.

BiLSTM+CRF. In this model, [23] feeds the outputs of BiLSTM into the CRF module to get the result of text sequence labeling. This model extracts target expressions from opinionated input sentences and categorizes each sentence.

BERT+CRF. It integrates BERT and CRF, fine-tuning after BERT pre-training, adding the CRF layer, and obtaining the sequence with a higher probability in the causal relation extraction based on sequence labeling.

BERT+BiLSTM. The outputs of the pre-trained-BERT model pass through a 3-layer BiLSTM as hidden states that may extract contextual information from a given sentence, and we obtain causal sequence labeling results at last.

BERT+BiLSTM+CRF [24] start by extracting text features with BERT, then apply BiLSTM to learn contextual information with BiLSTM and feed the outputs of the Bi-LSTM in CRF to obtain global optimum sequences with CRF.

Table 4. Results of single-task and multi-task experiments on E-CSB.

Task Type	Precision	Recall	F1
Single-	76.25	74.81	75.52
Multi-	77.54	76.32	76.93

5.3 Results and Discussion

Table 2 shows extraction results, which reveal that the BERT+BiLSTM+CRF model achieves limited improvements over BERT+BiLSTM since BERT is too strong to display minor differences between BiLSTM and CRF. Table 3 displays the causal strength classification experiment results, which show that all models perform better in E-CSB than in C-CSB, which may come from the scale of data. Furthermore, we conduct multi-task experiments on E-CSB in nearly the same setting as single-task experiments. The results of causal relationship extraction in the multi-task learning framework are in Table 4. The multi-task model outperforms single-tasking by 1.41 on the F1 metric, while it improves by 1.29 and 1.51 on precision and recall, respectively. Multi-task models perform better on all selected metrics than single-task models, demonstrating the significance of the causal strength classification task.

6   Conclusion and Future Work

In this paper, we propose a new Causal Strength Bank (CSB) dataset that includes causal strength information and introduce two novel tasks for Chinese and English causal strength classification. Our experiments demonstrate that the proposed dataset can improve causal relation extraction from texts. We also analyze the dataset from multiple perspectives to facilitate relevant research. However, due to its relatively small size, future work could involve expanding the dataset or creating additional benchmark datasets to further validate our findings. Our work contributes to the field of natural language processing by providing a new dataset for causal strength classification, and we hope it inspires further research in this area.

Acknowledgements. This work is supported by the National Natural Science Foundation of China under Grant (61976103), and the general foundation of the National University of Defense Technology under Grant (ZK22-11).

References

1. Mirza, P.: Extracting temporal and causal relations between events. In: Proceedings of the ACL 2014 Student Research Workshop, pp. 10–17 (2014)
2. Caselli, T., Vossen, P.: The event storyline corpus: a new benchmark for causal and temporal relation extraction. In: Proceedings of the Events and Stories in the News Workshop, pp. 77–86 (2017)
3. Hendrickx, I., et al.: Semeval-2010 task 8: multi-way classification of semantic relations between pairs of nominals. In: Proceedings of the 5th International Workshop on Semantic Evaluation, pp. 33–38 (2010)
4. Hashimoto, C., et al.: Toward future scenario generation: extracting event causality exploiting semantic relation, context, and association features. In: Proceedings of the 52nd Annual Meeting of the Association for Computational Linguistics (Volume 1: Long Papers), vol. 1, pp. 987–997 (2014)
5. Zhao, S., Liu, T., Zhao, S., Chen, Y., Nie, J.Y.: Event causality extraction based on connectives analysis. Neurocomputing **173**, 1943–1950 (2016)
6. Zhao, S., et al.: Constructing and embedding abstract event causality networks from text snippets. In: Proceedings of the Tenth ACM International Conference on Web Search and Data Mining, pp. 335–344. ACM (2017)

7. Liang, S., Zuo, W., Shi, Z., Wang, S., Wang, J., Zuo, X.: A multi-level neural network for implicit causality detection in web texts. Neurocomputing **481**, 121–132 (2022)
8. Mirza, P., Sprugnoli, R., Tonelli, S., Speranza, M.: Annotating causality in the TempEval-3 corpus. In: EACL 2014 Workshop on Computational Approaches to Causality in Language (CAtoCL), pp. 10–19. Association for Computational Linguistics (2014)
9. Ning, Q., Feng, Z., Wu, H., Roth, D.: Joint reasoning for temporal and causal relations. arXiv preprint arXiv:1906.04941 (2019)
10. Ding, X., Li, Z., Liu, T., Liao, K.: ELG: an event logic graph. arXiv preprint arXiv:1907.08015 (2019)
11. Luo, Z., Sha, Y., Zhu, K.Q., Hwang, S.W., Wang, Z.: Commonsense causal reasoning between short texts. In: Fifteenth International Conference on the Principles of Knowledge Representation and Reasoning (2016)
12. Sasaki, S., Takase, S., Inoue, N., Okazaki, N., Inui, K.: Handling multiword expressions in causality estimation. In: IWCS 2017 12th International Conference on Computational Semantics Short Papers (2017)
13. Yang, X., Obadinma, S., Zhao, H., Zhang, Q., Matwin, S., Zhu, X.: Semeval-2020 task 5: counterfactual recognition. In: Proceedings of the Fourteenth Workshop on Semantic Evaluation, pp. 322–335 (2020)
14. Zhang, Y., Yang, Q.: A survey on multi-task learning. arXiv preprint arXiv:1707.08114 (2017)
15. Akbik, A., Michael, T.: The weltmodell: a data-driven commonsense knowledge base. In: LREC, vol. 2, p. 5 (2014)
16. Church, K., Hanks, P.: Word association norms, mutual information, and lexicography. Comput. Linguist. **16**(1), 22–29 (1990)
17. Gordon, A.S., Bejan, C.A., Sagae, K.: Commonsense causal reasoning using millions of personal stories. In: Twenty-Fifth AAAI Conference on Artificial Intelligence (2011)
18. Devlin, J., Chang, M.W., Lee, K., Toutanova, K.: BERT: pre-training of deep bidirectional transformers for language understanding. arXiv preprint arXiv:1810.04805 (2018)
19. Mullen, T., Collier, N.: Sentiment analysis using support vector machines with diverse information sources. In: Proceedings of the 2004 Conference on Empirical Methods in Natural Language Processing, pp. 412–418 (2004)
20. Kim, Y.: Convolutional neural networks for sentence classification. In: Proceedings of the 2014 Conference on Empirical Methods in Natural Language Processing (EMNLP), pp. 1746–1751 (2014)
21. Zhou, P., Qi, Z., Zheng, S., Xu, J., Bao, H., Xu, B.: Text classification improved by integrating bidirectional LSTM with two-dimensional max pooling. arXiv preprint arXiv:1611.06639 (2016)
22. Lafferty, J., McCallum, A., Pereira, F.C.: Conditional random fields: probabilistic models for segmenting and labeling sequence data (2001)
23. Chen, T., Xu, R., He, Y., Wang, X.: Improving sentiment analysis via sentence type classification using BiLSTM-CRF and CNN. Expert Syst. Appl. **72**, 221–230 (2017)
24. Dai, Z., Wang, X., Ni, P., Li, Y., Li, G., Bai, X.: Named entity recognition using BERT BiLSTM CRF for Chinese electronic health records. In: 2019 12th International Congress on Image and Signal Processing, Biomedical Engineering and Informatics (CISP-BMEI), pp. 1–5. IEEE (2019)

Topological Graph Convolutional Networks Solutions for Power Distribution Grid Planning

Yuzhou Chen[1(✉)], Miguel Heleno[2], Alexandre Moreira[2], and Yulia R. Gel[3,4]

[1] Temple University, Philadelphia, USA
yuzhou.chen@temple.edu
[2] Lawrence Berkeley National Laboratory, Berkeley, USA
[3] The University of Texas at Dallas, Dallas, USA
[4] National Science Foundation, Alexandria, VA, USA

Abstract. With ever rising energy demands along with continuing proliferation of clean energy sources, the expanding analytic needs of the modern power sector can no longer be met by prevailing physical-based models and require new automatic solutions for planning, monitoring, and controlling tasks. In turn, artificial intelligence (AI) offers many necessary tools to develop such novel solutions. In this paper we take the first step towards bringing the utility of Topological Graph Neural Networks to power distribution grid planning and resilience quantification. We develop new Graph Convolutional Networks coupled with a zigzag topological layer for classification of distribution grid expansion plans. We also introduce bootstrap over the extracted zigzag persistence representations of the distribution grids which allows us to learn the most characteristic, or *hereditary* topological signatures over multiple graphs from the same family and, as a result, to improve classification performance both in terms of accuracy and stability. Our numerical experiments show that the new Bootstrapped Zigzag Persistence Based Graph Convolutional Networks (BZP-GCN) yields substantial gains in computational efficiency compared to the traditional methodology to assess the quality of investment and planning of distribution grids. Furthermore, BZP-GCN outperforms by a significant margin 6 state-of-the-art models in terms of classification accuracy.

Keywords: Graph neural networks · zigzag persistence · power distribution grid

1 Introduction

Extreme events, such as floods, windstorms or earthquakes have severely harm the power grid infrastructure, which resulted in long-term interruptions of electric power service [1]. Due to the potential security and socioeconomic consequences of these disaster-driven power outages, governments and policy makers

© The Author(s), under exclusive license to Springer Nature Switzerland AG 2023
H. Kashima et al. (Eds.): PAKDD 2023, LNAI 13935, pp. 123–134, 2023.
https://doi.org/10.1007/978-3-031-33374-3_10

started to include power grid resilience as part of regional and national adaptation plans [10]. When planning and regulating the power sector to accommodate these resilience objectives, a main challenge, particularly in distribution grids, is related with the High Impact Low Probability (HILP) nature of extreme events. Traditionally, security concerns in power distribution have been focused exclusively on reliability events, such equipment failures, maintenance, errors in system operation, etc. Due to the routine characteristic of these outages, reliability metrics are based on expected values (e.g., expected value of loss of load), and pay insufficient attention to HILP events due to their very small (almost negligible) probability of occurrence. To account for these events and effectively capture the power resilience, uncertainty realizations populating the right tail of the loss of load distribution should be taken into account, which can be done by employing metrics such as Value at Risk (VaR) or Conditional Value at Risk (CVaR) [24]. Another characteristic of reliability metrics is that they are designed for an *ex-post* statistical evaluation of the grid performance. In fact, when dealing with a large number of routine events, it is possible to say that *distribution grid A* is more reliable than *distribution grid B*, after observing several years of outages. However, in practice, the same principle cannot be applied to resilience against extreme events, again due to the HILP nature. Hence, in the process of developing (utilities) and approving (regulators) new resilience plans, we need new *ex-ante* methodologies of evaluating power grids. A possibility to perform these *ex-ante* evaluations of distribution grid expansion plans is through simulation-based methods [24]. However, for real size systems, these methods can take a prohibitive amount of time, as need to solve several optimization models to simulate the behaviour of the grid under multiple circumstances.

This opens an opportunity for artificial intelligence (AI) tools. Deep learning (DL) algorithms for automatic graph classification might be particularly suitable for this type of *ex-ante* applications, with the potential to dramatically decrease computational times in relation to simulation-based methods. Such reduction can be achieved by strategically extracting the most relevant topological information from distribution grid plans pre-computed plans as an alternative to observing and modelling detailed intra-hour stochastic events. Applications of such semi-supervised DL tools for graph classification, particularly, based on the graph neural network (GNN) architectures, have already been proven successful in a broad range of knowledge domains, from bioinformatics to social sciences to transportation systems [27]. Nonetheless, utility of GNNs for classification of power systems remains substantially under-explored. Most importantly it is still yet to be understood whether such GNN-based classification algorithms can properly replicate resilience metrics (such as the CVaR of the loss of load) and hence can be used in future applications to evaluate distribution grid plans. Motivated by these fundamental challenges at the intersection of AI and distribution grid planning, we introduce a novel approach to automatic classification of electricity system plans, namely, Graph Convolutional Networks (GCNs) with the power-based, bootstrapped zigzag topological layer. That is, we enhance GCNs by integrating the most characteristic, or *hereditary* topological signatures recorded over *multiple* power system plans. We extract such topological signatures by bridging the concept of zigzag persistence from algebraic topology with the notion of bootstrap inference from statistics. Zigzag persistence is the emerging methodology in computational topology, allowing us to simultaneously evaluate properties of graphs (or other topological spaces) with inclusions going in different directions [5,6]. Despite the recent results showing its utility in diverse areas of study such as ecology, biochemistry, and traffic analytics [9,15,26], zigzag persistence has never been applied for analysis of power grid networks and remains a largely unexplored concept in the data mining community. The key novelty of our contributions can be outlined as:

- We develop a novel classification algorithm for electricity system plans, based on a new GCN architecture with a bootstrapped zigzag topological layer – Bootstrapped Zigzag Persistence Based Graph Convolutional Networks (BZP-GCN).
- We propose a new methodology to learn the most characteristic, or *hereditary* topological signatures over multiple graphs from the same graph family, namely, *bootstrapped zigzag persistence representations*.
- In our expansive numerical experiments, we show that BZP-GCN yields substantial gains in computational efficiency compared to the stochastic optimization benchmarks from the power system community and also outperforms 6 state-of-the-art models from deep learning community by a significant margin in terms of graph classification accuracy.

2 Related Work

Graph Neural Networks for Power System Analysis. A wide variety of GNNs have been proposed in recent years for classification of non-Euclidean structures. However, applications of GNNs in power system analysis remain scarce for both transmission and distribution systems, with only few papers considering GNNs for distribution networks. Such representative approaches include recurrent GCN by [14] for multi-task transient stability assessment of power transmission systems, optimal power flow (OPF) optimization problem in transmission systems [23], node classification in transmission systems [8], and forecasting transmission system responses to contingencies [4]. One of such GNN applications to distribution systems include [7] who develop a GCN model for the task of fault location in distribution systems, and prove that GCN-based model is more robust to measurement errors compared with ML approaches. Compared to these existing techniques, our approach brings multiple new research directions. First, we represent graphical properties of power distribution grid by local topological signatures and define GCN based on *bootstrapped* zigzag persistence. This is a novel application of GCN not only in power system analysis but graph learning, in general. Our BZP-GCN is inspired by current GCNs but is carefully designed to capture local topological information from a bootstrapped sequence of *multiple* power distribution grids. Second, zigzag persistence and, generally, tools of topological data analysis, have never been applied to distribution power systems.

Methods for Distribution Grid Planning. The first phase of distribution grid planning is calculation of an expansion plan. Unlike the transmission and bulk power systems, in electrical distribution industry this process is often empirical, based on codes, electrical installation rules, and an ongoing discussion with the regulators. One of the recent approaches in the power system community is to improve this process via mathematical optimization. For instance, Nazemi et al. [22] presents a linear programming method to optimize investments in storage devices so as to make the distribution system more resilience when facing earthquakes. The optimization model of [19] aims to identify investments in generation and network expansion while considering reliability. Moreover, Munoz et al. [20] formulates a mixed-integer optimization problem to optimize the portfolio of investments in substations, lines, and transformers to decrease the loss of load costs. This class of models has the advantage of prescribing analytical optimal planning solutions for relatively small power networks under specific circumstances. However, due to their nature, these models struggle to provide generalizable and standardized industry solutions for large scale planning of distribution grids. The other family of grid planning methods are simulation-based. This aim at evaluating the performance of a system (or a future system) under different scenarios, to guarantee that a given grid planning solution – either obtained empirically or via mathematical optimization – is technically and economically feasible. Reliability indices are often assess through these simulation methods [3]. In resilience applications, advanced simulation approaches include Monte Carlo methods based on probabilistic models of the events combined with fragility curves to simulate network failures and their HILP consequences using CVaR of loss of load [12]. These simulations usually require the evaluating every single state of the system, including determining how the system would respond optimally to a loss of load in each time, across multiple scenarios. This process is cumbersome as solving a (non-parallelizable) sequence of optimization problems for large-scale systems various times requires a huge computational effort and a great deal of time. Within this context, we propose to replace this simulation-based evaluation with a topological GCN-based method that computes our resilience metric (CVaR) in a highly computational efficient manner. A fast and efficient evaluation method to assess the resilience of distribution grids, under potential candidate expansion plans, can hugely benefit regulators in their decision making process.

3   Methods

Our goal is to explore utility of GCN-based methods to classify distribution grid expansion plans in terms of resilience performance. In particular, our analysis aims at assessing the ability to approximate explicit risk-based resilience metrics, such as the CVaR of load loss, through the new topological descriptors of the distribution networks using GNNs. Let $\mathcal{G} = (\mathcal{V}, \mathcal{E}, A)$ denote a distribution grid, where $\mathcal{V}$ is a set of nodes, $\mathcal{E}$ is a set of edges, $A \in \mathbb{R}^{\mathcal{N} \times \mathcal{N}}$ is a symmetric adjacency matrix with $|\mathcal{V}| = \mathcal{N}$ nodes, and $X \in \mathbb{R}^{\mathcal{N} \times d}$ is a node feature matrix (here d is the dimension of node features). In a distribution grid, each node represents either a bus or a substation, and each edge is a distribution line between nodes. Distance among two nodes in $\mathcal{G}$ is denoted by A_{uv}, with $A_{uv} = 0$ if there exists no path connecting nodes u and v. Moreover, compared with graph theoretical analysis of electric distribution system, we make a clear distinction among substations (nodes of the distribution system that serve as a connection to the main transmission grid) and buses (nodes of the distribution system that may have a power load) in order to fully understand the relationship between the resiliency of a distribution system and node (bus or substation) attributes.

3.1   Base Resilience Evaluation Method

Methods that assess reliability and resilience are designed to evaluate the capability of the power grid to withstand multiple scenarios of outages while effectively supplying load. To account for pre-defined sets of scenarios of routine failures and HILP events, the resilience evaluation computes the distribution of total system loss of load (sum of loss of load across all nodes of the system) and then calculates the CVaR of this distribution. This process is usually carried out via Sequential Monte Carlo Simulation (SMCS) methods, which performs simulations of loss of load over annual time horizons. More specifically, to compute the annual CVaR of loss of load, we need to simulate the operation of the distribution system for several scenarios (e.g., 2000 scenarios), where each scenario corresponds to 365 d (one year) and the system is operated for each hour of each day. Each annual scenario, due to its respective failures, will then have an associated annual loss of load. Given the amounts of loss of load associated with all the scenarios, we can calculate the annual CVaR of loss of load. This procedure can take 24 h for a 54-Bus system as the one discussed in the case study to evaluate the annual CVaR of loss of load for a single candidate expansion plan. Hence, one of the main contributions of this paper is to replace the computationally expensive simulation by a method based on GCN that can quickly classify distribution grids according to their corresponding ranges of annual CVaR of loss of load.

3.2   Zigzag Persistent Homology

Persistent homology allows us to assess evolution of salient data shape patterns along various geometric dimensions. By shape here we broadly understand data properties which are invariant under continuous transformations, i.e., ones that do not alter "holes" in the data, e.g., crumpling, bending, and twisting. The main idea is to select some suitable scale parameter ϵ and then to study graph $\mathcal{G}$ not as a single object but as a nested sequence of graphs, or *filtration* $\mathcal{G}_1 \subseteq \ldots \subseteq \mathcal{G}_n = \mathcal{G}$, induced by monotonic changes of the parameter ϵ. For instance, suppose that $\mathcal{G}$ is an edge-weighted graph $(\mathcal{V}, \mathcal{E}, w)$, with $w : \mathcal{E} \mapsto \mathbb{R}$. Then, for each ϵ_j, $j = 1, \ldots, n$, we can set $\mathcal{G}_{\leq \epsilon_j} = (\mathcal{V}, \mathcal{E}, w^{-1}(-\infty, \epsilon_j])$, resulting in the induced edge-weighted filtration. Alternatively, for each ϵ_j, we can consider only induced subgraphs of $\mathcal{G}$ with maximal degree of ϵ_j, yielding a degree sublevel set filtration. (For more choices of graph filtrations see [13].) Armed with this construction, we can track which shape patterns such as independent components and cycles (dis)appear as scale ϵ varies, and also record their lifespans. Topological features with longer lifespan are said to persist and are likelier to deliver important information on the structural organization of $\mathcal{G}$. To make the process of pattern counting more systematic and efficient, we build an abstract simplicial complex $\mathbb{K}(\mathcal{G}_j)$ on each $\mathcal{G}_j$. Due to its computational costs, one of the most widely adopted choices here is Vietoris-Rips complex.

This framework allows us to extract and study the key shape descriptors from a single graph $\mathcal{G}$. However, *what if we observe not one but multiple graphs* $\{\mathcal{G}^1, \ldots, \mathcal{G}^{\mathcal{K}}\}$? *If these graphs belong to the same family* $\mathcal{G}$, *how can we track shape signatures which are not just individualistic but most characteristic, or "hereditary" for* $\mathcal{G}$? For instance, when we study cyber-physical networks which are designed according to the same engineering principles, we can expect that some properties will occur across many systems, thereby opening a path for transfer learning. We propose to address this problem using the concept of zigzag persistence and bootstrap. Based on quiver representations, zigzag persistence generalizes ordinary persistence to tracking characteristics of graphs (or other topological spaces) with inclusions going in different directions [5,6]. Here, we focus on assessing compatibility of persistent shape signatures across unions of graphs

$$\mathcal{G}^1 \to \mathcal{G}^1 \cup \mathcal{G}^2 \leftarrow \mathcal{G}^2 \to \mathcal{G}^2 \cup \mathcal{G}^3 \leftarrow \cdots \mathcal{G}^k \cup \mathcal{G}^{k+1} \leftarrow .$$

That is, we record indices in the above sequence at which topological features (dis)appear, for some given scale $\epsilon_\dagger$. If for the given $\epsilon_\dagger$ topological feature ρ (e.g., p-dimensional cycle where $0 \le p \le dim\mathbb{K}(\mathcal{G})$) is first recorded in $\mathbb{K}(\mathcal{G}^j)$, we say that the feature's birth is j, and if ρ first appears in $\mathbb{K}(\mathcal{G}^j \cup \mathcal{G}^{j+1})$, we record its birth as $j+1/2$. In turn, if ρ is last seen in $\mathbb{K}(\mathcal{G}^j)$, we record its death as j, while if it is last seen in $\mathbb{K}(\mathcal{G}^j \cup \mathcal{G}^{j+1})$, we say that its death is at $j+1/2$. Collecting births and deaths over the set $\mathfrak{J}$ of all observed topological features for a given $\epsilon_\dagger$, we get a zigzag persistent diagram (ZPD), i.e., a multiset $\mathcal{D}_{\epsilon_\dagger} = \{(b_\rho, d_\rho) \in \mathbb{R}^2 | b_\rho < d_\rho, \rho \in \mathfrak{J}\}$. Our hypothesis is that if we select multiple graph subsets of the same cardinality from the same graph family $\mathcal{G}$, e.g., $\{\mathcal{G}_{i_1}, \ldots, \mathcal{G}_{i_m}\}$, $\{\mathcal{G}_{k_1}, \ldots, \mathcal{G}_{k_m}\}$, and $\{\mathcal{G}_{l_1}, \ldots, \mathcal{G}_{l_m}\}$, we shall expect to observe that certain shape patterns (i.e., p-dimensional cycles) tend to manifest with similar persistence across all subsets of graphs. Such graph subsets can be selected using the statistical notion of bootstrap, while the extracted zigzag persistent features for each graph subset can be summarized via zigzag persistent images (i.e., the concept inspired by [9]).

Definition 1 (Bootstrapped Zigzag Persistence Image (B-ZPI)). *Let* $\mathcal{G}^*(b) = \{\mathcal{G}_{i_{b_1}}, \ldots, \mathcal{G}_{i_{b_m}}\} \in \mathcal{G}$ *be a bootstrapped subsample of graphs from the graph family* $\mathcal{G}$ *and* $b = \{1, \ldots, B\}$. *Let* $\epsilon_\dagger > 0$ *be a fixed scale and* $\mathcal{D}_{\epsilon_\dagger}(\mathcal{G}^*(b))$ *be the associated zigzag persistence diagram ZPD for* $\mathcal{G}^*(b)$. *Now, we first map the transformed ZPD* $T(\mathcal{D}_{\epsilon_\dagger}(\mathcal{G}^*(b)))$ *to an integrable function* $\phi : \mathbb{R}^2 \mapsto \mathbb{R}$ *which is a weighted sum of Gaussian functions* $\phi_{\mathcal{D}_{\epsilon_\dagger}(\mathcal{G}^*(b))}(z) = \sum_{z \in T(\mathcal{D}_{\epsilon_\dagger}(\mathcal{G}^*(b)))} f(\mu)g_\mu(z)$, *where* $z = (z_x, z_y) \in \mathbb{R}^2$, $f(\cdot)$ *is a non-negative weighting function, and* $g_\mu(\cdot)$ *is* a *Gaussian function with mean* $\mu \in \mathbb{R}^2$ *and variance* σ^2. *Then the bootstrapped zigzag persistence image B-ZPI for* $\mathcal{G}^*(b)$ *is obtained by integrating* $\phi_{\mathcal{D}_{\epsilon_\dagger}(\mathcal{G}^*(b))}(z)$ *over each grid box, i.e., B-ZPI*$^{(b)} = \iint \phi_{\mathcal{D}_{\epsilon_\dagger}(\mathcal{G}^*(b))}(z_x, z_y) dz_x dz_y$. *Finally, a collection*

$$\mathcal{B}_{\epsilon_\dagger} = \{\text{B-ZPI}^{(1)}, \text{B-ZPI}^{(2)}, \cdots, \text{B-ZPI}^{(B)}\}, \quad B \in \mathbb{Z}^+$$

is called a bootstrapped distribution of zigzag persistence images for a scale $\epsilon_\dagger$.

We can then use mean, median, quantiles and other parameters of $\mathcal{B}_{\epsilon_\dagger}$ for extracting the *most characteristic*, or *hereditary shape features* of the graph family $\mathcal{G}$ as well as for quantifying the associated uncertainty. Furthermore, $\mathcal{B}_{\epsilon_\dagger}$ can be used for formal statistical inference and hypothesis testing on particular p-dimensional cycles persistently re-occurring across $\{\mathcal{G}^*(1), \ldots, \mathcal{G}^*(B)\}$, which is a more fundamental mathematical problem which we leave for future research.

3.3 Bootstrapped Zigzag Image Representation Learning with Graph Convolutional Nets

Bootstrapped Zigzag Persistence Based Graph Convolutional Networks. Inspired by the message passing nature of GNNs, we propose a new model called Bootstrapped Zigzag Persistence Based Graph Convolutional Networks (BZP-GCN), because the bootstrapped zigzag persistence based graph convolutional layer is able to incorporate the sequential topological information from the sequence of power networks to the static power network representation. After ℓ-th iteration, the node representation in graph $\mathcal{G}_i$ can be defined as

$$h_{i,v}^{(\ell)} = f_{\text{MLP}} \left(\oplus \left(\text{AGG} \left(h_{i,u}^{(\ell-1)}, \forall u \in \mathcal{N}(v) \right), h_{i,v}^{(\ell-1)} \right) \right),$$

where $\oplus$ denotes the concatenation of two vectors, $\text{AGG}(\cdot)$ is the aggregation function that aggregates the output of each neighbor (e.g., *max*, *sum* and *average*), $\mathcal{N}(v)$ refers to 1-hop neighbors of node v, f_{MLP} denotes the multilayer perceptrons. Neighborhood vector $h_{i,v}^{(\ell)}$ incorporates node feature information from node v's neighborhood into the representation. Finally, we denote the signal convolved matrix as $H_i^{(\ell)} \in \mathbb{R}^{\mathcal{N} \times d_{out}}$ for graph $\mathcal{G}_i$, where d_{out} is the number of output channels. To encode the topological information of a sequence of data objects, we propose the bootstrapped zigzag persistence representation learning (BZPRL) module that uses CNN-based model f_{cnn} to learn topological features of bootstrapped distribution of zigzag persistence images, i.e., $\mathcal{B}_{\epsilon_\dagger, i} \in \mathbb{R}^{B \times p \times p}$. The operations can be formulated as

$$S_{i,(b)}^{(\ell)} = f_{\text{GMP}} \left(f_{\text{cnn}}^{(\ell)} \left(\text{B-ZPI}_i^{(b)} \right) \right), \qquad b = 1, 2, \ldots, B$$

where $\text{B-ZPI}_i^{(b)} \in \mathbb{R}^{1 \times p \times p}$ represents the b-th B-ZPI of resolution p for graph $\mathcal{G}_i$, and the global max pooling layer f_{GMP} is applied to get a representation vector of the image-level feature $S_{i,(b)}^{(\ell)} \in \mathbb{R}^{d_{out}}$. To fuse the multi-view information (i.e., bootstrapped zigzag persistence representations) and aggregate the individual learned zigzag persistence representation into a global representation, the final bootstrapped zigzag persistence representation is the average of B zigzag persistence representations $\bar{S}_i^{(\ell)} = \frac{1}{B} \sum_{b=1}^{B} S_{i,(b)}^{(\ell)}$. Next, to aggregate the graph structural information and topological-sequential information, we map the averaged bootstrapped zigzag persistence representation $\bar{S}_i^{(\ell)}$ to the representation vectors $H_i^{(\ell)}$. The aggregating operation can be formulated as $Q_i^{(\ell)} = H_i^{(\ell)} \bar{S}_i^{(\ell)}$. The core ideas behind the new graph representation are to (i) aggregate the local topological information from a sequence of power grid networks in bootstrapped zigzag persistence images and (ii) aggregate graph structural information from graph convolution, and topological information via bootstrapped zigzag persistence representation learning. Figure 1 provides an overview of the design of bootstrapped zigzag persistence-based graph convolutional layer.

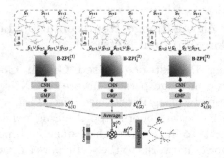

Fig. 1. The architecture of the BZP-GCN framework.

4   Experimental Studies

4.1   Datasets and Baselines

We consider learning CVaR of Annual Loss of Load through multi-class classification. Each of these classes represent a "degree" of system risk in relation to HILP events. The resulting classification could be then converted into regulatory planning standards, allowing to systematically compare resilience between different plans of the same grid and potentially across grids in similar geographical conditions [18]. Following a standard statistical practice [21], we perform binning into classes based on quantile ranges of annual CVaR of loss of load (kWh), i.e., 3 classes (i.e., low-, moderate-, and high-risk), 4 classes (i.e., low-, moderate-, middle-, and high-risk), and 5 classes (i.e., low-, moderate-, high-, very high-, and extreme high-robustness). The proposed methodology has been applied to two distribution system, namely 54-Bus System I and 54-Bus System II, which are modified versions of the 54-Bus system described in [19]. In the 54-Bus System I, we have 72 lines (50 existing and 22 candidate lines), 4 substation nodes and 50 load nodes. In the 54-Bus System II, we have 72 lines (52 existing and

Table 1. Average accuracy (%) comparison with baseline methods; ( ) denotes standard deviations.

Model	3 classes		4 classes		5 classes	
	54-Bus System I	54-Bus System II	54-Bus System I	54-Bus System II	54-Bus System I	54-Bus System II
RF [17]	73.1 (3.2)	69.9 (8.6)	70.2 (3.6)	63.3 (3.7)	63.5 (6.8)	61.3 (8.5)
WL subtree [25]	92.3 (3.4)	81.6 (9.3)	80.9 (2.3)	73.0 (7.3)	82.9 (6.2)	71.0 (9.8)
GCN [16]	91.9 (3.4)	81.1 (9.3)	85.0 (3.3)	72.2 (7.2)	81.2 (5.5)	71.7 (6.2)
Graph U-Net [11]	93.0 (3.5)	81.0 (9.0)	86.5 (3.7)	73.0 (6.1)	82.7 (5.1)	72.2 (9.9)
DCNN [2]	76.1 (7.0)	74.2 (8.1)	74.3 (5.9)	63.9 (5.1)	66.1 (10.2)	63.0 (9.9)
DGCNN [28]	92.1 (5.7)	80.1 (10.2)	88.6 (2.3)	72.6 (8.8)	83.2 (6.9)	72.2 (10.9)
BZP-GCN (ours)	***96.5 (2.8)**	**85.8 (8.0)**	**92.3 (3.0)**	**77.9 (8.6)**	*88.1 (5.0)	**76.9 (10.1)**

20 candidate lines), 2 substation nodes, 50 load nodes and 2 non-load nodes. To obtain our training data, we have generated several possible expansion plans for the two aforementioned systems (200 for the 54-Bus System I and 74 for 54-Bus System II). These expansion plans have been generated by selecting different subsets of the available candidate lines. Then, for each expansion plan of each system, we have simulated 2000 scenarios of annual operation, with hourly resolution. For each hour of each scenario, we have simulated independent Bernoulli trials for the availability of line segments of the distribution grid, considering the rate of routine failures (single-line failures) as 0.4 times per year and the rate of HILP failures (failures involving more than one line segment) as 0.01 times per year. By means of this simulation, we can attain the CVaR of annual loss of load for each expansion plan as described in Sect. 3.1. We compare our BZP-GCN model with widely used classification models, including (i) Random Forest [17] (RF); (ii) Weisfeiler-Lehman subtree kernel [25] (WL subtree kernel); (iii) Graph Convolutional Networks [16] (GCN); (iv) Graph U-Net [11]; (v) Diffusion-convolutional neural networks [2] (DCNN); (vi) Deep Graph CNN [28] (DGCNN).

Table 2. Ablation studies on 54-Bus System I (3 classes) and 54-Bus System II (3 classes) datasets.

Dataset	Architecture	Overall Accuracy
54-Bus System I	**BZP-GCN**	**96.5 (2.8)**
	BZP-GCN w/o BZPRL	**95.0 (3.9)
	ZP-GCN	*96.0 (3.5)
54-Bus System II	**BZP-GCN**	**85.8 (8.0)**
	BZP-GCN w/o BZPRL	***83.8 (10.5)
	ZP-GCN	**84.0 (9.1)

4.2 Experimental Settings

We use the Adam optimizer for 500 epochs to train BZP-GCN. For both 54-Bus Systems, BZP-GCN consists of 5 layers whose hidden feature dimension is 128, and each layer consists of two MLP blocks. The learning rate is 0.01 the dropout is set as 0.5, and the batch size is set as 32. In bootstrapped zigzag persistence representation learning, the CNN based model consists of 2 CNN layers. For the 54-Bus System I and 54-Bus System II, the filter size, kernel size, and stride is set to be 8, 2, 2 respectively. We set the size of global average pooling and global max pooling as 3×3. For bootstrapped distribution of zigzag persistence images, we set $B = 3$. The best results are in **bold** font. We also perform a one-sided two-sample t-test between the best result and the best performance achieved by the runner-up, where $*$, $**$, $* * *$ denote significant, statistically significant, highly statistically significant results, respectively. We implement our proposed BZP-GCN model using Python and Pytorch on NVIDIA GeForceX 3090 (24 GB memory). Our data and code is publicly available at https://github.com/bzpgcnpakdd/BZP-GCN.git.

4.3   Overall Results

The results on 54 bus systems, averaged over 10 cross-validation runs, are summarized in Table 1. As shown in Table 1 suggests the following key findings: (i) BZP-GCN surpasses all state-of-the-art baselines in terms of classification performance over all considered scenarios across all datasets; (ii) BZP-GCN delivers the relative gains with respect to the next best approach from 3.8% (for 3 classes of the 54-Bus System I and Graph U-Net as the competitor) to 6.7% (for 4 classes of the 54-bus System II and WL subtree as the competitor); (iii) BZP-GCN yields substantial reductions in computational costs with respect to the stochastic optimization methods adopted by the power system community and the *second lowest* running time among deep learning models. This proves BZP-GCN to be the most competitive approach for the automatic expansion plan classification both in terms of computational costs and accuracy among the considered ML tools; and (iv) as expected, performance of all models deteriorates with an increase of a number of classes; however, the number of classes is not a key aspect of this problem. In fact, in real world applications, it is expected that power resilience will be expressed in a small number of classes so that it can be incorporated into planning standards and regulatory proceedings.

4.4   Ablation Experiments

We now verify the effectiveness of the BZPRL module in BZP-GCN and the single zigzag persistence based graph convolutional networks (ZP-GCN, i.e., only considering one zigzag persistence image instead of bootstrapped zigzag persistence images) on 54-Bus System I (3 classes) and 54-Bus System II (3 classes). Table 2 shows that (i) BZP-GCN achieves the best performance in classification tasks on all datasets; (ii) BZPRL module shows a high utility for encoding topological information for graph representation learning; and (iii) bootstrapped zigzag persistence images enhance stability of the results. In addition, the results show that BZP-GCN equipped with BZPRL achieves statistically significant improvements over BZP-GCN w/o BZPRL and ZP-GCN under paired t-test.

5   Discussion

We have developed a new GCN-based architecture coupled with a zigzag topological layer for classification of distribution grid expansion plans which allows us to leverage the extracted most hereditary topological signatures over multiple distribution grid plans. The experiments have shown that the proposed BZP-GCN outperforms state-of-the-arts methods on distribution grid benchmarks. Moreover, the computational time of BZP-GCN to classify different grid plans (in the order of *seconds*) is encouraging, especially when compared with prevailing simulation based CVaR evaluation methodologies that may take several *hours* to compute. These findings indicate that BZP-GCN demonstrates a high potential to be successfully incorporated into the traditional expansion and planning algorithms to improve the computational times of the existing large scale simulation models for evaluation of distribution grid planning. As such, BZP-GCN forms a promising alternative to deliver automatic, computationally efficient, and standardized resilience analysis of the distribution networks, exclusively based on the grid topology and the grid assets. In the future, we will explore sensitivity of the BZP-GCN in face of operational plans that incorporate distribution assets (such as reclosers and sectionalizers) that can enhance resilience by introducing operational changes to the grid topology itself.

Acknowledgments. This work was supported by the Office of Naval Research (ONR) award N00014-21-1-2530. Any opinions, findings, conclusions, or recommendations expressed in this paper are those of the authors and do not necessarily reflect the views of ONR.

References

1. Amani, A.M., Jalili, M.: Power grids as complex networks: resilience and reliability analysis. IEEE Access **9**, 119010–119031 (2021)
2. Atwood, J., Towsley, D.: Diffusion-convolutional neural networks. In: NeurIPS. pp. 1993–2001 (2016)
3. Billinton, R., Billinton, J.: Distribution system reliability indices. In: IEEE Transactions on power Delivery. pp. 561–568 (1989). https://doi.org/10.1109/61.19247
4. Bush, B., Chen, Y., Ofori-Boateng, D., Gel, Y.: Topological machine learning methods for power system responses to contingencies. In: AAAI. vol. 35, pp. 15278–15285 (2021)
5. Carlsson, G.: Persistent homology and applied homotopy theory. In: Handbook of Homotopy Theory (2019)
6. Carlsson, G., De Silva, V.: Zigzag persistence. Found. Comput. Math. **10**(4), 367–405 (2010)
7. Chen, K., Hu, J., Zhang, Y., Yu, Z., He, J.: Fault location in power distribution systems via deep graph convolutional networks. IEEE J. Sel. Areas Commun. **38**(1), 119–131 (2019)
8. Chen, Y., Gel, Y., Avrachenkov, K.: LFGCN: Levitating over graphs with levy flights. In: ICDM (2020)
9. Chen, Y., Segovia-Dominguez, I., Gel, Y.R.: Z-GCNETs: Time zigzags at graph convolutional networks for time series forecasting. In: ICML (2021)

10. DOE: U.S. Department of Energy announces agency climate adaptation and resilience plan. DOE Press Release (2021)
11. Gao, H., Ji, S.: Graph U-Nets. In: ICML. pp. 2083–2092 (2019)
12. Gautam, P., Piya, P., Karki, R.: Resilience assessment of distribution systems integrated with distributed energy resources. In: IEEE Transactions on Sustainable Energy. pp. 1–1 (2020)
13. Hofer, C., Graf, F., Rieck, B., Niethammer, M., Kwitt, R.: Graph filtration learning. In: ICML. pp. 4314–4323 (2020)
14. Huang, J., Guan, L., Su, Y., Yao, H., Guo, M., Zhong, Z.: Recurrent graph convolutional network-based multi-task transient stability assessment framework in power system. IEEE Access **8**, 93283–93296 (2020)
15. Kim, W., Mémoli, F., Smith, Z.: Analysis of Dynamic Graphs and Dynamic Metric Spaces via Zigzag Persistence. In: Baas, N.A., Carlsson, G.E., Quick, G., Szymik, M., Thaule, M. (eds.) Topological Data Analysis. AS, vol. 15, pp. 371–389. Springer, Cham (2020). https://doi.org/10.1007/978-3-030-43408-3_14
16. Kipf, T.N., Welling, M.: Semi-supervised classification with graph convolutional networks. In: ICLR (2017)
17. Liaw, A., Wiener, M.: Classification and regression by randomforest. Rnews **2**(3), 18–22 (2002)
18. Ma, J., Zhang, J., Xiao, L., Chen, K., Wu, J.: Classification of power quality disturbances via deep learning. IETE Tech. Rev. **34**, 408–415 (2017)
19. Muñoz-Delgado, G., Contreras, J., Arroyo, J.: Multistage generation and network expansion planning in distribution systems considering uncertainty and reliability. IEEE Trans. Power Syst. **31**(5), 3715–3728 (2016)
20. Muñoz-Delgado, G., Contreras, J., Arroyo, J.M.: Distribution network expansion planning with an explicit formulation for reliability assessment. IEEE Trans. Power Syst. **33**(3), 2583–2596 (2018)
21. Nargesian, F., Samulowitz, H., Khurana, U., Khalil, E.B., Turaga, D.S.: Learning feature engineering for classification. In: IJCAI. pp. 2529–2535 (2017)
22. Nazemi, M., Moeini-Aghtaie, M., Fotuhi-Firuzabad, M., Dehghanian, P.: Energy storage planning for enhanced resilience of power distribution networks against earthquakes. IEEE Trans. Sustain. Energy **11**(2), 795–806 (2020)
23. Owerko, D., Gama, F., Ribeiro, A.: Optimal power flow using graph neural networks. In: ICASSP. pp. 5930–5934 (2020)
24. Poudel, S., Dubey, A., Bose, A.: Risk-based probabilistic quantification of power distribution system operational resilience. IEEE Syst. J. pp. 1–12 (2019)
25. Shervashidze, N., Schweitzer, P., Van Leeuwen, E.J., Mehlhorn, K., Borgwardt, K.M.: Weisfeiler-lehman graph kernels. J. Mach. Learn. Res. (JMLR) **12**(9) (2011)
26. Tymochko, S., Munch, E., Khasawneh, F.A.: Hopf bifurcation analysis using zigzag persistence. Algorithms **13**, 278 (2020)
27. Wu, Z., Pan, S., Chen, F., Long, G., Zhang, C., Philip, S.Y.: A Comprehensive Survey on Graph Neural Networks. IEEE Trans. Neural Netw. Learn, Syst (2020)
28. Zhang, M., Cui, Z., Neumann, M., Chen, Y.: An end-to-end deep learning architecture for graph classification. In: AAAI. vol. 32 (2018)

Label Distribution Learning with Discriminative Instance Mapping

Heng-Ru Zhang[✉][iD], Run-Ting Bai[iD], and Wen-Tao Tang[iD]

School of Computer Science, Southwest Petroleum University, ChengDu, China
zhanghrswpu@163.com

Abstract. Label distribution learning (LDL) is an effective tool to tackle label ambiguity since it allows one instance to be associated with multiple labels in different degrees. Therefore, the more complex but informative label space makes it challenging to directly model the relationship between original features and label distributions. In this paper, an algorithm called Label Distribution Learning with Discriminative Instance Mapping (LDLDIM) is proposed to select a discriminative instance pool (DIP) to map the original features into a more discriminative space. First, we design a criterion that incorporates label information to quantify the discriminative power of each instance. Second, we select several instances with the highest discriminative ability to construct the DIP, and map the instances to the discriminative space through the DIP. By exploiting label information, this criterion enables the selected DIP to ensure that instances that are close (far away) in label space remain close (far away) in the discriminative space. Finally, multiple regressions for prediction are trained on the label distributions and the new features that are obtained by distance mapping with DIP. Experiments and comparisons on 16 datasets illustrate that our algorithm outperforms 6 state-of-the-art LDL methods in most cases.

Keywords: Label distribution learning · Instance mapping · Discriminative ability

1 Introduction

In Single-Label Learning (SLL) and Multi-Label Learning (MLL), the hard labels are usually used to describe the instances, i.e., $0(-1)$ is for the irrelevant labels and 1 is for the relevant labels [21]. However, such a rough description cannot fully tackle the label ambiguity of the relative importance of each label to the instances [6].

Therefore, a paradigm named Label Distribution Learning (LDL) is proposed, which appropriately accommodates the different importance of each label to instances. Instead of investigating *"which label can describe the instance?"*, LDL further handles the label ambiguity problem by focusing on *"how much does each label describe the instance?"* [7]. Figure 1 reveals the difference between SLL, MLL and LDL. For SLL, the label of the image is identified as that of the largest semantic region (i.e., the building). For MLL, all relevant labels take the same value 1. For LDL, the values of all relevant labels together form a description vector whose sum is 1. Each element represents the relative importance of the corresponding label to the image.

© The Author(s), under exclusive license to Springer Nature Switzerland AG 2023
H. Kashima et al. (Eds.): PAKDD 2023, LNAI 13935, pp. 135–146, 2023.
https://doi.org/10.1007/978-3-031-33374-3_11

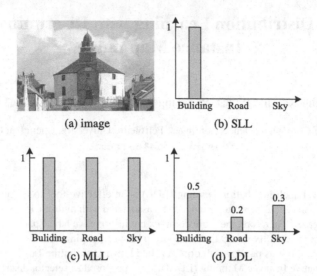

Fig. 1. A practical example of the difference between SLL, MLL and LDL.

Due to its unique label form, LDL can be used to tackle some informative real-world applications with more label ambiguity. Take the facial expression emotion recognition as an example, the expression of one person is frequently a mix of several basic emotions (e.g., happiness, sadness, surprise, fear, anger and disgust) [23]. Hence, the soft labels belongingness of LDL can model this kind of problem easily. During past few years, LDL has attracts a lot of attention and been mainly applied in many real-world tasks successfully, such as facial age estimation [8,10,20], head-pose estimation [9], beauty sensing [15], bone age assessment [3] and facial expression emotion recognition.

As mentioned earlier, LDL algorithms output a vector consisting of the description degree of each label to the sample. It further leads to a more complex but informative label space than SLL and MLL. Hence, many researchers tend to utilize label correlation. For example, LDLLC [11] encodes label correlation as a Pearson correlation coefficient to incorporate into the objective function. LDL-SCL [22] encodes local label correlations as additional features by computing instance distances from cluster centers. LDL-LDM [16] simultaneously mines global and local label correlations by learning label distribution manifold. Although, label correlations provide more information for the decision-making process of the LDL model. It does not directly enhance the discriminative power of the original features.

In this paper, we propose a mapping-based LDL method called LDLDIM to obtain a more discriminative feature space and learn label distributions based on it. Instead of adding customized regularization term or selecting some original features, we leverage the discriminative instances to map instances to a new features space. Therefore, we focus on selecting the optimal instances to build the DIP. Specifically, we design a criterion using the relationship between pairwise samples in label space. It can measure the discriminative power of the DIP, that is, to quantify the ability to map instances that

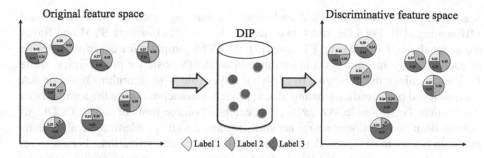

Fig. 2. Discriminative instance mapping. Each pie represents an instance, and the fan-shaped areas of different colors correspond to different labels, where the value represents the description degree of the corresponding label. Red circles represent the instances that are selected into the DIP. (Color figure online)

are close (far) in the label space to also close (far) in the discriminative space. Following the theoretical work [18], the criterion is further derived as a function that measures the discriminative ability of each instance. With this function, several instances with the highest discriminative ability are selected to construct the optimal DIP. Moreover, the representation of an instance in the discriminative space is the concatenation of its distances to all instances in the DIP. As shown in Fig. 2, after mapping by the DIP, instances with similar (dissimilar) label distributions will be close (far) to each other in the discriminative space.

The main contribution of this paper can be summarized as follows:

1) We propose an LDL algorithm with discriminative instance mapping to transform original features into a discriminative space to learn label distributions.
2) We design a criterion to measure the discriminative ability of candidate instance by exploiting the relationship of pair-wise label distributions.
3) We select several highest discriminative instances to construct the DIP to make all instances maximally distinguishable in the new mapping space.

The remainder of the paper is laid out as follows: in Sect. 2, we review some LDL algorithms that are proposed in recent years. In Sect. 3, the details of LDLDIM is described. In Sect. 4, the proposed algorithm is compared with several state-of-the-art LDL methods. In Sect. 5, we make a conclusion about LDLDIM.

2   Related Works

During the past few years, a lot of efforts have been made to develop LDL algorithms. These algorithms can be classified into three groups based on their main strategies: Problem Transformation (PT), Algorithm Adaptation (AA) and Specialized Algorithm (SA).

The algorithms based on PT are accomplished by transforming LDL tasks into a series of weighted single-label tasks. This kind of degradation-based strategy has been wildly applied in MLL and multi-instance learning, such as MLSVM [2] and MIBoosting [19]. For LDL tasks, two classic single-label classifiers SVM and Bayes are adopted, i.e., PT-SVM and PT-Bayes [7]. PT-SVM computes the description degree of each label by an improved implementation of Platt's posterior probabilities, while PT-Bayes achieve the same goal by applying Bayes rule. The algorithms based on AA are developed by extending existing SLL approaches to accommodate the form of label distribution. For example, AA-kNN [7] is extended on the basis of k-NN. DLDL [6] utilizes deep convolutional neural networks to tackle LDL problems. The algorithms based on SA try to design models that directly output label distributions. For example, Geng et al. proposed SA-IIS [7], which takes the maximum entropy model [1] as the output model, and adopts a strategy similar to Improved Iterative Scaling (IIS) [4] to complete the optimization.

Due to the label form of LDL is naturally more complex than that of SLL and MLL, it is difficult to achieve satisfactory results by directly constructing the feature-label mapping function like above algorithms. Therefore, some methods attempt to take label correlation into consideration. LDLLC [11] employs Pearson's correlation coefficients to measure the correlations between any two instances and use it to regularize the distance between corresponding parameter columns. However, Jia et al. [12,22] propose that local label correlations are more prevalent than global ones in most cases. They design a clustering based method to encode the local label correlations into additional features for each instance. Moreover, they further customize a loss term for these features to regularize the distance between predicted label distributions and the mean vectors of each cluster. Instead of investigating the label space, others tend to manipulate the feature space. LDLSF [14] applies l_1-regularization on the parameter matrix to make it sparse, and it can figure out which features are discriminative for which labels.

3 Approach

3.1 Notations

We use $\mathcal{X} = \{x_i\}_{i=1}^N \in \mathbb{R}^q$ to denote the q-dimensional input space and $\mathcal{Y} = \{y_1, y_2, \ldots, y_L\}$ as the L-dimensional label space, where N is the number of instances and L is that of labels. The instance $x_i \in \mathcal{X}$ is associated with a label distribution $D_i = [d_i^1, d_i^2, \ldots, d_i^L] \in \mathcal{D}$ and $\mathcal{D} = \{D_i\}_{i=1}^N$. Each d_i^j represents the description degree to x_i for a label y_j. The label distributions should satisfy the basic definition of LDL [7]: $\sum_{j=1}^L d_i^j = 1$, $d_i^j > 0$, which means the labels in $\mathcal{Y}$ can fully describe all instances. The discriminative instance pool (DIP) is represented by $\mathcal{P} \subset \mathcal{X}$, and $|\mathcal{P}| = k$.

3.2 Discriminative Instance Pool

We aim to find an optimal DIP $\mathcal{P}^* \subseteq \mathcal{X}$ which can be used to generate the new discriminative space. Therefore, $\mathcal{P}$ needs to have the ability to distinguish all instances through mapping [18]. Accordingly, we design an evaluation criterion $\mathcal{F}(\mathcal{P})$ to measure the distinguishing ability of $\mathcal{P}$, and the optimal DIP $\mathcal{P}^*$ can be found as:

$$\mathcal{P}^* = \arg\max_{\mathcal{P} \subseteq \mathcal{X}} \mathcal{F}(\mathcal{P}) \qquad s.t. \quad |\mathcal{P}| = k, \tag{1}$$

where $|\cdot|$ represents the cardinality of $\mathcal{P}$ and k is the number of selected instances.

In order to obtain DIP with the highest discriminative power, we consider two goals when constructing criterion $\mathcal{F}(\cdot)$. The first one is the connectivity between the instances that have similar label distributions. The second one is the separation between those that have dissimilar label distributions. That is to say, the DIP should ensure that the instances that are close (far away) in the label space remain close (far away) in the discriminative feature space. Consequently, the evaluation criterion $\mathcal{F}(\cdot)$ is formulated as:

$$\mathcal{F}(\mathcal{P}) = \sum_{i,j} \alpha(\boldsymbol{x}_i, \boldsymbol{x}_j) Q_{ij}, \tag{2}$$

where $\alpha(\boldsymbol{x}_i, \boldsymbol{x}_j)$ denotes the distance of $\boldsymbol{x}_i$ to $\boldsymbol{x}_j$ in the discriminative feature space and Q_{ij} is the corresponding element in the matrix $\boldsymbol{Q}$ which embeds the structural information of the label distributions. For universality, $\boldsymbol{Q}$ is defined as:

$$Q_{ij} = \begin{cases} \dfrac{-1}{|A| \cdot e^{dis(\boldsymbol{D}_i, \boldsymbol{D}_j)}}, \, dis(\boldsymbol{D}_i, \boldsymbol{D}_j) \leq \tau; \\[3mm] \dfrac{1}{|B| \cdot e^{dis(\boldsymbol{D}_i, \boldsymbol{D}_j)}}, \, dis(\boldsymbol{D}_i, \boldsymbol{D}_j) > \tau, \end{cases} \tag{3}$$

in which τ is a threshold for judging whether two label distributions is similar or not and $dis()$ represents the Euclidean distance. $|A|$ and $|B|$ denote the number of instance-pairs that have similar label distributions and dissimilar ones, i.e., $A = \{(i,j)|dis(\boldsymbol{D}_i, \boldsymbol{D}_j) \leq \tau\}$ and $B = \{(i,j)|dis(\boldsymbol{D}_i, \boldsymbol{D}_j) > \tau\}$. Accordingly, Eq. (2) implies the aforementioned two goals, i.e., the connectivity and separation.

We now give a more detailed form of $\alpha(\boldsymbol{x}_i, \boldsymbol{x}_j)$ as:

$$\alpha(\boldsymbol{x}_i, \boldsymbol{x}_j) = \|\boldsymbol{x}_i^{\mathcal{P}} - \boldsymbol{x}_j^{\mathcal{P}}\|_2^2 = \|\boldsymbol{I}\boldsymbol{x}_i^{\mathcal{X}} - \boldsymbol{I}\boldsymbol{x}_j^{\mathcal{X}}\|_2^2, \tag{4}$$

where $\boldsymbol{x}_i^{\mathcal{X}}$ is obtained by mapping based on all instances:

$$\boldsymbol{x}_i^{\mathcal{X}} = [dis(\boldsymbol{x}_i, \boldsymbol{x}_1), dis(\boldsymbol{x}_i, \boldsymbol{x}_2), \ldots, dis(\boldsymbol{x}_i, \boldsymbol{x}_N)]^{\mathsf{T}}. \tag{5}$$

The definition of $\boldsymbol{x}_i^{\mathcal{P}}$ is similar to that of $\boldsymbol{x}_i^{\mathcal{X}}$, but uses the DIP $\mathcal{P}$ as the mapping instance set:

$$\boldsymbol{x}_i^{\mathcal{P}} = [dis(\boldsymbol{x}_i, \boldsymbol{p}_1), dis(\boldsymbol{x}_i, \boldsymbol{p}_2), \ldots, dis(\boldsymbol{x}_i, \boldsymbol{p}_k)]^{\mathsf{T}}, \tag{6}$$

where $\boldsymbol{p}_i$ denotes the i-th instance in $\mathcal{P}$. $\boldsymbol{I} \in \mathbb{R}^{N \times N}$ is an diagonal indicator matrix whose diagonal elements indicate whether the corresponding instance belongs to the DIP. In other words, if $\boldsymbol{x}_i$ belongs to the DIP, then $I_{ii} = 1$, otherwise 0.

Substitute Eq. (4) into Eq. (2), the criterion $\mathcal{F}(\mathcal{P})$ is rewritten as:

$$\mathcal{F}(\mathcal{P}) = \sum_{i,j} \|\boldsymbol{I}\boldsymbol{x}_i^{\mathcal{X}} - \boldsymbol{I}\boldsymbol{x}_j^{\mathcal{X}}\|_2^2 Q_{ij} \tag{7}$$

In order to facilitate the subsequent derivation, a constant coefficient $\frac{1}{2}$ is added to Eq. (7), and the final formulation of the criterion is defined as:

$$\mathcal{F}(\mathcal{P}) = \frac{1}{2}\sum_{i,j}\|\boldsymbol{I}\boldsymbol{x}_i^{\mathcal{X}} - \boldsymbol{I}\boldsymbol{x}_j^{\mathcal{X}}\|_2^2 Q_{ij}$$

$$= \sum_{i,j}\boldsymbol{x}_i^{\mathcal{X}}\boldsymbol{I}^\mathsf{T}\boldsymbol{I}(\boldsymbol{x}_i^{\mathcal{X}})^\mathsf{T}Q_{i,j} - \sum_{i,j}\boldsymbol{x}_i^{\mathcal{X}}\boldsymbol{I}^\mathsf{T}\boldsymbol{I}(\boldsymbol{x}_j^{\mathcal{X}})^\mathsf{T}Q_{i,j}$$

$$= \sum_{i}\boldsymbol{x}_i^{\mathcal{X}}\boldsymbol{I}^\mathsf{T}\boldsymbol{I}(\boldsymbol{x}_i^{\mathcal{X}})^\mathsf{T}\sum_{j}Q_{i,j} - \sum_{i,j}\boldsymbol{x}_i^{\mathcal{X}}\boldsymbol{I}^\mathsf{T}\boldsymbol{I}(\boldsymbol{x}_j^{\mathcal{X}})^\mathsf{T}Q_{i,j}$$

$$= \sum_{i}\boldsymbol{x}_i^{\mathcal{X}}\boldsymbol{I}^\mathsf{T}\boldsymbol{I}(\boldsymbol{x}_i^{\mathcal{X}})^\mathsf{T}G_{i,i} - \sum_{i,j}\boldsymbol{x}_i^{\mathcal{X}}\boldsymbol{I}^\mathsf{T}\boldsymbol{I}(\boldsymbol{x}_j^{\mathcal{X}})^\mathsf{T}Q_{i,j} \qquad (8)$$

$$= tr(\boldsymbol{I}^\mathsf{T}\boldsymbol{V}^{\mathcal{X}}G(\boldsymbol{V}^{\mathcal{X}})^\mathsf{T}\boldsymbol{I}) - tr(\boldsymbol{I}^\mathsf{T}\boldsymbol{V}^{\mathcal{X}}Q(\boldsymbol{V}^{\mathcal{X}})^\mathsf{T}\boldsymbol{I})$$

$$= tr(\boldsymbol{I}^\mathsf{T}\boldsymbol{V}^{\mathcal{X}}(G - Q)(\boldsymbol{V}^{\mathcal{X}})^\mathsf{T}\boldsymbol{I})$$

$$= tr(\boldsymbol{I}^\mathsf{T}\boldsymbol{V}^{\mathcal{X}}Z(\boldsymbol{V}^{\mathcal{X}})^\mathsf{T}\boldsymbol{I})$$

$$= \sum_{\boldsymbol{p}_i\in\mathcal{P}}(\phi_i)^\mathsf{T}Z\phi_i,$$

where $\boldsymbol{V}^{\mathcal{X}} = [\boldsymbol{x}_1^{\mathcal{X}}, \boldsymbol{x}_2^{\mathcal{X}}, \ldots, \boldsymbol{x}_N^{\mathcal{X}}]$, G is a diagonal matrix generated from Q whose element $G_{i,i} = \sum_j Q_{ij}$, Z is a Laplacian matrix generalized from Q, $Z = [Z_{ij}]^{N \times N} = G - Q$. $\phi_i = [dis(\boldsymbol{p}_i, \boldsymbol{x}_1), dis(\boldsymbol{p}_i, \boldsymbol{x}_2), \ldots, dis(\boldsymbol{p}_i, \boldsymbol{x}_N)]^\mathsf{T} \in \{\mathbb{R}\}^{N \times 1}$.

Consequently, we can quantify the discriminative ability σ_i of each instance in $\mathcal{X}$:

$$\sigma_i = (\phi_i)^\mathsf{T}Z\phi_i. \qquad (9)$$

We sort the σ of each sample from high to low, and select the first k instances to construct the optimal DIP $\mathcal{P}$ that satisfies the following conditions:

$$\forall \boldsymbol{p}_i \in \mathcal{P}, \boldsymbol{x}_j \in \mathcal{X}, \sigma_i > \sigma_j, |\mathcal{P}| = k. \qquad (10)$$

3.3   Output Model

With the optimal DIP, we map instances from the original feature space to the discriminative space via the Eq. 6. $\boldsymbol{x}_i$ in discriminative space is represented as $\boldsymbol{x}_i^{\mathcal{P}} = [dis(\boldsymbol{x}_i, \boldsymbol{p}_1), dis(\boldsymbol{x}_i, \boldsymbol{p}_2), \ldots, dis(\boldsymbol{x}_i, \boldsymbol{p}_k)]^\mathsf{T}$. Furthermore, we utilize $\boldsymbol{V}^{\mathcal{P}} = [\boldsymbol{x}_1^{\mathcal{P}}, \boldsymbol{x}_2^{\mathcal{P}}, \ldots, \boldsymbol{x}_N^{\mathcal{P}}]$ to denote the mapped instance set. With $\boldsymbol{V}^{\mathcal{P}}$ and $\mathcal{D}$, a model consists of L regressions $(\mathcal{R}_1, \mathcal{R}_2, \ldots, \mathcal{R}_L)$ is trained, each of which is for one label.

For a test instance $\boldsymbol{x}_m$, we use the optimal $\mathcal{P}$ to map it into the discriminative space. The mapped vector is computed as $\boldsymbol{x}_m^{\mathcal{P}} = [dis(\boldsymbol{x}_m, \boldsymbol{p}_1), dis(\boldsymbol{x}_m, \boldsymbol{p}_2), \ldots, dis(\boldsymbol{x}_m, \boldsymbol{p}_k)]^\mathsf{T}$. With $\boldsymbol{x}_m^{\mathcal{P}}$ and the trained multiple regressions $\mathcal{R}_1, \mathcal{R}_2, \ldots, \mathcal{R}_L$, The i-th component of the predicted label distribution $[d_m^1, d_m^2, \ldots, d_m^L]$ can be obtained by:

$$d_m^i = \mathcal{R}_i(\boldsymbol{x}_m^{\mathcal{P}}). \qquad (11)$$

Algorithm 1 sets out the pseudocode of the LDLDIM. Among them, steps 1 to 8 are the process of obtaining the optimal DIP. Steps 9 to 12 are the process of mapping instances to discriminant space using DIP. Steps 13 to 15 are the process of training L regressors using the discriminative features and the true label distributions.

Algorithm 1. LDLDIM

Input: the input space $\mathcal{X}$, the label distributions $\mathcal{D}$, the number of labels L, the size of DIP k.
Output: trained regression models $\mathcal{R}_1, \mathcal{R}_2, \ldots, \mathcal{R}_L$, the DIP $\mathcal{P}$.
1: Initialize $\mathcal{P}, \boldsymbol{V}^{\mathcal{P}}, \sigma = \emptyset$.
2: Compute $\boldsymbol{Q}$ according to Eq. (3);
3: Compute $\boldsymbol{Z}$ according to Eq. (8);
4: **for** $(\boldsymbol{x}_i \in \mathcal{X})$ **do**
5:    Compute σ_i with Eq. (9);
6:    $\sigma = \sigma \cup \{\sigma_i\}$;
7: **end for**
8: Obtain the DIP $\mathcal{P}$ according to Eq. (10);
9: **for** $(\boldsymbol{x}_i \in \mathcal{X})$ **do**
10:    Compute $\boldsymbol{x}_i^{\mathcal{P}}$ according to Eq. (6);
11:    $\boldsymbol{V}^{\mathcal{P}} = \boldsymbol{V}^{\mathcal{P}} \cup \{\boldsymbol{x}_i^{\mathcal{P}}\}$;
12: **end for**
13: **for** $(i \in [1..L])$ **do**
14:    Train $\mathcal{R}_i$ using $\boldsymbol{V}^{\mathcal{P}}$ and $[d_1^i, d_2^i, \ldots, d_N^i]$;
15: **end for**
16: **return** $\mathcal{R}_1, \mathcal{R}_2, \ldots, \mathcal{R}_L$ and $\mathcal{P}$.

---

4 Experiments

4.1 Datasets

The experiments are undertaking on 10 yeast datasets (Yeast-alpha~Yeast-spoem),
1 genetic dataset (Human Gene), 3 facial expression datasets (Emotion6, SJAFFE,
SBU_3DFE), 1 image dataset (Natural Scene), and 1 movie rating dataset (Moive).
Table 1 shows their statistics and some detailed information can be found in [7] and
[13].

4.2 Evaluations

Following [22], 2 commonly used metrics are employed, one of which measures the
distance between two label distributions: *Euclidean* $\downarrow$, while the another measures the
similarity between them: *Intersection* $\uparrow$. Notably, $\downarrow$ indicates a lower value of the metrics, the better the performance, and $\uparrow$ indicates a higher value of the metrics, the better
the performance.

4.3 Parameter Analysis

For LDLDIM, there are two parameters that need to be manually set, namely the threshold τ for judging whether two label distributions are similar and the size of DIP k. In
order to make the above parameters have certain adaptability, we tend to set them in a
proportional way. The threshold is set to $r_1 \cdot \max\{dis(\boldsymbol{D}_i, \boldsymbol{D}_j)\}$ and size k is set to
$\lfloor r_2 \cdot N \rfloor$. Figure 3 shows the performance of the algorithm under different parameter
settings, each result is obtained using 5 times 5 cross validation ($5CV$). The results
are normalized to better exhibit the performance gap. It can be seen that each dataset

Table 1. Statistics of 16 LDL datasets, where N is number of instances, n is number of features, and L is number of classes.

Datasets	N	n	L
Yeast-alpha	2,465	24	18
Yeast-cdc	2,465	24	15
Yeast-cold	2,465	24	4
Yeast-diau	2,465	24	7
Yeast-dtt	2,465	24	4
Yeast-elu	2,465	24	14
Yeast-heat	2,465	24	6
Yeast-spo	2,465	24	6
Yeast-spo5	2,465	24	3
Yeast-spoem	2,465	24	2
Human Gene	30,542	36	68
Emotion6	1,980	168	7
SJAFFE	213	243	6
SBU_3DFE	2,500	243	6
Natural Scene	2,000	294	9
Movie	7,755	1,869	5

has different threshold requirements, but in general, the threshold needs to be small (r_1 is around 0.1). For DIP size, Yeast-alpha and Yeast-cdc only need a small DIP ($r_2 = 0.1$) to guarantee sufficient performance, which indicates that the feature-label relations of these two datasets is not particularly complicated. In contrast, that of Yeast-dtt is slightly more complicated, which calls for a larger DIP ($r_2 = 0.5$). Furthermore, Yeast-cold, SJAFFE, and SBU_3DFE have more complex feature-label relations and therefore require the largest DIP ($r_2 = 0.9$) for best results.

4.4 Baselines and Settings

We compare the proposed LDLDIM to 6 state-of-the-art algorithms, including AA-kNN [7], EDL [23], LDLSF [14], LDL-SCL [22], LDL-LDM [16], LDL-HR [17]. The parameter settings of above algorithms are set to the recommend ones in their corresponding papers.

4.5 Performance Comparison

The experiments are undertaking on 16 datasets with 2 metrics. Table 2 and Table 3 report the detailed experimental results of 7 comparing algorithms on all data sets, where the best performance among the comparing algorithms on each measure is marked in bold. Accordingly, LDLDIM ranks first on 12 out of 16 datasets under

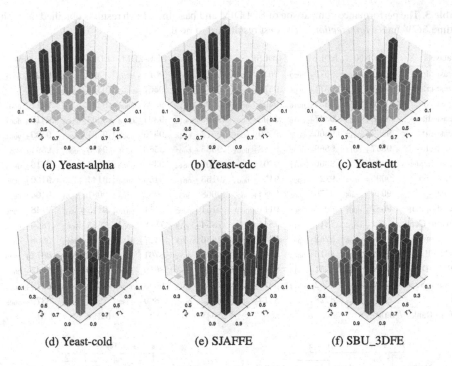

(a) Yeast-alpha   (b) Yeast-cdc   (c) Yeast-dtt

(d) Yeast-cold   (e) SJAFFE   (f) SBU_3DFE

Fig. 3. The performance of LDLDIM for different r_1 and r_2 (under *Intersection* ↑).

Table 2. The performance comparison of LDLDIM and baselines. Each result is obtained through 5 time $5CV$ under *Euclidean* ↓. The best results are in bold.

Datasets	AA-kNN	EDL	LDLSF	LDL-SCL	LDL-LDM	LDL-HR	LDLDIM
Yeast-alpha	.0279±.0006	.0260±.0011	.0235±.0001	.0234±.0003	.0235±.0001	.0412±.0003	**.0229±.0001**
Yeast-cdc	.0301±.0009	.0283±.0006	.0280±.0003	.0281±.0008	.0284±.0002	.0422±.0002	**.0277±.0001**
Yeast-cold	.0724±.0027	.0771±.0018	.0683±.0001	.0696±.0003	.0693±.0005	.0694±.0001	**.0671±.0009**
Yeast-diau	.0567±.0019	.0597±.0010	**.0529±.0001**	.0552±.0008	.0551±.0004	.0575±.0001	.0530±.0002
Yeast-dtt	.0512±.0019	.0508±.0022	.0480±.0001	.0493±.0017	.0489±.0002	.0486±.0001	**.0474±.0003**
Yeast-elu	.0297±.0010	.0289±.0005	.0279±.0005	.0279±.0005	.0282±.0001	.0365±.0001	**.0275±.0003**
Yeast-heat	.0624±.0020	.0629±.0016	.0593±.0001	.0596±.0016	.0599±.0002	.0609±.0001	**.0577±.0001**
Yeast-spo	.0879±.0030	.0843±.0029	.0822±.0019	**.0815±.0006**	.0830±.0001	.0830±.0001	.0816±.0002
Yeast-spo5	.1231±.0007	.1191±.0001	.1173±.0003	.1193±.0002	.1165±.0011	.1176±.0001	**.1147±.0002**
Yeast-spoem	.1291±.0001	.1274±.0001	.1230±.0001	.1229±.0002	.1257±.0018	.1251±.0001	**.1204±.0004**
Huamn Gene	.1036±.0044	.0887±.0021	.0864±.0001	**.0858±.0016**	.0866±.0002	.0943±.0001	.0860±.0001
Emotion6	.4551±.0134	.4502±.0029	**.4048±.0007**	.4111±.0018	.4349±.0030	.8897±.0072	.4109±.0014
SJAFFE	.1264±.0024	.1216±.0061	.1162±.0017	.1160±.0012	.1257±.0042	.1209±.0015	**.1157±.0019**
SBU_3DFE	.1633±.0007	.1510±.0018	.1485±.0001	.1459±.0001	.1433±.0022	.1423±.0003	**.1417±.0009**
Natural Scene	.4322±.0009	.5223±.0098	.4198±.0029	.4002±.0021	.4068±.0015	.4173±.0007	**.3750±.0007**
Movie	.1805±.0004	.1765±.0115	.1714±.0021	.1687±.0005	.2174±.0007	.1729±.0002	**.1649±.0005**
Mean Rank	6.38	5.69	2.63	3.00	4.13	4.94	1.25

Table 3. The performance comparison of LDLDIM and baselines. Each result is obtained through 5 time $5CV$ under *Intersection↑*. The best results are in bold.

Datasets	AA-kNN	EDL	LDLSF	LDL-SCL	LDL-LDM	LDL-HR	LDLDIM
Yeast-alpha	$.9561_{\pm.0012}$	$.9570_{\pm.0022}$	$.9619_{\pm.0005}$	$.9622_{\pm.0005}$	$.9615_{\pm.0002}$	$.9307_{\pm.0004}$	$\mathbf{.9625_{\pm.0001}}$
Yeast-cdc	$.9538_{\pm.0013}$	$.9571_{\pm.0008}$	$\mathbf{.9577_{\pm.0005}}$	$.9572_{\pm.0013}$	$.9567_{\pm.0003}$	$.9344_{\pm.0003}$	$.9576_{\pm.0001}$
Yeast-cold	$.9370_{\pm.0024}$	$.9332_{\pm.0016}$	$.9408_{\pm.0001}$	$.9396_{\pm.0013}$	$.9398_{\pm.0005}$	$.9344_{\pm.0003}$	$\mathbf{.9417_{\pm.0001}}$
Yeast-diau	$.9378_{\pm.0022}$	$.9347_{\pm.0010}$	$.9403_{\pm.0001}$	$.9395_{\pm.0011}$	$.9394_{\pm.0006}$	$.9365_{\pm.0001}$	$\mathbf{.9417_{\pm.0001}}$
Yeast-dtt	$.9557_{\pm.0017}$	$.9560_{\pm.0018}$	$.9580_{\pm.0001}$	$.9573_{\pm.0015}$	$.9576_{\pm.0002}$	$.9581_{\pm.0001}$	$\mathbf{.9589_{\pm.0001}}$
Yeast-elu	$.9557_{\pm.0014}$	$.9569_{\pm.0007}$	$.9588_{\pm.0001}$	$.9585_{\pm.0005}$	$.9580_{\pm.0001}$	$.9455_{\pm.0001}$	$\mathbf{.9591_{\pm.0001}}$
Yeast-heat	$.9368_{\pm.0018}$	$.9366_{\pm.0017}$	$.9401_{\pm.0001}$	$.9402_{\pm.0014}$	$.9396_{\pm.0003}$	$.9384_{\pm.0001}$	$\mathbf{.9416_{\pm.0001}}$
Yeast-spo	$.9096_{\pm.0034}$	$.9128_{\pm.0028}$	$.9154_{\pm.0001}$	$.9153_{\pm.0019}$	$.9156_{\pm.0006}$	$.9144_{\pm.0001}$	$\mathbf{.9160_{\pm.0002}}$
Yeast-spo5	$.9040_{\pm.0005}$	$.9070_{\pm.0001}$	$.9084_{\pm.0002}$	$.9068_{\pm.0001}$	$.9089_{\pm.0008}$	$.9081_{\pm.0001}$	$\mathbf{.9104_{\pm.0002}}$
Yeast-spoem	$.9087_{\pm.0001}$	$.9099_{\pm.0001}$	$.9111_{\pm.0001}$	$.9151_{\pm.0002}$	$.9111_{\pm.0013}$	$.9115_{\pm.0001}$	$\mathbf{.9148_{\pm.0002}}$
Huamn Gene	$.7451_{\pm.0036}$	$.7810_{\pm.0018}$	$.7844_{\pm.0005}$	$\mathbf{.7854_{\pm.0013}}$	$.7848_{\pm.0005}$	$.7555_{\pm.0001}$	$.7850_{\pm.0001}$
Emotion6	$.5494_{\pm.0008}$	$.5252_{\pm.0032}$	$\mathbf{.5857_{\pm.0008}}$	$.5836_{\pm.0018}$	$.5427_{\pm.0030}$	$.2442_{\pm.0072}$	$.5808_{\pm.0003}$
SJAFFE	$.8769_{\pm.0025}$	$.8816_{\pm.0063}$	$.8863_{\pm.0019}$	$.8861_{\pm.0011}$	$.8761_{\pm.0042}$	$.8706_{\pm.0018}$	$\mathbf{.8871_{\pm.0020}}$
SBU_3DFE	$.8466_{\pm.0006}$	$.8551_{\pm.0018}$	$.8569_{\pm.0001}$	$.8594_{\pm.0001}$	$.8619_{\pm.0017}$	$.8628_{\pm.0003}$	$\mathbf{.8647_{\pm.0008}}$
Natural Scene	$.5601_{\pm.0013}$	$.3662_{\pm.0074}$	$.5812_{\pm.0013}$	$.6170_{\pm.0011}$	$.6038_{\pm.0020}$	$.5802_{\pm.0013}$	$\mathbf{.6219_{\pm.0007}}$
Movie	$.8224_{\pm.0004}$	$.8228_{\pm.0094}$	$.8314_{\pm.0004}$	$.8322_{\pm.0015}$	$.7860_{\pm.0007}$	$.7912_{\pm.0002}$	$\mathbf{.8379_{\pm.0005}}$
Mean Rank	6.00	5.63	2.88	2.94	3.94	5.31	1.31

(a) *Euclidean ↓*   (b) *Intersection ↑*

Fig. 4. CD diagrams of the Bonferroni-Dunn test. Algorithms that have no connection to LDLDIM are considered to have significant performance differences from LDLDIM (CD = 2.014 at 0.05 significance level).

Euclidean ↓ and first on 13 out of 16 under *Intersection ↑*. For those datasets where LDLDIM do not achieve the best results, its performance is not far behind the top-ranked algorithms. The last row of each metric exhibits the mean ranks of each algorithm.

Furthermore, we apply the post-hoc Bonferroni-Dunn test [5] to verify whether the LDLDIM achieves competitive performance against the compared algorithms. The critical difference (CD) diagram at the 0.05 significance level is reported in Fig. 4. The mean rank of each algorithm is marked along the axis (lower ranks to the right). In addition, the algorithm whose mean rank is within one CD to that of the LDLDIM is connected with a thick line. It can be seen from the Fig. 4 that LDLDIM is significantly better than the algorithms AA-kNN, EDL, LDL-LDM and LDL-HR.

4.6 DIP Selection Result

To observe which instances are highly discriminative, we selected two datasets and visualized the sample distribution of their DIPs. For each dataset, the DIP size k is fixed to $\lfloor 0.1 \cdot n \rfloor$. Then, we calculate the discriminative power of each sample by the Eq. 9. Sorting from high to low discriminative ability, select the first k samples to construct DIP. After identifying all DIPs, we use Principal Component Analysis (PCA) to reduce the dimensionality of all instances to 2 and mark instances belonging to DIPs and others with different colors. As shown in the Fig. 5, the instances selected into DIP for each data set are marked in orange, and the remaining instances are marked in green. The results in the figure show that in different data sets, the closer to the edge of the data set is generally more discriminative.

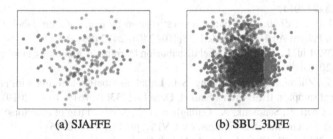

(a) SJAFFE   (b) SBU_3DFE

Fig. 5. DIP selection result on SJAFFE and SBU_3DFE. The instances in the DIP are marked in orange and the remaining instances are in green. (Color figure online)

5 Conclusion and Further Works

In this paper, a novel algorithm called LDLDIM that aims to map all instances into a more discriminative feature space is proposed. To obtain a suitable mapping, we design a criterion to quantify the discriminative power of instances. This criterion exploits the relationship information of pairwise instances in label space. Then we select a few optimal instances to build the DIP and map all the instances. Therefore, in the new discriminative feature space, instances with similar label distributions will be close to each other, which makes the instances easier to distinguish. Experimental results on 16 real-world datasets with 6 comparing algorithms exhibits the effectiveness of LDLDIM. In the future, we will try to come up with more specific measures of instance discriminative power, and will also improve the instance mapping method.

References

1. Berger, A., Della Pietra, S.A., Della Pietra, V.J.: A maximum entropy approach to natural language processing. Comput. Linguist. **22**(1), 39–71 (1996)
2. Boutell, M.R., Luo, J., Shen, X., Brown, C.M.: Learning multi-label scene classification. Pattern Recogn. **37**(9), 1757–1771 (2004)

3. Chen, C., Chen, Z., Jin, X., Li, L., Speier, W.F., Arnold, C.: Attention-guided discriminative region localization and label distribution learning for bone age assessment. IEEE J. Biomed. Health Inform. (2021)
4. Della Pietra, S., Della Pietra, V., Lafferty, J.: Inducing features of random fields. IEEE Trans. Pattern Anal. Mach. Intell. **19**(4), 380–393 (1997)
5. Demšar, J.: Statistical comparisons of classifiers over multiple data sets. J. Mach. Learn. Res. **7**, 1–30 (2006)
6. Gao, B.B., Xing, C., Xie, C.W., Wu, J., Geng, X.: Deep label distribution learning with label ambiguity. IEEE Trans. Image Process. **26**(6), 2825–2838 (2017)
7. Geng, X.: Label distribution learning. IEEE Trans. Knowl. Data Eng. **28**(7), 1734–1748 (2016)
8. Geng, X., Wang, Q., Xia, Y.: Facial age estimation by adaptive label distribution learning. In: ICPR, pp. 4465–4470 (2014)
9. Geng, X., Xia, Y.: Head pose estimation based on multivariate label distribution. In: CVPR, pp. 1837–1842 (2014)
10. Geng, X., Yin, C., Zhou, Z.H.: Facial age estimation by learning from label distributions. IEEE Trans. Pattern Anal. Mach. Intell. **35**(10), 2401–2412 (2013)
11. Jia, X., Li, W., Liu, J., Zhang, Y.: Label distribution learning by exploiting label correlations. In: AAAI (2018)
12. Jia, X., Li, Z., Zheng, X., Li, W., Huang, S.J.: Label distribution learning with label correlations on local samples. IEEE Trans. Knowl. Data Eng. **33**(4), 1619–1631 (2019)
13. Peng, K.C., Chen, T., Sadovnik, A., Gallagher, A.C.: A mixed bag of emotions: model, predict, and transfer emotion distributions. In: CVPR, pp. 860–868 (2015)
14. Ren, T., Jia, X., Li, W., Chen, L., Li, Z.: Label distribution learning with label-specific features. In: IJCAI, pp. 3318–3324 (2019)
15. Ren, Y., Geng, X.: Sense beauty by label distribution learning. In: IJCAI, pp. 2648–2654 (2017)
16. Wang, J., Geng, X.: Label distribution learning by exploiting label distribution manifold. IEEE Trans. Neural Networks Learn. Syst. **01**, 1–14 (2021)
17. Wang, J., Geng, X.: Learn the highest label and rest label description degrees. In: IJCAI, pp. 3097–3103 (2021)
18. Wu, J., Pan, S., Zhu, X., Zhang, C., Wu, X.: Multi-instance learning with discriminative bag mapping. IEEE Trans. Knowl. Data Eng. **30**(6), 1065–1080 (2018)
19. Xu, X., Frank, E.: Logistic regression and boosting for labeled bags of instances. In: PAKDD, pp. 272–281 (2004)
20. Zhang, H., Zhang, Y., Geng, X.: Practical age estimation using deep label distribution learning. Front. Comp. Sci. **15**(3), 1–6 (2021)
21. Zhang, M.L., Zhou, Z.H.: A review on multi-label learning algorithms. IEEE Trans. Knowl. Data Eng. **26**(8), 1819–1837 (2013)
22. Zheng, X., Jia, X., Li, W.: Label distribution learning by exploiting sample correlations locally. In: AAAI, vol. 32 (2018)
23. Zhou, Y., Xue, H., Geng, X.: Emotion distribution recognition from facial expressions. In: ACMMM, pp. 1247–1250 (2015)

Leveraging Generative Models for Combating Adversarial Attacks on Tabular Datasets

Jiahui Zhou[1], Nayyar Zaidi[1(✉)], Yishuo Zhang[1], Paul Montague[2], Junae Kim[2], and Gang Li[1]

[1] Deakin University, Burwood, VIC 3125, Australia
jhzhou@acm.org, {nayyar.zaidi,zhangyis,gang.li}@deakin.edu.au
[2] Defence Science and Technology Group, Canberra, Australia
{paul.montague,junae.kim}@defence.gov.au

Abstract. Artificial Neural Networks (ANN) models – a form of discriminative models – are the workhorse of deep learning research, and have resulted in a remarkable performance on a range of applications on a large variety of datasets. On tabular datasets, ANN models are preferable when learning from large quantities of data as non-parametric models such as Random Forest and XGBoost cannot be easily used due to their inherent in-core data processing (i.e., they require loading all the data in memory). The applicability and effectiveness of ANN models, however, come with a price. They have been shown to be susceptible to adversarial attacks, which can greatly compromise their performance and trust in their utilization. There has been a surge in research in developing effective defence strategies for adversarial attacks on ANN models, e.g., Madry, D2A3, etc. Recently, it has been shown that generative models are more robust to adversarial attacks than discriminative models. A natural question is – can generative models be used as a defence for discriminative models against adversarial attacks? This work addresses this question, where we study the power of generative models in warding off adversarial attacks for discriminative models. In this work, we propose an effective defence model – gD2A3 – that exploits the generative-discriminative equivalence of some ANN models. It uses the learned probabilities from a generative model to initialize the input layer parameters of a standard ANN model, and utilizes L_2 regularization of the input layer parameters as a defence mechanism. We show that our proposed model leads to better results than the state-of-the-art method D2A3 by conducting a thorough empirical study on a variety of datasets with two major adversarial attacks.

Keywords: Adversarial Attack · Tabular Data · Generative-Discriminative Models · Regularization

---

J. Zhou and N. Zaidi—Equal Contribution.

Supplementary Information The online version contains supplementary material available at https://doi.org/10.1007/978-3-031-33374-3_12.

H. Kashima et al. (Eds.): PAKDD 2023, LNAI 13935, pp. 147–158, 2023.
https://doi.org/10.1007/978-3-031-33374-3_12

1 Introduction

At the heart of deep learning is a parametric model in the form of an `Artificial Neural Network` (`ANN`), which is trained by optimizing a differentiable objective function. The error is propagated back through the network, and each parameter of the model is updated in an iterative gradient-descent optimization manner. This end-to-end training process, as it is known, is efficient as it can process notably large quantities of data in a strictly online or batch-processing manner. However, this gradient-based learning has a fundamental weakness – it opens the door to adversarial attacks [2]. The idea behind adversarial learning is that any malicious entity, with access to model parameters (weights), can obtain the respective gradients so that it can modify the input to achieve the desired output [8,13]. E.g., for an input $\mathbf{x}$ to a given model $f(\mathbf{x})$, $\mathbf{r}$ is an adversarial noise if $f(\mathbf{x} + \mathbf{r}) \neq f(\mathbf{x})$, where $|\mathbf{r}| \leq \epsilon$, for some (typically small and imperceptible) allowed perturbation size ϵ. Adversarial learning (attack and defence methods) is studied in great detail in the context of structured data such as images [4,9]. However, its application to tabular data is somewhat under-explored[1]. In the last few years, there has been a surge of research in developing effective defence strategies for adversarial attacks on `ANN` models, e.g., `Madry`, `D2A3`, etc. On tabular datasets, it is shown that `D2A3` leads to state-of-the-art results. Recently, it is shown that generative models are more robust to adversarial attacks than discriminative models. Can generative models be used as part of a defence strategy in discriminative models? If yes, how does this impact the results? We will address these questions in this work, as we propose a novel defence strategy that utilizes a novel combination of generative and discriminative learning, to achieve state-of-the-art results.

There is a long history of generative and discriminative models and learning in machine learning. Generative models optimize the log-likelihood objective function $(\mathrm{LL}(\theta) = \sum_{i=1}^{N} \mathrm{P}(y^{(i)}, \mathbf{x}^{(i)}))$, whereas discriminative models optimize conditional log-likelihood $(\mathrm{CLL}(\theta) = \sum_{i=1}^{N} \mathrm{P}(y^{(i)}|\mathbf{x}^{(i)}))$. Their equivalence in terms of the number of parameters to optimize is well established in the case of naive Bayes and logistic regression models [7]. There has been some work that discusses the robustness of generative and discriminative models [3] and claims that generative models are more robust to adversarial attacks than discriminative models. The work has been conducted in the context of image datasets. In fact, a variant of naive Bayes – the Latent Variable Model (`LVM`) model – is used in the study. We share the inspiration and insights from [3], as we exploit the use of generative models for adversarial defence. However, our work differs as our focus is on the integrating of generative models with discriminative models to ward off adversarial attacks on discriminative models.

Why Generative Models are Robust to Adversarial Attacks? – Before, we delve into why generative models are robust to adversarial attacks, let us take a step back and discuss the role of discretization in formulating a defence strategy. Note, the state-of-the-art defence strategy, `D2A3` [12], relies on the discretization

---

[1] By tabular data, we mean, a dataset in tabular format with discrete or categorical features such that the correlation among features is unknown.

of data at training time and adversarial training. For the sake of simplicity, we will only focus on the discretization part. The numeric input to the model is first discretized, and the discretized data is fed to the model for training. At testing time, the adversarial sample is first discretized and the discretized value is fed to the model. The effectiveness of this defence strategy stems from the bin number not changing between the original and adversarial samples. E.g., the age of a person says 23 if maliciously changed to 25 will be allocated to the same bin, say 3 as a result of discretization. However, if the bins are changed, for example, the value 29 is changed to 31, and the bin value is changed from say 3 to 4, this can result in altering the output of the model and hence performance degradation. Discretized data has limited degrees of freedom, which means that when an adversarial sample is discretized, there are two possibilities:

- First, it leads to a data point (feature values) that is already seen in the training data, and
- Second, it leads to a data point (feature values) that is different from the training data.

Now, let us analyse these two cases from the context of a generative model. A generative model aims to learn the parameters that quantify the distribution of any given data. It can be seen that in the former case, the distribution of the data (as parameterized by a generative model) is not changed, and the model already has learned how to handle even the adversarial datum. In the latter case, the probability of the data will be so small under the generative model that it will have minimal impact on the classification. In a nutshell, probabilities are less prone to changes in the data, resulting in making generative models more robust to adversarial attacks. Now that we have established the robustness of generative models, let us discuss how they can be incorporated during the learning of a discriminative model (especially deep ANNs).

A simple way to exploit the generative model's parameters (i.e., probabilities) is through pre-training of a generative model, followed by feeding these probabilities as an input to a discriminative model. The intuition is that even though the adversarial sample has resulted in changing the bin number, the resulting probability of obtaining that datum (or feature values) based on the generative model, can play its part. E.g., if the age attribute is changed to 98 which is then assigned to bin number 10 – now if P(Age == 10|other feature values) is used as an input to the discriminative model, it can derive the importance of that feature. If it is very small, say 0.00001, the impact of that adversarial manipulated feature is somewhat mitigated. Our proposed algorithm gD2A3 – *generative model inspired* D2A3 – relies on this strategy of inputting the log of probabilities as an input to the discriminative model. It also relies on regularization, which is commonly used for controlling for over-fitting in discriminative models. However, we will show that regularization in gD2A3 leads to a mechanism that can control the magnitude of the defence against any adversarial attack. E.g., a strong regularization can lead us to generative parameters, i.e., probabilities learned in the pre-training step.

The main contributions of this work are as follows:

- We have proposed a novel method – gD2A3 – that is based on utilizing the robustness of generative models for incorporating a defence mechanism in a discriminative model. It makes use of regularization to control the magnitude of defence during discriminative learning.
- For shallow models, we show that gD2A3 can result in an innovative combination of generative and discriminative learning. This is desirable, e.g., if one believes that an attack is happening, one can revert to a generative model during discriminative learning.
- We perform extensive analysis on 10 typical datasets and show that our method can lead to better results than the state-of-the-art method – D2A3.

The rest of this paper is organized as follows. We discuss some preliminary and related work in Sect. 2. The details of our proposed method are given in Sect. 3. We provide an experimental evaluation of our proposed method in Sect. 4. We conclude in Sect. 5 with pointers to future work.

2   Related Work and Preliminaries

Defence methods against adversarial attacks on tabular data are still limited in the current literature. A commonly used defence method for adversarial attacks on continuous data is `Madry` [5]. It leverages adversarial training to minimize the adversarial risk of the model. `TRADES` [11] is another commonly used defence method for continuous data which minimizes the regularized surrogate loss instead of directly training adversarially. The current state-of-the-art defence method for adversarial attacks on tabular data is `D2A3` [12]. The `D2A3` approach leverages discretization with adversarial training as the defence method and achieves extremely encouraging results.

Let us discuss the major attacks methods that are common for tabular datasets:

- `LowProFool` [1] is a white-box attack method in the tabular domain for generating imperceptible adversarial perturbations. It is based on minimizing the addition of (imperceptible) adversarial noise on the features via a gradient descent approach. The gradients of the adversarial noise are used to guide the updates towards a distinct target class to the clean sample. It is the state-of-the-art white-box attack method on tabular data [1].
- `DeepFool` [6] is another white-box attack method, and it works by adding adversarial noise to the clean sample by estimating the distance between the sample and the model decision boundary. The limitation of `DeepFool` is that the adversarial noise can be large.

Let us discuss the basics of generative and discriminative models, and delve into one formulation that innovatively combines the two models. A standard discriminative model is of the form of `logistic regression` (`LR`) – which takes the form

$$P_{\mathrm{LR}}(y|\mathbf{x}) = \frac{e^{\beta_y + \sum_{j=1}^{n} \beta_{x_j|y}}}{\sum_{c=1}^{C} e^{\beta_c + \sum_{j=1}^{n} \beta_{x_j|c}}}. \tag{1}$$

On the other hand, a standard generative model is `Naive Bayes` (NB) – which takes the form:

$$P_{NB}(y|\mathbf{x}) = \frac{\theta_y + \prod_{j=1}^{n} \theta_{x_j|y}}{\sum_{c=1}^{C} \theta_c + \prod_{j=1}^{n} \theta_{x_j|c}}. \tag{2}$$

Naive Bayes and logistic regression are generative and discriminative counterparts [7], i.e., there is equivalence and correspondence on the number of parameters but they differ in how these parameters are optimized. A notable work here is that of [10], which combines the two models i.e., naive Bayes and logistic regression in a single framework known as `WANBIA-C` – which takes the form:

$$P_{WC}(y|\mathbf{x}) = \frac{e^{\beta_y \log \theta_y + \sum_{j=1}^{n} \beta_{x_j|y} \log \theta_{x_j|y}}}{\sum_{c=1}^{C} e^{\beta_c \log \theta_c + \sum_{j=1}^{n} \beta_{x_j|c} \log \theta_{x_j|c}}}. \tag{3}$$

Equation 3 provides an excellent way of leveraging generative parameters during the learning of discriminative parameters. E.g., one can learn the generative parameters ($\boldsymbol{\theta}$) by maximizing the log-likelihood and fixing them. Later one can optimize the discriminative parameters ($\boldsymbol{\beta}$), by optimizing the traditional conditional log-likelihood. Our proposed gD2A3 method in this work is inspired by the `WANBIA-C` formulation, as it formalizes the integration of generative models in discriminative models that are deep (and not shallow like logistic regression).

3 Proposed Technique

In this section, let us discuss our proposed gD2A3 method for defending against adversarial attacks. As we discussed earlier, it is based on integrating the parameters of a generative model during the training of a discriminative model. In fact, we discussed one such integration in Sect. 2, when we discussed `WANBIA-C` – a formulation that is applicable to only shallow discriminative models. A natural question is: how can one exploit the `WANBIA-C` trick with deep discriminative models? Our proposed method, gD2A3, relies on the utilization of an embedding layer in a deep discriminative model (ANN). It learns a higher-order generative model such as a Bayesian Network in the form of k-dependence Bayesian Estimator classifier (KDB). The probabilities learned from the generative model are used to initialize the embedding layer. Note the embedding layer is not learnable, i.e., `learnable = False` is set during discriminative training. The output of the embedding layer is fed to the hidden layers of the ANN model, which constitutes the learnable part of the model during discriminative training. The output of the embedding layer is also regularized, to control the magnitude of the defence against any adversarial attack. In the following let us delve into the two salient features of gD2A3.

3.1 Pre-conditioning Layer

We denote the embedding layer in gD2A3 as a *pre-conditioning layer*. A pre-trained generative model of the form of KDB is utilized in this layer[2]. We have used KDB as a representative generative model because of the following reasons:

- Firstly, the model is easy to learn. The structure is based on mutual information between attributes and the class, as well as the conditional mutual information between two features and the class. The structure and parameters can be learned in two passes through the data.
- Secondly, the model incorporates naive Bayes (e.g., $k = 0$ in KDB is naive Bayes), and therefore, integration with shallow models (such as logistic regression) is straightforward.
- Finally, our conjecture is that the higher-order probabilities that KDB provides offer more robustness to adversarial attacks than plain naive Bayes.

In the following, let us formalize the form for our pre-conditioning layer. In KDB, the target feature denoted as Y is the parent of every feature. The feature X_i will take no more than k parents except the target feature Y, and it will have associated probabilities of the form:

$$P(X_i = x_i | \Pi(X_i) = \pi(X_i)), \tag{4}$$

where $\Pi(X_i)$ returns the parent features of X_i, and $\pi(X_i)$ denotes the values of each parent feature. E.g., if X_j, X_l and Y form the parents of X_i, then we have:

$$(\Pi(X_i) = \pi(X_i)) \equiv (X_j = x_j, X_l = x_l, Y = y). \tag{5}$$

Furthermore, for every value of feature X_i – if X_i takes k parents – we have an associated set of size Z_i of probabilities, where $Z_i = \prod_{f=1}^{k} |X_f|$. Here $|\cdot|$ denotes the cardinality of the feature. As an illustration, the feature EL (Education-Level) has two parents, say City, Age. Suppose, there is a total of 10 cities and 5 age groups in the dataset, this means that for every value of EL, we have an associated set of $Z_{EL} = 10 \times 5$ probabilities, which can be flattened to obtain a vector of size $Z_i = 50$. In other words, EL = 'Bachelor-Degree' or EL = 'Masters' or EL = 'PhD', is represented as a 50-dimensional vector of probabilities. Here, the size (Z_i) of the resultant vector is dependent on the value of k (i.e., how many parents a feature has). In the pre-conditioning layer of gD2A3, the embedding layer is of size Z_i for feature X_i, to establish a one-to-one correspondence of KDB probabilities with that of the embedding layer's parameters. The embedding layer parameters (β) are initialized with the log of the KDB probabilities as:

$$\beta_{X_i = x_i | X_j = x_j, \dots, Y = y} = \log P(X_i = x_i | X_j = x_j, \dots, Y = y) \tag{6}$$

As we described earlier, once the parameters of the embedding layer are initialized, they are frozen and not updated in the training process to follow.

---

[2] Note, the training time for KDB, typically with small order values of k due to the computational cost, i.e., $k = 0, k = 1$, is negligible compared to training the deep learning model.

3.2   Utilizing Generative Model During Discriminative Training

The pre-conditioning layer in gD2A3 is connected to a dense hidden layer leading to a deep ANN. One can, however, connect the output of the pre-conditioning layer to an output layer – this leads to a shallow model of the form as described in Eq. 3. In gD2A3, the output of the pre-conditioning layer is also regularized. However, instead of regularizing towards zero, it regularizes the parameters of the layer towards 1 as shown in the following equation:

$$\text{CLL}(\boldsymbol{\beta}) = \log(\text{P}(y|\mathbf{x})) + \lambda \|\boldsymbol{\beta}^{(1)} - 1\|^2, \tag{7}$$

where $\boldsymbol{\beta}^{(1)}$ represents the parameters associated with the first layer after the pre-conditioning layer in gD2A3. It can be seen that when λ is large, and hence $\boldsymbol{\beta}^{(1)}$ converges to 1, the input to the ANN is the output of the generative model (i.e., KDB probabilities). However, when λ is small, our model learns some set of weights on the input probabilities. Thus, λ can be seen as a mechanism to control the magnitude of the defence. A large value will result in resorting to the output of the generative model (a strong defence, but it can impact the performance if no attack is happening), whereas, a small value will let the ANN learn weights (a weak defence, but the performance can be better if no attack is happening).

3.3   Algorithm

Let us summarize the working of our proposed gD2A3 model:

- First, it learns a KDB model, and learns the structure as well as parameters (probabilities). The value of k is a hyper-parameter, default is 0. The probabilities are denoted as $\boldsymbol{\theta}$.
- Second, it initializes the weights of the pre-conditioning layer denoted as $\boldsymbol{\beta}^{(0)}$ with pre-trained KDB probabilities, i.e., $\boldsymbol{\beta}^{(0)} \leftarrow \boldsymbol{\theta}$. $\boldsymbol{\beta}^{(0)}$ is set to non-learnable in the discriminative training process to follow.
- Third, it optimizes an objective function of the form Eq. 7 optimizing parameters $\boldsymbol{\beta}^{(1)}, \dots \boldsymbol{\beta}^{(D)}$, where D denotes the number of layers in the deep ANN.
- Finally, the weights of the first layer $\boldsymbol{\beta}^{(1)}$ are regularized towards 1 to control the magnitude of the defence against adversarial attacks.

A simple illustration of this process is shown in Fig. 1.

4   Experimental Results

In this section, we empirically verify the effectiveness of our proposed gD2A3 model. We will test its efficacy for both shallow as well as deep models. By shallow, we mean a model that has no hidden layers – i.e., the output of the pre-conditioning layer is directly connected to the output layer, whereas a deep model has hidden layers present between the pre-conditioning and output layers.

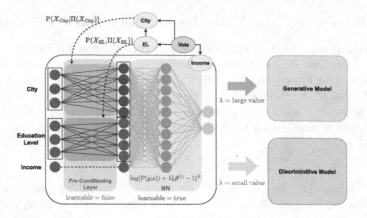

Fig. 1. Initialization of input layer based on pre-trained weights.

4.1 Experiment Setup

Dataset. We have used a total of 10 UCI classification datasets in our experiments. Out of the 10 datasets, 4 have more than 50K samples and are denoted as `Large`, whereas 3 datasets have between 5 and 50K samples and are denoted as `Medium`. The remaining 3 datasets have less than 5K samples and are denoted as `Small`. The statistics of the data are summarized in Table 1. We have used equal frequency discretization with 10 bins to convert numerical attributes to categorical attributes.

Adversarial Attack Setting and Evaluation Metric. We have made use of 2 commonly used white-box attack methods, which are common for tabular datasets, i.e. `Deepfool` and `LowProfool`. For each attack method, the architecture and the parameters of the target model are available, and the adversarial samples are directly generated. The maximum number of iterations for `Deepfool` and `LowProfool` is 50, which is the same as the default setting in their original implementation

Regarding the evaluation metrics, we have made use of an accuracy measure, which generally determines the level of resistance of a defence mechanism against an adversarial attack. The term `standard accuracy` is used for the normal case i.e., to quantify accuracy without an adversarial attack and the term `robust accuracy` is used to quantify accuracy for the case when an attack occurs.

4.2 Generative vs. Discriminative Models

Let us perform a preliminary experiment to compare the robustness of generative and discriminative models. We will use naive Bayes (`NB`) and logistic regression (`LR`) as examples of generative and discriminative models. We manipulated the data to perform two kinds of attacks, i.e., `LowProFool` and `DeepFool`. It can be seen from Table 2 that generative models are robust to adversarial attacks

Table 1. Description of datasets (m, n, c denotes the number of instances, features and classes respectively for each dataset).

Dataset	m	n	c	Size	Dataset	m	n	c	Size
Census-income	299285	41	2	Large	Magic	19020	10	2	Medium
SkinSegmentation	245057	3	2	Large	Page-blocks	5473	10	5	Medium
Higgs	98050	28	2	Large	Abalone	4177	8	3	Small
Connect-4	67557	42	3	Large	Vowel	990	13	11	Small
Adult	48842	14	2	Medium	Vowel-context	990	11	11	Small

Table 2. Comparison of Generative Model (NB) and Discriminative Model (LR) on all datasets.

Dataset	Model	Robust Accuracy	
	(no defence)	LowProFool	DeepFool
Census-income	LR	0.0422	0.0422
	NB	**0.739**	**0.739**
SkinSegmentation	LR	0.0562	0.0562
	NB	**0.869**	**0.869**
Higgs	LR	0.3235	0.3237
	NB	**0.5937**	**0.5937**
Connect-4	LR	0.2477	0.2477
	NB	**0.6845**	**0.6845**
Adult	LR	0.1297	0.13
	NB	**0.7925**	**0.7925**
Magic	LR	0.1547	0.1545
	NB	**0.6652**	**0.6495**
Page-blocks	LR	0.0192	0.0197
	NB	**0.828**	**0.746**
Abalone	LR	0.1902	0.8070
	NB	**0.7044**	**0.7066**
Vowel	LR	0.4444	0.5388
	NB	**0.7333**	**0.7333**
Vowel-context	LR	0.5611	0.4666
	NB	**0.7388**	**0.7388**

with a performance far superior to their discriminative counterpart, LR. This preliminary result is encouraging for our proposed defence strategy gD2A3, as it relies on the use of a generative model in its defence.

4.3 Analysis on Deep Model

In this work, we will use the state-of-the-art method D2A3 as the baseline. The deep model used in this work consists of 3 hidden layers with each layer consisting

Table 3. Performance Comparison for gD2A3 in Deep Model.

Models	λ	Standard Accuracy (Avg)	Robust Accuracy (Avg)	
			DeepFool	LowProFool
gD2A3	10^{-3}	0.8809	0.8673	0.8407
gD2A3	10^{+3}	**0.8841**	**0.8771**	**0.8482**
D2A3	–	0.8330	0.8210	0.8080

Table 4. Performance Comparison for gD2A3 in Shallow Model.

Models	λ	Standard Accuracy (Avg)	Robust Accuracy (Avg)	
			DeepFool	LowProFool
gD2A3	10^{-3}	**0.8180**	**0.8056**	**0.7950**
gD2A3	10^{+3}	0.8120	0.8053	0.7890
D2A3	–	0.7836	0.7494	0.7456

of 10 nodes. In particular, for both gD2A3 and D2A3, we use the same model structure, except that gD2A3 has an additional Pre-conditioning Layer. The magnitude of the defence is controlled by the regularization parameter λ, which is varied in this part of the experiment to investigate the trade-off between generative and discriminative models.

Table 3 shows this comparison of gD2A3 with D2A3, providing an average of results on 10 datasets. It is encouraging to see that our gD2A3 performs better than the current state-of-the-art D2A3 for both the Deepfool and LowProFool methods with the two regularization settings. Notably, gD2A3 performs better with higher λ, which suggests that it avails the output of the generative model as an effective defence against adversarial attacks. Comparing the gap between standard accuracy and robust accuracy, it can be seen that higher λ corresponds to a smaller gap between standard and robust accuracy. In summary, the results on gD2A3 with the deep model are extremely encouraging and suggest that the use of generative models can be an extremely effective defence against adversarial attacks on tabular data.

4.4 Analysis on Shallow Model

For sake of completeness. Let us see if our proposed formulation works for shallow models as well. For this experiment, we used an LR model with both gD2A3 and D2A3. Table 4 depicts the comparison of gD2A3 with D2A3 by reporting averaged results on 10 datasets. Like deep models, gD2A3 outperforms D2A3 on shallow models as well. However, one difference is that, gD2A3 now performs better with a smaller value of λ. This is an interesting result. We conjecture that shallow models could be inherently more robust to adversarial attacks than deep models, and so require far less reliance on generative parameters to ward-off attacks. A detailed analysis of shallow and deep learning models is warranted. However, it is outside the scope of this work and is left as future work. In summary, we can conclude from the results that our approach is as effective in providing defence against adversarial attacks on shallow models as it is on deep models.

4.5 Ablation Study on Data Size and Feature Order

We further conduct an ablation study on data size and feature-order (k in KDB) for gD2A3.

Table 5. Ablation Study on Data size for gD2A3.

Models	λ	Deep						Shallow					
		Large		Medium		Small		Large		Medium		Small	
		DF	LPF	DF	LPF	DF	LPF	DF	LPF	DF	LPF	DF	LPF
gD2A3	10^{-3}	0.8584	0.8494	0.9026	0.8936	0.8438	0.7764	0.8371	0.82	0.8971	0.8943	**0.6722**	**0.6621**
gD2A3	10^{+3}	**0.8599**	0.8539	**0.9039**	**0.8973**	**0.8733**	**0.7914**	**0.8376**	0.8168	**0.8978**	**0.8974**	0.6697	0.6433
D2A3	–	0.8571	0.8555	0.8932	0.8431	0.7006	0.7095	0.8358	0.8358	0.7885	0.7740	0.5616	0.5634

Table 6. Ablation Study on Feature-order, i.e., k in gD2A3.

Models	λ	Deep				Shallow			
		$k = 0$		$k = 1$		$k = 0$		$k = 1$	
		DF	LPF	DF	LPF	DF	LPF	DF	LPF
gD2A3	10^{-3}	0.8673	0.8407	**0.8720**	0.7537	**0.8056**	**0.7950**	0.8464	**0.8390**
gD2A3	10^{+3}	**0.8771**	**0.8482**	0.8718	**0.7711**	0.8053	0.7890	0.8444	0.8323
D2A3	–	0.8210	0.8080	0.8561	0.8224	0.7494	0.7456	0.7991	0.7722

Table 5 reports results on the deep and shallow models for three collections of datasets – Large, Medium and Small. It can be seen that gD2A3 works better with higher regularization λ value for most of the cases, except for small datasets with shallow models (as discussed in the last section). The results here are encouraging, as we can see that higher regularization can bring a stronger defence to the models.

Table 6 shows the comparison of the results for two values of k (KDB) in gD2A3. In general, for both $k = 0$ and $k = 1$, gD2A3 wins with higher regularization for the deep model and lower regularization for the shallow model. For the deep model with $k = 1$, the dimension is large for most of the data, and therefore affords more opportunity for adversarial attacks for LowProFool – which targets the features specifically.

5   Conclusion and Future Work

In this work, we studied if generative models can be used as a defence mechanism against adversarial attacks on discriminative models such as deep ANNs. Our proposed formulation, gD2A3, leverages Bayesian Networks as a generative model and proposes a Pre-conditioning Layer. This pre-conditioning layer incorporates the weights learned in the pre-trained generative model. The first layer of the ANN after the pre-conditioning layer relies on regularization to control the magnitude of the defence by pulling the input towards that of the generative model's output. It results in blending generative models in discriminative training. We show empirically on 10 datasets that gD2A3 leads to better results on the majority of datasets than the existing state-of-the-art method D2A3. For future work, we will focus on the following directions:

- Currently, the generative models (probabilities) are static – that is, with new adversarial examples, they are not updated. As we mentioned earlier, the

real benefit of generative models will stem from the fact that they can adapt to a newly forged adversarial distribution, and therefore, a moving average estimate of probabilities is extremely desirable.
– We are interested in studying the robustness of gD2A3 to adversarial attacks based on the network depth (shallow vs. deep). Currently, it is not clear if the attack's success rate depends on the depth of the model.

Acknowledgement. This research is funded through a **Defence Science and Technology Group** DAIRNet grant.

References

1. Ballet, V., Renard, X., Aigrain, J., Laugel, T., Frossard, P., Detyniecki, M.: Imperceptible adversarial attacks on tabular data. arXiv preprint arXiv:1911.03274 (2019)
2. Kurakin, A., Goodfellow, I., Bengio, S.: Adversarial machine learning at scale. arXiv preprint arXiv:1611.01236 (2016)
3. Li, Y.: Are generative classifiers more robust to adversarial attacks? CoRR abs/1802.06552 (2018). https://arxiv.org/abs/1802.06552
4. Liu, X., Hsieh, C.: From adversarial training to generative adversarial networks. CoRR abs/1807.10454 (2018), https://arxiv.org/abs/1807.10454
5. Madry, A., Makelov, A., Schmidt, L., Tsipras, D., Vladu, A.: Towards deep learning models resistant to adversarial attacks. arXiv preprint arXiv:1706.06083 (2017)
6. Moosavi-Dezfooli, S.M., Fawzi, A., Frossard, P.: Deepfool: a simple and accurate method to fool deep neural networks. In: CVPR (2016)
7. Ng, A., Jordan, M.: On discriminative vs. generative classifiers: a comparison of logistic regression and Naive Bayes. In: Dietterich, T., Becker, S., Ghahramani, Z. (eds.) Advances in Neural Information Processing Systems, vol. 14. MIT Press (2001). https://proceedings.neurips.cc/paper/2001/file/7b7a53e239400a13bd6be6c91c4f6c4e-Paper.pdf
8. Qiu, S., Liu, Q., Zhou, S., Wu, C.: Review of artificial intelligence adversarial attack and defense technologies. Appl. Sci. **9**(5), 909 (2019)
9. Schott, L., Rauber, J., Brendel, W., Bethge, M.: Robust perception through analysis by synthesis. CoRR abs/1805.09190 (2018), https://arxiv.org/abs/1805.09190
10. Zaidi, N.A., Carman, M.J., Cerquides, J., Webb, G.I.: Naive-bayes inspired effective pre-conditioner for speeding-up logistic regression. In: 2014 IEEE International Conference on Data Mining. pp. 1097–1102 (2014). https://doi.org/10.1109/ICDM.2014.53
11. Zhang, H., Yu, Y., Jiao, J., Xing, E., El Ghaoui, L., Jordan, M.: Theoretically principled trade-off between robustness and accuracy. In: ICML (2019)
12. Zhou, J., Zaidi, N., Zhang, Y., Li, G.: Discretization inspired defence algorithm against adversarial attacks on tabular data. In: Gama, J., Li, T., Yu, Y., Chen, E., Zheng, Y., Teng, F. (eds.) PAKDD 2022. LNCS, vol. 13281, pp. 367–379. Springer, Cham (2022). https://doi.org/10.1007/978-3-031-05936-0_29
13. Zhou, M., Wu, J., Liu, Y., Liu, S., Zhu, C.: Dast: data-free substitute training for adversarial attacks. In: Proceedings of the IEEE/CVF Conference on Computer Vision and Pattern Recognition, pp. 234–243 (2020)

Weak Correlation-Based Discriminative Dictionary Learning for Image Classification

Huang-Kai Zhang and Li Zhang$^{(\boxtimes)}$ (ID)

School of Computer Science and Technology, Soochow University,
Suzhou 215006, China
20225227027@stu.suda.edu.cn, zhangliml@suda.edu.cn

Abstract. Currently, representation-based learning is widely used in image classification because of its good mathematical interpretability. However, when the number of training samples available is small, the performance of the general representation-based model is bad. For this reason, we propose a new weak correlation-based discriminative dictionary learning (WCDDL) method, which learns a weakly correlated class-specific structured dictionary by narrowing the correlation between each sub-dictionary representation. WCDDL can reduce the impact of sparse training samples in decreasing the classification accuracy to a certain extent. Experimental results show that our proposed method can achieve better classification performance compared to existing representation-based algorithms even when training samples are sparse.

Keywords: Dictionary learning · Image classification · Weak correlation

1 Introduction

Image classification has long been a core task in the field of computer vision, with the aim of distinguishing different classes of images based on their features. In recent years, image classification algorithms have received a lot of attention because of their usefulness in many application areas (e.g., face recognition [21], handwritten digit recognition [15], hyperspectral remote sensing imagery classification [10]). Presently, a large number of image classification algorithms have been proposed. The mainstream classification algorithms can be broadly classified into two categories: non-parametric and parametric methods. The former is mainly representation-based, say the nearest neighbor classifier, and the latter consists of a lot of methods, including support vector machine (SVM) and neural network (NN). This study focuses on representation-based methods for their mathematical interpretability.

Supported by the Natural Science Foundation of the Jiangsu Higher Education Institutions of China under Grant No. 19KJA550002, by the Six Talent Peak Project of Jiangsu Province of China under Grant No. XYDXX-054, and by the Priority Academic Program Development of Jiangsu Higher Education Institutions.

As the earliest representation-based method, the nearest neighbor classifier considers the simply representation for a given point using its nearest neighbors and ignores the degree of representation. Therefore, researchers have developed more representation-based methods, such as sparse representation-based classifier (SRC) [8] and collaborative representation-based classifier (CRC) [19]. Generally, representation-based methods linearly represent a test sample using the given training samples and classify it by measuring the residuals between it and its class representations. During the representation process, SRC uses each class of training samples to linearly represent test samples, whereas CRC adopts all training samples to represent test samples. When the number of training samples in each class is small, the SRC representation can lead to a degradation of classification performance. However, CRC can avoid this situation well, which makes researchers prefer CRC. Therefore, some CRC extensions have been proposed, such as collaborative-competitive representation-based classifier (CCRC) [18], double competitive constraints-based collaborative representation for classification (DCCRC) [6], weighted discriminative collaborative competitive representation (WDCCR) [5], and Probabilistic CRC (ProCRC) [4]. Although the classification performance of CRC-based methods seems to be good, their computational time increases as the amount of data increases. To maintain the classification performance and reduce the computational effort, it is considered to learn the information features of samples through dictionaries and apply them in the classification method.

Back in 2006, Aharon et al. [1] proposed a dictionary learning method, the K-SVD algorithm, which is a generalization of K-means. Although K-SVD trains a dictionary with superior performance and performs well in image recovery and image compression, it is not suitable for image classification tasks because it does not utilize the label information of data. In view of the shortcoming, Zhang et al. [20] proposed a discriminative K-SVD (D-KSVD) algorithm based on K-SVD. D-KSVD adds a new label term to the original objective function by introducing label information, which makes it possible to maintain the performance of the dictionary while making it applicable to image classification tasks. Jiang et al. [9] proposed label consistent KSVD (LC-KSVD) by adding a new label consistency constraint (discriminating sparse coding errors) to the objective function of D-KSVD, associating each dictionary atom in the dictionary with its corresponding specific label and forcing samples in the same class to have similar sparse representations.

The dictionaries learned by above methods are all shared ones, which can adequately capture the main features needed for facial images when the intra-class variation of facial images is small. However, the representation capability of the shared dictionary would be decreased when images in the same class have a great intra-class distance owing to various factors during the shooting process. To remedy it, class-specific dictionary learning algorithms have been proposed, including Fisher discrimination-based dictionary learning (FDDL) [17], discriminative dictionary learning via Fisher discrimination K-SVD [22], and probabilistic collaborative dictionary learning [12]. Class-specific dictionary learning can learn dictionaries with better representation performance for sufficient data with large intra-class distances; thus, the uncertainty of dictionary atoms will increase if a few training samples are used to learn the complete information of each class, which eventually leads to the degradation of classification performance.

In this paper, we propose a new weak correlation-based discriminative dictionary learning (WCDDL) method. WCDDL first learns a structured dictionary by all training samples to ensure that the structured dictionary has a good reconstruction performance for test samples and then effectively associates each sub-dictionary with its corresponding class using the properties of class-specific structured dictionaries. In this way, each class-specific sub-dictionary has a good reconstruction ability for the training samples of that class. To address the issue of weak dictionary discriminative ability brought by sparse training samples in each class, WCDDL incorporates a term, called the weak correlation term, which is used to weaken the correlation between sub-dictionaries. Different from the methods in [11,21] increasing the number of training samples to preserve the classification performance, WCDDL increases the inter-class distance by applying the weak correlation term so that a small number of samples can distinguish classes.

2   Proposed Method

This section presents WCDDL. Before explaining our proposed algorithm, we describe its learning framework. The goal of WCDDL is to learn a structured dictionary based on a given data set and apply the well-learned dictionary to classify unseen data points.

Let the given training sample set be $H = \{(\mathbf{y}_1, \ell_1), \ldots, (\mathbf{y}_n, \ell_n)\}$, where $\mathbf{y}_i \in \mathcal{R}^m$ is the ith training sample, $\ell_i \in \{1, \ldots, C\}$ is the label of $\mathbf{y}_i$, and n and C are the numbers of samples and classes. For the cth class, we denote its sample matrix as $\mathbf{Y}_c = [\mathbf{y}_{c_1}, \ldots, \mathbf{y}_{c_{n_c}}] \in \mathcal{R}^{m \times n_c}$, where $c_i \in \{1, \ldots, n\}$, and n_c is the number of training samples in the cth class. Note that $n = \sum_{c=1}^{C} n_c$. Then, the entire training sample matrix is $\mathbf{Y} = [\mathbf{Y}_1, \ldots, \mathbf{Y}_C] \in \mathcal{R}^{m \times n}$. Let $\mathbf{D} = [\mathbf{D}_1, \ldots, \mathbf{D}_C] \in \mathcal{R}^{m \times (r \times C)}$ be the structured training dictionary, where $\mathbf{D}_c \in \mathcal{R}^{m \times r}$ is the sub-dictionary corresponding to the cth class, and r is the number of atoms in the sub-dictionaries.

During the classification procedure, a test sample $\mathbf{y} \in \mathcal{R}^m$ can be represented by a linear combination of atoms of the structured dictionary $\mathbf{D}$. That is $\mathbf{y} \approx \mathbf{D}\mathbf{x} = \sum_{c=1}^{C} \mathbf{D}_c \mathbf{x}^c$, where $\mathbf{x} = [\mathbf{x}^1, \ldots, \mathbf{x}^C]^T \in \mathcal{R}^{(r \times C) \times 1}$ is the coefficient vector of the dictionary $\mathbf{D}$ for sample $\mathbf{y}$, and $\mathbf{x}^c = [x_1^c, \ldots, x_r^c] \in \mathcal{R}^r$ is the coefficients vector of the cth sub-dictionary for the sample $\mathbf{y}$.

2.1 Dictionary Learning Algorithm

In our WCDDL, the structured dictionary can be obtained by solving the following optimization problem:

$$f(\mathbf{D}, \mathbf{X}) = \min_{(\mathbf{D}, \mathbf{X})} r(\mathbf{Y}, \mathbf{D}, \mathbf{X}) + \lambda g(\mathbf{D}, \mathbf{X}) \tag{1}$$

where $\mathbf{X} = [\mathbf{x}_1, \ldots, \mathbf{x}_n] \in \mathcal{R}^{(r \times C) \times n}$ is the coefficient matrix with respect to the training sample matrix $\mathbf{Y}$, and $\mathbf{x}_i$ is the coefficient vector of sample $\mathbf{y}_i$, $r(\mathbf{Y}, \mathbf{D}, \mathbf{X})$ is the reconstruction term, $g(\mathbf{D}, \mathbf{X})$ is the weak correlation term, and $\lambda > 0$ is the regularization term.

The purpose of reconstruction term $r(\mathbf{Y}, \mathbf{D}, \mathbf{X})$ is to learn a structured dictionary that extracts feature information from samples. In [17], $r(\mathbf{Y}, \mathbf{D}, \mathbf{X})$ is defined as

$$r(\mathbf{Y}, \mathbf{D}, \mathbf{X}) = \sum_{c=1}^{C} r(\mathbf{Y}_c, \mathbf{D}, \mathbf{X}_c) \tag{2}$$

where $r(\mathbf{Y}_c, \mathbf{D}, \mathbf{X}_c)$ is used to learn the sub-dictionary of the cth class, and $\mathbf{X}_c$ is the coefficient matrix with respect to $\mathbf{Y}_c$. Further, this reconstruction term can be decomposed as

$$r(\mathbf{Y}, \mathbf{D}, \mathbf{X}) = \sum_{c=1}^{C} \left\{ \|\mathbf{Y}_c - \mathbf{D}\mathbf{X}_c\|_2^2 + \|\mathbf{Y}_c - \mathbf{D}_c\mathbf{X}_c^c\|_2^2 + \sum_{j=1, j \neq c}^{C} \|\mathbf{D}_j\mathbf{X}_c^j\|_2^2 \right\} \tag{3}$$

where $\mathbf{X}_c^c$ is the coefficient matrix with respect to $\mathbf{Y}_c$ for the cth class. In (3), the first term $\|\mathbf{Y}_c - \mathbf{D}\mathbf{X}_c\|_2^2$ allows us to learn a dictionary that represents each sample well approximately; the second term $\|\mathbf{Y}_c - \mathbf{D}_c\mathbf{X}_c^c\|_2^2$ uses the sub-dictionary $\mathbf{D}_c$ to represent samples in the cth class as much as possible provided that the trained dictionary can represent the test samples approximately; the third term $\sum_{j=1, j \neq c}^{C} \|\mathbf{D}_j\mathbf{X}_c^j\|_2^2$ improves the discriminative property of the second term by reducing the representation ability of classes except the cth class.

To compensate for the decrease of dictionary representation ability induced by small training samples, we introduce the weak correlation term [5], which has the following form:

$$g(\mathbf{D}, \mathbf{X}) = \sum_{c=1}^{C} g(\mathbf{D}_c, \mathbf{X}_c)$$

$$= \sum_{c=1}^{C} \sum_{j=1, j \neq c}^{C} \|\mathbf{D}_c\mathbf{X}_c^c + \mathbf{D}_j\mathbf{X}_c^j\|_2^2 \tag{4}$$

$$= \sum_{c=1}^{C} \sum_{j=1, j \neq c}^{C} \|\mathbf{D}_c\mathbf{X}_c^c\|_2^2 + \|\mathbf{D}_j\mathbf{X}_c^j\|_2^2 + 2(\mathbf{D}_c\mathbf{X}_c^c)^T(\mathbf{D}_j\mathbf{X}_c^j)$$

By (4), we can see that minimizing $(\mathbf{D}_c\mathbf{X}_c^c)^T(\mathbf{D}_j\mathbf{X}_c^c)$ is equivalent to minimizing $g(\mathbf{D}, \mathbf{X})$, which can be regarded as the correlation between dictionary representations of class c and class j. In other words, the minimization of $g(\mathbf{D}, \mathbf{X})$ is to weaken the correlation between dictionary representations. When this term keeps shrinking, the correlation between dictionary representations of class c and class j would also keep decreasing. In this way, we can make a distinction between classes only by extracting a small amount of category information when the training samples are sparse.

2.2 Solution

Now, we consider the solution to (1), using an alternating iterative method that is to alternatively update the structured dictionary $\mathbf{D}$ and the sparse coefficient matrix $\mathbf{X}$.

First, we fix the structured dictionary $\mathbf{D}$ and update the coefficient matrix $\mathbf{X}_c$ class by class. The specific objective function used to update $\mathbf{X}_c$ is as follows:

$$
\begin{aligned}
f(\mathbf{X}_c) = \min_{\mathbf{X}_c}\ & \|\mathbf{Y}_c - \mathbf{D}\mathbf{X}_c\|_2^2 + \|\mathbf{Y}_c - \mathbf{D}_c\mathbf{X}_c^c\|_2^2 \\
& + \sum_{j=1, j\neq c}^{C} \|\mathbf{D}_j\mathbf{X}_c^j\|_2^2 + \lambda \sum_{j=1, j\neq c}^{C} \|\mathbf{D}_c\mathbf{X}_c^c + \mathbf{D}_j\mathbf{X}_c^j\|_2^2
\end{aligned}
\tag{5}
$$

In order to solve (5), we use a soft threshold function [14] for updating $\mathbf{X}_c$, which is widely used in sparse signal reconstruction tasks. The soft threshold function has the form as follows:

$$
S(\beta) = sign(\beta) \odot (|\beta| - \lambda_1 w)_+
\tag{6}
$$

where the variable β could be a scalar, vector, or matrix, $\odot$ denotes the element multiplication of two vectors or matrices, $\lambda_1 > 0$ is the weight factor, $w \in \mathcal{R}$ is the threshold parameter that controls the magnitude of each change in β, $sign(\cdot)$ is the sign function, and $(\cdot)_+ = \max(\cdot, 0)$. In addition, we need to find the partial derivative of $f(\mathbf{X}_c)$ with respect to $\mathbf{X}_c$. That is

$$
\begin{aligned}
\frac{\partial f(\mathbf{X}_c)}{\partial \mathbf{X}_c} = & - 2\mathbf{D}^T(\mathbf{Y}_c - \mathbf{D}\mathbf{X}_c) - 2\mathbf{D}_c^T(\mathbf{Y}_c - \mathbf{D}_c\mathbf{X}_c^c) + 2 \sum_{j=1, j\neq c}^{C} \mathbf{D}_j^T\mathbf{D}_j\mathbf{X}_c^j \\
& + 2 \sum_{j=1, j\neq c}^{C} (\mathbf{D}_c^{cT} + \mathbf{D}_c^{jT})(\mathbf{D}_c\mathbf{X}_c^c + \mathbf{D}_j\mathbf{X}_c^j)
\end{aligned}
\tag{7}
$$

Let $\beta = \mathbf{X}_c - \frac{\lambda}{2}\frac{\partial f(\mathbf{X}_c)}{\partial \mathbf{X}_c}$. We use the soft threshold function (6) to iteratively update the coefficient matrix for each class with the following equation:

$$
\mathbf{X}_c^t = S(\mathbf{X}_c^{t-1} - \frac{\lambda}{2}\frac{\partial f(\mathbf{X}_c^{t-1})}{\partial \mathbf{X}_c^{t-1}}), \quad c = 1, \dots, C
\tag{8}
$$

where t is the current iteration number, and $\mathbf{X}_c^{t-1}$ is the coefficient matrix of the cth class generated in the $(t-1)$th iteration. We keep iterations until $\mathbf{X}_c$ converges or t reaches the predetermined maximum number of iterations.

After updating the coefficient matrix $\mathbf{X}$, we fix it unchanged and update the structured dictionary $\mathbf{D}$, for which we also use a class-by-class update scheme. The optimization problem with respect to only $\mathbf{D}$ is

$$f(\mathbf{D}_c) = \min_{\mathbf{D}_c} \ \|\mathbf{Y}_c - \mathbf{D}\mathbf{X}_c\|_2^2 + \|\mathbf{Y}_c - \mathbf{D}_c\mathbf{X}_c^c\|_2^2 + \sum_{j=1,j\neq c}^{C} \|\mathbf{D}_j\mathbf{X}_c^j\|_2^2$$

$$+ \lambda \sum_{j=1,j\neq c}^{C} \|\mathbf{D}_c\mathbf{X}_c^c + \mathbf{D}_j\mathbf{X}_c^j\|_2^2 \tag{9}$$

Let $\mathbf{Y}' = \mathbf{Y}_c - \sum_{j=1,j\neq c}^{C} \mathbf{D}_j\mathbf{X}_c^j$, so (9) can be rewritten as

$$f(\mathbf{D}_c) = \min_{\mathbf{D}_c} \ \|\mathbf{Y}' - \mathbf{D}_c\mathbf{X}_c^c\|_2^2 + \|\mathbf{Y}_c - \mathbf{D}_c\mathbf{X}_c^c\|_2^2 + \sum_{j=1,j\neq c}^{C} \|\mathbf{D}_j\mathbf{X}_c^j\|_2^2$$

$$+ \lambda \sum_{j=1,j\neq c}^{C} \|\mathbf{D}_c\mathbf{X}_c^c + \mathbf{D}_j\mathbf{X}_c^j\|_2^2 \tag{10}$$

Similarly, the partial derivative of $f_{\mathbf{D}_c}$ with respect to $\mathbf{D}_c$ must be calculated for optimizing (10). We make the partial derivative of $f_{\mathbf{D}_c}$ with respect to $\mathbf{D}_c$ equal to zero and then obtain

$$\frac{\partial f(\mathbf{D}_c)}{\partial \mathbf{D}_c} = - 2\mathbf{X}_c^{cT}(\mathbf{Y}' - \mathbf{D}_c\mathbf{X}_c^c) - 2\mathbf{X}_c^{cT}(\mathbf{Y}_c - \mathbf{D}_c\mathbf{X}_c^c)$$

$$+ 2\lambda \sum_{j=1,j\neq c}^{C} \mathbf{X}_c^{cT}(\mathbf{D}_c\mathbf{X}_c^c + \mathbf{D}_j\mathbf{X}_c^j) = 0 \tag{11}$$

Since each column in the coefficient matrix $\mathbf{X}_c^{cT}$ is linearly independent, it is a column-full rank matrix. By arranging (11), we have

$$\mathbf{Y}' - \mathbf{D}_c\mathbf{X}_c^c + \mathbf{Y}_c - \mathbf{D}_c\mathbf{X}_c^c - \lambda \sum_{j=1,j\neq c}^{C} (\mathbf{D}_c\mathbf{X}_c^c + \mathbf{D}_j\mathbf{X}_c^j) = 0 \tag{12}$$

Rearranging (12), we can further obtain

$$\frac{2\mathbf{Y}_c - (\lambda+1)\sum_{j=1,j\neq c}^{C} \mathbf{D}_j\mathbf{X}_c^j}{2+\lambda C} = \mathbf{D}_c\mathbf{X}_c^c \tag{13}$$

Let $\mathbf{Y}'' = \frac{2\mathbf{Y}_c - (\lambda+1)\sum_{j=1,j\neq c}^{C} \mathbf{D}_j\mathbf{X}_c^j}{2+\lambda C}$. Then (13) can be rewritten in the following form:

$$\mathbf{Y}'' = \mathbf{D}_c\mathbf{X}_c^c \tag{14}$$

We use the K-SVD method to get $\mathbf{D}_c$ in (14) and then obtain the structured dictionary $\mathbf{D}$.

We keep repeating iterations to calculate the above formula until the coefficient matrix and dictionary converge or satisfy our maximum number of iterations.

2.3   Classification Algorithm

After we obtain the well-trained structured dictionary $\mathbf{D}$, we consider the label prediction for a given test sample. Let $\mathbf{y}_{test} \in \mathcal{R}^m$ be an arbitrary test sample. To estimate its class label, we first compute the coefficient vector $\mathbf{x}_{test}$ according to the structured dictionary $\mathbf{D}$. The objective function for solving the coefficient vector is as follows:

$$\mathbf{x}_{test} = \arg\min_{\mathbf{x}} \left\{ \|\mathbf{y}_{test} - \mathbf{D}\mathbf{x}\|_2^2 + \lambda_1 \|\mathbf{x}\|_1 \right\} \tag{15}$$

To solve (15), we use the orthogonal matching tracking algorithm (OMP) [13] to find the optimal solution vector $\mathbf{x}_{test}$.

To predict the label of $\mathbf{y}_{test}$, we make the dictionary representation $\mathbf{D}_c \mathbf{x}_{test}^c$ of $\mathbf{y}_{test}$ as the residual operation to obtain the representation residual e_c of the cth class. The residual of the cth class is calculated as follows:

$$e_c = \|\mathbf{y}_{test} - \mathbf{D}_c \mathbf{x}_{test}^c\|_2^2 + \lambda_2 \|\mathbf{x}_{test} - \mathbf{m}_c\|_2^2 \tag{16}$$

where $0 < \lambda_2 < 1$ is the weight factor, $\mathbf{m}_c$ is the mean vector of coefficient vectors of the cth class. In (16), the first term $\|\mathbf{y}_{test} - \mathbf{D}_c \mathbf{x}_{test}^c\|_2^2$ is mainly used to calculate the residuals, and the second term $\|\mathbf{x}_{test} - \mathbf{m}_c\|_2^2$ is used to estimate how similar the coefficient vector is to the representation coefficient vector of cth class. If $\|\mathbf{x}_{test} - \mathbf{m}_c\|_2^2$ is small, then the representation coefficient vector of $\mathbf{y}_{test}$ is close to the representation coefficient vector of the cth class sample. Finally, the label of the test sample $\mathbf{y}_{test}$ is predicted by

$$\ell_{test} = \arg\min_{c=1,\dots,C} e_c \tag{17}$$

3   Experiments

The goal of this section is to validate the feasibility and efficiency of the proposed algorithm. First, we briefly introduce the facial datasets used for experiments and then, compare the classification accuracy of different representation-based models on different datasets, including K-SVD [1], D-KSVD [20], FDDL [17], LC-KSVD [9], and Label embedded dictionary learning (LEDL) [16].

3.1   Datasets

In our experiments, two public facial datasets are introduced here, Yale [3,7] and ORL [2], which are widely used to validate representation-based algorithms. The description of these datasets is given as follows:

- **Yale**: This database is a facial dataset that contains 165 face images from 15 people, each with 11 face images. These human images have different facial expressions: center light, with glasses, happy, left light, without glasses, normal, right light, sad, sleepy, surprised, and blinking. Fig 1(a) shows some specific examples of the Yale dataset.

- **ORL**: The ORL database is also a facial dataset that contains 400 face images from 40 people, each with 10 face images. These images were acquired at different times of day, under different lighting, with different facial expressions (eyes open/closed, smiling/not smiling) and facial details (with/without glasses). Figure 1(b) provides some specific examples from the ORL dataset.

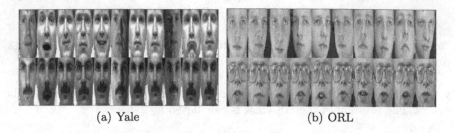

(a) Yale   (b) ORL

Fig. 1. Some image samples from (a) Yale and (b) ORL databases.

3.2   Performance Comparison

For comparison, we randomly divide both Yale and ORL datasets into a training set and a test set. Compared models are first trained on the training set and then tested on the test set to provide the classification performance. Each dataset is divided in the following way: p images from each class are taken as the training samples and the remaining images in this class are regarded as the test samples, where p takes values in the set $\{2, 3, 4, 5, 6, 7, 8\}$ and the divided dataset is called pTrain. The randomness of division may have a certain influence on experimental results. To eliminate the randomness, we perform 50 random division for each pTrain and report the average results over 50 trails.

Classification Performance. Table 1 presents experimental results on Yale, where the highest values among compared methods are highlighted in bold. First, We can clearly observe that the performance of all compared algorithms gradually decreases when the number of training samples decreases, which is inevitable. Second, we can see that WCDDL has a good performance on seven divided datasets compared to other algorithms. When the number of training samples is small, WCDDL performs much better than other methods. For example, the proposed WCDDL has a classification accuracy of 72.6% on 2Train in which each class consists of only two samples for training. In this case, our method improves the accuracy by 10.4% compared to the second-best algorithm FDDL. For the case of 8Train, WCDDL is 6.9% higher in accuracy than the second-best LEDL. Relatively speaking, the improvement of WCDDL on smaller datasets is more obvious. Note that WCDDL is designed on the basis of FDDL.

In detail, our method replaces the discriminative coefficient term in FDDL with the weak correlation term. Findings indicate that WCDDL is much better than FDDL, which means that the weak correlation term works well.

Table 2 shows experimental results on ORL, where the highest values among the six methods are in bold. Similar conclusions can be obtained by experimental results in Table 2. More training samples induce better classification performance, and the classification accuracy of WCDDL is also higher than other representation-based algorithms under different training set divisions. On the ORL dataset, WCDDL ranks first in seven divisions, followed by LEDL. With the increase of training samples, the gap between WCDDL and LEDL in accuracy is closing all the time. In other words, the advantage of WCDDL is that it deals with sparse training samples better.

Table 1. Mean accuracy obtained by compared methods on Yale with various divided datasets.

Divided dataset	K-SVD	D-KSVD	LC-KSVD	FDDL	LEDL	WCDDL
2Train	58.1 ± 3.9	48.0 ± 6.2	41.9 ± 6.2	62.2 ± 4.7	24.0 ± 5.0	$\mathbf{72.6 \pm 3.8}$
3Train	62.9 ± 4.1	52.5 ± 5.4	42.9 ± 4.9	66.4 ± 3.6	54.0 ± 5.1	$\mathbf{78.6 \pm 2.8}$
4Train	66.4 ± 3.9	56.1 ± 5.6	66.8 ± 3.7	69.7 ± 3.4	63.7 ± 4.3	$\mathbf{86.3 \pm 2.6}$
5Train	69.1 ± 4.6	57.8 ± 6.2	70.4 ± 4.7	73.9 ± 3.5	64.9 ± 4.7	$\mathbf{89.8 \pm 2.9}$
6Train	80.7 ± 4.4	57.4 ± 6.4	77.1 ± 3.7	74.0 ± 3.9	78.9 ± 4.7	$\mathbf{91.2 \pm 3.0}$
7Train	82.6 ± 4.8	57.5 ± 7.5	82.5 ± 4.7	75.7 ± 4.9	82.6 ± 5.1	$\mathbf{91.7 \pm 2.4}$
8Train	82.3 ± 4.9	60.0 ± 7.2	82.8 ± 5.0	77.7 ± 5.1	84.8 ± 5.5	$\mathbf{91.7 \pm 3.5}$

Table 2. Mean accuracy obtained by compared methods on ORL with various divided datasets.

Divided dataset	K-SVD	D-KSVD	LC-KSVD	FDDL	LEDL	WCDDL
2Train	72.8 ± 3.1	48.4 ± 3.7	63.1 ± 2.5	71.0 ± 2.8	74.8 ± 3.5	$\mathbf{79.3 \pm 3.1}$
3Train	79.3 ± 2.4	56.1 ± 4.5	79.5 ± 2.7	75.7 ± 2.2	82.6 ± 2.0	$\mathbf{86.1 \pm 2.0}$
4Train	88.1 ± 2.2	55.6 ± 4.4	87.8 ± 1.9	78.2 ± 2.2	88.8 ± 1.8	$\mathbf{91.9 \pm 1.6}$
5Train	91.1 ± 2.0	52.6 ± 5.1	90.8 ± 2.0	79.8 ± 2.8	91.8 ± 1.8	$\mathbf{94.2 \pm 1.7}$
6Train	92.6 ± 2.2	56.7 ± 4.4	92.4 ± 2.2	80.5 ± 3.1	92.8 ± 2.2	$\mathbf{95.3 \pm 1.8}$
7Train	93.7 ± 1.8	57.7 ± 6.5	94.0 ± 1.8	81.5 ± 3.2	94.4 ± 1.7	$\mathbf{96.5 \pm 1.5}$
8Train	94.2 ± 2.3	72.7 ± 6.6	94.0 ± 2.5	82.9 ± 3.3	94.8 ± 2.3	$\mathbf{96.5 \pm 1.7}$

Representation Visualization. Owing to the relationship between WCDDL and FDDL, we further compare them by observing their structured dictionary and reconstructed images. After training a model, we can obtain the structured dictionary $\mathbf{D}$ and coefficient matrix $\mathbf{X}$ and then reconstruct unseen images with $\mathbf{D}$ and X.

Figures 2(a) and 2(b) are the dictionary and reconstructed samples generated by FDDL, respectively. By observing Fig. 2, we can see that the structured dictionary trained by FDDL extracts the overall facial features; thus, so FDDL achieves good visualization results in the subsequent reconstruction of images.

Figure 3(a) shows the first two classes of atoms of the structured dictionary obtained by WCDDL, and Fig. 3(b) plots the first two classes of samples reconstructed by WCDDL. By comparing the dictionary and reconstructed images obtained by FDDL and WCDDL, we find that the visualization result of WCDDL is not good as that of FDDL. The proposed WCDDL aims to learn the distinguished features of different classes and reduce the correlation between dictionary representations of different classes. Thus, the samples we reconstruct are very close in the same class and differ more between classes.

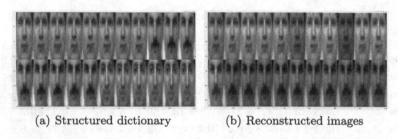

    (a) Structured dictionary   (b) Reconstructed images

Fig. 2. Visualization of FDDL on Yale, (a) structured dictionary **D**, and (b) reconstructed images.

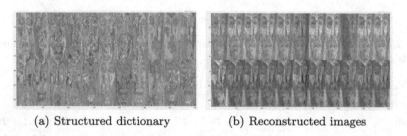

    (a) Structured dictionary   (b) Reconstructed images

Fig. 3. Visualization of WCDDL on Yale, (a) structured dictionary **D**, and (b) reconstructed images.

4   Conclusion

In this study, image classification is implemented by the proposed WCDDL. To solve the issue of poor dictionary classification capability induced by sparse training samples, we introduce a weak correlation term to reduce the relation between any two classes. To evaluate WCDDL, we conduct extensive experiments using two common facial datasets. Experimental results show the classification advantage of our approach in particular for sparse training samples. On the Yale dataset, WCDDL is higher than the second-best method by 10.4% when there are only two training samples for each class. At the same time, we also validate the efficiency of the weak correlation term by comparing FDDL and WCDDL. Our work can be not only applicable to face classification, but also be extended to other classification tasks, such as scene classification and handwriting recognition.

Of course, there still have some issues with our current approach. For example, the dictionary update in WCDDL takes a two-stage approach, where the iterative update for the coefficient matrix is time-consuming. Thus, we may take a simpler and more efficient update approach to improve the overall performance of the model in the future. Therefore, how to further improve the classification performance and reduce the computational time when the training sample set is sparse will be our future research goal.

References

1. Aharon, M., Elad, M., Bruckstein, A.: K-SVD: an algorithm for designing overcomplete dictionaries for sparse representation. IEEE Trans. Signal Process. **54**(11), 4311–4322 (2006)
2. Cai, D., He, X., Han, J., Zhang, H.J.: Orthogonal Laplacianfaces for face recognition. IEEE Trans. Image Process. **15**(11), 3608–3614 (2006)
3. Cai, D., He, X., Hu, Y., Han, J., Huang, T.: Learning a spatially smooth subspace for face recognition. In: Proceedings of the IEEE Conference on Computer Vision and Pattern Recognition (CVPR'07) (2007)
4. Cai, S., Zhang, L., Zuo, W., Feng, X.: A probabilistic collaborative representation based approach for pattern classification. In: Proceedings of the IEEE Conference on Computer Vision and Pattern Recognition, pp. 2950–2959 (2016)
5. Gou, J., Wang, L., Yi, Z., Yuan, Y., Ou, W., Mao, Q.: Weighted discriminative collaborative competitive representation for robust image classification. Neural Netw. **125**, 104–120 (2020)
6. Gou, J., et al.: Double competitive constraints-based collaborative representation for pattern classification. Comput. Electr. Eng. **84**, 106632 (2020)
7. He, X., Yan, S., Hu, Y., Niyogi, P., Zhang, H.J.: Face recognition using Laplacianfaces. IEEE Trans. Pattern Anal. Mach. Intell. **27**(3), 328–340 (2005)
8. Iliadis, M., Spinoulas, L., Berahas, A.S., Wang, H., Katsaggelos, A.K.: Sparse representation and least squares-based classification in face recognition. In: 2014 22nd European Signal Processing Conference (EUSIPCO), pp. 526–530 (2014)
9. Jiang, Z., Lin, Z., Davis, L.S.: Label consistent K-SVD: learning a discriminative dictionary for recognition. IEEE Trans. Pattern Anal. Mach. Intell. **35**(11), 2651–2664 (2013)
10. Li, W., Du, Q.: A survey on representation-based classification and detection in hyperspectral remote sensing imagery. Pattern Recogn. Lett. **83**, 115–123 (2016)
11. Liu, Z., Pu, J., Wu, Q., Zhao, X.: Using the original and symmetrical face training samples to perform collaborative representation for face recognition. Optik **127**(4), 1900–1904 (2016)
12. Lv, S., Liang, J., Di, L., Yunfei, X., Hou, Z.: A probabilistic collaborative dictionary learning-based approach for face recognition. IET Image Proc. **15**(4), 868–884 (2021)

13. Pati, Y., Rezaiifar, R., Krishnaprasad, P.: Orthogonal matching pursuit: recursive function approximation with applications to wavelet decomposition. In: Proceedings of 27th Asilomar Conference on Signals, Systems and Computers, vol. 1, pp. 40–44 (1993)
14. Rosasco, L., Verri, A., Santoro, M., Mosci, S., Villa, S.: Iterative projection methods for structured sparsity regularization (2009)
15. Sasao, T., Horikawa, Y., Iguchi, Y.: Handwritten digit recognition based on classification functions. In: 2020 IEEE 50th International Symposium on Multiple-Valued Logic (ISMVL), pp. 124–129. IEEE (2020)
16. Shao, S., Xu, R., Liu, W., Liu, B.D., Wang, Y.J.: Label embedded dictionary learning for image classification. Neurocomputing **385**, 122–131 (2020)
17. Yang, M., Zhang, L., Feng, X., Zhang, D.: Fisher discrimination dictionary learning for sparse representation. In: 2011 International Conference on Computer Vision, pp. 543–550. IEEE (2011)
18. Yuan, H., Li, X., Xu, F., Wang, Y., Lai, L.L., Tang, Y.Y.: A collaborative-competitive representation based classifier model. Neurocomputing **275**, 627–635 (2018)
19. Zhang, L., Yang, M., Feng, X.: Sparse representation or collaborative representation: which helps face recognition? In: 2011 International Conference on Computer Vision, pp. 471–478. IEEE (2011)
20. Zhang, Q., Li, B.: Discriminative K-SVD for dictionary learning in face recognition. In: 2010 IEEE Computer Society Conference on Computer Vision and Pattern Recognition, pp. 2691–2698. IEEE (2010)
21. Zhang, Y., Liu, W., Fan, H., Zou, Y., Cui, Z., Wang, Q.: Dictionary learning and face recognition based on sample expansion. Appl. Intell. **52**(4), 3766–3780 (2022)
22. Zheng, H., Tao, D.: Discriminative dictionary learning via fisher discrimination K-SVD algorithm. Neurocomputing **162**, 9–15 (2015)

Data-dependent and Scale-Invariant Kernel for Support Vector Machine Classification

Vinayaka Vivekananda Malgi[✉], Sunil Aryal[✉], Zafaryab Rasool,
and David Tay

Deakin University, Geelong, Waurn Ponds, VIC 3216, Australia
vinayaka.malgi@outlook.com, sunil.aryal@deakin.edu.au

Abstract. Kernel similarity function allows a Support Vector Machine (SVM) classifier to learn the maximum margin hyperplane in a higher dimensional space where two classes are linearly separable without explicitly mapping the data. Most existing kernel functions (e.g., RBF) use spatial positions of two data instances in the input space to compute their similarity. These kernels are data distribution independent and sensitive to data representation (i.e., units/scales used to measure/express data). Since this can be unknown in many real-world applications, a careful selection of a suitable kernel is required for a given problem. In this paper, we present a new kernel function based on probability data mass that is both data-dependent and scale-invariant. Our empirical results show that the proposed SVM kernel outperforms popular existing kernels.

Keywords: SVM classification · Kernel functions · Data-dependent kernel · Scale-invariant kernel · Information-theoretic similarity

1 Introduction

Support Vector Machine (SVM) classifier learns the separating hyperplane that maximises the margin between data points belonging to different classes [3]. Because data in many real-world applications have complex structures and classes are often not linearly separable in the input space, the idea of *'kernel trick'* allows learning the maximum margin hyperplane in a higher dimensional space where the classes are linearly separable without explicitly projecting the data. To achieve this, it requires special type of measures/functions to compute pairwise similarity of data. They are called kernel similarity functions, often referred to as 'kernels' in the SVM classification context. Most commonly used kernels such as Radial Basis Function (RBF) and Laplacian use spatial distance of two points in the input space to compute their kernel similarity. As discussed by Aryal et al. (2020) [2], such a distance-based notion of similarity may not be effective in real-world problems as it is:

V. V. Malgi and S. Aryal—They contributed equally to this work.

H. Kashima et al. (Eds.): PAKDD 2023, LNAI 13935, pp. 171–182, 2023.
https://doi.org/10.1007/978-3-031-33374-3_14

1. *Data-independent*: The similarity of two instances solely depends on their spatial positions and it is not affected by the distribution of other data. Unit distance between two points has the same degree of similarity everywhere in the space regardless of the density distribution. Psychologists [6] have argued that the two points in a sparse region are considered more similar than the other two points with the same geometric distance but located in dense regions. For example, two Caucasian persons are judged as less similar in Europe than in Asia because there are many Caucasian people in Europe compared to those in Asia, i.e., density is higher.

2. *Sensitive to data representation*: Because geometric model relies on the spatial positions of data points in the input space, they are sensitive to how data are represented/expressed. The distance between the two points can change significantly if the same data is represented/expressed differently by non-linear scaling. In real-world applications, data come from various sources and can be measured/expressed in different forms. For example, sample variability can be measured as standard deviation (σ) or variance (σ^2) and credit risk of customers can be measured as Income-to-Debt ratio or Debt-to-Income ratio. When data are given for analysis, mostly we are given only data values (numbers) and we may not know how they are represented, let alone the most appropriate representation.

These are the reasons as to why a kernel function that works well in one problem or dataset does not work well for others. Thus, kernel function has to be selected carefully for a given problem.

Recently, some data-dependent (dis)similarity measures [1,2,5,9] have been proposed. Among them, only the m_p-dissimilarity [2] is fully data-dependent and invariant to the change in data representation. It is data-dependent because the similarity of two objects in each attribute/feature is estimated as the probability data mass between them. Two instances in dense regions are less similar or more dissimilar to each other than the two instances with the same spatial distance but located in sparse (low-density) regions. It captures the essence of the human perceptual notion of similarity suggested by psychologists. It is robust to data representation as it does not use feature values directly in the similarity calculations, it just uses the number of data points between the two data points under consideration in each feature. As the ranking is preserved or reversed even in the case of non-linear scaling of data, data mass between any two points is not changed. It has been shown to produce better and more consistent results across a wide range of datasets in the k-nearest neighbors (k-NN) classification and content-based information retrieval problems (CBIR) [2]. However, in its current form, it cannot be used as a kernel function in the SVM classification because the self-similarity of data instances is not constant.

In this paper, we extend a variant of m_p-dissimilarity, namely m_0-dissimilarity, into a kernel function and use it in the SVM classification framework. We call the new kernel function as **P**robability **M**ass-based **K**ernel (PMK). Like m_0-dissimilarity, PMK is both data-dependent and scale-invariant. Our empirical results show that PMK consistently outperforms widely used state-of-the-art (SoTA) data-dependent and data-independent kernel functions, making it the optimal choice kernel function in practical real-world problems, particularly in domains where data are captured from various sources using different devices/sensors.

2   PMK: Probability Mass-based Kernel

2.1   Notations and Preliminaries

Let D be a collection of N labelled training instances, where each instance $\mathbf{x} = \langle x_1, x_2, \cdots, x_M \rangle$ is represented by an M-dimensional vector of its values of the M selected features; and L be a N-dimensional vector of their class labels. In this paper, we focus on numeric data, i.e., each $x_i \in \mathbb{R}$ ($\mathbb{R}$ is a real domain). In the training process, a SVM classifier learns the maximum margin hyperplane separating different classes using the similarity matrix of all $(\mathbf{x}, \mathbf{y})$ pairs in D based on a kernel function $K(\mathbf{x}, \mathbf{y})$ and label vector L. In the testing phase, to predict a class label of an unseen test instance $\mathbf{q}$, the vector of its kernel similarities with training instances in D is used. Thus, the choice of kernel function $K(\mathbf{x}, \mathbf{y})$ is central to learning a good SVM classifier.

2.2 m_0-dissimilarity

The m_0-dissimilarity of $\mathbf{x}$ and $\mathbf{y}$ is estimated as [2]:

$$m_0(\mathbf{x}, \mathbf{y}) = \frac{1}{M} \sum_{i=1}^{M} \log\left(\frac{|R_i(\mathbf{x}, \mathbf{y})|}{N} \right) \tag{1}$$

where, $R_i(\mathbf{x}, \mathbf{y}) = [\min(x_i, y_i), \max(x_i, y_i)]$ is a region covering $\mathbf{x}$ and $\mathbf{y}$ in dimension i; and $|R_i(\mathbf{x}, \mathbf{y})| = |\{\mathbf{z} \in D : \min(x_i, y_i) \leq z_i \leq \max(x_i, y_i)\}|$, where $| \cdot |$ represents cardinality.

It is data-dependent because $|R_i(\mathbf{x}, \mathbf{y})|$ depends on how many other instances fall between x_i and y_i. It is scale-invariant because the number of data points falling between x_i and y_i does not change due to linear or non-linear scaling of data as the ranking is either preserved or reversed. Because, (i) $|R_i(\mathbf{x}, \mathbf{x})|$ and $|R_i(\mathbf{y}, \mathbf{y})|$, and (ii) $|R_i(\mathbf{x}, \mathbf{x})|$ and $|R_j(\mathbf{x}, \mathbf{x})|$, can be different depending on the probability masses at x_i, y_i, and x_j; the self-dissimilarity of data instances based on m_0-dissimilarity are not constant, i.e., $m_0(\mathbf{x}, \mathbf{x})$ and $m_0(\mathbf{y}, \mathbf{y})$ can be different. However, self-dissimilarity is minimal for any instance $\mathbf{x}$, i.e., $m_0(\mathbf{x}, \mathbf{x}) \leq m_0(\mathbf{x}, \mathbf{y})$ for $\forall \mathbf{y} \neq \mathbf{x}$. Because of non-constant self-(dis)similarity, it cannot be used as a kernel function in SVM.

It is computationally expensive to compute $|R_i(\mathbf{x}, \mathbf{y})|$ as it requires a range search, especially in the case where either or both of them are unseen and N is large. However, as suggested in [2], it can be approximated quickly by converting a continuous-valued domain in each dimension i into an ordinal discrete domain by discretizing the range of data values into $b \ll N$ intervals/bins $(h_{i,1}, h_{i,2}, \cdots, h_{i,b})$. By storing the frequency of each bin from D in the preprocessing step, $|R_i(\mathbf{x}, \mathbf{y})|$ can be approximated quickly based on the frequencies of the bins where $\mathbf{x}$ and $\mathbf{y}$ falls and bins in between as $|R_i(\mathbf{x}, \mathbf{y})| = \sum_{a=l}^{u} |h_{i,a}|$, where $h_{i,l}$ and $h_{i,u}$ are the bins in which $\min(x_i, y_i)$ and $\max(x_i, y_i)$ fall. In each dimension i, data frequencies between all pairs of bins can be pre-computed and stored as a $b \times b$ matrix. Then, $|R_i(\mathbf{x}, \mathbf{y})|$ can be computed as a table look-up by finding bins where they fall. Because Equal-Width Discretisation (EWD), where bins are of the same width, is sensitive to unit/scales of data and outliers, Equal-Frequency Discretisation (EFD), where bins have the same frequency where possible, is used. As it may not be possible in practice to have the same frequency in each bin because of duplicate values, bins may have different frequencies.

2.3   Proposed New Kernel Function

Based on m_0-dissimilarity (Eq. 1), we define a new **P**robability **M**ass-based **K**ernel similarity function, referred to as 'PMK' in short, as:

$$K_{PMK}(\mathbf{x}, \mathbf{y}) = \frac{2 * m_0(\mathbf{x}, \mathbf{y})}{m_0(\mathbf{x}, \mathbf{x}) + m_0(\mathbf{y}, \mathbf{y})} \tag{2}$$

Unlike $m_0(\mathbf{x}, \mathbf{y})$ which is a measure of dissimilarity, $K_{PMK}(\mathbf{x}, \mathbf{y})$ is a similarity of $\mathbf{x}$ and $\mathbf{y}$ and it is in the range of $[0, 1]$. The self-similarity of data instances is always maximal and the constant of 1. Now it can be used as a kernel function to learn a SVM classifier. Like $m_0(\mathbf{x}, \mathbf{y})$, it is both data-dependent and invariant to units/scales of data.

The proposed kernel function has a probabilistic interpretation. It can be viewed as the multi-dimensional extension of Lin's Information Theoretical Similarity Measure [7] for ordinal data, where the similarity of two one-dimensional ordinal values x and y is estimated as:

$$s_{lin}(x, y) = \frac{2 \times \log \sum_{z=\min(x,y)}^{\max(x,y)} P(z)}{\log P(x) + \log P(y)} \tag{3}$$

where, $P(x)$ is the probability of x and it is estimated using the frequency of value x, $f(x)$, as $P(x) = f(x)/N$; and $\sum_{z=\min(x,y)}^{\max(x,y)} P(z)$ is the probability mass in the range between and including x and y, $R(x, y)$.

In the literature, the similarity of two instance $\mathbf{x}$ and $\mathbf{y}$ in a multidimensional space using Lin's approach is estimated by aggregating their similarities in each dimension based on Eq. 3:

$$s_{lin}(\mathbf{x}, \mathbf{y}) = \frac{1}{M} \sum_{i=1}^{M} \frac{2 \times \log \sum_{z_i=\min(x_i,y_i)}^{\max(x_i,y_i)} P(z_i)}{\log P(x_i) + \log P(y_i)} \tag{4}$$

Aryal et al. (2020) [2] used Lin's similarity measure defined in Eq. 4 in multidimensional continuous spaces using the same discretisation as done for m_0-dissimilarity discussed above. The similarity of $\mathbf{x}$ and $\mathbf{y}$ in each dimension i is estimated based on Eq. 3 using probability masses in one-dimensional regions $R_i(\mathbf{x}, \mathbf{y})$, $R_i(\mathbf{x}, \mathbf{x})$, and $R_i(\mathbf{y}, \mathbf{y})$. Lin's measure is applied in each dimension separately and the one-dimensional similarities are aggregated to compute the final

Table 1. An example of data distribution in two dimensions of a multi-dimensional dataset [2]

Dim	$Inst1$	$Inst2$	$Inst3$	$Inst4$	$Inst5$	$Inst6$	$Inst7$	$Inst8$	$Inst9$	$Inst10$
	.	.	.	.	.	.	.	.	.	.
	.	.	.	.	.	.	.	.	.	.
i	2	2	1	1	1	1	1	1	1	1
j	2	2	2	2	2	2	2	2	1	1
	.	.	.	.	.	.	.	.	.	.
	.	.	.	.	.	.	.	.	.	.

similarity in the multidimensional space. A more natural extension of Lin's approach in multidimensional space would be to define an M-dimensional region $R(\mathbf{x},\mathbf{y})$ that encloses $\mathbf{x}$ and $\mathbf{y}$ which has the length of $R_i(\mathbf{x},\mathbf{y})$ in each dimension i, and estimate the probability mass in the region. The similarity of $\mathbf{x}$ and $\mathbf{y}$ can be estimated as:

$$s'_{lin}(\mathbf{x},\mathbf{y}) = \frac{\log P(R(\mathbf{x},\mathbf{y}))}{\log P(R(\mathbf{x},\mathbf{x})) + \log P(R(\mathbf{y},\mathbf{y}))} \tag{5}$$

To have a reasonable estimate of $P(R(\mathbf{x},\mathbf{y}))$ in a high-dimensional space, a large amount of data is required. It is not realistic in many application domains. However, we can get a good approximation of it with Naive Bayesian assumption that dimensions are independent as $P(R(\mathbf{x},\mathbf{y})) \approx \prod_{i=1}^{M} P(R_i(\mathbf{x},\mathbf{y}))$ and Eq. 5 can be written as:

$$s'_{lin}(\mathbf{x},\mathbf{y}) = \frac{\log \prod_{i=1}^{M} P(R_i(\mathbf{x},\mathbf{y}))}{\log \prod_{i=1}^{M} P(R_i(\mathbf{x},\mathbf{x})) + \log \prod_{i=1}^{M} P(R_i(\mathbf{y},\mathbf{y}))} \tag{6}$$

Replacing the log of products with the sum of logs, $P(R_i(\mathbf{x},\mathbf{y})) = |R_i(\mathbf{x},\mathbf{y})|/N$ and using Eq. 1, Eq. 6 results in the PMK defined in Eq. 2.

One fundamental difference between PMK and the traditional Lin's approach (Eq. 4) is that the similarity of $\mathbf{x}$ and $\mathbf{y}$ in dimensions where $x_i = y_i$ is always 1 in the latter irrespective of $P(x_i)$, whereas that is not the case in PMK as it considers the $\log P(x_i)$ in the calculation. Considering $P(x_i)$ can be useful because sharing rare values can provide more information about the similarity of $\mathbf{x}$ and $\mathbf{y}$ than sharing a very frequent value. To understand this, let's look at an example dataset shown in Table 1 [2], $Inst1$ and $Inst2$ have the same values in dimensions i and j, but their value in dimension i is less common (has lower probability) than their value in dimension j. In the case of Lin's approach as used in the literature, their similarities in dimensions i and j contribute equally to the overall similarity. But, as suggested by psychologists, they provide different amount of information about the similarity of $Inst1$ and $Inst2$. Common rare value in dimension i provides more information compared to sharing frequent value in dimension j. Many instances can have the same value in many

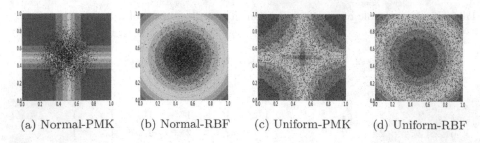

(a) Normal-PMK   (b) Normal-RBF   (c) Uniform-PMK   (d) Uniform-RBF

Fig. 1. Contour plots of kernel similarities of points in a 2-dimensional space with the centre (0.5, 0.5) under Normal and Uniform data distributions.

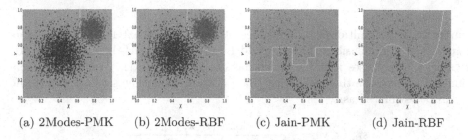

(a) 2Modes-PMK   (b) 2Modes-RBF   (c) Jain-PMK   (d) Jain-RBF

Fig. 2. Decision boundaries of SVM classifiers using PMK and RBF kernel in two 2-dimensional two-class synthetic datasets with different class densities.

dimensions in high-dimensional problems as data often lies in a low-dimensional manifold. Therefore, measure like PMK that considers $P(x_i)$ can perform better than those which do not.

To demonstrate the difference between data-dependent and data-independent kernel similarity, we present the contour plots of kernel similarities of points in a 2-dimensional space to the centre (0.5, 0.5) using PMK and RBF kernels given two datasets of 5000 points generated from normal and uniform distributions in Fig. 1. Dark orange represents the regions of high similarity, while dark blue represents the region of low similarity. It shows the data-dependent behaviour of PMK as it adapts the contours to underlying data distribution. But, the most widely used RBF kernel produces exactly the same contour regardless of underlying data distribution.

To show the effect of PMK in SVM classification, we present decision boundaries learned by SVM using PMK and RBF kernels in two 2-dimensional two-class synthetic datasets with classes of different density distribution in Fig. 2. In both datasets, the decision boundaries of RBF are pushed towards sparse classes. As a result, it maximises correct predictions for the dense classes, more samples from sparse classes are misclassified. However, in the case of PMK, the decision boundaries are pushed towards dense classes in both cases. This result suggests that PMK can better differentiate classes with varying densities rather than favouring dense class.

2.4   Positive Definiteness of the PMK Kernel

The basic notion in non-linear SVM is the existence of a mapping of the input pattern into a higher dimensional space, that more readily allow for a shattering (separation) of the different classes. In practice the mapping need not be explicitly defined, but is implicit via the kernel function. For a given kernel function $K(\mathbf{x}, \mathbf{y})$ to correspond to a higher dimensional mapping, it must be positive definite and satisfies the Mercers condition, i.e. for any arbitrary function $f(\mathbf{x})$, the following condition must be satisfied

$$\int d\mathbf{x} \int d\mathbf{y} \; f(\mathbf{x}) K(\mathbf{x}, \mathbf{y}) f(\mathbf{y}) \geq 0 \tag{7}$$

Unless the kernel function is relatively simple, e.g. polynomial, Gaussian, it is not analytically tractable to show if (7) is satisfied or not. This is certainly the case with our proposed PMK kernel, which is dependent on the data distribution. We adopted an experimental approach to show that the kernel is positive definite by checking the eigenvalues of the pairwise kernel similarity matrix (aka Gram matrix). We randomly generated datasets with 5000 points from three types of distributions: (i) normal; (ii) uniform; and (ii) a mixture of five Gaussians. We considered spaces with two, five and ten dimensions. For each distribution and dimensionality, we generated three different datasets resulting in 27 synthetic datasets. We also tested for Gram matrices of real-world datasets as shown in Table 2. In all cases, with both synthetic and real-world datasets, eigenvalues are non-negative.

3   Empirical Evaluation

We used 21 real-world datasets sourced from the UCI Machine Learning Repository [4] having varying sizes (from 699 to 43680), varying numbers of classes (from 2 to 100) and varying numbers of attributes (10 to 6826). The properties of these datasets are provided in the first four columns of Table 2. For each dataset, a 10-fold cross-validation was conducted and reported the average accuracy over 10 folds. Support Vector Classifier (SVC) of the Scikit-Learn Machine Learning Library [8] was used as an SVM classifier.

3.1   Comparison of SVM with PMK and Other Kernels

We compared PMK with five state-of-the-art (SoTA) data-independent and data-dependent kernel functions: (i) Isolation Kernel (IK); (ii) Radial Basis Function (RBF); (iii) Laplacian (Lap.); (iv) Lin's measure as defined in the literature by aggregating Lin's similarity in each dimension (Lin); and (v) Laplacian kernel on the rank transformation of data (Lap.R). Lap.R is considered because using kernels on the rank transformation of data is the simplest way of making them invariant to the change in data representation. As rank transformation of data can be computationally expensive, we discretised data as done in PMK and

Table 2. Average classification accuracy of SVM classifier over a 10-fold cross-validation. The best average accuracy and best average rank are boldfaced. N:#Instances, M: #Dimensions, and K: #Classes in a dataset.

Data[Ref]	N	M	K	PMK	IK	RBF	Lap	Lin	Lap.R
Breast Cancer [4]	699	10	2	**0.800**	0.727	0.799	0.768	0.675	0.767
Gtzan [2]	1000	230	10	0.782	0.697	**0.799**	0.783	0.650	0.774
Hba [2]	1500	187	15	**0.780**	0.665	0.720	0.761	0.710	0.765
Steel Plate [4]	1941	27	7	0.751	0.777	0.745	**0.778**	0.724	0.717
Rejafada [4]	1996	6826	2	**0.982**	0.958	0.956	0.966	0.952	0.954
Mfeat [4]	2000	649	10	**0.990**	0.980	0.986	0.983	0.927	0.937
Cardio [4]	2126	23	3	0.993	0.992	**0.995**	0.989	0.779	0.925
Hydraulic [4]	2205	43680	4	**1.000**	0.996	0.723	0.554	0.954	0.993
Segment [4]	2310	19	7	0.980	0.976	0.977	**0.986**	0.959	0.820
Fbis [2]	2463	2000	17	**0.890**	0.690	0.800	0.849	0.670	0.660
Madelon [4]	2600	500	2	**0.625**	0.555	0.590	0.606	0.520	0.622
Malware [4]	2955	1087	4	**0.995**	0.985	0.978	0.993	0.962	0.994
Page Blocks [4]	5473	10	5	**0.987**	0.977	0.971	0.974	0.900	0.888
First Order [4]	6118	51	5	0.565	**0.580**	0.562	0.535	0.425	0.520
Satimage [4]	6435	36	7	0.926	0.907	0.917	**0.927**	0.900	0.888
Musk [4]	6598	166	2	**0.998**	0.990	0.994	0.992	0.846	0.941
Taiwan Bank [4]	6819	96	2	0.967	0.969	0.970	**0.972**	0.934	0.950
Isolet [4]	7797	617	26	0.970	**0.977**	0.975	0.572	0.787	0.962
Corel [2]	10000	67	100	**0.503**	0.415	0.400	0.492	0.494	0.150
Ismis [4]	12495	191	6	0.938	0.921	**0.974**	0.972	0.939	0.221
Gas sensory [4]	13910	128	6	0.994	**0.996**	0.993	0.995	0.991	0.976
Avg. Acc.				**0.876**	0.841	0.847	0.830	0.795	0.783
Avg. Rank.				**1.799**	3.427	3.094	2.714	5.189	4.713

used bin ranks. Among the contending kernels: (i) Lin and Lap.R are both data-dependent and robust; (ii) IK is data-dependent but not robust; and (iii) RBF and Lap are neither data-dependent nor robust. It is interesting to note that though Lin and Lap.R are data-distribution dependent when $x_i \neq y_i$, they do not consider data distribution in the case of $x_i = y_i$. In a way, they are partially data-dependent.

The SVM cost parameter 'C' and parameters of kernel functions are tuned in each train-test fold through five-fold cross-validation: C in $\{0.01, 0.1, 10, 100\}$; Number of Bins (b) for PMK, Lin and Lap.R in $\{25, 50, 75, 100, (\log_2 N + 1)\}$; Subsample size ($\psi$) for IK in $\{2^m | m = 2, 3, 4, 5, 6, 7, 8\}$; and γ for RBF, Lap and Lap.R in $\{0.01, 0.1, 1, 10, 100\}$. The forest size parameter in IK was set to the default value of 100 as suggested by the authors of [9]. Because IK is a random method, for each fold, we did 10 runs and took the average accuracy.

Table 3. Average classification accuracy of SVM with PMK and other SoTA classifiers. RF, XGB, LMNN and ITML did not complete due to the 'out of memory' errors (n/a) on three large/high-dimensional datasets.

Data	SVM_{pmk}	RF	XGB	$LMNN_{knn}$	$ITML_{svm}$	$LMNN_{svm}$
Breast Cancer	0.800	0.963	**0.986**	0.966	0.958	0.962
Gtzan	**0.782**	0.729	0.734	0.725	0.700	0.740
Hba	**0.780**	0.725	0.731	0.677	0.704	0.741
Steel Plate	0.751	**0.768**	0.763	0.721	0.720	0.752
REJAFADA	**0.982**	0.976	0.993	0.961	0.957	0.956
Mfeat	**0.990**	0.987	0.981	0.985	0.969	0.984
Cardio	**0.993**	0.905	0.908	0.775	0.910	0.920
Hydraulic	**1.000**	n/a	n/a	n/a	n/a	n/a
Segment	**0.980**	0.975	0.977	0.955	0.921	0.968
Fbis	**0.890**	0.823	0.847	0.710	0.560	0.557
Madelon	0.625	0.733	**0.843**	0.553	0.474	0.605
Malware	**0.995**	0.989	0.992	0.972	0.977	0.974
Page Blocks	**0.987**	0.980	0.972	0.966	0.958	0.965
First Order	0.565	**0.585**	0.583	0.538	0.542	0.500
Satimage	**0.926**	0.916	0.918	0.911	0.895	0.924
Musk	**0.998**	0.977	0.985	0.976	0.996	0.997
Taiwan Bank	**0.967**	0.925	0.822	0.811	0.923	0.930
Isolet	**0.970**	0.966	0.734	0.957	0.967	0.952
Corel	0.503	0.502	**0.611**	0.424	0.350	0.250
Ismis	**0.938**	n/a	n/a	n/a	n/a	n/a
Gas sensory	**0.996**	n/a	n/a	n/a	n/a	n/a

The average classification accuracies over a 10-fold cross-validation run of SVC classifiers with different kernels are provided in Table 2. PMK produced the best classification results in 11 datasets followed by Lap. in 4 datasets, and RBF and IK in 3 datasets each. Lin and Lap.R did not produce the best result in any dataset. The average accuracy and the average rank show that PMK produced better and more consistent results across different datasets with the best average rank of 1.799 followed by Lap. (2.714) and RBF (3.094).

3.2 Comparison of SVM with PMK Against Other SoTA Classifiers

Table 3 reports the classification results of SVM with PMK and the SoTA classification algorithms of Random Forest (RF), eXtreme Gradient Boosting (XGB), and Large Margin Nearest Neighbor ($LMNN_{knn}$). One way of making SVM with data-independent kernels (e.g., RBF kernel) data-dependent is to transform data into some latent space using data-dependent techniques like metric learning and

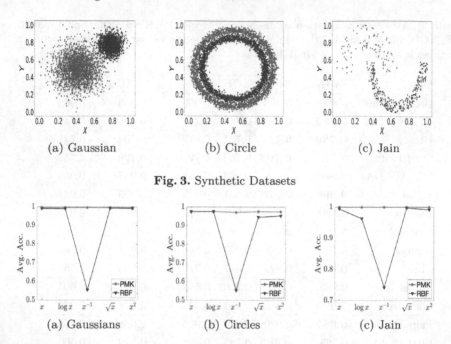

Fig. 3. Synthetic Datasets

Fig. 4. Average classification accuracy of PMK and RBF kernels over a 10-fold classification with different transformations of data.

use SVM in the latent space. We used SVM classification with RBF kernel on latent spaces resulted by Large Margin Nearest Neighbor (LMNN) and Information Theoretic Metric Learning (ITML), represented as $ITML_{svm}$ and $LMNN_{svm}$ in Table 3, respectively. The results show PMK had the best accuracy results in 16 datasets followed by XGB in 3 datasets and RF in 2 datasets. For SVM using PMK and in cases of LMNN and ITML with SVM using RBF kernel, we tuned the cost and kernel parameters as mentioned earlier. For RF and XGB, we tuned the number of estimators and learning rate parameter in the range of {100, 200, 400, 600, 800, 1000} and {0.01, 0.08, 0.1, 0.2, 0.3}, respectively. For the metric learning method of LMNN (both SVM and k-NN), we tuned the number of neighbors (k) in the range of {5, 10, 15, 20, 25, 50}.

3.3   Robustness Towards Scales of Measurement

To understand the robustness of PMK with respect to the change in units/scales used to represent data, we evaluated the performances of PMK and RBF in SVM classification in three two-class two-dimensional synthetic datasets namely Gaussian, Circle, and Jain, shown in Fig. 3. We created four variants of each dataset using non-linear scaling of data based on logarithm, inverse, square root and square, where each feature value x was transformed to $\log x$, x^{-1}, $\sqrt{x}$ and x^2 respectively. Since, x^{-1} and $\log x$ are not defined for $x = 0$, all the transformations were applied on $x' = c(x + \delta)$ as discussed in [2,5], where $\delta = 0.0001$ and

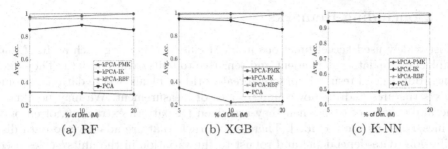

Fig. 5. Average classification accuracy of RF, XGB and k-NN classifiers in Hydraulic dataset with reduced dimensions.

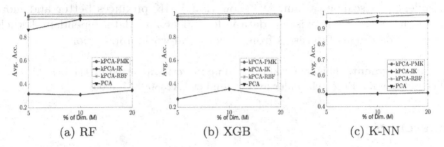

Fig. 6. Average classification accuracy of RF, XGB and k-NN classifiers in Rejafada dataset with reduced dimensions.

$c = 100$. Data values in both dimensions are normalised to be in the range of $[0, 1]$ before and after the transformations. The average accuracy over a 10-fold cross-validation run of PMK and RBF on the three synthetic datasets and their scaled variants are shown in Fig. 4. As shown in the figure, PMK performed consistently (i.e., same average accuracy was obtained for all five variants) due to its scale-invariant characteristic, while the accuracies of RBF fluctuated when data representations were changed and it performed poorly for the case of x^{-1} in particular. Since RBF uses Euclidean distance, the similarity between two objects is affected by the non-linear scaling of the data.

3.4   Dimensionality Reduction Using Kernel PCA

We also evaluated the effectiveness of the proposed PMK kernel for dimensionality reduction using kernel PCA (kPCA). We used Hydraulic and Rejafada, two datasets with dimensionalities more than 5000. We reduced the number of dimensions to 5%, 10% and 20% of the original input dimensions and used three classifiers - RF, XGB and K-NN ($k = 10$). The average classification accuracies over a 10-fold cross-validation run of the three classifiers using reduced dimensions based on kPCA with PMK, IK and RBF and simple PCA are shown in Fig. 5 and Fig. 6. In all cases, dimensionality reduction based on kPCA using PMK produced better classification results and kPCA using IK produced the worst classification results.

4   Concluding Remarks

Most widely used kernel functions in SVM classifier learning such as RBF and Laplacian are data-independent and sensitive to units/scales of data. They may not produce good results in practical real-world problems, where data have complex structures and unknown units/scales of measurement. We may not know the units/scales of data when they are given for pattern extraction, often only values/numbers are provided. Therefore, kernels that are adaptive to data distribution (data-dependent) and robust to the variation in the units/scales used to represent data (scale-variant) are preferred. In this paper, we present one such kernel based on probability data mass called PMK which is both data-dependent and scale-invariant. We show that PMK produces better and more consistent results than existing data-independent and data-dependent kernels across a wide range of datasets from various real-world applications.

Acknowledgment. This material is based upon work supported by the U.S Air Force Office of Scientific Research under award number FA2386-20-1-4005.

References

1. Aryal, S., Ting, K.M., Haffari, G., Washio, T.: Mp-dissimilarity: a data dependent dissimilarity measure. In: 2014 IEEE International Conference on Data Mining, pp. 707–712. IEEE (2014)
2. Aryal, S., Ting, K.M., Washio, T., Haffari, G.: A comparative study of data-dependent approaches without learning in measuring similarities of data objects. Data Min. Knowl. Disc. **34**(1), 124–162 (2020)
3. Cristianini, N., Shawe-Taylor, J.: An introduction to support vector machines and other Kernel-based learning methods. Cambridge University Press (2000)
4. Dua, D., Graff, C.: UCI machine learning repository. http://archive.ics.uci.edu/ml. University of california, Irvine, CA. School Inf. Comput. Sci. **25**, 27 (2019)
5. Fernando, T.L., Webb, G.I.: SimUSF: an efficient and effective similarity measure that is invariant to violations of the interval scale assumption. Data Min. Knowl. Disc. **31**(1), 264–286 (2017)
6. Krumhansl, C.L.: Concerning the applicability of geometric models to similarity data: the interrelationship between similarity and spatial density. Psychol. Rev. **85**(5), 445–463 (1978)
7. Lin, D., et al.: An information-theoretic definition of similarity. In: International Conference on Machine Learning (ICML), pp. 296–304 (1998)
8. Pedregosa, F., et al.: Scikit-learn: machine learning in Python. J. Mach. Learn. Res. **12**, 2825–2830 (2011)
9. Ting, K.M., Zhu, Y., Zhou, Z.H.: Isolation kernel and its effect on SVM. In: Proceedings of the 24th ACM SIGKDD International Conference on Knowledge Discovery & Data Mining, pp. 2329–2337 (2018)

Enhancing Robustness of Prototype with Attentive Information Guided Alignment in Few-Shot Classification

Tae-Hyung Kim[1], Woo-Jeoung Nam[2], and Seong-Whan Lee[1(✉)]

[1] Department of Artificial Intelligence, Korea University, Seoul, South Korea
{th_kim,sw.lee}@korea.ac.kr
[2] School of Computer Science and Engineering, Kyungpook National University,
Daegu, South Korea
nwj0612@knu.ac.kr

Abstract. In this paper, we carefully revisit the issues of conventional few-shot learning: i) gaps in highlighted features between objects in support and query samples, and ii) losing the explicit local properties due to global pooled features. Motivated by them, we propose a novel method to enhance robustness in few-shot learning by aligning prototypes with abundantly informed ones. As a way of providing more information, we smoothly augment the support image by carefully manipulating the discriminative part corresponding to the highest attention score to consistently represent the object without distorting the original information. In addition, we leverage word embeddings of each class label to provide abundant feature information, serving as the basis for closing gaps between prototypes of different branches. The two parallel branches of explicit attention modules independently refine support prototypes and information-rich prototypes. Then, the support prototypes are aligned with superior prototypes to mimic rich knowledge of attention-based smooth augmentation and word embeddings. We transfer the imitated knowledge to queries in a task-adaptive manner and cross-adapt the queries and prototypes to generate crucial features for metric-based few-shot learning. Extensive experiments demonstrate that our method consistently outperforms existing methods on four benchmark datasets.

Keywords: Few-shot classification · Data augmentation · Attention mechanism

1 Introduction

There are various fields of computer vision tasks in real life [18,26,34], and deep neural networks (DNNs) have achieved tremendous performance in those fields [10,17] e.g., classification, segmentation, and object detection with the help of abundant large-scale annotated datasets. However, they are not always available due to the high cost of human annotations and the lack of the data itself. To address the issues, few-shot learning (FSL) aims to categorize samples from novel classes given only a few labeled samples by mimicking human abilities [2,31].

© The Author(s), under exclusive license to Springer Nature Switzerland AG 2023
H. Kashima et al. (Eds.): PAKDD 2023, LNAI 13935, pp. 183–194, 2023.
https://doi.org/10.1007/978-3-031-33374-3_15

Metric-based prototype learning has shown promising results in FSL [28,29, 31]. CNN-based feature extractors embed images into a common feature space. Distance metrics are then applied to match query embeddings with prototypes of support embeddings and assign a category of the closest prototype to classify queries. Theoretically, the robust matching procedure aims to capture relevant features between query samples and prototypes.

Recent approaches employ the self-attention mechanism to capture relevant features [5,37]. However, global pooled features that follow the last global average pooling layer of the CNN feature extractor induce spatially implicit attention and result in traditional limitations of losing important spatial information and being sensitive to object poses in low-data regimes. Moreover, cluttered backgrounds and significant intra-class variations shift global pooled features of the same category further away in the metric space [20,33,38], leading to unrobust prototype matching. Several works [13,16] remove the global pooling layer to preserve explicit spatial features. Nevertheless, a discrepancy problem presented as a gap of emphasized features between query and support samples prevents the model from focusing on relative features and learning robust prototypes [19].

Some current approaches utilize available class label information to obtain crucial features [3,36,39]. These approaches make it possible to reduce the noise of prototypes by giving consistency to support samples. However, such augmentation is not available for query samples as they are considered test samples, resulting in extending fundamental differences between feature spaces of semantic and visual modality. Thereby, significant gaps between query and modality mixed support samples should be reduced.

In this paper, we propose Attentive Information Guided Alignment (AIGA) that aims to achieve model robustness by closing the gap among information differences of network branches. The overall model has three branches that handle information-rich prototypes, support prototypes, and queries. For the information-rich branch, working as a supervisor, attention-based smooth augmentation makes a variation of crucial common parts of objects, mitigating the dissension between query and support samples. In addition, we leverage auxiliary semantic knowledge by utilizing class embeddings to compensate for the limited data of FSL. Including the support branch, we utilize an explicit attention module that learns spatial and channel information of objects. The gap among prototypes is closed by alignment losses that distill knowledge from information-rich prototypes to non-augmented ones. Queries adapt to non-augmented prototypes with rich augmentation knowledge in a task-adaptive manner.

The main contribution of our work can be summarized as follows:

- We propose AIGA for robust prototype learning in metric-based few-shot classification. Our model aims to reduce the gap among information differences during the training procedure, e.g., attention-based smooth augmentation and auxiliary semantic knowledge.
- We propose to utilize the parallel explicit spatial and channel attention modules with a prototype alignment mechanism and to transfer the knowledge task-adaptively from prototype to query feature, leading to cross-adapting between refined prototypes and query features.
- Extensive experiments and ablation studies verify the effectiveness of AIGA.

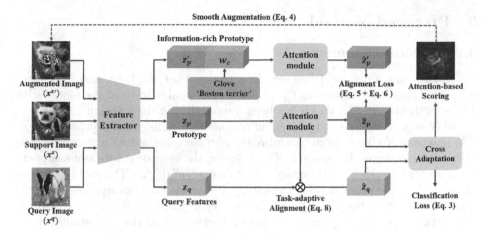

Fig. 1. The overall architecture of AIGA. $x^{s'}$, x^s and x^q pass through the shared feature extractor and are generated as z'_p, z_p and z_q. Class embeddings are concatenated with z'_p and explicit attention modules generate $\hat{z}'_p$ and $\hat{z}_p$. They are aligned by $\mathcal{L}_{aln}$ and z_q is task-adaptively refined to $\hat{z}_q$. Finally, $\hat{z}_p$ and $\hat{z}_q$ are cross-adapted, resulting in the few-shot classification loss $\mathcal{L}_{fsl}$. $\otimes$ denotes an element-wise product operation.

2   Related Work

FSL approaches fall into two main streams, i) optimization-based methods and ii) metric-based methods. Optimization-based approaches aim to learn better initialized models that can quickly adapt to novel samples with a few gradient steps [7,24,27]. We follow metric-based approaches that aim to match queries with prototypes representing the ground-truth labels of queries in the metric space by capturing relevant features of query and support samples [28,29,31].

To capture the relevant features, FEAT [37] proposes an embedding adaptation as a set-to-set transformation based on Transformer [30], but only in support samples, not query samples. CAD [5] extends FEAT to adapt support and query samples to each other. A single shared attention module replaces query values with each other to mutually reweight the features through attention scores. CAN [13] and RENet [16] generate the cross-correlational maps to produce adaptively refined spatial features. Other works present feature map reconstruction tasks [33,38] and mutual nearest neighbors [20] to fully utilize local features.

Some FSL approaches leverage auxiliary semantic knowledge to compensate for the lack of data. AM3 [35] and SEGA [36] utilize word embeddings of class labels to generate visual-semantic prototypes. ECKPN [3] utilizes label embeddings as class-level knowledge and combines them with corresponding visual knowledge via message-passing to guide the inference of query samples. ArL [39] concurrently learns relative object similarity with semantic soft labels and prediction of the absolute semantic label to better leverage semantic knowledge.

3   Proposed Method

3.1   Problem Formulation

Given two split datasets: {base set: D_b and novel set: D_n}, C_b and C_n classes for each dataset are disjoint as $C_b \cap C_n = \emptyset$. The goal is to train the model with D_b generalizable to D_n, which has been unseen during the training.

In N-way K-shot meta-training, an episode that mimics the test process randomly selects N classes from C_b and K support samples and K' non-overlapping query samples for each class from D_b. Thereby, the episode consists of a support set $S = \{(x_i^s, y_i^s)\}_{i=1}^{NK}$ and a query set $Q = \{(x_j^q, y_j^q)\}_{j=1}^{NK'}$. The meta-test goes through the same procedure for C_n and D_n. Following the Prototypical Networks [28], we extract features z_i^s, z_j^q from images x_i^s, x_j^q through feature extractor f_θ and obtain a prototype p_c representing each class c, and the probability distribution of x_j^q using a distance function $d(\cdot)$ as follows:

$$p_c = \frac{1}{|S_c|} \sum_{(x_i^s, y_i^s) \in S_c} f(x_i^s), \quad c = 1, ..., N, \tag{1}$$

$$p_\theta(y_j = c | x_j^q) = \frac{\exp(-d(f_\theta(x_j^q), p_c)}{\sum_k \exp(-d(f_\theta(x_j^q), p_k))}. \tag{2}$$

The basic loss $\mathcal{L}_{fsl}$ for training the classification model is:

$$\mathcal{L}_{fsl} = \frac{1}{|Q|} \sum_{(x_i, y_i) \in Q} -\log p_\theta(y_i | x_i). \tag{3}$$

3.2   Query-Prototype Cross-Adaptation

We introduce cross-adaptation between queries and prototypes to reduce the gap among branches and provide attention priority for augmentation. CAD [5] proposed self-attention-based cross-attention for all query and support samples. Although relevant features could be captured between query and support samples, it stacks the gap of differences between highlighted features, hindering the robustness of the model.

Instead, we leverage FEAT [37] to generate task-specific features of queries and prototypes. FEAT utilizes self-attention only to prototypes for this purpose, but we also apply it to queries to obtain each task-relevant feature of the prototypes and queries. Then, we leverage cosine similarity to obtain relevant features between the prototypes and queries to reduce the gap mentioned above and enhance the robustness.

3.3   Information-Rich Prototypes

Attention-based Smooth Augmentation. We generate smooth-augmented support samples based on attention scores to capture the relevant areas of the

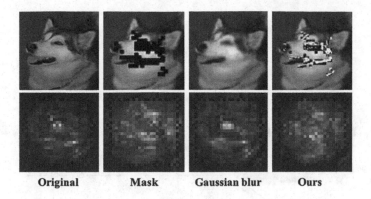

Original   **Mask**   **Gaussian blur**   **Ours**

Fig. 2. Attention-based smooth augmentation. The first row shows the augmentation results and the second row represents attention maps. Ours complements the shortcomings of previous methods, leading to augmentation object-friendly rather than coarse.

object. By manipulating areas with high attention scores, we try not to focus only on a part of the object and not to interfere with robust prototype learning.

We sort the attention scores of self-attention of prototypes in cross-adaptation and augment the support samples in the order of the top $n\%$. Specifically, the temporary attention mask $\bar{M}$ consists of $[0,1]$ and is thresholded by the top $n\%$ attention scores. However, if $\bar{M}$ is directly multiplied by the image, there is an issue that information is significantly different from neighboring pixels due to the loss of information in the masked part [19], as shown in the second column of Fig. 2. This vulnerability of DNNs to perturbations has been widely researched to prevent unpredictable phenomena in the field of explainability and robustness [9,22]. As another way of perturbation, Gaussian blur masking alleviates dramatic differences among pixels. However, it is insufficient to provide additional effects as it is similar to the original attention map.

Therefore, we introduce smooth augmentation that complements both methods by subtracting the Gaussian blurred sample from the original one with attention-based threshold to obtain the augmented support sample $x^{s\prime}$. This augmentation relieves information in over-emphasized areas of the image. It spreads attention to the entire area of the target object, forming a more natural boundary with the surrounding pixels than simply masking with zero.

Auxiliary Semantic Knowledge. In the circumstance of a limited dataset, recent works [3,36,39] utilize class embeddings to feed additional information. While our approach is similar in terms of leveraging semantic knowledge to enrich information, we utilize it to train the superior branch that plays the role of the advisor during alignment. We extract class embeddings $w_c \in \mathbb{R}^{c\prime}$ from the pre-trained word embedding model and concatenate w_c along channels of support

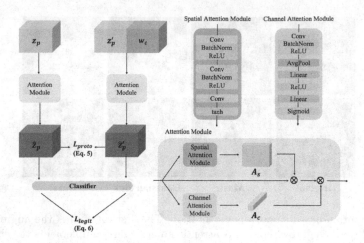

Fig. 3. Illustration of the explicit attention module with prototype alignment mechanism. Detailed descriptions are in the related section.

feature maps. The process for enriching advisor prototypes is as follows:

$$z'_p = \tau(\frac{1}{K} \sum_{x^s \in S_c} f(x^s - (\Psi(x^s) \odot \bar{M})), w_c), \qquad (4)$$

where $\tau(\cdot)$, $\Psi(\cdot)$, and $\odot$ denote the concatenation, Gaussian blur, and element-wise product operation, respectively.

3.4 Explicit Spatial and Channel Attention Module

Explicit attention modules consist of two branches for information-rich and support prototypes. Each branch consists of explicit spatial and channel attention modules, as shown in Fig. 3.

We build a stack of convolutional layers for the explicit spatial attention modules that emphasize salient regions from a spatial point of view. The channels of prototypes are compressed in one dimension to generate spatial attention maps denoted as A_s and A'_s for the support and information-rich prototype. We also introduce the channel attention module to find which patterns are informative in given features. A convolutional layer and SELayer [14] squeeze spatial dimensions and generate A_c and A'_c. Then a refined support prototype is finally obtained:

$$\hat{z}_p = z_p \odot A_s \odot A_c. \qquad (5)$$

A refined information-rich prototype $\hat{z}'_p$ is obtained in the same way.

We design a prototype alignment mechanism consisting of two losses to align support prototypes with information-rich prototypes so that the former mimics the latter knowledge. The prototype feature alignment loss L_{proto} reduces the gap between prototypes by fitting the probability distribution of each feature as follows:

$$\mathcal{L}_{proto} = -\frac{\exp(\hat{z}_p)}{\sum \exp(\hat{z}_p)} \log \frac{\exp(\hat{z}'_p)}{\sum \exp(\hat{z}'_p)}. \tag{6}$$

In addition, the logit alignment loss L_{logit} aligns the global logit of prototypes via a shared global classifier f_{cls} that classifies all classes in the base set:

$$\mathcal{L}_{logit} = -\frac{\exp(f_{cls}(\hat{z}_p))}{\sum \exp(f_{cls}(\hat{z}_p))} \log \frac{\exp(f_{cls}(\hat{z}'_p))}{\sum \exp(f_{cls}(\hat{z}'_p))}, \tag{7}$$

so the final prototype alignment loss is defined as: $L_{aln} = L_{proto} + L_{logit}$.

Task-Adaptive Query Alignment. We align query features to support prototypes in a task-adaptive manner. Intuitively thinking, it is desirable to directly align query features on information-rich prototypes to obtain rich augmentation knowledge. However, there is a risk of distortion due to differences between visual-semantic and visual-only features. Thus, we indirectly transfer the knowledge to query features using the channel attention of support prototypes:

$$\hat{z}_q = z_q \odot \mathrm{A}_c. \tag{8}$$

We feed this newly adapted feature as input to cross-adaptation and recursively operate during the training.

4 Experimental Evaluation

4.1 Datasets

We utilize four benchmarks of few-shot classification for a fair evaluation. **mini-ImageNet** [31] is a subset of ImageNet [6] consisting of 100 classes with 600 images per class. According to the standard setting [24], we split 100 classes into 64 base, 16 validation, and 20 novel classes, respectively. **tieredImageNet** [25] is a larger hierarchical subset of ImageNet with 20/6/8 superclasses consisting of 351/97/160 subclasses for base, validation, and novel, respectively. **CUB-200-2011** [32] is a fine-grained dataset of 200 classes of birds containing a total of 11788 images. We split the dataset into 100/50/50 classes, following [12]. **CIFAR-FS** [1] consists of 64/16/20 classes with 600 images per class.

4.2 Implementation Details

We adopt ResNet-12 [10] as the feature extractor without taking the last global pooling layer. It provides feature maps $z \in \mathbb{R}^{6 \times 6 \times 640}$ that preserve spatial information. Using the pre-trained word embedding model GloVe [23], we obtain word embeddings of class labels with 300 dimensions. For N-way K-shot meta-learning, we set $N = 5, K = 1, 5$. The model is trained for 200 epochs for meta-training with 600 tasks per epoch. Each task contains randomly sampled 15 query samples per class. For evaluation, we report the average classification accuracy with 95% confidence intervals of 600 randomly sampled test episodes.

Table 1. Few-shot classification accuracy on miniImageNet and tieredImageNet in the 5-way k-shot setting. † denotes the methods which leverage semi-supervised learning.

Method	Backbone	miniImageNet		tieredImageNet	
		1-shot	5-shot	1-shot	5-shot
MathcingNet [31]	ResNet-12	63.08 ± 0.80	75.99 ± 0.60	68.50 ± 0.92	80.60 ± 0.71
ProtoNet [28]	ResNet-12	62.39 ± 0.21	80.53 ± 0.14	68.23 ± 0.23	84.03 ± 0.16
Cosine [4]	ResNet-12	55.43 ± 0.81	77.18 ± 0.61	61.49 ± 0.91	82.58 ± 0.30
CAN [13]	ResNet-12	63.85 ± 0.48	79.44 ± 0.34	69.89 ± 0.51	84.23 ± 0.37
Boosting [8]†	WRN-28-10	64.03 ± 0.46	80.68 ± 0.33	70.53 ± 0.51	84.98 ± 0.36
AM3 [35]	ResNet-12	65.30 ± 0.49	78.10 ± 0.36	69.08 ± 0.47	82.58 ± 0.31
S2M2 [21]	ResNet-34	63.74 ± 0.18	79.45 ± 0.12	–	–
DeepEMD [38]	ResNet-12	65.91 ± 0.82	82.41 ± 0.56	71.16 ± 0.87	86.03 ± 0.58
FEAT [37]	ResNet-12	66.78 ± 0.20	82.05 ± 0.14	70.80 ± 0.23	84.79 ± 0.16
FRN [33]	ResNet-12	66.45 ± 0.19	82.83 ± 0.13	72.06 ± 0.22	86.89 ± 0.14
RENet [16]	ResNet-12	67.60 ± 0.44	82.58 ± 0.30	71.61 ± 0.51	85.28 ± 0.35
ECKPN [3]	ResNet-12	70.48 ± 0.38	85.42 ± 0.46	73.59 ± 0.45	88.13 ± 0.28
PTN [15]†	WRN-28-10	82.66 ± 0.97	88.43 ± 0.67	**84.70 ± 1.14**	89.14 ± 0.71
SEGA [36]	ResNet-12	69.04 ± 0.26	79.03 ± 0.18	72.18 ± 0.30	84.28 ± 0.21
CAD [5]	ResNet-12	77.56 ± 0.72	87.68 ± 0.57	77.55 ± 0.74	90.73 ± 0.54
HCTransformers [11]	ViT-S	74.74 ± 0.17	89.19 ± 0.13	79.67 ± 0.20	91.72 ± 0.11
Ours	ResNet-12	**82.79 ± 1.04**	**91.37 ± 0.75**	83.51 ± 0.96	**93.02 ± 0.71**

Table 2. Few-shot classification accuracy on CUB-200-2011 and CIFAR-FS in the 5-way k-shot setting. † denotes the methods which leverage semi-supervised learning.

Method	Backbone	CUB-200-2011		CIFAR-FS	
		1-shot	5-shot	1-shot	5-shot
MathcingNet [31]	ResNet-12	71.87 ± 0.85	85.08 ± 0.57	–	–
ProtoNet [28]	ResNet-12	66.09 ± 0.92	82.50 ± 0.58	66.09 ± 0.92	82.50 ± 0.58
Cosine [4]	ResNet-34	68.00 ± 0.83	84.50 ± 0.51	60.39 ± 0.28	72.85 ± 0.65
Boosting [8]†	WRN-28-10	–	–	73.60 ± 0.30	86.00 ± 0.20
S2M2 [21]	ResNet-34	72.92 ± 0.83	86.55 ± 0.51	62.77 ± 0.23	75.75 ± 0.13
DeepEMD [38]	ResNet-12	75.65 ± 0.83	88.69 ± 0.50	–	–
FEAT [37]	ResNet-12	73.27 ± 0.22	85.77 ± 0.14	–	–
FRN [33]	ResNet-12	83.55 ± 0.19	92.92 ± 0.10	–	–
RENet [16]	ResNet-12	79.49 ± 0.44	91.11 ± 0.24	74.51 ± 0.46	86.60 ± 0.32
ECKPN [3]	ResNet-12	77.43 ± 0.54	92.21 ± 0.41	79.20 ± 0.40	91.00 ± 0.50
SEGA [36]	ResNet-12	84.57 ± 0.22	90.85 ± 0.16	78.45 ± 0.24	86.00 ± 0.20
CAD [5]	ResNet-12	82.95 ± 0.67	90.80 ± 0.51	79.97 ± 0.72	94.13 ± 0.41
Ours	ResNet-12	**89.03 ± 0.89**	**93.56 ± 0.54**	**87.01 ± 0.98**	**94.40 ± 0.61**

4.3 Quantitative Assessment

We first report the results for miniImageNet and tieredImageNet in Table 1. We observe considerable performance gains without pre-training strategies or larger backbones e.g., WRN-28-10 and ViT-S. We achieve the new state-of-the-art compared to existing methods, except for the 1-shot evaluation on tieredImageNet. However, PTN employs semi-supervised learning that utilizes a number of unlabeled data. It compensates for the lack of data in FSL, which signif-

Table 3. Ablation on the effect of each module on miniImageNet.

| Augmentation | | | | | |
Mask	Blur	Smooth	Semantic	1-shot	5-shot
–	–	–	–	78.46 ± 0.93	88.11 ± 0.77
✓	–	–	–	79.22 ± 0.95	88.63 ± 0.78
–	✓	–	–	79.57 ± 1.05	88.93 ± 0.77
–	–	✓	–	80.14 ± 1.00	89.51 ± 0.73
✓	–	–	✓	81.46 ± 1.06	90.12 ± 0.77
–	✓	–	✓	81.92 ± 1.06	90.69 ± 0.74
–	–	✓	✓	**82.79 ± 1.04**	**91.37 ± 0.75**

icantly improves performance, especially in the 1-shot case with dramatically fewer data. Indeed, with the exception of PTN, our method records state-of-the-art with large margins even in 1-shot evaluation on all datasets including CUB-200-11 and CIFAR-FS in Table 2. In addition, we outperform the methods leveraging semantic knowledge e.g., AM3, ECKPN, and SEGA.

5 Ablation Study

5.1 Analysis for Explicit Modules

The first row in Table 3 is the result of applying only explicit attention modules and the cross-adaptation without augmentation and the alignment mechanism. It's slightly inferior to ours but still outperforms existing methods. We observe that explicit attention modules and task-adaptive query alignment greatly help cross-adaptation via spatial implicit self-attention. Augmentation improves performance, especially when leveraging auxiliary semantic knowledge, as shown in rows 5–7. In particular, the improvement at 1-shot is more remarkable than 5-shot. Since the data available in the 1-shot setting is much smaller, the effect is further maximized when supplemented with additional data. In other words, it is justifiable and effective to utilize semantic knowledge when data is scarce. Our method, the last row, reports the best performance. Thus, smoothing the most distinct regions in support samples to focus the model on the whole part of the target object and distilling knowledge into query samples to align them with prototypes are very effective for FSL.

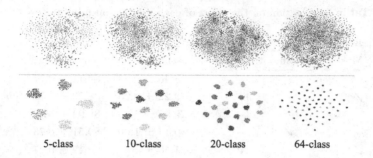

Fig. 4. t-SNE visualization on miniImageNet at 5-way 5-shot.

5.2 Effect of Augmentation

Here, we compare the effects of different augmentation methods. In Fig. 2, the attention map of the original image highlights only certain parts of the target object, making it difficult to match query and support samples. Therefore, we mask the area with the high attention score as shown in (b), capturing the whole part of the object better. But sharp differences from surrounding pixels hinder performance, as shown in Table 3. Instead of masking, we apply Gaussian blur to that area, resulting in (c). However, there is a limit to performance improvement because the highlighted area of the original image is only slightly wider. Our method (d) yields significant performance by subtracting Gaussian blurred regions from the original image, smoothing the sharp gap between the augmented portion and surrounding pixels.

5.3 Visualization

The first and second rows of Fig. 4 represent the support and query features after the feature extractor and cross-adaptation, respectively. The left three are novel classes, and the other one is all base classes. Our features are well-clustered regardless of base and novel classes.

6 Conclusion

In this paper, we propose a novel method: Attentive Information Guided Alignment (AIGA), with enrichment mechanisms of the advisor branch and alignment among branches. Attention-based smooth augmentation and auxiliary semantic knowledge help to generate information-rich prototypes. They contain more discriminative knowledge with explicit attention modules to enhance robustness. Afterward, we align support prototypes with information-rich ones via prototype alignment mechanism and also align query samples with support prototypes to share the rich knowledge in a task-adaptive manner. Queries and support prototypes cross-adapt to capture relevant features for robust matching. We experimentally validate the effectiveness of AIGA and achieve superior performance.

Acknowledgement. This work was supported by Institute of Information & communications Technology Planning & Evaluation (IITP) grant funded by the Korea government(MSIT) (No. 2019-0-00079, Artificial Intelligence Graduate School Program(Korea University) and No. 2022-0-00984, Development of Artificial Intelligence Technology for Personalized Plug-and-Play Explanation and Verification of Explanation and No.2019-0-01371, Development of brain-inspired AI with human-like intelligence).

References

1. Bertinetto, L., Henriques, J.F., Torr, P., Vedaldi, A.: Meta-learning with differentiable closed-form solvers. In: ICLR (2018)
2. Bülthoff, H.H., Lee, S.W., Poggio, T., Wallraven, C., et al.: Biologically Motivated Computer Vision: Second International Workshop, BMCV 2002, Tübingen, Germany, November 22–24, 2002, Proceedings, vol. 2525. Springer Science & Business Media (2002)
3. Chen, C., Yang, X., Xu, C., Huang, X., Ma, Z.: ECKPN: explicit class knowledge propagation network for transductive few-shot learning. In: CVPR, pp. 6596–6605 (2021)
4. Chen, W.Y., Liu, Y.C., Kira, Z., Wang, Y.C.F., Huang, J.B.: A closer look at few-shot classification. In: ICLR (2018)
5. Chikontwe, P., Kim, S., Park, S.H.: Cad: co-adapting discriminative features for improved few-shot classification. In: CVPR, pp. 14554–14563 (2022)
6. Deng, J., Dong, W., Socher, R., Li, L.J., Li, K., Fei-Fei, L.: ImageNet: a large-scale hierarchical image database. In: CVPR, pp. 248–255 (2009)
7. Finn, C., Abbeel, P., Levine, S.: Model-agnostic meta-learning for fast adaptation of deep networks. In: ICML, pp. 1126–1135 (2017)
8. Gidaris, S., Bursuc, A., Komodakis, N., Pérez, P., Cord, M.: Boosting few-shot visual learning with self-supervision. In: ICCV, pp. 8059–8068 (2019)
9. Goodfellow, I.J., Shlens, J., Szegedy, C.: Explaining and harnessing adversarial examples. arXiv preprint arXiv:1412.6572 (2014)
10. He, K., Zhang, X., Ren, S., Sun, J.: Deep residual learning for image recognition. In: CVPR, pp. 770–778 (2016)
11. He, Y., et al.: Attribute surrogates learning and spectral tokens pooling in transformers for few-shot learning. In: CVPR, pp. 9119–9129 (2022)
12. Hilliard, N., Phillips, L., Howland, S., Yankov, A., Corley, C.D., Hodas, N.O.: Few-shot learning with metric-agnostic conditional embeddings. arXiv preprint arXiv:1802.04376 (2018)
13. Hou, R., Chang, H., Ma, B., Shan, S., Chen, X.: Cross attention network for few-shot classification. In: NeurIPS (2019)
14. Hu, J., Shen, L., Sun, G.: Squeeze-and-excitation networks. In: CVPR, pp. 7132–7141 (2018)
15. Huang, H., Zhang, J., Zhang, J., Wu, Q., Xu, C.: PTN: a poisson transfer network for semi-supervised few-shot learning. In: AAAI, pp. 1602–1609 (2021)
16. Kang, D., Kwon, H., Min, J., Cho, M.: Relational embedding for few-shot classification. In: ICCV, pp. 8822–8833 (2021)
17. Krizhevsky, A., Sutskever, I., Hinton, G.E.: ImageNet classification with deep convolutional neural networks. In: NeurIPS (2012)
18. Lee, M.S., Yang, Y.M., Lee, S.W.: Automatic video parsing using shot boundary detection and camera operation analysis. Pattern Recogn. **34**(3), 711–719 (2001)

19. Li, J., Wang, Z., Hu, X.: Learning intact features by erasing-inpainting for few-shot classification. In: AAAI, pp. 8401–8409 (2021)
20. Liu, Y., Zheng, T., Song, J., Cai, D., He, X.: DMN4: few-shot learning via discriminative mutual nearest neighbor neural network. In: AAAI, pp. 1828–1836 (2022)
21. Mangla, P., Kumari, N., Sinha, A., Singh, M., Krishnamurthy, B., Balasubramanian, V.N.: Charting the right manifold: manifold mixup for few-shot learning. In: WACV, pp. 2218–2227 (2020)
22. Nam, W.J., Lee, S.W.: Gradient hedging for intensively exploring salient interpretation beyond neuron activation. arXiv preprint arXiv:2205.11109 (2022)
23. Pennington, J., Socher, R., Manning, C.D.: Glove: global vectors for word representation. In: EMNLP, pp. 1532–1543 (2014)
24. Ravi, S., Larochelle, H.: Optimization as a model for few-shot learning. In: ICLR (2016)
25. Ren, M., et al.: Meta-learning for semi-supervised few-shot classification. In: ICLR (2018)
26. Roh, M.C., Kim, T.Y., Park, J., Lee, S.W.: Accurate object contour tracking based on boundary edge selection. Pattern Recogn. **40**(3), 931–943 (2007)
27. Rusu, A.A., et al.: Meta-learning with latent embedding optimization. In: ICLR (2018)
28. Snell, J., Swersky, K., Zemel, R.: Prototypical networks for few-shot learning. In: NeurIPS (2017)
29. Sung, F., Yang, Y., Zhang, L., Xiang, T., Torr, P.H., Hospedales, T.M.: Learning to compare: relation network for few-shot learning. In: CVPR, pp. 1199–1208 (2018)
30. Vaswani, A., et al.: Attention is all you need. In: NeurIPS (2017)
31. Vinyals, O., et al.: Matching networks for one shot learning. In: NeurIPS, pp. 3630–3638 (2016)
32. Wah, C., Branson, S., Welinder, P., Perona, P., Belongie, S.: The caltech-ucsd birds-200-2011 dataset (2011)
33. Wertheimer, D., Tang, L., Hariharan, B.: Few-shot classification with feature map reconstruction networks. In: CVPR, pp. 8012–8021 (2021)
34. Xi, D., Podolak, I.T., Lee, S.W.: Facial component extraction and face recognition with support vector machines. In: Proceedings of Fifth IEEE International Conference on Automatic Face Gesture Recognition, pp. 83–88. IEEE (2002)
35. Xing, C., Rostamzadeh, N., Oreshkin, B., Pinheiro, O.P.O.: Adaptive cross-modal few-shot learning. In: NeurIPS (2019)
36. Yang, F., Wang, R., Chen, X.: Sega: semantic guided attention on visual prototype for few-shot learning. In: WACV, pp. 1056–1066 (2022)
37. Ye, H.J., Hu, H., Zhan, D.C., Sha, F.: Few-shot learning via embedding adaptation with set-to-set functions. In: CVPR, pp. 8808–8817 (2020)
38. Zhang, C., Cai, Y., Lin, G., Shen, C.: DeepEMD: few-shot image classification with differentiable earth mover's distance and structured classifiers. In: CVPR, pp. 12203–12213 (2020)
39. Zhang, H., Koniusz, P., Jian, S., Li, H., Torr, P.H.: Rethinking class relations: absolute-relative supervised and unsupervised few-shot learning. In: CVPR, pp. 9432–9441 (2021)

Clustering

Clustering

An Improved Visual Assessment with Data-Dependent Kernel for Stream Clustering

Baojie Zhang[1], Yang Cao[2(✉)], Ye Zhu[2], Sutharshan Rajasegarar[2], Gang Liu[3], Hong Xian Li[2], Maia Angelova[2], and Gang Li[2]

[1] Xi'an Shiyou University, Shaanxi 710065, China
bjzhang@tulip.academy
[2] Deakin University, Burwood, VIC 3125, Australia
charles.yang@ieee.org,
{sutharshan.rajasegarar,hong.li,maia.a,gang.li}@deakin.edu.au
[3] Harbin Engineering University, Harbin 150001, China
liugang@hrbeu.edu.cn

Abstract. The advances of 5G and the Internet of Things enable more devices and sensors to be interconnected. Unlike traditional data, the large amount of data generated from various sensors and devices requires real-time analysis. The data objects in a stream will change over time and only have a single access. Thus, traditional methods no longer meet the needs of fast exploratory data analysis for continuously generated data. Cluster tendency assessment is an effective method to determine the number of potential clusters. Recently, there are methods based on Visual Assessment of cluster Tendency (VAT) proposed for visualising cluster structures in streaming data using cluster heat maps. However, those heat maps rely on Euclidean distance that does not consider the data distribution characteristics. Consequently, it would be difficult to separate adjacent clusters of varied densities. In this paper, we discuss this issue for the latest inc-siVAT method, and propose to use a data-dependent kernel method to overcome it for clustering streaming data. Extensive evaluation on 7 large synthetic and real-world datasets shows the superiority of kernel-based inc-siVAT over 4 recently published state-of-the-art online and offline clustering algorithms.

Keywords: Cluster tendency assessment · VAT · Isolation kernel · Clustering · Data stream

1 Introduction

The advanced development of 5G and the Internet of Things enables more devices and sensors to be interconnected, and those devices continuously generate massive and high-speed data streams that pose a challenge for real-time data processing and analysis, due to their dynamicity, velocity and heterogeneity [18]. There are lots of applications based on streaming data analysis such as stock market modelling, network traffic monitoring, and system outage detection [7].

© The Author(s), under exclusive license to Springer Nature Switzerland AG 2023
H. Kashima et al. (Eds.): PAKDD 2023, LNAI 13935, pp. 197–209, 2023.
https://doi.org/10.1007/978-3-031-33374-3_16

Clustering, as an unsupervised learning algorithm, aims to partition data objects into different groups such that similar objects are in the same group. It can be used to automatically summarize the streaming data to extract meaningful hidden patterns and insights. Traditional clustering methods usually work on a finite amount of data as batch models. However, objects in streaming data will change over time and only have a single access. Thus, traditional methods no longer meet the needs of fast exploratory data analysis for streaming data.

Recently, streaming data clustering has become a hot research topic taking into account restrictions of computational memory and execution time. Many traditional clustering methods have been modified to use the two-phase framework, i.e., online and offline, to deal with data streams [2]. For example, DenStream [6] is based on DBSCAN [9] and StreamKM++ [1] is an extension of k-means++ [3]. However, determining the number of clusters in a data set is a crucial problem for most clustering methods. It is also important to show the trajectory of cluster changes in a data stream.

Visual Assessment of cluster Tendency (VAT) [4] and the improved VAT (iVAT) [25] create a reordered dissimilarity matrix for visualisation or a cluster heat map to show the possible clusters as dark blocks along the diagonal. In the heat map, each pixel reflects the dissimilarity value between two objects, i.e., the higher the similarity value, the darker pixel is. The number of dark blocks along the diagonal in the heat map represents the number of possible clusters in the dataset. There are many VAT-inspired methods proposed to visualise cluster structures in high-volume, high-velocity and high-dimensionality streaming data. Scalable iVAT (siVAT) [10] relies on an intelligent sampling scheme to retain a small size of samples. inc-iVAT/dec-iVAT [14] are incremental and decremental algorithms for visualizing the evolving cluster structures in data streams. Presently, inc-siVAT [18] is developed to visualise long data streams based on an incremental maxmin random sampling (MMRS) algorithm that can dynamically update the intelligent samples on the fly.

Nevertheless, existing VAT-based methods usually generate cluster heat maps relying on Euclidean distance that does not take the data distribution characteristics into account. If there are clusters close to each other with varied densities, it may be difficult to separate adjacent clusters, i.e., some objects from one cluster fall inside a block of another cluster. Thus, it is imperative to utilise a more adaptive dissimilarity measure for VAT-based methods such that different clusters are shown in separated blocks with similar dark colours.

A large number of existing studies have shown that the kernel approach can improve the performance of distance and density-based clustering algorithms, such as spectral clustering [12] and kernel DBSCAN [17]. In this paper, we investigate the density bias issue in VAT methods and propose to use a data-dependent kernel for improving the VAT method on clustering large streaming data. The contributions of this paper include:

– Investigating the drawbacks of using Euclidean distance in inc-siVAT [18] for clustering large data streams. Given a cluster heat map, the dense clusters tend to appear with darker blocks, which are easier to be identified and extracted. This brings about challenges when different clusters with varying densities exist in a data set. This results in a kind of density-bias issue that prevails in most distance or density-based clustering algorithms [27], i.e., having a bias towards dense clusters (in showing them with more darker colour blocks in the heat map, as opposed to less denser clusters) when using Euclidean distance.

- Proposing an improved iVAT framework utilising a newly developed data-dependent kernel method, i.e., Isolation kernel (IK) [17,22], to significantly improve both the quality of the cluster heat map visualisation and the intelligent sampling scheme. In order to meet the requirement of real-time update, we propose to map the new objects into a finite IK feature space and then use Maximum Mean Discrepancy (MMD) method [5] for the dissimilarity calculation. In addition, we apply an adaptive cluster extraction strategy to effectively identify the local meaningful clusters from the cluster heat map.
- Extensively evaluating the performance of IK-based inc-siVAT, which we call as inc-IKiVAT[1], on two synthetic and five real datasets. The results show the superiority of inc-IKiVAT over 4 recently published state-of-the-art online and offline clustering algorithms.

This paper is organised as follows: Sect. 2 provides an introduction about the VAT methods and their extension to streaming data. Section 3 shows the proposed algorithm utilising IK to overcome the drawbacks of inc-siVAT. Section 4 shows the empirical evaluation results, followed by a conclusion in Sect. 5.

2   Relate Work

Many clustering algorithms need the number of clusters to be pre-specified as input for performing clustering on a given dataset. Cluster tendency assessment is an effective method to determine the number of potential clusters from the dataset. Bezdek et al. [4] have proposed the first visual assessment of tendency (VAT) algorithm that produces a heat map based on a dissimilarity matrix. Given a dataset $O = \{o_1, \ldots, o_n\}$, $D_n = [d_{ij}]$ is a n by n pairwise dissimilarity matrix between all $o_i, o_j \in O$, and d_{ij} can be calculated using a dissimilarity measure. VAT algorithm uses a modified Prim's algorithm, i.e., a standard minimum spanning tree (MST) algorithm, to reorder a pairwise dissimilarity matrix D_n to D_n^*, and then presents each value in D_n^* as a grey pixel in a cluster heat map. In a heat map, the darker colour indicates a lower dissimilarity or higher similarity between two objects. Therefore, different clusters can be shown as separated dark blocks along the diagonal of the heat map. Since the VAT methods rely on a minimum spanning tree algorithm, the k aligned clusters can be identified simply by cutting the largest $k - 1$ edge [13].

To improve the contrast between different clusters, Wang et al. [25] proposed an iVat to convert the distance value in D_n^* with a path-based distance, i.e.,:

$$D_{ij}'^* = \min_{p \in p_{ij}} \max_{1 \leq h \leq |p|} D_{p[h]p[h+1]}^*. \tag{1}$$

---

[1] The code of `inc-IKiVAT` is on https://github.com/charles-cao/inc-IKiVAT.

where $p \in p_{ij}$ is an acyclic path in the set of all acyclic paths between o_i and o_j.

However, neither of the above two methods can be used in large datasets due to the high computational complexity of $\mathcal{O}(n^3)$. In order to overcome this weakness, scalable iVAT (siVAT) [10] used a maxmin random sampling (MMRS) approach as an intelligent sampling scheme to construct a VAT image from small representative samples. In addition, Kumar et al. [14] proposed both incremental and decremental processing for the iVAT method, named inc-iVAT/dec-iVAT. They utilise a dynamic incremental deletion mechanism based on a sliding window model. When the window size is fixed to n', inc-iVAT/dec-iVAT provides visualization on the last n' objects in the input stream with a continuous deletion and insertion process.

In order to visualise evolving structures in a long data stream, inc-siVAT [18] is developed to visualise long data streams based on MMRS that can dynamically update the intelligent samples on the fly. The MMRS begins by finding k samples that are furthest from each other in the first data chunk, then the remaining samples are grouped with their nearest sample, before finally selecting n' smart samples using a proportional random sampling from each group. inc-siVAT incrementally updates the smart sample to reflect changes in the iVAT image for visualising a data stream.

Nevertheless, all the above VAT-based methods generate cluster heat maps relying on Euclidean distance that does not take the data distribution characteristics into account. We will discuss this issue and propose the use of a data-dependent kernel method to overcome it in the next section.

3   Kernel-Based inc-siVAT

The similarity measure (i.e., Euclidean distance) used in inc-siVAT could produce a misleading cluster heat map that can not reflect the structure of the given dataset, especially when there are adjacent clusters with varied densities. There are two key consequences:

(i) A dark block of a dense cluster in the heat map would contain objects from adjacent sparse clusters. This is because when clusters are close to each other, boundary objects from a sparse cluster are likely to belong to an adjacent dense cluster using the nearest neighbour linking process [23].

(ii) The colour of a sparse cluster block shown in the cluster heat map is much lighter than that of a dense cluster block. The boundaries between dense and sparse cluster blocks become blurred, since the pairwise distances between two objects in these regions are similar.

Figure 1 shows an example on a synthetic dataset with four Gaussian clusters of different variances. When using Euclidean distance to generate the cluster heat map on all objects, inc-siVAT will produce the cluster heat map shown in Fig. 1b, in which the sparsest cluster located at the left top region is split into multiple dark sub-blocks. Two dense clusters located at the right bottom region likely belong to a large single block. Although inc-siVAT uses a path-based distance

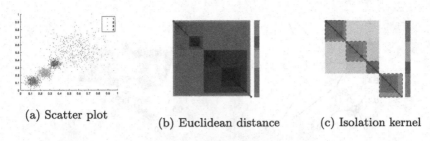

(a) Scatter plot

(b) Euclidean distance

(c) Isolation kernel

Fig. 1. (a) is the scatter plot of a synthetic dataset with four Gaussian clusters. (b) and (c) shows the cluster heat maps generated by inc-siVAT using two dissimilarity measures on the dataset shown in (a). On each cluster heat map, the blue boxes represent the extracted clusters based on MST. The objects from the same cluster have the same colour and each colour represents a ground truth cluster. The colour bar represents the true label for each row (as a data object). AMI of the extracted clusters in (b) and (c) are 0.61 and 0.93, respectively. (Color figure online)

measure to produce a much sharper heat map, it still suffers from extracting sparse clusters from indistinct blocks. This is a kind of density-bias issue existing in most distance or density-based clustering algorithms [27], i.e., having a bias towards dense clusters when using Euclidean distance. Thus, it is imperative to utilise an adaptive dissimilarity measure for VAT-based methods, such that different clusters are shown in separated blocks with similar dark colours.

In the last two decades, many kernel methods are developed to improve the performance of distance and density-based clustering [12]. Recently, Isolation kernel (IK), as a novel data-dependent similarity measure, has been proposed to improve density-based clustering [17], kernel regression [20], and time series anomaly detection [19]. In this paper, we propose to use IK for improving inc-siVAT on clustering large streaming data.

IK has two crucial properties: (i) two objects in a sparse region are more similar than two objects of equal inter-point distance in a dense region. This property enables a distance/density-based clustering algorithm to better separate complex clusters with varied densities; and (ii) IK has a finite-dimensional feature map with binary features that enables an efficient similarity calculation for streaming data.

The key idea of IK is to use a space partitioning strategy $\mathbb{H}_\psi(O)$ to split the whole data space into ψ non-overlapping partitions based on a random sample of ψ objects from the given dataset O. The similarity between any two objects $x, y \in O$ is the probability of how likely x and y can be split into the same partition under all partitions, which can be defined as

$$\mathcal{K}_\psi(x, y|O) = \mathbb{E}_{\mathbb{H}_\psi(O)}[\mathbb{1}(x, y \in \theta[z] \mid \theta[z] \in H)], \qquad (2)$$

where $\mathbb{1}(\cdot)$ is an indicator function, and $\theta[z]$ is an isolating partition.

Here we implement the partitioning H using a Voronoi diagram generated from the random subset $S \subset O$ with $|S| = \psi$ [17]. The similarity can be estimated

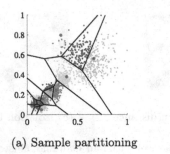
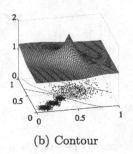

(a) Sample partitioning   (b) Contour

Fig. 2. Demonstration of Isolation kernel ($\psi = 16$) on the synthetic data used in Fig. 1: (a) An example partitioning H in which red points are the 16 subsample objects. (b) Contours with reference to point (0.5, 0.5).

using the *Monte Carlo* method, i.e., $\mathcal{K}_\psi(x, y|O)$ is the percentage of both x, y falling into the same cell over t independently generated Voronoi diagrams.

Figure 2a illustrates a partitioning result on the synthetic dataset used in Fig. 1 with the subsample sizes $\psi = 16$. It can be seen from the result that a dense region is likely to be split into smaller cells than a sparse region. Thus, objects in the spare region are more similar to $\mathbf{x}$ than points in the dense region to $\mathbf{x}$ with equal distance, as shown in Fig. 2b. As a result, IK can overcome the density-bias issue of inc-siVAT to produce a much clearer cluster heat map, as shown in Fig. 1c.

Moreover, Isolation kernel has a finite feature map defined as follows [21]:

Definition 1. *For point* $\mathbf{x} \in \mathbb{R}^d$, *the feature mapping* $\Phi : \mathbf{x} \rightarrow \{0, 1\}^{t \times \psi}$ *of* κ_ψ *is a binary vector that represents the partitions in all the partitioning* $H_i \in \mathbb{H}_\psi(D)$, $i = 1, \ldots, t$; *where* $\mathbf{x}$ *falls into only one of* ψ *hyperspheres in each partitioning* H_i.

To meet the requirement of real-time smart sample update in inc-siVAT, we propose to continuously map new objects into the finite IK feature map for fast dissimilarity calculation. Based on the kernel trick, the Maximum Mean Discrepancy (MMD) [5] between two points x and y is their Euclidean distance in the feature space, which can be derived as follow:

$$
\begin{aligned}
||\Phi(x) - \Phi(y)||^2 &= ||\Phi(x)||^2 + ||\Phi(y)||^2 - 2\langle \Phi(x), \Phi(y) \rangle \\
&= tK_\psi(x, x|D) + tK_\psi(y, y|D) - 2tK_\psi(x, y|D) \\
&= 2t(1 - K_\psi(x, y|D)) \propto -K_\psi(x, y|D)
\end{aligned}
\tag{3}
$$

Therefore, to kernelise inc-siVAT, we only need to use IK features of new objects as the input for inc-siVAT and then leave all procedures of inc-siVAT unchanged. The overall time complexity of IK-based inc-siVAT remains the same as the original inc-siVAT algorithm, since the IK feature mapping has linear complexity of $\mathcal{O}(nt\psi)$ [21]. Note that the random subset S used for building IK can be chosen from the first data chunks or be updated over different chunks. In the following experiments, S are from the first 2,000 objects for large streams.

To efficiently detect the local dark blocks in a cluster heat map generated by inc-siVAT, we propose a method called *Clusters Extraction from the cluster heat Map* (CEM) based on the local and adjacent comparison. CEM considers each row from the heat map as a vector of grey pixels, and then splits the top $k-1$ paired adjacent rows with the largest dissimilarity scores. Then the k split sub-regions on the cluster heat map indicates k clusters.

4   Experiment and Analysis

In this section, we use 2 synthetic and 5 real-world datasets to evaluate IK-based inc-siVAT. First, we compared different dissimilarity measures and cluster extraction methods for inc-siVAT. Then, we show the superiority of `inc-IKiVAT` (the IK-based inc-siVAT with the cluster extraction method CEM) over 4 recent state-of-the-art clustering methods.

The properties of the 7 datasets are shown in Table 1. Figure 3 shows scatter plots of 2 synthetic datasets, which contain clusters of varied densities. Mnist is from *UCI Machine Learning Repository* [8]. ImageNet-10, STL-10 and CIFAR-10 are TensorFlow datasets, and we utilised a recently proposed unsupervised deep learning method [15] to extract the feature representation from these images as the input for different clustering algorithms. Before the experiments, Min-Max normalisation is used for each dataset to make all the feature values between 0 and 1. We randomly permuted the order of points in each dataset and streamed points sequentially for online clustering algorithms.

4.1   Evaluation of inc-siVAT Variations

We compare the clustering results of inc-siVAT with two different dissimilarity measures, i.e., *Euclidean distance* and *Isolation Kernel*, along with two cluster extraction methods, i.e., *MST* [13] and our proposed method *CEM*. We searched the best parameter values for each algorithm within a reasonable range. The IK method has parameter $t = 200$ and $\psi \in \{2^1, 2^2, \ldots, 2^5\}$. inc-siVAT parameters range are $k\prime \in \{10, 20, 30\}$, $np \in \{150, 200, 250, \ldots, 400\}$, where $k\prime$ is the desired number of maximin (distinguished) points and np is an approximate (desired) size of MMRS sample. For cluster extraction methods, the number of clusters is $k \in \{2, 3, \ldots, 50\}$. For each clustering algorithm, We report the best result in Adjusted Mutual Information (AMI) [24] over 10 independent trials.

Table 1 shows the best clustering performance of the different variations of inc-siVAT on 7 datasets. Regarding cluster extraction, CEM achieves better than or the same performance as MST on all 7 datasets. Furthermore, IK-based clustering results are better than those based on Euclidean distance with any cluster extraction method.

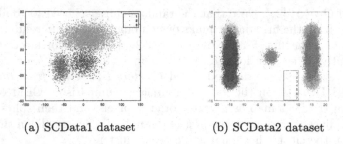

(a) SCData1 dataset   (b) SCData2 dataset

Fig. 3. Synthetic datasets scatter plots.

Table 1. Best performance of different inc-siVAT clustering methods in AMI on 7 datasets. For each dataset, the underlined and the boldfaced are the best performers within the same measure and over all measures, respectively.

Datasets	Euclidean distance			Euclidean distance		Isolation kernel	
	#Objects	#Features	#Clusters	MST	CEM	MST	CEM
SCData1	7,000	2	3	0.75	<u>0.84</u>	0.92	**1.00**
SCData2	100,000	2	7	0.86	**0.96**	0.92	**0.97**
Mnist	10,000	784	10	0.07	<u>0.24</u>	0.39	**0.57**
Imagenet10	13,000	128	10	0.84	<u>0.89</u>	0.88	**0.94**
Stl10	13,000	128	10	0.66	<u>0.70</u>	0.67	**0.72**
Dogs	19,500	128	15	0.36	<u>0.44</u>	0.38	**0.53**
Cifar10	60,000	128	10	0.61	<u>0.71</u>	0.70	**0.74**
avg.				0.59	<u>0.68</u>	0.69	**0.78**

We visualize the result of different IK-based and Euclidean distance-based inc-siVAT on *SCData1* and *Imagenet10* datasets as shown in Fig. 4. It's clear that IK significantly improves the cluster contrast in the cluster heat map produced by inc-siVAT and the cluster in each block is purer than using distance, as shown in the colour bars. Therefore, IK enables both cluster extraction methods to identify clusters much more precisely.

Moreover, we investigate the parameter sensitivity of IK-based inc-siVAT using two datasets as a demonstration, shown in Fig. 5. The results indicate that IK-based inc-siVAT usually performs well with $\psi = 16$ or 32.

4.2 Comparison with Different Clustering Algorithms

In this subsection, we select *IK-based inc-siVAT with CEM* (inc-IKiVAT) to compare with 4 recently published clustering algorithms as follows. OnCAD [7] is an online clustering algorithm, while other three are offline clustering algorithms.

- SEC [16] conducts spectral clustering on the co-association matrix in order to find the consensus partition and is robust to incomplete basic partitions.

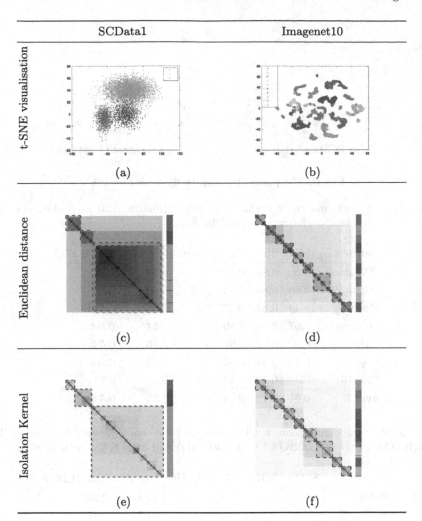

Fig. 4. (a) and (b) show the t-SNE visualisation results on imagenet10 and stl10 datasets, respectively. (c) - (f) show the clustering results of inc-siVAT using two dissimilarity measures on the two datasets. On each heat map, the blue and green boxes represent the extracted clusters using the MST and CEM, respectively. The colour bar represents the true cluster label for each row (as a data object). (Color figure online)

- OnCAD [7] is an online clustering algorithm that discovers the temporal evolution of clusters and anomalies in real-time.
- SGL [11] is a k-means-like clustering algorithm based on a graph learning framework to detect more meaningful clusters.
- GraphLSHC [26] is a large-scale hypergraph spectral clustering that provides cost-effective solutions to solve the eigenvector problem.

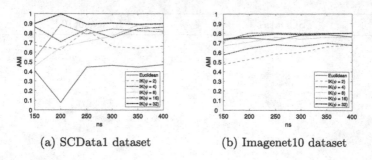

(a) SCData1 dataset   (b) Imagenet10 dataset

Fig. 5. The parameter sensitivity with $k\prime = 30$.

Table 2. Best performance of 5 clustering algorithms in AMI on 7 datasets. The boldfaced are the best performers on each dataset.

Datasets	SEC	SGL	GraphLSHC	OnCAD	inc-IKiVAT
SCData1	0.50	0.81	0.52	0.71	**1.00**
SCData2	0.66	0.86	0.84	0.62	**0.97**
Mnist	0.49	0.63	**0.77**	0.07	0.57
Imagenet10	0.67	0.88	0.89	0.14	**0.94**
Stl10	0.50	0.66	0.67	0.16	**0.72**
Dogs	0.26	0.51	0.51	0.36	**0.53**
Cifar10	0.49	**0.75**	0.74	0.39	0.74
avg.	0.51	0.73	0.71	0.35	**0.78**

Table 3. Runtime comparison in CPU seconds on a machine with four cores CPU (Intel(R) Core(TM) i5-6300HQ CPU @ 2.30 GHz) and 8GB RAM with Matlab.

Dataset	SEC	SGL	GraphLSHC	OnCAD	inc-IKiVAT
SCData1	1.82	8.48	0.12	1.68	12.94
SCData2	38.19	123.53	2.68	41.77	267.01
Mnist	130.95	17.16	0.55	23.16	34.64
Imagenet10	17.00	19.85	0.27	34.50	35.10
Stl10	23.12	19.87	0.52	11.26	37.89
Dogs	52.86	37.45	1.12	47.10	56.53
Cifar10	201.38	80.89	1.93	32.95	141.52

Table 2 shows that `inc-IKiVAT` achieves the highest average AMI score of 0.78, and is better than other cluster extraction methods on all datasets except Mnist and Cifar10. The run time of different algorithms in Table 3. `inc-IKiVAT` has the linear time complexity, e.g., the size of SCdata2 grows tenfold over Mnist, but the run time of `inc-IKiVAT` just grows about eight-fold.

5 Conclusions

We discover that the key drawback of inc-siVAT is due to Euclidean distance used as the dissimilarity measure for generating the cluster heat map. When a streaming dataset has adjacent clusters of varied densities, the dark block in the cluster heat map may contain sample objects from different clusters, and the cluster boundaries may become unclear. This is because the sampling and reordering processes in inc-siVAT do not consider the data local density distributions.

In order to overcome this shortcoming, we propose to use a data-dependent method, i.e., Isolation kernel, for dissimilarity calculation. We also demonstrate that, with the adaptive local cluster extraction method CEM, IK-based inc-siVAT can produce significantly better clustering results than state-of-the-art online and offline clustering algorithms. In the future, we will develop a more effective Isolation kernel method to fit the evolving distribution changes.

Acknowledgements. This work is supported by Natural Science Foundation of Heilongjiang Province under grant number LH2021F015, National Foreign Cultural and Educational Expert Project under grant number G2021180008L.

References

1. Ackermann, M.R., Märtens, M., Raupach, C., Swierkot, K., Lammersen, C., Sohler, C.: Streamkm++ a clustering algorithm for data streams. J. Exp. Algorithmics (JEA) **17**, 2–1 (2012)
2. Aggarwal, C.C., Philip, S.Y., Han, J., Wang, J.: A framework for clustering evolving data streams. In: Proceedings 2003 VLDB Conference, pp. 81–92. Elsevier (2003)
3. Arthur, D., Vassilvitskii, S.: K-means++: The advantages of careful seeding. In: Proceedings of the Eighteenth Annual ACM-SIAM Symposium on Discrete Algorithms, SODA 2007, pp. 1027–1035. Society for Industrial and Applied Mathematics, USA (2007)
4. Bezdek, J. C. Hathaway, R.J.: Vat: a tool for visual assessment of (cluster) tendency. In: International Joint Conference on Neural Networks (2002)
5. Borgwardt, K.M., Gretton, A., Rasch, M.J., Kriegel, H.P., Schölkopf, B., Smola, A.J.: Integrating structured biological data by Kernel Maximum Mean Discrepancy. Bioinformatics **22**(14), 49–57 (2006)
6. Cao, F., Estert, M., Qian, W., Zhou, A.: Density-based clustering over an evolving data stream with noise. In: Proceedings of the 2006 SIAM International Conference on Data Mining, pp. 328–339. SIAM (2006)
7. Chenaghlou, M., Moshtaghi, M., Leckie, C., Salehi, M.: Online clustering for evolving data streams with online anomaly detection. In: Phung, D., Tseng, V.S., Webb, G.I., Ho, B., Ganji, M., Rashidi, L. (eds.) PAKDD 2018. LNCS (LNAI), vol. 10938, pp. 508–521. Springer, Cham (2018). https://doi.org/10.1007/978-3-319-93037-4_40
8. Dua, D., Graff, C.: UCI machine learning repository (2017). http://archive.ics.uci.edu/ml

9. Ester, M., Kriegel, H.P., Sander, J., Xu, X., et al.: A density-based algorithm for discovering clusters in large spatial databases with noise. In: kdd, vol. 96, pp. 226–231 (1996)
10. Havens, T.C., Bezdek, J.C., Palaniswami, M.: Scalable single linkage hierarchical clustering for big data. In: 2013 IEEE Eighth International Conference on Intelligent Sensors, Sensor Networks and Information Processing, pp. 396–401. IEEE (2013)
11. Kang, Z., Lin, Z., Zhu, X., Xu, W.: Structured graph learning for scalable subspace clustering: from single view to multiview. IEEE Trans. Cybernetics (2021)
12. Kang, Z., Peng, C., Cheng, Q., Xu, Z.: Unified spectral clustering with optimal graph. In: Proceedings of the AAAI Conference on Artificial Intelligence, vol. 32 (2018)
13. Kumar, D., Bezdek, J.C., Rajasegarar, S., Leckie, C., Palaniswami, M.: A visual-numeric approach to clustering and anomaly detection for trajectory data. Vis. Comput. **33**(3), 265–281 (2017)
14. Kumar, D., Bezdek, J.C., Rajasegarar, S., Palaniswami, M., Leckie, C., Chan, J., Gubbi, J.: Adaptive cluster tendency visualization and anomaly detection for streaming data. ACM Trans. Knowl. Discovery Data (TKDD) **11**(2), 1–40 (2016)
15. Li, Y., Hu, P., Liu, Z., Peng, D., Zhou, J.T., Peng, X.: Contrastive clustering. In: 2021 AAAI Conference on Artificial Intelligence (AAAI) (2021)
16. Liu, H., Wu, J., Liu, T., Tao, D., Fu, Y.: Spectral ensemble clustering via weighted k-means: theoretical and practical evidence. IEEE Trans. Knowl. Data Eng. **29**(5), 1129–1143 (2017)
17. Qin, X., Ting, K.M., Zhu, Y., Lee, V.C.: Nearest-neighbour-induced isolation similarity and its impact on density-based clustering. In: Proceedings of the AAAI Conference on Artificial Intelligence, vol. 33, pp. 4755–4762 (2019)
18. Rathore, P., Kumar, D., Bezdek, J.C., Rajasegarar, S., Palaniswami, M.: Visual structural assessment and anomaly detection for high-velocity data streams. IEEE Trans. Cybernetics **51**(12), 5979–5992 (2021)
19. Ting, K.M., Liu, Z., Zhang, H., Zhu, Y.: A new distributional treatment for time series and an anomaly detection investigation. Proc. VLDB Endowment **15**(11), 2321–2333 (2022)
20. Ting, K.M., Washio, T., Wells, J., Zhang, H., Zhu, Y.: Isolation kernel estimators. Knowledge and Information Systems, pp. 1–29 (2022)
21. Ting, K.M., Wells, J.R., Washio, T.: Isolation kernel: the x factor in efficient and effective large scale online kernel learning. Data Min. Knowl. Disc. **35**(6), 2282–2312 (2021)
22. Ting, K.M., Xu, B.C., Washio, T., Zhou, Z.H.: Isolation distributional kernel: a new tool for kernel based anomaly detection. In: Proceedings of the 26th ACM SIGKDD International Conference on Knowledge Discovery & Data Mining, pp. 198–206 (2020)
23. Ting, K.M., Zhu, Y., Carman, M., Zhu, Y., Washio, T., Zhou, Z.H.: Lowest probability mass neighbour algorithms: relaxing the metric constraint in distance-based neighbourhood algorithms. Mach. Learn. **108**(2), 331–376 (2019)
24. Vinh, N.X., Epps, J., Bailey, J.: Information theoretic measures for clusterings comparison: is a correction for chance necessary? In: Proceedings of the 26th Annual International Conference on Machine Learning, pp. 1073–1080 (2009)
25. Wang, L., Nguyen, U.T., Bezdek, J.C., Leckie, C.A., Ramamohanarao, K.: ivat and avat: enhanced visual analysis for cluster tendency assessment. In: Pacific-Asia Conference on Knowledge Discovery and Data Mining, pp. 16–27. Springer (2010)

26. Yang, Y., Deng, S., Lu, J., Li, Y., Gong, Z., U, L.H., Hao, Z.: Graphlshc: towards large scale spectral hypergraph clustering. Inf. Sci. **544**, 117–134 (2021)
27. Zhu, Y., Ting, K.M., Carman, M.J., Angelova, M.: Cdf transform-and-shift: an effective way to deal with datasets of inhomogeneous cluster densities. Pattern Recogn. **117**, 107977 (2021)

Selecting the Number of Clusters K with a Stability Trade-off: An Internal Validation Criterion

Alex Mourer[1,4], Florent Forest[2,4(✉)], Mustapha Lebbah[3], Hanane Azzag[2], and Jérôme Lacaille[4]

[1] SAMM, Université Paris 1 Panthéon Sorbonne, Paris, France
[2] LIPN (CNRS UMR 7030), Université Sorbonne Paris Nord, Villetaneuse, France
f@florentfo.rest
[3] DAVID lab, Université de Versailles/Paris-Saclay, Versailles, France
[4] Safran Aircraft Engines, Moissy-Cramayel, France

Abstract. Model selection is a major challenge in non-parametric clustering. There is no universally admitted way to evaluate clustering results for the obvious reason that no ground truth is available. The difficulty to find a universal evaluation criterion is a consequence of the ill-defined objective of clustering. In this perspective, clustering stability has emerged as a natural and model-agnostic principle: an algorithm should find stable structures in the data. If data sets are repeatedly sampled from the same underlying distribution, an algorithm should find similar partitions. However, stability alone is not well-suited to determine the number of clusters. For instance, it is unable to detect if the number of clusters is too small. We propose a new principle: a good clustering should be stable, and within each cluster, there should exist no stable partition. This principle leads to a novel clustering validation criterion based on between-cluster and within-cluster stability, overcoming limitations of previous stability-based methods. We empirically demonstrate the effectiveness of our criterion to select the number of clusters and compare it with existing methods. Code is available at https://github.com/FlorentF9/skstab.

Keywords: clustering · model selection · stability · internal validation

1 Introduction

Clustering is an unsupervised learning technique aiming at discovering structure in unlabeled data. It can be defined as the "partitioning of data into groups (a.k.a. clusters) so that similar [...] elements share the same cluster and the

---

A. Mourer and F. Forest—Equal contribution. Supported by ANRT CIFRE grants and Safran Aircraft Engines.

Supplementary Information The online version contains supplementary material available at https://doi.org/10.1007/978-3-031-33374-3_17.

H. Kashima et al. (Eds.): PAKDD 2023, LNAI 13935, pp. 210–222, 2023.
https://doi.org/10.1007/978-3-031-33374-3_17

members of each cluster are all similar" [3]. These goals are contradictory because of the non-transitivity of similarity: if A is similar to B, and B is similar to C, A is not necessarily similar to C. Since clustering is an ill-posed problem, it cannot be properly solved using this definition, and algorithms often optimize only one of its aspects. For instance, K-means only guarantees that dissimilar objects are separated, and on the other hand, single linkage clustering only guarantees that similar objects will end up in the same cluster. As a consequence, model selection is a major challenge in non-parametric clustering.

In the sample-based framework adopted in this work, model selection assesses whether partitions found by an algorithm correspond to meaningful structures of the underlying distribution, and not just artifacts of the algorithm or sampling process [29,30]. Practitioners need to evaluate clustering results in order to select the best parameters for an algorithm (e.g. the number of clusters K) or choose between different algorithms. Plenty of evaluation methods exist in literature, but they usually incorporate strong assumptions on the geometry of clusters or on the underlying distribution.

There is a need for a general, model-agnostic evaluation method. Clustering stability has emerged as a principle stating that"to be meaningful, a clustering must be both *good* and the only *good* clustering of the data, up to small perturbations. Such a clustering is called stable. Data that contains a stable clustering is said to be clusterable" [24]. Hence, a clustering algorithm should discover stable structures in the data. In statistical learning terms, if data sets are repeatedly sampled from the same underlying distribution, an algorithm should find similar partitions. As we do not have access to the data-generating distribution, perturbed data sets are obtained either by sampling or injecting noise into the original data. Stability seems to be an elegant principle, but there are still severe limitations in practice. For instance, stability does not necessarily depend on clustering outcomes but can be solely related to properties of the data such as symmetries [6]. As outlined in [34], there exist various protocols to estimate stability. Unfortunately, a thorough study that evaluates them in practice is lacking.

Contributions. We propose a method for quantitatively and visually assessing the presence of structure in clustered data. The main contributions of our work can be stated as follows:

- To our knowledge, this is the first large-scale empirical study on clustering stability analysis.
- A novel definition of clustering is proposed, based on between-cluster and within-cluster stability. Based on this definition, we introduce Stadion, the stability difference criterion, along with an interpretable visualization tool, called *stability paths*.
- We show that additive noise perturbation is reliable, and a methodology to determine the amount of perturbation is proposed.
- We assess the ability of Stadion to select the number of clusters K on a vast collection of data sets and compare it with state-of-the-art methods.

2   Related Work

Internal clustering indices measure the quality of a clustering when ground-truth labels are unavailable. Most criteria rely on a combination of between-cluster and within-cluster distances. Between-cluster distance measures how distinct clusters are dissimilar, while within-cluster distance measures how elements belonging to the same cluster are similar. Unfortunately, this incorporates a prior on the geometry of clusters [10, 11, 13, 14, 27, 28, 32].

Stability analysis for clustering validation is a long-established technique. It can be traced back as far as 1973 [31] and from there has drawn increasing attention [4–6, 8, 22, 34]. Some works concluded that stability is not a well-suited tool for model selection [29]. In the general case, stability can only detect if the number of clusters is too large for the K-means algorithm (see Fig. 1). A partition with too few clusters is indeed stable, except for perfectly symmetric distributions. More accurately, these works proved that the asymptotic stability of risk-minimizing clustering algorithms, as sample size grows to infinity, only depends on whether the objective function has one or several global minima.

Albeit significant theoretical efforts, few empirical studies have been conducted. Each study focuses on specific practical implementations of stability, but as mentioned in [7, 34], a thorough study comparing all protocols in practice does not exist and a more objective evaluation of these results is warranted.

3   Clustering Stability

A data set $\mathbf{X} = \{\mathbf{x}_1, \ldots, \mathbf{x}_N\}$ consists in N independent and identically distributed (i.i.d.) samples, drawn from a data-generating distribution $\mathcal{P}$ on an underlying space $\mathcal{X}$. Formally, a clustering algorithm $\mathcal{A}$ takes as input the data set $\mathbf{X}$, some parameter $K \geq 1$, and outputs a clustering $\mathcal{C}_K = \{C_1, \ldots, C_K\}$ of $\mathbf{X}$ into K disjoint sets. Thus, a clustering can be represented by a function $\mathbf{X} \to \{1, \ldots, K\}$ assigning a label to every point of the input data set. Some algorithms can be extended to construct a partition of the entire underlying space. This partition is represented by an extension operator function $\mathcal{X} \to \{1, \ldots, K\}$ (e.g. for center-based algorithms, we compute the distance to the nearest center).

Let $\mathbf{X}$ and $\mathbf{X}'$ be two data sets drawn from the same distribution and note $\mathcal{C}_K$ and $\mathcal{C}'_K$ their respective clusterings. Let s be a similarity measure such that $s(\mathcal{C}_K, \mathcal{C}'_K)$ measures the agreement between the two clusterings. Then, for a given sample size N, the stability of a clustering algorithm $\mathcal{A}$ is defined as the expected similarity between $\mathcal{C}_K$ and $\mathcal{C}'_K$ on different data sets $\mathbf{X}$ and $\mathbf{X}'$, sampled from the same distribution $\mathcal{P}$,

$$\mathrm{Stab}(\mathcal{A}, K) := \mathbb{E}_{\mathbf{X}, \mathbf{X}' \sim \mathcal{P}^N} \left[s(\mathcal{C}_K, \mathcal{C}'_K) \right]. \tag{1}$$

This quantity is unavailable in practice, as we have a finite number of samples, so it needs to be estimated empirically. Various methods have been devised to estimate stability using perturbed versions of $\mathbf{X}$.

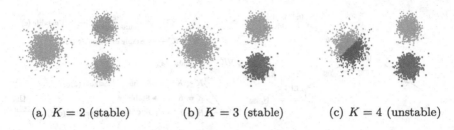

(a) $K = 2$ (stable)   (b) $K = 3$ (stable)   (c) $K = 4$ (unstable)

Fig. 1. Example data set with three clusters. The labels correspond to the K-means clustering result for $K = 2$, 3 and 4. K-means is stable even if K is too small.

The first methods used in literature are based on resampling the original data set (splitting in half [31], subsampling [8], bootstrapping [15,16], jackknife [35], etc.). Another method consists in adding random noise either to the data points [25] or to their pairwise distances [1,33]. For high-dimensional data, alternatives are random projections or randomly adding or deleting variables [31]. Once the perturbed data sets are generated, there are several ways to compare the resulting clusterings. With noise-based methods, it is possible to compare the clustering of the original data set (reference clustering) with the clusterings obtained on perturbed data sets, or to compare only clusterings obtained on the latter. With sampling-based methods, we can compare overlapping subsamples on data points where both clusterings are defined [15], or compare clusterings of disjoint subsamples (using for instance an extension operator or a supervised classifier to transfer labels from one sample to another [22]). Finally, possible similarity measures include external indices such as the ARI [15,36].

Before discussing in details the mechanisms of stability, we introduce a trivial example to illustrate its main issue: it cannot detect in general whenever K is too small. Consider the example presented in Fig. 1 with three clusters. On any sample from such a distribution, as soon as we have a reasonable amount of data, K-means with $K = 2$ always constructs the solution separating the left cluster from the two right clusters. Consequently, it is stable despite $K = 2$ being the wrong number of clusters. This situation was pointed out in [6].

In the case of algorithms that minimize an objective function (e.g. center-based or spectral), two different sources of instability have been identified [34]. First, *jittering* is caused by assignment changes at cluster boundaries after perturbation. Therefore, strong jitter is produced when a cluster boundary cuts through high-density regions. Second, *jumping* refers to the algorithm ending up in different local minima. The most important cause of jumping is initialization. Another cause is the presence of several global minima of the objective function. This happens if there are perfect symmetries in the distribution, which is very unlikely in real-world data. Examples are provided as supplementary material.

However, practitioners mainly use algorithms with consistent initialization strategies. For instance with K-means, we keep the best trial over a large number of runs and use the K-means++ heuristic. This initialization tends to make K-means deterministic, differently from the random initialization proposed in

214   A. Mourer et al.

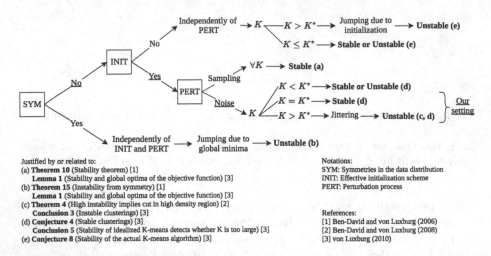

Fig. 2. Diagram explaining sources of instability in different settings, based on theoretical results for K-means, with large sample size, assuming $K \ll N$ and the underlying distribution has $K^\star$ well-separated clusters that can be represented by K-means. We consider no symmetries, effective initialization and noise-based perturbation, thus instability (due to jittering) arises when K is too large, and when K is too small whenever cluster boundaries are in high-density regions.

[9,34], which allows jumping to occur whenever $K > K^\star$, where $K^\star$ is the true number of clusters. Throughout this work, we consider a setting with large enough sample size, without perfect symmetries and with consistent initialization, that we deem to be realistic. Thus, we do not rely on jumping as the main source of instability even when $K > K^\star$, and rather rely on jittering. As a consequence, we need a perturbation process that produces jittering. We settle for noise-based perturbation, because as soon as N is reasonably large, resampling methods become trivially stable whenever there is a single global minimum [6,34]. We summarize important results in the diagram Fig. 2 and provide a simple example where sampling methods such as [8,22] fail in the supplementary material. To conclude, in our setting, a noise-based perturbation process causes jittering, enabling stability to indicate whenever K is too large. On the other hand, stability cannot in general detect when K is too small. In order to overcome this limitation, we introduce the concept of *within-cluster stability*.

4   Between-cluster and Within-Cluster Stability

A clustering algorithm applied with the same parameters to perturbed versions of a data set should find the same structure and obtain similar results. The stability principle described by (1) relies on between-cluster boundaries and we thus call it *between-cluster stability*. Therefore, it cannot detect structure within clusters. In Fig. 1, $K = 2$ is stable, whereas one cluster contains two sub-clusters. This sub-structure cannot be detected by between-cluster stability alone. Obviously, this implies that stability is unable to decide whether a data set is clusterable or not (i.e. when $K^* = 1$), which is a severe limitation. For this very reason, we introduce a second principle of *within-cluster stability*: clusters should not be composed of several sub-clusters. This implies the absence of stable structures inside any cluster. In other words, any partition of a cluster should be unstable. The combination of these two principles leads to a new definition of a clustering:

Definition 1. *A clustering is a partitioning of data into groups so that the partition is stable, and within each cluster, there exists no stable partition.*

A clustering should have a high between-cluster stability and a low within-cluster stability. Despite its apparent simplicity, implementing this principle is a difficult task. As seen in the last section, between-cluster stability can be estimated in many ways. On the other hand, within-cluster stability is a challenging quantity to define and estimate. We propose a method to estimate both quantities, and then we detail and discuss our choices.

4.1 Stadion: A Novel Stability-Based Validity Index

Let $\{\mathbf{X}_1, \ldots, \mathbf{X}_D\}$ be D perturbed versions of the data set obtained by adding random noise to the original data set $\mathbf{X}$. Between-cluster stability of algorithm $\mathcal{A}$ with parameter K estimates the expectation (1) by the empirical mean of the similarities s between the reference clustering $\mathcal{C}_K = \mathcal{A}(\mathbf{X}, K)$ and the clusterings of the perturbed data sets,

$$\text{Stab}_\text{B}(\mathcal{A}, \mathbf{X}, K) := \frac{1}{D} \sum_{d=1}^{D} s\left(\mathcal{A}(\mathbf{X}, K), \mathcal{A}(\mathbf{X}_d, K)\right). \tag{2}$$

Since s is a similarity measure, this quantity needs to be maximized. In order to define within-cluster stability, we need to assess the presence of stable structures inside each cluster. To this aim, we propose to *cluster again* the data within each cluster of $\mathcal{C}_K$. Formally, let $\Omega \subset \mathbb{N}^*$ be a set of numbers of clusters. The k-th cluster in the reference clustering is noted C_k, its number of elements N_k. Within-cluster stability of algorithm $\mathcal{A}$ is defined as

$$\text{Stab}_\text{W}(\mathcal{A}, \mathbf{X}, K, \Omega) := \sum_{k=1}^{K} \left(\frac{1}{|\Omega|} \sum_{K' \in \Omega} \text{Stab}_\text{B}(\mathcal{A}, C_k, K')\right) \times \frac{N_k}{N}. \tag{3}$$

As a good clustering is unstable within each cluster, this quantity needs to be minimized. Hence, we propose to build a new validity index combining between-cluster and within-cluster stability. A natural choice is to maximize the difference between both quantities. We call this index *Stadion*, standing for *stability difference criterion*:

$$\text{Stadion}(\mathcal{A}, \mathbf{X}, K, \Omega) := \text{Stab}_\text{B}(\mathcal{A}, \mathbf{X}, K) - \text{Stab}_\text{W}(\mathcal{A}, \mathbf{X}, K, \Omega). \tag{4}$$

The same partition $\mathcal{C}_K = \mathcal{A}(\mathbf{X}, K)$ is used in both terms of (4). Thus, Stadion evaluates the stability of an algorithm w.r.t. a reference partition.

How to perturb data? We consider the setting in Fig. 2 that is deemed to be real-istic. Neither jumping nor jittering will occur if data are perturbed by sampling processes, as soon as there is enough data. Therefore, only noise-based pertur-bation is considered here. Among them, we adopt the ε-Additive Perturbation (ε-AP) with Gaussian or uniform noise, assuming variables are scaled to zero mean and unit variance. The number of perturbations D can be kept very low and still gives reliable estimates (an analysis on the influence of D is conducted in the supplementary material.

How to choose ε? A central trade-off has to be taken into account when per-turbing the data set. If the noise level ε is too strong, we might alter the very structure of the data. We propose to circumvent this issue by *not* choosing a single value for ε, but a grid of values. By gradually increasing ε from 0 to a value $\varepsilon_{\max}$, we obtain what we call a *stability path*, i.e. the evolution of stability as a function of ε. This method has one crucial advantage: it allows to compare partitions for different values of ε without the necessity of choosing one. How-ever, it comes with two drawbacks: setting both the fineness and the maximum value of the grid. In our experiments, the fineness does not play a major role in the results. A straightforward method to fix a maximum value $\varepsilon_{\max}$ beyond which comparisons are not meaningful anymore is as follows. The perturbation corresponding to $\varepsilon_{\max}$ is meant to destroy the cluster structure of the original data. This corresponds to the value where the data are no longer clusterable, i.e. $K = 1$ becomes the solution with the best Stadion value. A first guess at $\varepsilon_{\max} = \sqrt{p}$ (where p is the data dimension) works well in practice. We found that visualizing the stability paths (see Fig. 3) greatly helps interpreting the structures found by an algorithm, hence improving the *usefulness* of results.

How to compare partitions? The similarity measure s chosen to compare two partitions is the ARI. Note that it is used to compare cluster assignments and not the ground-truth labels. Its value is in $[0, 1]$, thus Stadion has a value in $[-1, 1]$, with 1 corresponding to the best clustering and -1 to the worst. A total of 16 different similarity measures (such as the NMI) were compared (results of this study are in supplementary material).

How to aggregate the Stadion path? To compute a scalar validity index for model selection, the Stadion path must be aggregated on the noise strength ε from 0 to $\varepsilon_{\max}$. Two aggregation strategies, the maximum (Stadion-max) and the mean (Stadion-mean), are evaluated in our experiments.

The within-cluster stability is governed by the parameter Ω, which detects stable structures inside clusters of $\mathcal{C}_K$. As these are unknown, averaging several different values in Ω gives a better estimate. In absence of sub-clusters, all par-titions will be unstable because cluster boundaries will be placed in high-density regions. For the opposite reason, in presence of sub-clusters, at least some par-titions will result in higher stability, thus increasing the within-cluster stability. The analysis of influence conducted in supplementary material shows that Ω has low impact on Stadion results and can be set easily.

An important assumption behind our implementation of within-cluster stability is that for non-clusterable structures (e.g. uniform noise), the algorithm will place cluster boundaries in high-density regions to produce instability through jittering. This encompasses a wide range of algorithms such as center-based, spectral or Ward linkage clustering which, for the sake of saving cost, would cut through dense clouds of points. If this requirement is not fulfilled, further studies are needed to determine whether this method will work.

5   Experiments

5.1   A Simple Example with Stability Paths

We begin by illustrating our method with K-means and uniform ε-AP on the example discussed previously (see Fig. 1). Figure 3 displays between-cluster stability, within-cluster stability and Stadion as a function of the noise strength ε. For reasonable amounts of noise, the solutions $K = 1$, $K = 2$ and $K = 3$ are all perfectly stable, showing the insufficiency of between-cluster stability alone to indicate whenever K is too small. The solutions for $K \geq 4$ cut through the clusters and are thus unstable due to jittering. However, the solutions for $K = 1$ and $K = 2$ both have high within-cluster stability, caused by the presence of sub-clusters, which is not the case for $K \geq 3$. By computing a difference, our criterion Stadion combines this information and is able to indicate the correct number of clusters ($K = 3$) by selecting the Stadion path with the highest maximum or mean value. Through its formulation, Stadion is acting as a *stability trade-off*. The stability paths also give additional insights about the data structure. For example, we can read from the between-cluster stability path how the clusters successively merge together as ε increases. Finally, the last graph (called stability trade-off plot) represents Stadion-mean for different values of K.

5.2   Benchmark of Clustering Validation Methods

Methodology. Importantly, we aim at evaluating clustering validation methods and not the clustering algorithms themselves. Thus, we evaluate methods on a large collection of 73 artificial benchmark data sets, most of them extensively used in literature, with a guaranteed known ground-truth cluster structure. Most data sets are available in [2,17] and all data will be shared after publication. It was ensured that the evaluated algorithms are able to obtain good solutions (i.e., reasonably close to the ground-truth clustering), for some optimal parameter K. The data sets also provide various difficulty levels by varying the numbers, sizes, variances, shapes of clusters and noisy, close-by or overlapping clusters.

To compare the different validation methods, we first report the number of data sets where each method found $K^\star$, which we refer to as the number of *wins*. However, only checking whether $K^\star$ is selected is not always related to the goodness of the partition w.r.t. ground-truth, as the algorithm does not necessarily succeed in finding a good partition into $K^\star$ clusters.

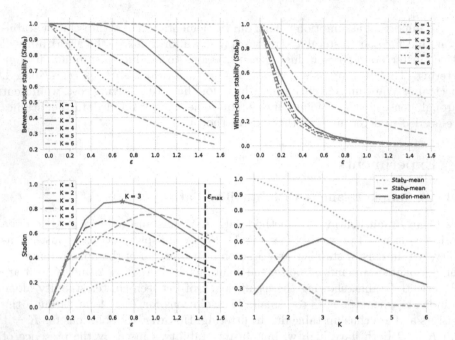

Fig. 3. Between-cluster stability paths (top left), within-cluster stability paths (top right), Stadion paths (bottom left) and stability trade-off curve (bottom right) for K-means on the data set shown in Fig. 1, for $K \in \{1 \ldots 6\}$. ε is the amplitude of the uniform noise perturbation. The best solution $K = 3$ is selected either by taking the maximum or by averaging Stadion over ε until $\varepsilon_{\max}$. The trade-off plot represents the averaged Stadion, between- and within-cluster stability as a function of K.

Thus, we also compute the ARI between the selected partition and the ground-truth. Let us note $\mathcal{Y}_{K^*} = \{Y_1, \ldots, Y_{K^*}\}$ the ground-truth partition. The performance of each method is assessed by computing $\mathrm{ARI}(\mathcal{Y}_{K^*}, \mathcal{C}_{\hat{K}})$, where $\hat{K}$ is the estimated number of clusters. In order to compare methods over multiple data sets, we compute the average ranks, denoted $\overline{R_{\mathrm{ARI}}}$. Since data sets have different difficulties, their results are not comparable and simply reporting an average would be meaningless [12]. Thus, comparing their ranks is a more sound and fair solution.

In this benchmark, three algorithms are considered: K-means, Gaussian Mixture Models (GMM) and Ward hierarchical clustering. For K-means, two versions of Stadion are evaluated: the first one using the stability computation described in Sect. 4.1 (referred to as the *standard* version), and the second one with an approximation using the extension operator (referred to as the *extended* version). As seen in Sect. 3, an extension operator extends a clustering to new data points. K-means extends naturally by computing the Euclidean distance to centers. Hence, instead of re-running K-means for each perturbation, we directly predict the cluster assignments of perturbed data points. GMM allows a similar extension, by assigning points to the cluster with the highest posterior probability. It is the only version considered due to GMM's computational cost.

Table 1. Benchmark results on 73 data sets for K-means, Ward and GMM. Average rank of the ARI between the selected clustering and ground-truth clustering ($\overline{R}_{ARI}$) and number of times the ground-truth K^* was selected (wins).

Method	K-means $\overline{R}_{ARI}$	wins	Ward $\overline{R}_{ARI}$	wins	GMM $\overline{R}_{ARI}$	wins
K^* (Oracle)	8.11	73	4.77	73	5.05	73
Stadion-max	**7.46**	50	**5.25**	**54**	–	–
Stadion-mean	7.70	51	5.80	49	–	–
Stadion-max (extended)	7.58	**56**	–	–	**5.59**	**56**
Stadion-mean (extended)	8.09	48	–	–	6.79	43
BIC	–	–	–	–	6.45	48
Wemmert-Gancarski [13]	8.33	53	5.40	**54**	5.77	52
Silhouette [28]	9.55	46	6.47	45	7.01	45
Lange [22]	10.18	45	6.53	51	6.99	48
Davies-Bouldin [11]	10.21	40	6.45	41	7.29	34
Ray-Turi [27]	10.28	37	6.97	40	7.68	33
Hennig [19]	10.72	37	–	–	–	–
Calinski-Harabasz [10]	11.44	41	7.14	39	7.43	37
Gap statistic (B) [32]	11.49	29	–	–	–	–
X-means [26]	11.56	28	–	–	–	–
Dunn [14]	13.09	26	7.77	33	7.92	34
Hofmeyr [21]	13.20	30	–	–	–	–
Xie-Beni	13.30	22	7.61	34	8.19	28
Gap statistic (A) [32]	13.57	26	–	–	–	–
G-means [18]	13.74	24	–	–	–	–
Ben-Hur [8]	14.34	20	7.86	31	8.85	28
SpecialK [20]	17.07	19	–	–	–	–

Table 1 summarizes results for each algorithm and validation method. We evaluated $K \in \{1, \ldots, 60\}$. For Stadion, we used uniform noise, $D = 10$, $\Omega = \{2, \ldots, 10\}$ and $s = $ ARI. We also evaluate the partitions obtained with the ground-truth K^* and a selection of widely used clustering validation indices (when applicable) [13], the Gap statistic [32] (with alternative versions A and B implemented in [23]), BIC, X-means [26], G-means [18], the Hennig procedure [19], stability methods [8,22], and the recent SpecialK [20]. For SpecialK, we used the default parameters indicated by the authors, but the assumptions made by the method failed on 11 data sets, explaining the poor results. Unfortunately, other methods like dip-means or skinny-dip did not have easy-to-use available implementations and were not included in this study.

Stadion-max achieves the best results overall. On K-means, it is even ranked higher than the Oracle in terms of ARI. The second-best performing index is Wemmert-Gancarski (WG). It was shown in [1] that agglomerative clustering is not robust to noise, which explains inferior Stadion results with Ward. Moreover, results are slightly biased in favor of the indices that are only valid for $K \geq 2$, unlike Stadion that will select $K = 1$ on non-clusterable distributions.

6   Conclusion

In this paper, we tackled some of limitations of cluster stability for model selection. Our contribution is twofold. First, stability can be well estimated through additive noise perturbation, solving the limitations of sampling-based stability methods. Second, we introduce the Stadion (stability difference criterion), a novel criterion acting as a trade-off between traditional stability and within-cluster stability. Furthermore, our method to control the amount of perturbation provides an interpretable visualization called stability paths.

We evaluated Stadion and methods of the literature on 73 clustering benchmark data sets. Performance is superior or on par with internal clustering indices that were designed with specific cluster geometries in mind, while relying on more general assumptions. This comes at a computational cost, requiring to run the algorithm many times. Nevertheless, studies have shown that it can be drastically reduced by down-sizing the hyperparameters with negligible impact on performance. Moreover, most theoretical results used here were derived for K-means, and more work is needed to extend these concepts to other algorithms.

Altogether, model selection remains a challenge and there is yet no theory nor a methodology that can fulfill this task perfectly. We proposed an empirical method showing interesting results along with hints to a theoretical background that could be established in future work. We hope that it will spark much-needed interest in the research community to further advance this field.

References

1. Balcan, M.F., Liang, Y.: Clustering under perturbation resilience. SIAM J. Comput. (2016)
2. Barton, T.: https://github.com/deric/clustering-benchmark
3. Ben-David, S.: Clustering-what both theoreticians and practitioners are doing wrong. In: Thirty-Second AAAI Conference on Artificial Intelligence (2018)
4. Ben-David, S., Pál, D., Simon, H.U.: Stability of k-means clustering. In: International Conference on Computational Learning Theory (2007)
5. Ben-David, S., Von Luxburg, U.: Relating clustering stability to properties of cluster boundaries. In: 21st Annual Conference on Learning Theory, COLT 2008 (2008)
6. Ben-David, S., Von Luxburg, U., Pál, D.: A sober look at clustering stability. In: International Conference on Computational Learning Theory (2006)
7. Ben-David, S., Reyzin, L.: Data stability in clustering: a closer look. Theoretical Computer Science (2014)

8. Ben-Hur, A., Elisseeff, A., Guyon, I.: A stability based method for discovering structure in clustered data. Pacific Symposium on Biocomputing (2002)
9. Bubeck, S., Meila, M., Luxburg, U.V.: How the initialization affects the stability of the k-means algorithm. ESAIM - Probability and Statistics (2012)
10. Caliński, T., Harabasz, J.: A dendrite method for cluster analysis. Commun. Stat.(1974)
11. Davies, D.L., Bouldin, D.W.: A cluster separation measure. IEEE Trans. Pattern Anal. Mach. Intell. (1979)
12. Demšar, J.: Statistical comparisons of classifiers over multiple data sets. J. Mach. Learn. Res. (2006)
13. Desgraupes, B.: ClusterCrit: clustering indices. CRAN Package (2013)
14. Dunn, J.C.: Well-Separated clusters and optimal fuzzy partitions. J. Cybern. (1974)
15. Falasconi, M., Gutierrez, A., Pardo, M., Sberveglieri, G., Marco, S.: A stability based validity method for fuzzy clustering. Pattern Recogn. (2010)
16. Fang, Y., Wang, J.: Selection of the number of clusters via the bootstrap method. Comput. Stat. Data Anal. **56**(3), 468–477 (2012)
17. Gagolewski, M., Bartoszuk, M., Cena A.G.: A new, fast, and outlier-resistant hierarchical clustering algorithm (2016)
18. Hamerly, G., Elkan, C.: Learning the k in k-means. In: NIPS (2004)
19. Hennig, C.: Cluster-wise assessment of cluster stability. Comput. Stat. Data Anal. **52**(1), 258–271 (2007)
20. Hess, S., Duivesteijn, W.: K is the magic number - inferring the number of clusters through nonparametric concentration inequalities. In: EMCL-PKDD (2019)
21. Hofmeyr, D.P.: Degrees of freedom and model selection for k-means clustering. arXiv preprint arXiv:1806.02034 (2018)
22. Lange, T., Roth, V., Braun, M.L., Buhmann, J.M.: Stability-based validation of clustering solutions. Neural Comput. (2004)
23. Maechler, M., Rousseeuw, P., Struyf, A., Hubert, M., et al.: Package 'cluster' (2013)
24. Meila, M.: How to tell when a clustering is (approximately) correct using convex relaxations. In: Advances in Neural Information Processing Systems (2018)
25. Möller, U., Radke, D.: A cluster validity approach based on nearest-neighbor resampling. In: Proceedings - International Conference on Pattern Recognition (2006)
26. Pelleg, D., Moore, A.: X-means: extending k-means with efficient estimation of the number of clusters. In: International Conference on Machine Learning (2000)
27. Ray, S., Turi, R.: Determination of number of clusters in k-means clustering and application in colour image segmentation. In: Proceedings of the 4th International Conference on Advances in Pattern Recognition and Digital Techniques (1999)
28. Rousseeuw, P.J.: Silhouettes: A graphical aid to the interpretation and validation of cluster analysis. J. Comput. Appl. Math. (1987)
29. Shamir, O., Tishby, N.: Cluster stability for finite samples. In: Advances in Neural Information Processing Systems (2007)
30. Smith, S.P., Dubes, R.: Stability of a hierarchical clustering. Pattern Recogn. (1980)
31. Strauss, J.S., Bartko, J.J., Carpenter, W.T.: The use of clustering techniques for the classification of psychiatric patients. British J. Psychiatry (1973)
32. Tibshirani, R., Walther, G., Hastie, T.: Estimating the number of clusters in a data set via the gap statistic. J. Royal Stat. Soc. Ser. B (2001)
33. Vijayaraghavan, A., Dutta, A., Wang, A.: Clustering stable instances of euclidean k-means. In: Advances in Neural Information Processing Systems (2017)
34. Von Luxburg, U.: Clustering stability: an overview. Found. Trends® Mach. Learn. (2010)

222   A. Mourer et al.

35. Yeung, K.Y., Haynor, D.R., Ruzzo, W.L.: Validating clustering for gene expression data. Bioinformatics (2001)
36. Zhao, Q., Xu, M., Fränti, P.: Extending external validity measures for determining the number of clusters. Intell. Syst. Design Appl. (2011)

Adaptive View-Aligned and Feature Augmentation Network for Partially View-Aligned Clustering

Xianchao Zhang[1], Mengyan Chen[1], Jie Mu[1], and Linlin Zong[1,2]([✉])

[1] Key Laboratory for Ubiquitous Network and Service Software of Liaoning Province, School of Software, Dalian University of Technology, Dalian 116620, China
`{xczhang,llzong}@dlut.edu.cn`, `{mengyan_chen,jiem}@mail.dlut.edu.cn`
[2] State Key Laboratory for Novel Software Technology, Nanjing University, Nanjing 210023, China

Abstract. As an important task in multi-view clustering, partially view-aligned clustering has attracted increasing attention in recent years. However, previous algorithms have two limitations: (1) they manually calculate the fixed alignment matrix based on Euclidean distance and use the fixed matrix for common feature expression. The manual fixed alignment matrix fails to adequately reflect the similarity of the training data; (2) the process of learning features is isolated from the downstream clustering task, thus learned features are unsuitable for the clustering scenario. In this paper, we propose an adaptive view-aligned and feature augmentation network (**AFAN**) to tackle these two issues. First, we propose an adaptive view-aligned module to calculate the alignment matrix with the self-attention mechanism. The calculated alignment matrix can capture data similarity by jointly learning data features and view alignment. Second, we introduce a self-augmentation strategy to encourage the learned features of the same cluster to be crowded together. Extensive experimental results show that **AFAN** outperforms state-of-the-art approaches on four benchmark datasets.

Keywords: Partially View-aligned · Clustering · Deep Learning

1 Introduction

Partially view-aligned clustering aims to solve the data inconsistency problem caused by a portion of data misalignment among views. The problem is called

This work was supported in part by the National Natural Science Foundation of China under Grant 61972065, Grant 62006034; in part by the Natural Science Foundation of Liaoning Province under Grant 2021-BS-067; in part by the Social Science Planning Foundation of Liaoning Province under Grant L21CXW003; in part by the State Key Laboratory of Novel Software Technology, Nanjing University under Grant KFKT2022B41; and in part by the Dalian High-level Talent Innovation Support Plan under Grant 2021RQ056.

H. Kashima et al. (Eds.): PAKDD 2023, LNAI 13935, pp. 223–235, 2023.
https://doi.org/10.1007/978-3-031-33374-3_18

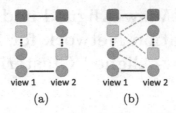

view 1   view 2   view 1   view 2

(a)   (b)

Fig. 1. Partially view-aligned problem (PVP) and the solution of PVP.

partially view-aligned problem (PVP) [4]. Figure 1(a) depicts the PVP, where different shapes indicate different categories, different colors indicate different samples and the solid black line indicates the prior alignment relationships between the samples. For two views data, only part of the data is aligned, and the correspondence of the other data is unaligned. For example, the samples represented by blue and orange are aligned and they are connected across views with the solid black line. The samples represented by yellow and green are unaligned. For the PVP, we could conduct multi-view clustering on the partial data directly, but the interaction of unaligned multi-view data is limited, which will decrease the performance of partially view-aligned clustering. A better strategy is to align all the data before clustering. Since the goal of the clustering task is to classify the data into the correct category, the solution of PVP only needs to align one sample in one view with samples from the same category in other views. As shown in Fig. 1(b), taking the yellow rectangle in view 1 as an example, the solution of PVP designs the alignment strategy to establish the yellow rectangle alignment relationships with the blue and yellow rectangles in view 2.

A few traditional methods have been proposed to alleviate the negative effects caused by PVP [6,23]. For example, Lampert et al. [6] proposed a weakly-paired maximum covariance analysis (WMCA) to handle weakly paired data. These methods have achieved promising results, but they are shallow models that cannot handle high-dimensional and complex data.

Deep view-aligned strategies use deep neural networks to capture the feature of high-dimensional and complex data. They are based on Euclidean distance to manually compute the alignment matrix, which is separated from the network training. Specifically, PVC [4] calculates the fixed alignment matrix based on Euclidean distance and then applies the fixed alignment matrix in the training phase. However, these methods still face two issues: (1) they manually compute the fixed alignment matrix based on Euclidean distance and use the fixed matrix for common feature expression. The manual fixed alignment matrix is separated from the training process and may not be the optimal alignment matrix for aligning the data; (2) the process of learning features is isolated from the downstream clustering task, thus learned features are unsuitable for the clustering scenario.

In this paper, we propose an adaptive view-aligned and feature augmentation network (**AFAN**) to tackle these two issues. The AFAN framework is shown in Fig. 2. First, AFAN employs an adaptive view-aligned module (AVAM) to adaptively calculate the alignment matrix with the self-attention mechanism. Compared with manual calculation, the alignment matrix obtained from AVAM can capture data similarity by jointly learning data features and view alignment. Second, AFAN exploits a self-augmentation strategy to encourage the learned features belonging to the same cluster to be crowded together. Such crowded features are clustering-friendly [22]. Specifically, we perform feature augmentation on the constructed nearest neighbor graph, thus encouraging the obtained sample features to be suitable for the clustering scenario. The main contributions of this work are summarized as follows:

- We propose a new partially view-aligned clustering method that uses an adaptive view-aligned module to adaptively calculate the alignment matrix with the self-attention mechanism. Our method solves the limitation that Euclidean distance-based alignment matrix fails to adequately reflect the similarity of the training data.
- We introduce a self-augmentation strategy to encourage the learned features belonging to the same class to be crowded together. Such crowded features are clustering-friendly and are more likely to be assigned to the correct clusters.
- Extensive experiment results on four benchmark datasets have demonstrated that our method is effective and consistently outperforms state-of-the-art approaches.

2   Related Work

Multi-view Clustering. With the increase of information collection methods, data can be collected from different sources, which is called multi-view data. Multi-view clustering (MVC) can handle multi-view data and make full use of the consistency and complementary information between different views. Existing multi-view clustering methods [12,13,19,21] are based on two ideal assumptions: completeness of data and correspondence of views. However, in the real world, it is quite difficult for multi-view data to satisfy the above assumptions simultaneously, which leads to partial multi-view clustering problems.

Partial Multi-view Clustering. Partial multi-view clustering is divided into two categories, one solves the partially data-missing problem (PDP) [4] and the other solves the partially view-aligned problem (PVP).

PDP occurs when some data is missing in some views of multi-view data. In recent years, many deep methods have been proposed to solve PDP [14,15,17]. These approaches utilize the remaining information in the views and eliminate the negative effects of missing views.

Unlike PDP, PVP means that only a portion of the data is aligned across views and the alignment relationships of the other data is unknown. Some traditional methods have been proposed to solve PVP. Lampert et al. [6] proposed

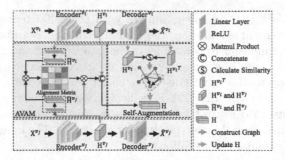

Fig. 2. Framework of the proposed adaptive view-aligned and feature augmentation network(AFAN). AFAN consists of three components: multi-view autoencoder, adaptive view-aligned module(AVAM) and self-augmentation strategy. We input the aligned data X^{v_i} and shuffled data X^{v_j} into the specific autoencoders to obtain the potential representations H^{v_i} and H^{v_j}. Then, AVAM computes the alignment matrix adaptively using $\widetilde{H}^{v_i}$ and $\widetilde{H}^{v_j}$ obtained by transforming H^{v_i} and H^{v_j}. We align and concatenate H^{v_i} and H^{v_j} to obtain the sample common representation H according to the alignment matrix. The self-augmentation strategy uses H^{v_i} and $(H^{v_i})^T$ to construct the nearest neighbor graph to enhance H. $\widehat{X}^{v_i}$ and $\widehat{X}^{v_j}$ are reconstructed data.

weakly-paired maximum covariance analysis (WMCA) to overcome the limitation of weakly paired data. Zhang et al. [23] proposed a small number of inter-view constraints instead of mapping to obtain mutual information between views. Deep partially view-aligned methods that capture high-dimensional features of the data are in their infancy. PVC [4] implements the differentiable agent of the Hungarian algorithm to establish the alignment relationships of unaligned data. MvCLN [18] proposed a noise-robust contrastive loss. After the MvCLN is trained, the Euclidean distance matrix is calculated as the alignment matrix.

Compared to existing deep methods, our approach adaptively learns the alignment matrix throughout the whole network training process and generates cluster-oriented features for downstream clustering tasks.

3   Proposed Method

In this section, we describe our proposed adaptive view-aligned and feature augmentation network(AFAN). As shown in Fig. 2, AFAN consists of three components, i.e., multi-view autoencoder, adaptive view-aligned module (AVAM), and self-augmentation strategy. In the following sections, we first introduce the formulation definition of PVP in Sect. 3.1 and then elaborate on the design of the three parts in Sect. 3.2, Sect. 3.3 and Sect. 3.4.

3.1   Problem Formulation

Given a multi-view dataset with V views and n samples $\left\{X^{(v)}\right\}_{v=1}^{V}$, where $X^{(v)} = \left\{x_1^{(v)}, x_2^{(v)}, \ldots, x_n^{(v)}\right\}^T$. The multi-view data $X^{(v)}$ with PVP is divided into two parts: aligned data $A^{(v)}$ and unaligned data $U^{(v)}$. Specifically, $X^{(v)} = \{A^{(v)}, U^{(v)}\}$. We train the network using the aligned data $A^{(v)}$. For any two views v_i and v_j, $A^{(v_i)}$ keeps the original arrangement, $A^{(v_j)}$ is randomly shuffled to simulate the unaligned data, and the alignment relationships between $A^{(v_i)}$ and $A^{(v_j)}$ is known. Our AFAN learns from the aligned data $A^{(v)}$ and then outputs the alignment matrix of all data $X^{(v)}$ to establish the cross-view alignment relationships. Our goal can be expressed by the following formula:

$$E^{(v_i)}\left(X^{(v_i)}\right) = W^{(v_i,v_j)} * E^{(v_j)}\left(X^{(v_j)}\right), \tag{1}$$

where $E(\cdot)$ is the mapping of the encoder, $*$ represents matmul product. $=$ is an operator, which denotes that the sample features in views v_i and v_j are aligned with correct correspondence via $W^{(v_i,v_j)}$. $W^{(v_i,v_j)} \in R^{n \times n}$ is the alignment matrix, which reveals the data correspondence between views v_i and v_j.

3.2   Multi-view Autoencoder

To extract view-specific features, we employ view-specific encoders and decoders. Inspired by under-complete autoencoder [16], we design the view-specific encoder $E^{(v)}$ with fully connected layers and the corresponding decoder $D^{(v)}$ for each view. This approach not only preserves the unique information of each view but also solves the problem of inconsistent view dimensions.

3.3   Adaptive View-Aligned Module

To automatically learn the alignment relationships among views, we design the adaptive view-aligned module (AVAM). AVAM uses self-attention to adaptively compute the alignment matrix. The alignment matrix can capture data similarity by jointly learning data features and view alignment.

To take advantage of the prior knowledge of the aligned data $A^{(v)}$, we train the network with the aligned data $A^{(v)}$. Taking two views v_i and v_j as an example, we pass unchanged data $A^{(v_i)}$ and shuffled data $A^{(v_j)}$ through the view-specific encoders described above to obtain latent representations $H_a^{(v_i)} = E^{(v_i)}\left(A^{(v_i)}\right)$ and $H_a^{(v_j)} = E^{(v_j)}\left(A^{(v_j)}\right)$, where $H_a^{(v_i)}$ and $H_a^{(v_j)} \in R^{n_a \times d}$, n_a is the number of samples in $A^{(v)}$, d is the feature dimension. Then, we transform $H_a^{(v_i)}$ and $H_a^{(v_j)}$ to obtain $\widetilde{H}_a^{(v_i)} = H_a^{(v_i)} * w_Q$ and $\widetilde{H}_a^{(v_j)} = H_a^{(v_j)} * w_K$ respectively, where $w_Q \in R^{d \times d_k}$ and $w_K \in R^{d \times d_k}$ are learnable parameters. The adaptive alignment matrix $W_a^{(v_i,v_j)}$ between views v_i and v_j is defined as follows:

$$W_a^{(v_i,v_j)} = \text{attention}(v_i, v_j) = \text{softmax}\left(\frac{\widetilde{H}_a^{(v_i)}\left(\widetilde{H}_a^{(v_j)}\right)^T}{\sqrt{d_k}}\right), \tag{2}$$

where $W_a^{(v_i,v_j)} \in R^{n_a \times n_a}$ denotes the alignment relationships between views v_i and v_j. $W_a^{(v_i,v_j)}$ is a score matrix, whose (i,j)-th entry measures the alignment ratio given by the i-th instance representation in view v_i to the j-th instance representation in view v_j.

Hence, to jointly learn the sample features and view alignment of V views, we propose the following loss function L_1:

$$
L_1 = \sum_{v_i=1}^{V} \left\| A^{(v_i)} - D^{(v_i)} \left(E^{(v_i)} \left(A^{(v_i)} \right) \right) \right\|_2^2
$$
$$
+ \sum_{v_i \neq v_j} \left\| E^{(v_i)} \left(A^{(v_i)} \right) - W_a^{(v_i,v_j)} * E^{(v_j)} \left(A^{(v_j)} \right) \right\|_2^2, \tag{3}
$$

where the former reconstruction loss preserves the local structure and prevents distortion of the embedding space. The latter loss minimizes the difference between view v_i and the linearly varying view v_j to make the representations of the same data in different views as similar as possible.

To further learn the prior alignment relations of aligned data $A^{(v)}$, we design the loss function L_2 to measures the difference between the learned alignment matrix $W_a^{(v_i,v_j)}$ and the ground-truth alignment matrix $W_{gt}^{(v_i,v_j)} \in \{0,1\}$:

$$
L_2^{(v_i,v_j)} = \| W_a^{(v_i,v_j)} - W_{gt}^{(v_i,v_j)} \|_2^2, \tag{4}
$$

By synthesizing the above objectives L_1 and L_2, our AFAN is designed to minimize the following objective function:

$$
L = L_1 + \mu \sum_{v_i \neq v_j} L_2^{(v_i,v_j)}, \tag{5}
$$

where $\mu > 0$ is a trade-off hyper-parameter.

For multiple views, we need to select one view as the anchor and compute the alignment matrix W between other shuffled views and the anchor view respectively. e.g., we take view v_i as the anchor and compute the alignment matrix of view v_i and shuffled view $v_j, v_j \in \{1, 2, \ldots, V\}, v_j \neq v_i$ to make view v_j align to view v_i.

3.4   Self-augmentation Strategy

To learn cluster-oriented sample features, we introduce a self-augmentation strategy. The self-augmentation strategy maps samples to a feature space where sample features belonging to the same cluster crowd together.

We apply the self-augmentation strategy to all data $X^{(v)}$ to enhance all sample representations. Similarly, taking views v_i and v_j as an example, we take the unchanged data $X^{(v_i)}$ and the shuffled data $X^{(v_j)}$ through the multi-view encoders to obtain the latent representations $H^{(v_i)}$ and $H^{(v_j)}$, and then compute the all data alignment matrix $W^{(v_i,v_j)}$. After that, we use the alignment matrix $W^{(v_i,v_j)}$ to concatenate $H^{(v_i)}$ and $H^{(v_j)}$ as the common representation:

$$
H = H^{(v_i)} \| W^{(v_i,v_j)} H^{(v_j)}, \tag{6}
$$

where H is the common representation of the multi-view data $X^{(v)}$, $\|$ denotes concatenate operation.

Note that in this concatenate method, view v_i acts as an anchor, and then view v_j is aligned to it through linear changes. So we use $H^{(v_i)}$ to build the k-nearest neighbor graph, where vertex refers to sample features. Specifically, calculate the cosine similarity between two samples and construct the graph S:

$$S_{i,j} = \begin{cases} \dfrac{\left(H^{(v_i)}\right)_{i,:}\left(H^{(v_i)}\right)_{j,:}^T}{\left\|\left(H^{(v_i)}\right)_{i,:}\right\|_2 \left\|\left(H^{(v_i)}\right)_{j,:}\right\|_2}, i \neq j \\ 0, i = j \end{cases} \tag{7}$$

where $\left(H^{(v_i)}\right)_{i,:}$ and $\left(H^{(v_i)}\right)_{j,:}$ denote the i-th and j-th row of $H^{(v_i)}$, $S_{i,j}$ represents feature similarity between the i-th and j-th samples, $\|\cdot\|_2$ is l_2-norm.

Then, we leave the largest k values for each row and column in S and apply normalization on it to get the adjacency matrix $E = D^{-\frac{1}{2}}SD^{-\frac{1}{2}}$. D is the degree diagonal matrix and $D_{ii} = \sum_j S_{i,j}$. After that, we use E as the Laplacian matrix in GCN [7] to propagate feature information. The i-th row sample common representation in H is updated as follows:

$$H_{i,:}^* = \sum_{j=1}^{n} (\alpha I + E)_{i,j}^\tau H_{j,:}, \tag{8}$$

where H^* is the updated common representation. I is the identity matrix, α and τ are hyper-parameters. α controls the balance of neighbors' representations and the self-ones. τ decides the number of times for feature propagation.

Next, use k-means to cluster H^* to get the final clustering results. The training process of our AFAN is summarized in Algorithm 1.

4 Experiments

4.1 Experimental Settings

Datasets. We experiment on four multi-view datasets. **Caltech101-20** [8,12] contains 2386 images from 20 classes. We use HOG and GIST features as two views. **Scene-15** [3] consists of 15 categories and 4485 images. We use two features, i.e. PHOG and GIST features. **LUse-21**[1] consists of 2100 samples from 21 different categories. We choose CENTRIST and GIST features for our experiments. **Leaves** [10] contains the binary images of 1584 leaf samples of 99 classes. We choose Shape and Margin features for experiments.

For the above four multi-view datasets, we randomly divide them into two parts $\left\{A^{(v)}, U^{(v)}\right\}_{v=1}^{V}$ with the equal size, where $A^{(v)}$ retains alignment relationships across views to represent aligned data, and $U^{(v)}$ are randomly shuffled to simulate unaligned data.

---

[1] http://weegee.vision.ucmerced.edu/datasets/landuse.html.

Algorithm 1: Training process of AFAN

Input: Dataset $\left\{X^{(v)}\right\}_{v=1}^{V} = \left\{A^{(v)}, U^{(v)}\right\}_{v=1}^{V}$; Batch size m; Hyper-parameters μ, k, α and τ.

Output: Clustering results *label*.

1: **Initialization**: Pre-train the multi-view encoders and decoders with aligned data $A^{(v)}$ by Eq. (3).
2: **while** not converged **do**
3:     Choose a batch of data from unchanged data $A^{(v_i)}$ and shuffled data $A^{(v_j)}$.
4:     Calculate loss L by Eq. (5) and the alignment matrix of this batch by Eq. (2).
5:     Update network parameters via back-propagation.
6: **end while**
7: Feed all multi-view data $X^{(v)}$ into trained AFAN and obtain data feature $H^{(v)}$.
8: Calculate all data alignment matrix W by Eq. (2) and calculate common representation H by Eq. (6).
9: Update feature H to get H^* by Eq. (8).
10: Obtain clustering results *label* by perform k-means on H^*.

Compared Methods. We compare AFAN with the following baselines: (1) Multi-view clustering methods: CCA [2], DCCA [1], DCCAE [17], LMSC [20], MVC-DMF [24], SwMC [11], AE^2-Nets [21]. (2) Partially view-aligned clustering methods: PVC [4] and MvCLN [18].

In the comparison process, only PVC, MvCLN and our AFAN can solve PVP, so we use the same two comparison schemes as PVC: (1) for multi-view clustering methods that cannot handle partially view-aligned data, we first use PCA to map the original data to the latent space, then apply the Hungarian algorithm in the latent space to align the unaligned data, and finally execute these multi-view clustering methods on the aligned data. For PVC, MvCLN and our AFAN, we perform directly on partially view-aligned data. (2) we directly execute these multi-view methods that cannot handle partially view-aligned data on fully-aligned data.

We adopt widely used accuracy (ACC), normalized mutual information (NMI), and F-measure(F-mea) to measure the performance of all methods.

Settings. For all baselines, we fine-tune them according to the parameter ranges and experimental details described in the original paper until the performance is optimal and stable. For our proposed AFAN, we use the Adam optimizer [5] with the learning rate of 10^{-3} with weight decay 10^{-5}. The batch size is fixed to 128. For four datasets, we experimentally set the best pre-train epoch from {500, 1000, 1500, 2000} and seek μ in the range {0.001, 0.01, 0.1, 1, 10, 100 1000}. The encoder dimensions are set to D-1024-1024-20 for Caltech101-20, LUse-21, and Leaves, and D-2048-1024-128 for Scene-15, where D is the dimension of input data. We set the decoder with a symmetrical structure and adopt ReLU as the activation function.

Table 1. Clustering performance of ACC(%), NMI(%) and F-mea(%), and the best results are in bold.

Aligned	Method	Caltech101-20			Scene-15			LUse-21			Leaves		
		ACC	NMI	F-mea	ACC	NMI	F-mea	ACC	NMI	F-mea	ACC	NMI	F-mea
Fully	CCA	39.52	57.53	33.90	34.38	37.37	29.07	21.43	29.27	15.31	26.54	76.20	18.39
	DCCA	41.95	60.72	35.89	35.88	39.90	30.08	30.95	**45.93**	**27.83**	32.10	**83.22**	24.99
	DCCAE	44.17	60.83	**42.11**	**36.68**	**40.56**	30.11	27.62	44.21	19.58	27.78	80.96	21.07
	LMSC	31.56	32.17	22.18	33.60	32.98	31.09	**32.67**	36.99	22.50	49.56	72.12	35.55
	MVC-DMF	**59.72**	**62.76**	38.16	29.70	29.72	26.45	25.19	32.57	15.00	39.67	68.08	26.29
	SwMC	52.68	56.87	34.96	27.47	35.71	24.55	14.24	16.92	10.37	**67.17**	82.68	33.62
	AE^2	30.09	32.31	19.11	36.16	39.98	**34.14**	13.46	10.56	10.51	41.90	73.18	**41.30**
Partially	CCA	**42.20**	**60.84**	**36.15**	32.46	**35.81**	26.00	11.90	25.36	8.23	24.07	76.71	18.88
	DCCA	23.60	30.64	21.44	**33.24**	34.27	**28.94**	**32.86**	**47.28**	**28.66**	26.54	76.03	22.65
	DCCAE	31.68	38.61	26.22	30.28	33.31	27.11	22.38	37.77	18.26	29.63	**79.20**	**25.55**
	LMSC	26.03	35.65	20.27	22.16	16.46	20.00	18.57	16.40	10.15	**32.89**	60.50	19.56
	MVC-DMF	19.74	21.58	17.43	21.54	19.31	19.78	10.31	4.80	5.82	11.52	42.24	1.49
	SwMC	38.94	30.14	21.09	18.73	21.30	15.85	11.52	10.12	9.17	20.01	31.43	2.43
	AE^2	29.97	47.64	26.16	29.52	28.43	26.45	12.68	9.95	10.56	28.18	60.51	23.16
Partially	PVC	48.07	65.39	55.03	37.32	39.33	33.05	30.52	52.50	25.09	47.66	70.86	47.12
	MvCLN	44.92	55.61	32.07	37.00	39.40	34.76	28.00	32.66	26.41	38.35	71.44	37.30
	AFAN(ours)	**51.38**	**69.19**	**58.47**	**38.93**	**40.15**	**38.11**	**36.00**	**58.92**	**30.26**	**57.77**	**84.11**	**54.19**

4.2 Experimental Results and Analysis

The clustering results are listed in Table 1. Partially represents the comparison scheme (1): the multi-view clustering methods are performed on the dataset preprocessed by the Hungarian algorithm, and the partially view-aligned clustering methods are performed on the partially view-aligned data. Fully represents the comparison scheme (2): the multi-view clustering methods are directly performed on the fully view-aligned data. Note that the Caltech101-20 and Scene-15 datasets are the same as PVC, we directly use the experimental results in PVC on these two datasets.

We observe that: (1) in the first comparison scheme, our AFAN significantly outperforms other baselines in all four datasets. Particularly, our AFAN outperforms the best baseline by 3.31%, 3.8%, 3.44% (Caltech101-20), 1.61%, 0.75%, 3.35% (Scene-15), 5.48%, 6.42%, 3.85% (LUse-21), and 10.11%, 12.67%, 7.07% (Leaves) on the ACC, NMI and F-mea metrics. This verifies that our AFAN can perform effective alignment and clustering. There are two reasons. On the one hand, compared with the method based on Euclidean distance to calculate the alignment matrix, AFAN can capture the data similarity in the training process to adaptively calculate the alignment matrix. On the other hand, the data features learned by AFAN through the self-augmentation strategy are suitable for clustering scenarios. (2) in the second comparison scheme, our method still achieves competitive results even though the input data for the baselines are fully view-aligned data, while ours is partially view-aligned data.

Table 2. Ablation studies on Scene-15. "✓" (yes) or "×" (no) mean that we use or do not use the self-augmentation strategy.

Self-augmentation	ACC(%)	NMI(%)	F-mea(%)
×	37.77	39.67	36.93
✓	**38.9**	**40.15**	**38.11**

4.3 Ablation Studies

In this section, we conduct ablation studies to validate the effectiveness of our self-augmentation strategy on the Scene-15 dataset.

We first compare the results of our proposed AFAN with and without the self-augmentation strategy on ACC, NMI and F-mea. As shown in Table 2, The self-augmentation strategy improves the clustering performance to a certain extent, which indicates that it is crucial and beneficial for our proposed AFAN.

Then, we employ the t-SNE [9] technique to visualize the sample embeddings obtained by our AFAN with and without the self-augmentation strategy. As shown in Fig. 3, (a) is the visualization result without the self-augmentation strategy, and (b) is the visualization result with the self-augmentation strategy. The sample embeddings from the same cluster in (a) are scattered, while they are compact in (b). We can see that the self-augmentation strategy maps the data into a clustering-friendly space where distance-based clustering algorithms (e.g., k-means) are able to classify the data more easily.

 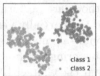

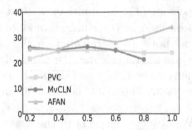

(a) Result without the self-augmenta-tion strategy.   (b) Result with the self-augmentation strategy.

Fig. 3. Visualization results of sample embeddings obtained by our AFAN with and without the self-augmentation strategy with t-SNE on Scene-15 dataset.

Fig. 4. Clustering F-measure(%) of PVC, MvCLN and AFAN on LUse-21 dataset with varying alignment ratio. (MvCLN does not work with alignment ratio = 1.0).

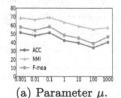 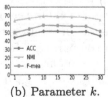 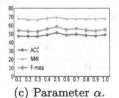 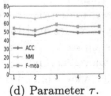

(a) Parameter μ.   (b) Parameter k.   (c) Parameter α.   (d) Parameter τ.

Fig. 5. Clustering performance of AFAN for different u, k, α and τ on Caltech101-20 dataset.

4.4   Parameter Analysis

The Balance Hyper-parameter μ. As can be seen in Eq. (5), AFAN introduces a hyper-parameter μ. We set μ to vary in the range {0.001, 0.01, 0.1, 1, 10, 100, 1000}. Figure 5(a) shows that AFAN obtains stable clustering results in the range of μ from 0.001 to 0.1. Finally, we select $\mu = 0.1$ in this range.

The Hyper-parameters k, α and τ of the Self-augmentation Strategy. To investigate the influence of k in the k-nearest neighbor graph, we report the performance of AFAN by increasing k from 1 to 30 with an interval of 5. The results in Fig. 5(b) show that AFAN performs stably when k ranges into [10, 25], which indicates that our AFAN is robust to the parameter k.

AFAN introduces hyper-parameters α and τ in Eq. (8). We fix the k as 10 and report the performance with a set of α values in {0.1, 0.2, 0.3, 0.4, 0.5, 0.6, 0.7, 0.8, 0.9, 1.0} and τ values in {1, 2, 3, 4, 5}. The performance of AFAN is shown in Fig. 5(c) and 5(d). We can observe that both α and τ are too large or too small to be inappropriate. Specifically, for α, AFAN achieves the best performance when neighbor representations and self-representation are equally divided, i.e., $\alpha = 0.5$. For τ, when τ is too small, only the information of the nearest neighbors can be obtained. When τ is too large, negative information of samples from other clusters is received, which affects the clustering results. Finally, we set $\alpha = 0.5$ and $\tau = 3$ as the hyper-parameter values of the AFAN.

4.5   Alignment Ratio Analysis

In this section, we investigate the performances of PVC, MvCLN and our AFAN with different alignment ratios. Figure 4 shows clustering F-measure(%) of these methods on the LUse-21 dataset with varying alignment ratios. We observe that: (1) the performance of our AFAN generally increases with the alignment rate. (2) AFAN outperforms PVC and MvCLN basically at all alignment ratio settings.

5   Conclusion

In this paper, we propose an adaptive view-aligned and feature augmentation network(AFAN) to solve the partially view-aligned problem, which avoids the limitations of existing partially view-aligned methods. By utilizing an adaptive view-aligned module, AFAN can adaptively compute the alignment matrix with the self-attention mechanism. The calculated alignment matrix can capture data similarity by jointly learning data features and view alignment. By introducing a self-augmentation strategy, AFAN encourages the learned features belonging to the same class to be crowded together. Such features are clustering-friendly and improve the clustering performance. Extensive experiment results on four benchmark datasets verify the effectiveness of AFAN. In the future, we will consider extending our approach to other tasks with both missing and unaligned data.

References

1. Andrew, G., Arora, R., Bilmes, J.A., Livescu, K.: Deep canonical correlation analysis. In: ICML, vol. 28, pp. 1247–1255 (2013)
2. Chaudhuri, K., Kakade, S.M., Livescu, K., Sridharan, K.: Multi-view clustering via canonical correlation analysis. In: ICML, vol. 382, pp. 129–136 (2009). https://doi.org/10.1145/1553374.1553391
3. Fei-Fei, L., Perona, P.: A bayesian hierarchical model for learning natural scene categories. In: CVPR, pp. 524–531 (2005). https://doi.org/10.1109/CVPR.2005.16
4. Huang, Z., Hu, P., Zhou, J.T., Lv, J., Peng, X.: Partially view-aligned clustering. In: NeurIPS (2020)
5. Kingma, D.P., Ba, J.: Adam: A method for stochastic optimization. In: ICLR (2015)
6. Lampert, C.H., Krömer, O.: Weakly-paired maximum covariance analysis for multimodal dimensionality reduction and transfer learning. In: Daniilidis, K., Maragos, P., Paragios, N. (eds.) ECCV 2010. LNCS, vol. 6312, pp. 566–579. Springer, Heidelberg (2010). https://doi.org/10.1007/978-3-642-15552-9_41
7. Li, J., Liu, G.: Few-shot image classification via contrastive self-supervised learning. CoRR abs/2008.09942 (2020)
8. Li, Y., Nie, F., Huang, H., Huang, J.: Large-scale multi-view spectral clustering via bipartite graph. In: AAAI, pp. 2750–2756 (2015)
9. Van der Maaten, L., Hinton, G.: Visualizing data using t-sne. J. Mach. Learn. Res. **9**(11) (2008)
10. Mallah, C., Cope, J., Orwell, J., et al.: Plant leaf classification using probabilistic integration of shape, texture and margin features. Signal Processing, Pattern Recognition and Applications **5**(1), 45–54 (2013)
11. Nie, F., Li, J., Li, X.: Self-weighted multiview clustering with multiple graphs. In: IJCAI, pp. 2564–2570 (2017). https://doi.org/10.24963/ijcai.2017/357
12. Peng, X., Huang, Z., Lv, J., Zhu, H., Zhou, J.T.: COMIC: multi-view clustering without parameter selection. In: ICML, vol. 97, pp. 5092–5101 (2019)
13. Vinokourov, A., Shawe-Taylor, J., Cristianini, N.: Inferring a semantic representation of text via cross-language correlation analysis. In: NIPS, pp. 1473–1480 (2002)
14. Wen, J., Zhang, Z., Xu, Y., Zhang, B., Fei, L., Xie, G.: Cdimc-net: cdeep incomplete multi-view clustering network. In: IJCAI, pp. 3230–3236 (2020). https://doi.org/10.24963/ijcai.2020/447
15. Wen, J., Zhang, Z., Zhang, Z., Wu, Z., Fei, L., Xu, Y., Zhang, B.: Dimc-net: deep incomplete multi-view clustering network. In: Proceedings of the 28th ACM International Conference on Multimedia, pp. 3753–3761 (2020). https://doi.org/10.1145/3394171.3413807
16. Xie, J., Girshick, R.B., Farhadi, A.: Unsupervised deep embedding for clustering analysis. In: ICML, vol. 48, pp. 478–487 (2016)
17. Xu, C., Guan, Z., Zhao, W., Wu, H., Niu, Y., Ling, B.: Adversarial incomplete multi-view clustering. In: IJCAI, pp. 3933–3939 (2019). https://doi.org/10.24963/ijcai.2019/546
18. Yang, M., Li, Y., Huang, Z., Liu, Z., Hu, P., Peng, X.: Partially view-aligned representation learning with noise-robust contrastive loss. In: CVPR, pp. 1134–1143. Computer Vision Foundation/IEEE (2021)
19. Yin, M., Huang, W., Gao, J.: Shared generative latent representation learning for multi-view clustering. In: AAAI, pp. 6688–6695 (2020)

20. Zhang, C., Hu, Q., Fu, H., Zhu, P., Cao, X.: Latent multi-view subspace clustering. In: CVPR, pp. 4333–4341 (2017). https://doi.org/10.1109/CVPR.2017.461
21. Zhang, C., Liu, Y., Fu, H.: Ae2-nets: autoencoder in autoencoder networks. In: CVPR, pp. 2577–2585 (2019)
22. Zhang, X., Mu, J., Liu, H., Zhang, X.: Graphnet: graph clustering with deep neural networks. In: ICASSP, pp. 3800–3804 (2021)
23. Zhang, X., Zong, L., Liu, X., Yu, H.: Constrained nmf-based multi-view clustering on unmapped data. In: AAAI, pp. 3174–3180. AAAI Press (2015)
24. Zhao, H., Ding, Z., Fu, Y.: Multi-view clustering via deep matrix factorization. In: Singh, S., Markovitch, S. (eds.) AAAI, pp. 2921–2927 (2017)

Data Mining Processes and Pipelines

Continuously Predicting the Completion of a Time Intervals Related Pattern

Nevo Itzhak[1](✉), Szymon Jaroszewicz[2,3], and Robert Moskovitch[1]

[1] Software and Information Systems Engineering,
Ben-Gurion University of the Negev, Beer Sheva, Israel
nevoit@post.bgu.ac.il, robertmo@bgu.ac.il
[2] Institute of Computer Science, Polish Academy of Sciences, Warsaw, Poland
szymon.jaroszewicz@ipipan.waw.pl
[3] Faculty of Mathematics and Information Science,
Warsaw University of Technology, Warsaw, Poland

Abstract. In various domains, such as meteorology or patient data, events' durations are stored in a database, resulting in symbolic time interval (STI) data. Additionally, using temporal abstraction techniques, time point series can be transformed into STI data. Mining STI data for frequent time intervals-related patterns (TIRPs) was studied in recent decades. However, for the first time, we explore here how to continuously predict a TIRP's completion, which can be potentially applied with patterns that end with an event of interest, such as a medical complication, for its prediction. The main challenge in performing such a completion prediction occurs when the time intervals are coinciding, but not finished yet, which introduces an uncertainty in the evolving temporal relations, and thus on the TIRP's evolution process. In this study, we introduce a new structure to overcome this challenge and several continuous prediction models (CPMs). In the segmented CPM (SCPM), the completion probability depends only on the pattern's STIs' starting and ending points, while a machine learning-based CPM (CPML) incorporates the duration between the pattern's STIs' beginning and end times. Our experiment shows that overall, CPML based on an ANN performed better than the other CPMs, but CPML based on NB or RF provided the earliest predictions.

Keywords: continuous prediction · early prediction · temporal patterns

1 Introduction

Frequent temporal patterns, whether given by a domain expert or discovered by mining, were used already for temporal knowledge discovery, clustering, or classification [5,7]. Being able to continuously estimate, in real-time, whether a temporal pattern would fully occur, while its components are being revealed, is desirable, and can be useful in various applications, such as event prediction. Estimating the probability of the last pattern's component (e.g., an event of interest) occurrence, is of great interest. For example, predicting the death of a patient in the intensive care unit (ICU), based on continuous data, consisting of a temporal pattern that was observed in the data ending with death.

© The Author(s), under exclusive license to Springer Nature Switzerland AG 2023
H. Kashima et al. (Eds.): PAKDD 2023, LNAI 13935, pp. 239–251, 2023.
https://doi.org/10.1007/978-3-031-33374-3_19

In many real-life data science problems, in which data are gathered from various sources, the multivariate temporal data are heterogeneous. Some variables may be sampled regularly but at different frequencies (e.g. sensor measurements) and some variables irregularly (e.g. variables measured manually or event-driven). Other temporal variables may be represented by events that may or may not have varying duration. In this study, we propose to employ the use of *temporal abstraction* [6, 10] to transform the entire heterogeneous multivariate temporal data into meaningful symbolic time interval series. A *symbolic time interval (STI)* is a triplet of a start time, end time, and a symbol from an ordered alphabet.

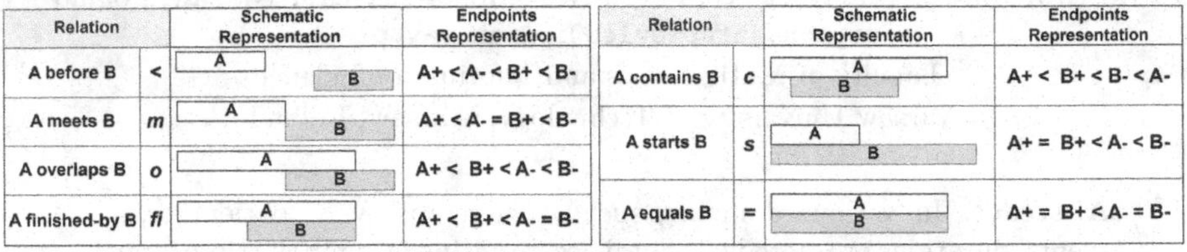

Relation		Schematic Representation	Endpoints Representation	Relation		Schematic Representation	Endpoints Representation
A before B	<	A / B	A+ < A- < B+ < B-	A contains B	c	A / B	A+ < B+ < B- < A-
A meets B	m	A / B	A+ < A- = B+ < B-	A starts B	s	A / B	A+ = B+ < A- < B-
A overlaps B	o	A / B	A+ < B+ < A- < B-				
A finished-by B	fi	A / B	A+ < B+ < A- = B-	A equals B	=	A / B	A+ = B+ < A- = B-

Fig. 1. Allen's seven temporal relations between a pair of STIs.

From STI data, frequent time intervals-related patterns can be discovered [4,10], which were shown in the past to be useful for knowledge discovery and as features for classification and prediction [5,7,11]. A *time intervals-related pattern (TIRP)* is comprised of a series of STIs and a set that defines all Allen's temporal relations (Fig. 1) between each of the pairs of STIs. For example, a pattern from time intervals data may be that hospitalized patients with COVID-19 frequently start with symptoms of "fever" and "cough," and a week later also begin experiencing shortness of breath, in which case the symptoms have not ended at the ICU admission time. Note that a TIRP's definition does not include the STIs' durations and their durations can vary among different instances. Using a frequent TIRP that ends with an event of interest, such as a patient's death, may allow for real-time continuous prediction of the completion of the TIRP's instance and of the occurrence of the event of interest. For example, the TIRP illustrated in Fig. 2 is defined by three STIs and three temporal relations, where the last STI, C, is considered as the event of interest.

Predicting a TIRP's completion is challenging due to the TIRP's instances variability which is reflected by the varying duration of the STIs, as noted earlier, and the varying duration of the gaps between the instances' STIs (e.g., in Fig. 2, between STI B and C). We define for the first time, as far as we know, the problem of predicting continuously the completion of a TIRP, and introduce novel models for the TIRP's continuous completion prediction.

The contributions of the paper are the following:

1. Defining the problem of continuous prediction of a TIRP's completion.
2. Introducing two novel methods for continuous prediction of a TIRP's completion.
3. A rigorous evaluation on real-life datasets, including new metrics to evaluate the continuous prediction of a TIRP's completion model.

2   Background

One of the forms of temporal abstraction is *state abstraction,* in which based on given cutoffs the time point values are categorized into symbols, and when adjacent time points have the same symbol, they are concatenated into a symbolic time interval. Several methods were proposed in the literature to learn the cutoffs from the data, such as equal width discretization (EWD), symbolic aggregate approximation (SAX) [6], and more [10].

A *symbolic time interval (STI)* $I = (s, e, sym)$, is a triplet of start time $s \in \mathbb{R}_{\geq 0}$, end time $e \in \mathbb{R}_{\geq 0}$, $e \geq s$ and a symbol *sym* $(sym \in \Sigma)$ from an ordered alphabet Σ. A *time intervals-related pattern (TIRP)* Q is defined as a pair $Q = (IS, R)$, where $IS = \{I^1, ..., I^k\}$ is a series of k STIs and $R = \{r(I^i, I^j) : 1 \leq i < j \leq k\}$ is a set that defines all Allen's temporal relations (Fig. 1) between each of the $(k^2 - k)/2$ pairs of STIs in IS.

Given STI series data, TIRPs can be discovered for which several TIRP mining methods were proposed in the past two decades [4,10,11], most of which use Allen's temporal relations [1] that include seven temporal relations between a pair of STIs, as shown in Fig. 1. A TIRP is called frequent if its vertical support exceeds a predefined minimum threshold. Given a database DB of $|DB|$ unique entities (e.g., patients), the *vertical support* $VS(DB, Q)$ of a TIRP Q is defined as the cardinality of the set DB^Q of distinct entities within which Q holds at least once, divided by $|DB|$ (the total number of entities in DB), $VS(DB, Q) = |DB^Q|/|DB|$.

Frequent TIRPs are typically used as features for temporal data classification or prediction [5], as proposed first in [11]. Liu et al. [7] suggested a TIRPs-semantic-based probabilistic framework for STI data that can be used to answer varied semantic-level queries in a unified way, such as predicting future activities given observed ones. To the best of our knowledge, no previous study has investigated the task of continuous prediction of a TIRP's completion.

3   Methods

A model M predicts a TIRP Q's completion, given a database DB, by estimating the probability of observing the remaining part of Q, given its observed part at time t_c. An estimation is provided at each *current* time point t_c, and changes as a given Q's instance evolves over time. The database DB comprises $|DB|$ entities (e.g., patients), where each entity contains a series of STIs. We assume that in a specific STI series, STIs with the same symbols can not overlap.

Let p_{t_c} denote a prefix representing the observed part of Q at t_c, and s_{t_c} denote a suffix representing the remaining part of Q at t_c that is expected to occur. Thus, to estimate the Q's completion probability, at time point t_c, $Pr(Q \mid t_c)$, the following simple model can be used, which typically represents the confidence of a rule in sequential patterns:

$$Pr(Q \mid t_c) = Pr(s_{t_c} \mid p_{t_c}) = \frac{Pr(p_{t_c}, s_{t_c})}{Pr(p_{t_c})} = \frac{Pr(Q)}{Pr(p_{t_c})}. \qquad (1)$$

The calculation in Formula 1 answers the question: "Out of all the times we saw p_{t_c}, how many times was it followed by s_{t_c} (i.e., Q has unfolded to completion)?" Thus, the number of times each p_{t_c} of Q occurs in the database and the number of times p_{t_c} is followed by s_{t_c} should be counted. Since the database DB comprises multiple entities, and each entity contains a lexicographically ordered STI series, instances of Q and p_{t_c} may be discovered more than once in a single entity. Each such instance is *counted separately* in the computation.

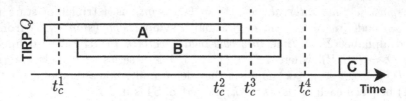

Fig. 2. TIRP Q's completion probability is estimated at any time point (e.g., t_c^2).

Applying Formula 1 to a relatively simple example sheds light on the challenges that arise while continuously predicting TIRP's completion. In Fig. 2, a TIRP $Q = \{A \text{ overlaps } B, A \text{ before } C, B \text{ before } C\}$ is shown, together with four time points t_c^1, t_c^2, t_c^3, and t_c^4 chosen to illustrate partial instances of the pattern to demonstrate various types of challenges. Calculating the numerator $Pr(Q)$ is quite straightforward and is done by counting the number of times that A *overlaps B* is followed by A *before C* and B *before C* (i.e., Q). However, calculating the denominator $Pr(p_{t_c})$ is more challenging in some cases.

At t_c^4, the denominator $Pr(p_{t_c^4})$ is equal to the probability of seeing A *overlaps B*, which results in no uncertainty. Similar computations can be carried out at time points t_c^1 and t_c^3, but the situation is more complex since the time points are located after a starting point and before an ending point of an STI. Thus since an STI that is not finished yet is involved, p_{t_c} and s_{t_c} cannot be described with Allen's temporal relations. Instead, we need to use a different representation based on STIs' *tieps* (Definition 1).

Definition 1. (tiep) A *time interval endpoint* is a triplet $(t, type, sym)$ consisting of a time stamp $t \in \mathbb{R}_{\geq 0}$, an endpoint type, which can be either starting (+) or ending (-), and a symbol ($sym \in \Sigma$) from an ordered alphabet Σ.

Example 1. In Fig. 1, for an STI $A = (A_s, A_e, \text{``}A''\text{''})$, the starting and ending *tieps* are defined respectively as $A+ = (A_s, +, \text{``}A''\text{''})$ and $A+ = (A_e, -, \text{``}A''\text{''})$.

A total order on *tieps* (Definition 1) is defined based on their time stamps, which are real numbers. Thus, the *tieps* can be used in inequalities defining temporal relations, while their structure will be exploited in the following sections.

At t_c^1, the $p_{t_c^1}$ is "A that has started but not ended yet," and thus, $Pr(A+)$ denotes the probability of seeing STI A that has started in DB. In the learning stage, in the database DB, each STI has its starting and ending *tieps*, and thus, the probability $Pr(A+)$ equals the probability of seeing STI A in DB. Similarly, at t_c^3, the $p_{t_c^3}$ can be represented by the following inequality between the STIs' *tieps*: $A+ < B+ < A-$.

However, in the learning stage, since STI B has started but not ended yet, its ending *tiep* $B-$ has to satisfy $t_c^3 < B-$ in database DB. Thus, the prefix's *tiep* ordering should be extended to $p_{t_c^3} = A+ < B+ < A- < B-$ to represent that STI B ended after B ended. The extended $p_{t_c^3}$ is equivalent to Allen's temporal relation A *overlaps* B (see Fig. 1). Uncertainty occurs at time point t_c^2, since $p_{t_c^2}$ includes STIs A and B that have already started but not yet ended (i.e., $t_c^2 < A-$ and $t_c^2 < B-$), it results in uncertainty about which temporal relation between A and B will finally unfold. Three different temporal relations are possible: *overlaps*, *contains*, or *finished-by*, which should be considered and used in Formula 1.

3.1   The Unfinished Coinciding STIs Challenge

Definition 2. (Unfinished STI) An *unfinished STI* I^* at time t_c is an STI whose starting *tiep* I^*+ satisfies $0 \le I^*+ \le t_c$ and whose ending *tiep* I^*- satisfies $t_c < I^*-$.

Throughout the text, the asterisk (*) will indicate that an STI is unfinished. The start time of an unfinished STI is known at time t_c, but its end time is not. In fact, it is censored: we only know that it is later than t_c.

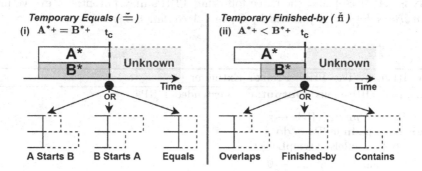

Fig. 3. The evolving temporary temporal relations.

A pair of unfinished coinciding STIs A^* and B^* may evolve into three possible temporal relations. The logic follows from the *tieps* representation of Allen's temporal relations that is presented in Fig. 1. Figure 3.i shows that in the case of the *temporary equals* ($\doteq$) temporal relation, their temporal relation may eventually evolve into A *starts* B, or B *starts* A, or stay at A equals B. The reason that A *starts* B and B *starts* A cannot be distinguished at t_c is that the exact temporal relation is determined by their end times, which are not yet known. Similarly, the *temporary finished-by* ($\tilde{\text{fi}}$) temporal relation shown in Fig. 3.ii, may eventually evolve into three possible Allen's temporal relations: A *overlaps* B, A *contains* B, or stay at A *finished-by* B.

3.2   TIRP-Prefixes

A TIRP can be represented by a series of starting and ending *tieps* instead of a series of STIs.

To maintain the conjunction of pairwise temporal relations among the STIs, the set of *tieps* has to be transformed into a sorted tieps' series, based on Allen's *tieps* representation (Fig. 1, right column).

A TIRP is divided into TIRP-prefixes (Def. 3) that are part of the TIRP's evolution process, which are created based on sub-sequences of the TIRP's *tieps*. In each TIRP-prefix, since the temporal relation between two unfinished STIs is uncertain, the temporary temporal relation $\check{r}$ is used to express the disjunction of possible final temporal relations based on the unfinished coinciding STIs challenge logic explained in Fig. 3.

Definition 3. (TIRP-prefix) Let Q be a TIRP of length k. A *TIRP-prefix* $\check{Q}$ of Q is defined as a pair $\check{Q} = (\check{IS}, \check{R})$, where $\check{IS}$ is a lexicographical ordered STI series of $\check{k} \leq k$ finished ($\check{IS}_f$) and unfinished ($\check{IS}_*$) STIs: $\check{IS} = \check{IS}_f \cup \check{IS}_*$, and $\check{R}$ is the set of all the temporal relations between each of the pairs of STIs in $\check{IS}$: $\check{R} = \check{R}_f \cup \check{R}_*$, where $\check{R}_f = \{r(I^i, I^j) : 1 \leq i < j \leq \check{k} \land \neg(I^i \in \check{IS}_* \land I^j \in \check{IS}_*)\}$ and $\check{R}_* = \{\check{r}(I^{*,i}, I^{*,j}) : 1 \leq i < j \leq \check{k} \land I^{*,i} \in \check{IS}_* \land I^{*,j} \in \check{IS}_*\}$.

Example 2. In Fig. 2, at t_c^2, it is known that $p_{t_c^2} = A^*{+} < B^*{+}$, thus the TIRP-prefix is $\{A^* \text{ fi } B^*\}$ and the three following TIRPs may potentially evolve into: $\{A \text{ overlaps } B\}$ or $\{A \text{ finished-by } B\}$ or $\{A \text{ contains } B\}$.

Algorithm 1. The TIRP-Prefix's Extender

Input: px - TIRP-prefix; **Output**: epx - extended TIRPs

```
1: unfPairs ← px.Ř*; eRels ← ∅
2: for each pa in unfPairs do
3:     eRels ← eRels ∪ tempLogic(pa)
4: cmb = comb(eRels); epx ← ∅
5: for each c in cmb do
6:     cand ← cmb ∪ px.Řf
7:     if validTransitionTable(cand) then
8:         epx ← epx ∪ cand
9: return epx
```

In the TIRP-Prefix's Extender algorithm (Algorithm 1), given a TIRP-prefix *px*, generates a set *epx* of all possible TIRPs that can evolve from *px*. The TIRP-prefix's temporary disjunctions of temporal relations (*unfPairs*, line 1) are set based on the rules presented in Fig. 3 by using the *tempLogic* function (lines 2-3). Then, the *comb* function generates a set of all the possible temporal relations that can be evolved given $\check{R}_*$, which is stored in *cmb*. Then, each *cmb* is joined with the TIRP-prefix non-temporary disjunctions of temporal relations $\check{R}_f$ and assigned to *cand*. Each *cand* represents a TIRP that can evolve from *px*.

However, *cand* can be a pattern with combinations of temporal relations that contradict each other. For example, A^* *overlaps* B^* and B^* *overlaps* C^*, the temporal relation between A^* and C^* can not be the relations *finished-by* or *contains*, but only *overlaps*. Thus, Allen's transition table [1] is used to reduce the number of generated candidates by avoiding impossible patterns (line 7).

Lastly, the potential evolved TIRPs *epx*'s instances have to be detected in the STI database DB by using rather a STIs based [10] or sequence-based [4] representation.

3.3 Continuous Prediction Models (CPMs)

In this section we will present the following two continuous TIRP completion prediction models:

SCPM. The predicted TIRP's completion probability changes only at time points where *tieps* appear. As discussed in detail when Formula 1 was explained, the probability of TIRP Q completion is $Pr(Q)/Pr(p_{t_c})$, where p_{t_c} is a TIRP-prefix of Q at time t_c.

CPML. The durations between consecutive TIRP-prefixes' *tieps* were used as features. For that, naive Bayes (NB) [9], random forest (RF) [2], and artificial neural network (ANN) [8] classifiers were used.

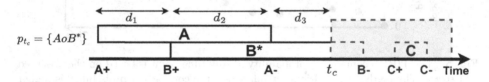

Fig. 4. Time durations d_1, d_2, and d_3 are based on *tieps* A+, B+, A-, and t_c, which are used as features for the classifiers to perform the TIRP's completion prediction.

Records for the classifier were created to represent the evolution of a TIRP over time. Multiple records were used as input for the classifier to include all the time stamps for each evolving TIRP-prefix instance. The TIRP's duration elements that were not observed until time point t_c are set to zero. Each instance's record target was set to whether the instance was finally unfolded into a TIRP's completion or not. For example, in Fig. 4, the TIRP-prefix instance {A *overlaps* B^*} is represented by the time durations between the four consecutive *tieps* ($f_1 = [A+,B+], f_2 = [B+,A-], f_3 = [A-,B-]$, and $f_4 = [B-,C+]$), which are used as features ($[f_1, f_2, f_3, f_4]$) for the classifier. For the instance at t_c, the record values are $f_1 = d_1$, $f_2 = d_2$, $f_3 = d_3$, and $f_4 = 0$.

The parameters for the models were selected after testing the performance of each considered combination. The parameters that performed best are the following: RF - maximum depth of 5, using bootstrap with 100 trees in the forest; ANN - two 50-neurons hidden layers, a maximum of 20 epochs, a batch size of 16, learning rate of 0.001 with gradually decreasing with early stopping, and with the ReLU activation function. NB, RF, and ANN were implemented with Python 3.6 and the Scikit-Learn package (scikit-learn.org) version 0.22.1. For parameters we did not specify, we used the package defaults.

3.4   Early Warning Strategies

Early warning strategies are used to decide that the TIRP will likely unfold once there is a high likelihood of the completion of the TIRP, based on the CPMs' estimated probabilities.

A decision that the TIRP will be unfolded is made immediately after the probability exceeds the prediction decision threshold (e.g., the gray point in Fig. 5) or when the threshold was consistently exceeded for some pre-defined time τ (e.g., the blue point in Fig. 5, which is defined with τ of three time stamps).

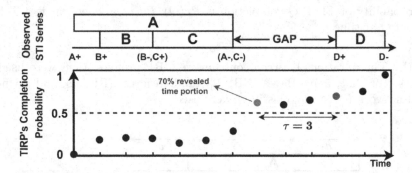

Fig. 5. The TIRP's completion probability at any time point is based on the observed STI series. The prediction decision is made when the completion probability is higher than the threshold, which is the horizontal dashed line (0.5 in this case), and a time delay τ has been passed.

4   Evaluation

Our goal was to evaluate the effectiveness of using continuous prediction models (CPMs) in predicting a TIRP's completion. The main *research questions* for this study were:

RQ1. Which CPM performs better, in terms of prediction performance and earliness, in predicting the completion of a TIRP?

RQ2. Which value of τ performs best, in terms of prediction performance and earliness, in predicting the completion of a TIRP?

4.1 Datasets

We evaluated the proposed models using real-life medical and non-medical datasets: cardiac surgical patients (CSP) [12], acute hypertensive episodes (AHE) [5], diabetes (DBT) [10], and elderly first injury fall (EFIF) [3] datasets. The events of interest were defined as the first occurrence of the following: CSP - cardiac index lower than 2.5 $L/min/m^2$, AHE - the target onset, DBT - HbA1C greater than 9%, and EFIF - first fall in elderly with a severe or moderate injury. Table 1 summarizes the main parameters of each dataset: entities (e.g., patients) number (#Ent), variables number (#Var), entities' maximum number of timestamps (#Timestamps), time granularity (Granularity), entities with the event of interest (#EntEvent), where the values in parentheses represent the percentage of #EntEvent out of #Ent, and the averaged different number of discovered TIRPs that ended with the events of interest (#TIRPs).

Table 1. The evaluation datasets' parameters

Name	#Ent	#Var	#Timestamps	Granularity	#EntEvent	#TIRPs
CSP	329	13	720	minutes	115 (35%)	257
AHE	1,000	4	238	hours	500 (50%)	246
DBT	1,710	12	24	months	239 (14%)	256
EFIF	823	15	144	weeks	121 (15%)	529

4.2 Experimental Setup

The models were evaluated on the ability to predict the completion of a TIRP that ended with an event of interest. Being able to predict continuously the completion of a TIRP, means it is possible to predict an event of interest. The entities' demographic data were not used, and only the time-based data were used for the continuous prediction. All the datasets were abstracted into STI series using SAX [6] with three symbols per variable. Then, TIRPs were discovered from the STI data using the KarmaLego algorithm [10] using Allen's seven temporal relations [1]. The patterns were discovered using 15% minimal vertical support from the entities that contained the event of interest.

Only the discovered patterns that ended with the event of interest were used. Yet, all entities, with or without the event of interest, were used to learn the model's parameters. TIRP-prefixes can be detected more than once in an entity's records, in which case each detected instance of the TIRP or its TIRP-prefixes were considered separately. The TIRP-prefixes instances from the training set were used to learn the model. To evaluate the models, all instances that started with the TIRP's earliest *tieps*, were used in the experiments. The events' STIs were considered instantaneous events, and the beginning of the event of interest was considered as the TIRP's completion.

Since each TIRP's completion was based on a different number of detected TIRP-prefixes' instances, the imbalance ratio differed between the patterns. We ran the experiments with ten-fold cross-validation, using target stratification. The instances of a TIRP and its TIRP-prefixes of the same entity appeared exclusively in the same fold.

Evaluation Metrics. To evaluate the models' performance, a receiver operating characteristics (ROC) curve was calculated, together with the corresponding area under the curve (AUROC). The decisions were made based on a prediction decision threshold, that was varied between 0 to 1, and the decision time delay τ was varied using 0, 1, 2, and 3 time units. The ROC was created by varying the prediction decision thresholds between 0 to 1 (Fig. 5).

The *revealed time portion (RTP)* refers to the percentage of the instance that is revealed, at the time of the decision, relative to the entire instance's duration (start till its end time – known retrospectively). The revealed time is in percentage, since each instance, even of the same TIRP, may have a different duration. For example, in Fig. 5 the grey dot's RTP is 70%, where $D+$ is considered as the event of interest and thus the instance's end.

4.3   Experiments and Results

Due to the limited space, each experiment is described first, followed by its results. The results are based on a total of 1,288 different TIRPs (Table 1). Each point on the charts represents the mean performance of the different TIRPs in each dataset, including confidence intervals of 95%.

Preliminary Analysis. In this experiment, we present the performance in retrospect, analyzing the decision accuracy if the decision was made according to any of the RTPs (rather than according to the model threshold decision). Additionally, we wanted to understand whether there is an ideal RTP for a decision, or the more it reveals is better. The results were computed for all TIRP-prefixes instances for each RTP, based on the completion probability at this point.

Figure 6 presents the mean AUROC at various TIRP-prefixes instances' RTP. The charts show that the more the instance is revealed, the more accurate predictions are provided by the models. RF and ANN perform best on all the datasets, except the EFIF dataset, in which the SCPM performs best above 40% revealed portion, while the CPMs perform worse. Thus, instance unfoldment can be predicted better when more information is revealed, but there is no optimal stage. However, early predictions are also desirable, so there is a trade-off, as we demonstrate in Fig. 8.

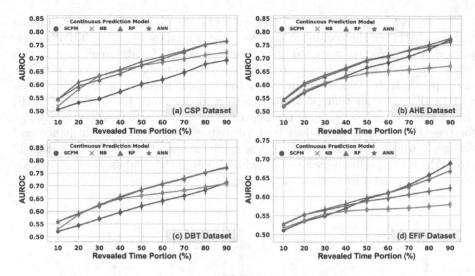

Fig. 6. The more the instance is revealed, the more accurate the predictions are. Overall, RF and ANN perform best on all the datasets, except the EFIF dataset.

Continuous TIRP's Completion Prediction. In this experiment, we evaluated the models' ability to estimate the TIRPs' instances' completion, where the results were computed for all TIRP-prefixes' instances based on the decisions made by using the early warning strategies.

To answer RQ1, Fig. 7 presents the mean results of the models in predicting the TIRPs' instances completion with different values of τ. The ANN performed significantly best, except on the EFIF, in which SCPM performed best. This implies that the duration distributions between TIRP-prefixes' consecutive *tieps* are similar between instances that ended with and without the event of interest on the EFIF dataset, and this may explain why SCPM performed better. The NB performed worst in all datasets.

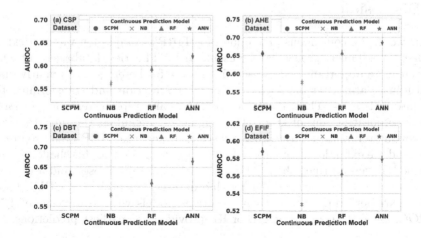

Fig. 7. ANN performed better than the other models with an average of 1.5% AUROC.

To answer RQ1 and RQ2, Fig. 8 presents the AUROC versus the mean instances' RTP of the corresponding decisions, for cases when the TIRP's completion was correctly predicted (true positive cases), for different models and values of τ, which are represented by five different colors and four different sizes, respectively. While NB provided the earliest predictions for CSP, AHE, and EFIF, its prediction performances were poor. Also, for CSP, and EFIF, the SCPM, and ANN provided the latest but most accurate predictions. Except for the DBT dataset, there is a trade-off between prediction performance and earliness, where more accurate models need more time to make decisions. It strengthens the preliminary analysis, which showed the models were more accurate as time passed for each instance (Fig. 6).

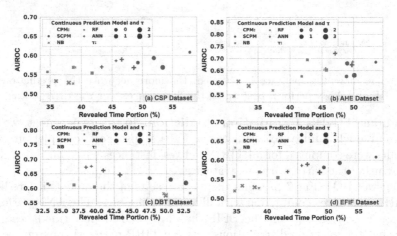

Fig. 8. More accurate models need more time to make decisions. Overall, SCPM provided the latest predictions, and NB and RF provided the earliest predictions.

5   Discussion

In this work, the continuous prediction of a TIRP's completion was studied for the first time. This approach can be useful with STI series databases, but what makes it more important and impactful is its use for heterogeneous multivariate longitudinal data, after employing temporal abstraction and transforming the data into STI series. Thus, it can be applied to any type of temporal variable, while incorporating any of them. The challenges, including the uncertainty of the evolving temporal relations, were discussed, and the TIRP-prefix representation and the extender algorithm (Algorithm 1) to overcome this challenge were described. Based on that, the SCPM and CPML were proposed and a rigorous evaluation was performed on four real-life datasets. Overall, the CPML based on an ANN performed better than the other models with an average of 1.5% AUROC, but CPML based on NB or RF provided the earliest predictions. For future work, we intend to gear this methodology for event prediction, by applying it with multiple instances of various types of TIRPs that end with the event of interest.

Acknowledgments. Nevo Itzhak was funded by the Israeli Ministry of Science and Technology Jabotinsky scholarship grant #3-16643.

References

1. Allen, J.F.: Maintaining knowledge about temporal intervals. Commun. ACM **26**(11), 832–843 (1983)
2. Breiman, L.: Random forests. Mach. Learn. **45**(1), 5–32 (2001)
3. Dvir, O., Wolfson, P., Lovat, L., Moskovitch, R.: Falls prediction in care homes using mobile app data collection. In: Michalowski, M., Moskovitch, R. (eds.) AIME 2020. LNCS (LNAI), vol. 12299, pp. 403–413. Springer, Cham (2020). https://doi.org/10.1007/978-3-030-59137-3_36
4. Harel, O., Moskovitch, R.: Complete closed time intervals-related patterns mining. In: Proceedings of the AAAI Conference on Artificial Intelligence, vol. 35, pp. 4098–4105 (2021)
5. Itzhak, N., Pessach, I.M., Moskovitch, R.: Prediction of acute hypertensive episodes in critically ill patients. Artif. Intell. Med. **139**, 102525 (2023)
6. Lin, J., Keogh, E., Wei, L., Lonardi, S.: Experiencing SAX: a novel symbolic representation of time series. Data Min. Knowl. Disc. **15**(2), 107–144 (2007). https://doi.org/10.1007/s10618-007-0064-z
7. Liu, L., et al.: A framework of mining semantic-based probabilistic event relations for complex activity recognition. Inf. Sci. **418**, 13–33 (2017)
8. McCulloch, W.S., Pitts, W.: A logical calculus of the ideas immanent in nervous activity. Bull. Math. Biophys. **5**(4), 115–133 (1943)
9. Minsky, M.: Steps toward artificial intelligence. Proc. IRE **49**(1), 8–30 (1961)
10. Moskovitch, R., Shahar, Y.: Fast time intervals mining using the transitivity of temporal relations. Knowl. Inf. Syst. **42**(1), 21–48 (2015)
11. Patel, D., Hsu, W., Lee, M.L.: Mining relationships among interval-based events for classification. In: Proceedings of the 2008 ACM SIGMOD International Conference on Management of Data, pp. 393–404. ACM (2008)
12. Verduijn, M., Sacchi, L., Peek, N., Bellazzi, R., de Jonge, E., de Mol, B.A.: Temporal abstraction for feature extraction: a comparative case study in prediction from intensive care monitoring data. Artif. Intell. Med. **41**(1), 1–12 (2007)

Interactive Pattern Mining Using Discriminant Sub-patterns as Dynamic Features

Arnold Hien[1,2], Samir Loudni[2(✉)], Noureddine Aribi[3], Abdelkader Ouali[1], and Albrecht Zimmermann[1]

[1] Normandie Univ., UNICAEN, CNRS - UMR GREYC, Caen, France
{abdelkader.ouali,albrecht.zimmermann}@unicaen.fr
[2] TASC (LS2N-CNRS), IMT Atlantique, 4 rue Alfred Kastler, 44307 Nantes, France
{arnold.hien,samir.loudni}@imt-atlantique.fr
[3] Lab. LITIO, University of Oran1, 31000 Oran, Algeria
aribi.noureddine@univ-oran1.dz

Abstract. Recent years have seen a shift from a pattern mining process that has users define constraints before-hand, and sift through the results afterwards, to an interactive one. This new framework depends on exploiting user feedback to learn a quality function for patterns. Existing approaches have a weakness in that they use static pre-defined low-level features, and attempt to learn independent weights representing their importance to the user. As an alternative, we propose to work with more complex features that are derived directly from the pattern ranking imposed by the user. Those features are used to learn weights to be aggregated with low-level features and help to drive the quality function in the right direction. Experiments on UCI datasets show that using higher-complexity features leads to the selection of patterns that are better aligned with a hidden quality function while being competitively fast when compared to state-of-the-art methods.

1 Introduction

Constraint-based pattern mining is a fundamental data mining task, extracting locally interesting patterns to be either interpreted directly by domain experts, or to be used as descriptors in downstream tasks, such as classification or clustering. Since the publication of the seminal paper [1], two problems have limited the usability of this approach: 1) how to translate user preferences and background knowledge into constraints, and 2) how to deal with the large result sets that often number in the thousands or even millions of patterns. Replacing the original support-confidence framework with other quality measures [14] does not address the pattern explosion. Post-processing results via condensed representations still typically leaves many patterns, while pattern set mining [11] just pushes the problem further down the line.

In recent years, research on *interactive pattern mining* has proposed to alter the mining process itself: instead of specifying constraints once, mining a result set, and then post-processing it, interactive pattern mining performs an iterative loop [12]. This loop involves three repeating main steps: (1) pattern extraction in which a relatively small set of patterns is extracted; (2) interaction in which the user expresses his preferences w.r.t.

H. Kashima et al. (Eds.): PAKDD 2023, LNAI 13935, pp. 252–263, 2023.
https://doi.org/10.1007/978-3-031-33374-3_20

those patterns; (3) preference learning in which the expressed preferences are translated into a quality assessment function for mining patterns in future iterations.

The most recent proposal to dealing with the question of finding interesting patterns involves the user, via interactive pattern mining [12] often involving sampling [3], with LETSIP [4] one of the end points of this development. Other interactive methods have been proposed, APLE [5], another approach based on active preference learning to learn a linear ranking function using RANKSVM [8], and IPM [3], an MCMC-based interactive sampling framework. However, existing approaches have a short-coming: to enable preference learning, they represent patterns by independent descriptors, such as included items or covered transactions, and expect the learned function, usually a regression or multiplicative weight model, to handle relations.

In this paper, we propose a new interactive pattern mining approach that introduces more complex class of descriptors for *explainable ranking*, thereby allowing to capture the importance of item interactions. These descriptors exploit the concept of **discriminating sub-patterns**, which separate patterns that are given low rank by the user from those with high rank. By temporarily adding those descriptors, we can learn weights for them, which are then apportioned to involved items without blowing up the feature space. Results on UCI datasets show favourable trade-offs in quality-time of learning.

2 Preliminaries

Pattern Mining. Given a database $\mathcal{D}$, a language $\mathcal{L}$ defining subsets of the data and a selection predicate q that determines whether an element $\phi \in \mathcal{L}$, the task is to find the theory $\mathcal{T}(\mathcal{L}, \mathcal{D}, q) = \{\phi \in \mathcal{L} \mid q(\mathcal{D}, \phi) \text{ is true}\}$. A well-known pattern mining task is *frequent itemset mining* [1]. Let $\mathcal{I}$ be a set of n *items*, an *itemset* (or pattern) X is a non-empty subset of $\mathcal{I}$. The language of itemsets corresponds to $\mathcal{L}_{\mathcal{I}} = 2^{\mathcal{I}} \backslash \emptyset$. A transactional dataset $\mathcal{D}$ is a bag (or multiset) of transactions over $\mathcal{I}$, where each *transaction t* is a subset of $\mathcal{I}$, i.e., $t \subseteq \mathcal{I}$; $\mathcal{T} = \{1, ..., m\}$ a set of m *transaction* indices. An itemset X *occurs* in a transaction t, iff $X \subseteq t$. The *cover* of X in $\mathcal{D}$ is the bag of transactions in which it occurs: $\mathcal{V}_{\mathcal{D}}(X) = \{t \in \mathcal{D} \mid X \subseteq t\}$. The *support* of X in $\mathcal{D}$ is the size of its cover: $sup_{\mathcal{D}}(X) = |\mathcal{V}_{\mathcal{D}}(X)|$.

Learning from Preferences. An algorithmic template of the *Mine, Interact, Learn, Repeat* framework is listed in Algorithm 1. The interactive process proceeds iteratively for some reasonable number of iterations T, which depends on the task at hand. Let $\Phi : \mathcal{L}_{\mathcal{I}} \to \mathbb{R}$ denote the true, unobserved preferences function of the user. The algorithm maintains an internal estimate φ^t of the true function, where $t \in [T]$ is the iteration index. At each iteration, it selects a query $\mathcal{X}^t$ to be posed to the user. The user's feedback is then used (possibly along with all the feedback received so far) to compute a new estimate φ^{t+1} of Φ. Key questions concerning instantiations of the Mine, interact, learn, repeat framework include 1) feature representations of patterns to be ranked and the feedback format, 2) learning user's preferences from feedback, 3) mining with learned preferences, and crucially, 4) selecting the patterns to show to the user.

a) User Interaction & Pattern Representations. User feedback w.r.t. patterns takes the form of providing a total order over a (small) set of patterns [4, 12], called a query. User feedback $\{X_1 \succ X_3 \succ X_2\}$, for instance, indicates that the user prefers X_1 over

---

Algorithm 1: Template algorithm for interactive pattern mining

---

1 **In**: Dataset $\mathcal{D}$, set of patterns $\mathcal{X}$
2 **Parameters**: Query size k, number of steps T, feature pattern representations $\mathcal{F}$
3 **Out**: φ : Ranking function

4 **begin**
5     $\mathcal{U} \leftarrow \emptyset, \varphi^0 \leftarrow$ initial function estimates
6   **for** $t = 1, 2 \ldots T$ **do**
7   select a query $\mathcal{X}^t$ based on φ^{t-1} ▷ Mine
8   ask query $\mathcal{X}^t$ to the user and get feedback $\widehat{\mathcal{R}}^t$ ▷ Interact
9        $\mathcal{U} \leftarrow \mathcal{U} \cup \widehat{\mathcal{R}}^t$, compute φ^t based on φ^{t-1} and $\mathcal{U}$ ▷ Learn φ
10   **return** φ;

X_3, which they prefer over X_2 in turn, and so on. Pattern representations determines how the user characterizes patterns of interest to him. Patterns are commonly represented using a vector of *static features* (also called **descriptors**) $\mathcal{F} = \langle F_1, \ldots, F_n \rangle$. The description of a pattern X w.r.t. $\mathcal{F}$ is given by the vector $\mathbf{P} = \langle \mathbf{P_1}, \ldots, \mathbf{P_n} \rangle$, where $\mathbf{P_i}$ is the value associated to F_i. Examples of features include numerical descriptors like $Len(X) = |X|/|\mathcal{I}|$, $Freq(X) = sup_{\mathcal{D}}(X)/|\mathcal{D}|$, in which case the corresponding $\mathbf{P_i}$ is truly in $\mathbb{R}$, or binary descriptors $Items(i, X) = [i \in X]$; and $Trans(t_i, X) = [X \subseteq t_i]$, where $[.]$ denotes the Iverson bracket, leading to (partial) feature vectors $\in \{0,1\}^{|\mathcal{I}|}$ and $\in \{0,1\}^{|\mathcal{D}|}$, respectively.

b) Learning from Feedback. Evaluating patterns in terms of quality function is a very natural way of representing preferences. In the object preferences scenario [9], such a function is a mapping $\varphi : \mathcal{X} \rightarrow \mathbb{R}$ that assigns a score $\varphi(X)$ to each pattern X and, thereby, induces an order on $\mathcal{X}$. As in [4], we use a parametrized logistic function to measure the interestingness/quality of a given pattern X: $\varphi_{logistic}(X; w, A) = A + \frac{1-A}{1+e^{-\mathbf{w}_{\mathcal{F}} \cdot \mathbf{P}}}$, where $\mathbf{P}$ the afore-mentioned description of a pattern X, $\mathbf{w}_{\mathcal{F}}$ the weight vector associated to descriptors $\mathcal{F}$ reflecting feature contributions to pattern interestingness, and A is a parameter that controls the range of the interestingness measures, i.e. $\varphi_{logistic} \in (A, 1)$. However, setting feature weights manually is tedious, thus we present in Sect. 3 an algorithm that learns the weights based on easy-to-provide feedback with respect to patterns. Given a user feedback $\mathcal{U} = \{X_1 \succ X_3 \succ X_2\}$ which is translated into pairwise rankings $\{(X_1 \succ X_3), (X_1 \succ X_2), \ldots\}$, each ranked pair $X_i \succ X_j$ corresponds to a classification example $(\mathbf{P}_i - \mathbf{P}_j, +)$ of a training dataset. We use Stochastic Coordinate Descent (SCD) [13] for minimizing logistic loss stemming from this training dataset, and use the learned weights in $\varphi_{logistic}$.

3 DiSPaLe: Discriminating Sub-Pattern Feature Learning

We present DISPALE, an instantiation of the framework described by Algorithm 1, which exploits more complex descriptors in combination with static low-level features to learn logistic functions. These new descriptors are learned from discriminating sub-patterns. The sequel describes how discriminating sub-pattern are extracted from the user-defined pattern ranking and how they are used in the learning component of DIS-PALE (see Algorithm 2). Table 1b summarizes the different notations used in this paper.

Algorithm 2: DISPALE **(Di**scriminating **S**ub-**Pa**ttern feature **L**earning)

---

1 **In**: Dataset $\mathcal{D}$, set of patterns $\mathcal{X}$
2 **Parameters** : Query size k, number of steps T, range A, query retention ℓ, feature pattern representations $\mathcal{F}$
3 **Out**: φ : Ranking function;
4 **begin**
5 $\quad \mathcal{U} \leftarrow \emptyset, w_{\mathcal{F}}^0 \leftarrow \mathbf{0}, \mathcal{X}^0 \leftarrow \emptyset, \varphi^0 = \varphi_{logistic}(w_{\mathcal{F}}^0, A)$
6 $\quad$ **for** $t = 1, 2 \ldots T$ **do**
7 $\qquad \mathcal{X}^t \leftarrow \text{TOP}(\mathcal{X}^{t-1}, \ell) \cup (\text{SAMPLEPATTERNS}(\mathcal{D}, \varphi^{t-1}) \times (k - \ell))$
8 $\qquad \hat{\mathcal{R}}^t \leftarrow \text{RANK}(\mathcal{X}^t), \ disc \leftarrow \text{MINEDISCRIMINATING}(\mathcal{X}^t, \hat{\mathcal{R}}^t), \ \mathcal{U} \leftarrow \mathcal{U} \cup \hat{\mathcal{R}}^t$
9 $\qquad \langle w_{\mathcal{F}}^t, w_{\mathcal{F}_{disc}}^t \rangle \leftarrow \text{LEARNWEIGHTS}(\mathcal{U}, \mathcal{F} \cup \mathcal{F}_{disc})$
10 $\qquad w_{\mathcal{F}}^t \leftarrow \text{UPDATEWEIGHTS}(w_{\mathcal{F}}^t, w_{\mathcal{F}_{disc}}^t)$
11 $\qquad \varphi^t \leftarrow \varphi_{logistic}(w_{\mathcal{F}}^t, A)$
12 $\quad$ **return** φ^T

3.1 Towards More Expressive and Learnable Pattern Descriptions

Features for pattern representation involve indicator variables for included items or sub-graphs, or for covered transactions [3,4], or pattern length, etc. The issue is that such features are treated as if they were independent, whether in the logistic function mentioned above, or multiplicative functions [3,12]. While this allows to identify pattern components that are *globally* interesting for the user, it is impossible to learn relationships such as "the user is interested in item i_1 if item i_3 is present but item i_4 is absent". In addition, the pattern elements whose inclusion is indicated are defined before-hand, and the user feedback has no influence on them. We therefore propose to learn more expressive features in order to improve the learning of user preferences. In this work, we propose to consider *discriminating sub-patterns* that better capture (or explain) these preferences. Those features exploit ranking-correlated patterns, *i.e.*, patterns that influence the user ranking either by allowing some patterns to be well ranked or the opposite.

a) Interclass Variance. As explained above, our goal is to mine sub-patterns that discriminate between patterns that have been given a high user ranking and those that received a low one. An intuitive way of modelling this problem consists of considering the numerical ranks given to individual patterns as numerical labels and the mining setting as akin to regression. We are not aiming to build a full regression model but only to mine an individual pattern that correlates with the numerical label. For this purpose, we use the interclass variance measure as proposed by [10].

Definition 1. *Let $\mathcal{X}$ be a query, $\hat{\mathcal{R}}$ the user ranked patterns, $\mathcal{X}_Y$ the subset of patterns X in $\mathcal{X}$ containing the sub-pattern Y, and $\overline{\mathcal{X}}_Y = \mathcal{X} - \mathcal{X}_Y$. The interclass variance of the sub-pattern y is defined by: $ICV(Y, \hat{\mathcal{R}}) = |\mathcal{X}_Y| \cdot (\mu(\mathcal{X}) - \mu(\mathcal{X}_Y))^2 + |\overline{\mathcal{X}}_Y| \cdot (\mu(\mathcal{X}) - \mu(\overline{\mathcal{X}}_Y))^2$, where $\mu(\mathcal{X}) = \frac{1}{|\mathcal{X}|} \cdot \sum_{X \in \mathcal{X}} r(X)$, and $r(X)$ is the rank of X in $\hat{\mathcal{R}}$.*

b) Extracting Discriminating Sub-patterns. To find the sub-pattern $Y \subseteq X \in \mathcal{X}$ with the greatest interclass variance ICV, we systematically search the pattern space spanned by the items involved in patterns of the user's query $\mathcal{X}$. Semantically, this is the sub-pattern whose presence in one or more patterns X has influenced their ranking. So, if $Y \subseteq X$, we can say that the ranking of X at the $r(X)^{th}$ position in $\hat{R}$ is more likely to be explained by the presence of sub-pattern Y.

Algorithm 3: Extracting discriminating sub-patterns

---

1 **Function** $MineDiscriminating(\mathcal{X}, \widehat{R})$
2     $ICV_{max} \leftarrow 0, disc \leftarrow \emptyset$
3     $I_{\mathcal{X}} \leftarrow \{i \in X \mid X \in \mathcal{X}\}$
4     $\mathcal{S} \leftarrow I_{\mathcal{X}}$
5   **For each** $i \in I_{\mathcal{X}}$ **do**
6   **If** $ICV(i, \widehat{R}) \geq ICV_{max}$ **then**
7           $ICV_{max} \leftarrow ICV(i, \widehat{R}), disc \leftarrow \{i\}$

8   **For each** $Y \in \mathcal{S}$ **do**
9   **While** $(\exists i \in I_{\mathcal{X}} \wedge \exists X \in \mathcal{X} \ st. \ Y \cup \{i\} \subseteq X \wedge i \notin Y)$ **do**
10   **If** $ICV(Y \cup \{i\}, \widehat{R}) \geq ICV_{max}$ **then**
11             $ICV_{max} \leftarrow ICV(Y \cup \{i\}, \widehat{R}), disc \leftarrow Y \cup \{i\}, \mathcal{S} \leftarrow \mathcal{S} \cup disc$

12   **return** $disc$

Algorithm 3 implements the function MINEDISCRIMINATING (see Algorithm 2, line 8), which learns the best discriminating pattern as a descriptor. Its accepts as input the query $\mathcal{X}$ and the ranked patterns $\widehat{R}$ by the user, and returns the sub-itemset with the highest ICV. Its starts by computing the ICV of all items of the patterns in $\mathcal{X}$ (loop 5–7). Then, it iteratively combines the items to form a larger and finer-grained discriminating sub-pattern (loop 8-11). Obviously, before combining sub-itemsets, we should ensure that the resulting sub-pattern belongs to an existing pattern $X \in \mathcal{X}$ (line 9). If such a sub-pattern exists, we update the value of ICV_{max}, we save the best discriminating sub-pattern computed so far and we update with $disc$ the set of sub-itemsets that can be extended for further improvements (lines 10-11). Finally, the best discriminating pattern is returned at line 12.

Example 1. Consider a dataset with items $\mathcal{I} = \{1, \ldots, 7\}$. Let's consider a user query $\mathcal{X} = \{X_1, X_2, X_3, X_4\}$, with $X_1 = \{5, 7\}, X_2 = \{2, 7\}, X_3 = \{1\}, X_4 = \{4\}$ and let $\widehat{R} = \{X_2 \succ X_1 \succ X_3 \succ X_4\}$. For $Y = \{2\}$, we have $\mathcal{X}_Y = \{X_2\}, \overline{\mathcal{X}}_Y = \{X_1, X_3, X_4\}, \mu(\mathcal{X}_Y) = 1, \mu(\overline{\mathcal{X}}_Y) = 3$ and $\mu(\mathcal{X}) = 2.5$. Applying definition 1 gives $ICV(2, \widehat{R}) = 3$. After the first loop of Algorithm 3, $ICV_{max} = 4$ and $disc = \{7\}$.

3.2 Discriminating Sub-patterns as Descriptors

Exploiting discriminating sub-patterns as a new descriptors in DISPALE brings meaningful knowledge to consider during an interactive preference learning. In fact, this sub-pattern correlated with the user's ranking emphasizes the items of interest related to his ranking. Now, we describe how these discriminating patterns can be used in order to improve the learning function $\varphi_{logistic}$ for patterns.

A direct way of exploiting discriminating sub-patterns consists of adding them as new descriptors to the initial features $\mathcal{F}$ during the iterations. However, this will increase the size of $\mathcal{F}$, introduces additional cost and most probably leads to over-fitting and generalization issues of the learning function $\varphi_{logistic}$. Instead, we propose to use the discriminating sub-pattern $disc$ extracted at each iteration as a **temporary descriptor** F_{disc} that can be added to $\mathcal{F}$ in order to learn a weight $w_{F_{disc}}$ (see Algorithm 2, line 9). We propose three types of discriminating descriptors:

- F_{disc_X}: a binary descriptor used to assess the presence/absence of $disc$ in a pattern $X \in \mathcal{X}$; its associated value $\mathbf{P_{disc_X}} = \begin{cases} 1 \text{ if } disc \subseteq X \\ 0 \text{ otherwise} \end{cases}$
- F_{disc_T}: a numerical descriptor representing the frequency of the discriminating sub-pattern $disc$; its associated value $\mathbf{P_{disc_T}} = \begin{cases} sup_{\mathcal{D}}(disc)/|\mathcal{D}| \text{ if } disc \subseteq X \\ 0 \text{ otherwise} \end{cases}$
- F_{disc_I}: a numerical descriptor representing the relative size of the discriminating sub-pattern $disc$; its associated value $\mathbf{P_{disc_I}} = \begin{cases} |disc|/|\mathcal{I}| \text{ if } disc \subseteq X \\ 0 \text{ otherwise} \end{cases}$

By denoting $\mathcal{F}_{disc}$ the set of discriminating descriptors added to $\mathcal{F}$, we obtain the following temporary vector of descriptors: $\langle \underbrace{F_1, \ldots, F_n}_{\mathcal{F}}, \underbrace{F_{disc_X}, F_{disc_T}, F_{disc_I}}_{\mathcal{F}_{disc}} \rangle$.

Given the user-defined pattern ranking $\widehat{\mathcal{R}}^t$ at iteration t on query $\mathcal{X}^t$, we learn two weight vectors : the weight vector $w_{\mathcal{F}}^t$ associated to $\mathcal{F}$ and the weight vector $w_{\mathcal{F}_{disc}}^t$ associated to $\mathcal{F}_{disc}$. As descriptors $\mathcal{F}_{disc}$ are added temporary, the weights learned for $w_{\mathcal{F}_{disc}}^t$ are used back to update the weights $w_{\mathcal{F}}^t$ in order to be exploited for the next iteration. This new learning schema can be summarized as follows (see Fig. 1 in [7]):

(i) each pattern $X \in \mathcal{X}^t$ is converted into a vector $\mathbf{P} = \langle \mathbf{P_1}, \ldots, \mathbf{P_n}, \mathbf{P_{disc_X}}, \mathbf{P_{disc_T}}, \mathbf{P_{disc_I}} \rangle$, where $\mathbf{P_i}$ is the value associated to a feature/descriptor $F_i \in \mathcal{F} \cup \mathcal{F}_{disc}$.
(ii) new weights $w_{F_i}^t$ are learned for each descriptor $F_i \in \mathcal{F} \cup \mathcal{F}_{disc}$. The learned weights $w_{\mathcal{F}_{disc}}^t$ are then used back to update the weights $w_{\mathcal{F}}^t$ using a multiplicative weight method (see Algorithm 2, line 10).
(iii) finally, a new estimate φ^t is computed using the new $w_{\mathcal{F}}^t$ (see Algorithm 2, line 11).

3.3 Updating the Weights of Feature Pattern Representations

Let $disc$ be the discriminating sub-pattern extracted from the query $\mathcal{X}^t$. We propose two rules to update the weight vector $w_{\mathcal{F}}^t$ from the weight vector $w_{\mathcal{F}_{disc}}^t$:

- for *binary features* $F_i \in \mathcal{F}$ (items and transactions):
    o $F_i \equiv items(i, disc), w_{F_i}^t = f^{ag}(w_{F_i}^t, w_{F_{disc_X}}^t)$
    o $F_i \equiv Trans(t_i, disc) \wedge t_i \in \mathcal{V}_{\mathcal{D}}(disc), w_{F_i}^t = f^{ag}(w_{F_i}^t, w_{F_{disc_X}}^t)$
- for *numerical features* $F_i \in \mathcal{F}$ (frequency, length, ...):
    o $F_i \equiv Freqency : w_{F_i}^t = f^{ag}(w_{F_i}^t, w_{F_{disc_T}}^t)$
    o $F_i \equiv Lenght : w_{F_i}^t = f^{ag}(w_{F_i}^t, w_{F_{disc_I}}^t)$

The Multiplicative Weights Method [2] is a simple idea which has been repeatedly discovered in fields as diverse as Machine Learning, and Optimization. The setting for this method is the following: A decision maker (DM) has a choice of n decisions, and needs to repeatedly make a decision. The method assigns initial weights to the DM, and updates these weights multiplicatively and iteratively according to the feedback of how well an expert performed. Following this idea, we propose, at each iteration, to update the feature weights $w_{\mathcal{F}}^t$ by multiplying them with factors which depend on the learned weights of discriminating descriptors in that iteration. Intuitively, this updating scheme tend to focus higher weight on features that better explain patterns ranked by the user in the long run, thus increasing the probability of being present in patterns of the next iterations. We propose to multiplicative updating rules:

Table 1. θ represents an absolute value.

Dataset	$|\mathcal{D}|$	$|\mathcal{I}|$	θ	$\#\mathcal{P}$
Anneal	812	89	660	149 331
Chess	3 196	75	2 014	155 118
German	1 000	110	350	161 858
Heart-cleveland	296	95	115	153 214
Hepatitis	137	68	35	148 289
Kr-vs-kp	3 196	73	2 014	155 118
Lymph	148	68	48	146 969
Mushroom	8 124	112	813	155 657
Soybean	630	50	28	143 519
Vote	435	48	25	142 095
Zoo-1	101	36	10	151 806

(a) Dataset Characteristics.

Notation	Significance
$t \in [T]$	Iteration index
$\mathcal{X}^t$	User query
$\hat{R}^t$	User-defined feedback on $\mathcal{X}^t$
$\mathcal{F}$	Vector of feature representations of patterns
$\mathcal{F}_{disc}$	Vector of discriminating descriptors of patterns
$\mathbf{P}$	Pattern description w.r.t ($\mathcal{F} \cup \mathcal{F}_{disc}$)
$disc$	Discriminating sub-pattern extracted from $\hat{R}^t$
F_{disc_X}	Binary descriptor related to the presence/absence of $disc$ in X
F_{disc_T}	Numerical descriptor related to the frequency of $disc$
F_{disc_I}	Numerical descriptor related to the relative size $disc$
$w_{\mathcal{F}}^t$	Weight vector associated to static features $\mathcal{F}$
$w_{\mathcal{F}_{disc}}^t$	Weight vector associated to dynamic features $\mathcal{F}_{disc}$
η	Regularization parameter
$\varphi_{logistic}$	Learned logistic function

(b) Notations.

- by a *linear factor*: $f^{ag}(w_{F_i}^t, w_{F_{disc}}^t) = w_{F_i}^t \times (1 + \eta \cdot w_{F_{disc}}^t)$
- by an *exponential factor*: $f^{ag}(w_{F_i}^t, w_{F_{disc}}^t) = w_{F_i}^t \times \exp^{\eta \cdot w_{F_{disc}}^t}$

where $\eta \in]0, \frac{1}{2}]$ is regularization parameter used to control the increase in weights resulting from this update. In our experiments (see Sect. 4), we compare both updating rules for learning weights.

Example 2. Let us consider Example 1 and $disc = \{7\}$. Let us assume that $\mathcal{F}$ represents items and frequency features and $\mathcal{F}_{disc} = \langle F_{disc_X}, F_{disc_T} \rangle$. Suppose that $Freq(X_1) = 0.54$, $Freq(X_3) = 0.36$ and $Freq(disc) = 0.63$. According to $\mathcal{F}$, $X_1 = \{5, 7\}$ is represented by the vector $\mathbf{P}_1 = \langle 0, 0, 0, 0, 1, 0, 1, 0.54 \rangle$, while $X_3 = \{1\}$ by $\mathbf{P}_3 = \langle 1, 0, 0, 0, 0, 0, 0, 0.36 \rangle$. Using additionally features $\mathcal{F}_{disc}$, we obtain the new vector $\mathbf{P}_1 = \langle 0, 0, 0, 0, 1, 0, 1, 0.54, \mathbf{1}, \mathbf{0.63} \rangle$ since $disc \subset X_1$. Similarly, for $X_3 = \{1\}$, $\mathbf{P}_3 = \langle 0, 0, 0, 0, 1, 0, 1, 0.36, \mathbf{0}, \mathbf{0} \rangle$. Let $t = 1$, to learn the weights $w_{\mathcal{F}}^1$ and $w_{\mathcal{F}_{disc}}^1$, the user's feedback is translated into pairwise rankings and distances between vectors $\mathbf{P}_i$ for each pair are calculated (see Sect. 2). After the learning step, we obtain $w_{\mathcal{F}}^1 = \langle -0.33, 0.99, 0, -0.99, 0.33, 0, 1.33, 0.15 \rangle$ and $w_{\mathcal{F}_{disc}}^1 = \langle 1.33, 0.84 \rangle$. Using the linear factor with $\eta = 0.2$, we update the weight $w_{F_7}^1$ associated to item 7 (since $7 \in disc$) and the weight $w_{F_{disc_T}}^1$ associated to frequency as follows: $w_{F_7}^1 = w_{F_7}^1 \times (1 + \eta \cdot w_{F_{disc_X}}^1) = 1.68$; $w_{F_{disc_T}}^1 = w_{F_{disc_T}}^1 \times (1 + \eta \cdot w_{disc_T}^1) = 0.175$. After the updating step, the resulting weight vector $w_{\mathcal{F}}^1 = \langle -0.33, 0.99, 0, -0.99, 0.33, 0, \mathbf{1.68}, \mathbf{0.175} \rangle$.

4 Experiments

a) Evaluation Methodology and Pattern Selection. To experimentally evaluate our approach DISPALE, we emulate user feedback using a (hidden) quality measure Φ, which is not known to the learning algorithm. We follow the same protocol used in [4]: for each dataset, a set $\mathcal{P}$ of frequent patterns is mined without prior user knowledge. We assume that there exists a user ranking $\hat{R}$ on the set $\mathcal{P}$, derived from Φ, i.e. $X \succ Y \Leftrightarrow \Phi(X) > \Phi(Y)$. Thus, the task is to learn a logistic function $\varphi_{logistic}$ to sample frequent patterns which approximates Φ. We use surprisingness $surp$ as Φ, where $surp(X) = \max\{sup_{\mathcal{D}}(X) - \prod_{i=1}^{|X|} sup_{\mathcal{D}}(\{i\}), 0\}$.

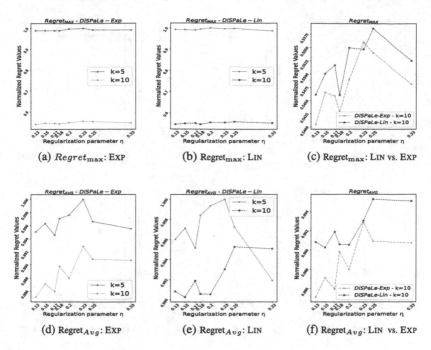

Fig. 1. Effects of the parameter η of DISPALE on $Regret_{max}$, $Regret_{Avg}$ and $\ell = 0$. Results are aggregated over data sets and the three feature combinations (I, IT, ILFT). Regret values are normalized to the range [0,1] based on the maximum regret value.

To compare the performance of the different approaches, we use *cumulative regret*, which is the difference between the ideal value of a certain measure M and its observed value, summed over iterations for each dataset. At each iteration t, we evaluate the regret of ranking pattern X_i by Φ as follows: we compute the percentile rank $pct.rank(X_i)$ by Φ of each pattern $X_i \in \mathcal{X}^t$ ($1 \leq i \leq k$) as $pct.rank(X_i) = (DI + \frac{DE}{2})/|\mathcal{P}|$ where $DI = |Y \in \mathcal{P}, \Phi(Y) < \Phi(X_i)|$ and $DE = |Y \in \mathcal{P}, \Phi(Y) = \Phi(X_i)|$. The percentile rank over all patterns of $\mathcal{X}^t$ measures the ability of the learnable function $\varphi_{logistic}$ to extract interesting patterns, i.e. patterns X_i for which $\Phi(X_i)$ is higher. Thus, the ideal value is 1 (e.g., the highest possible value of Φ over all frequent patterns has the percentile rank of 1). The regret is then defined as $1 - M_{(1 \leq i \leq k)}(pct.rank(X_i))$ where $M \in \{max, Avg\}$ and $k = |\mathcal{X}^t|$. We repeat each experiment 10 times with different random seeds; the average cumulative regret is reported. We ensure that all compared methods are sampled on the same pattern bases at each iteration.

For the evaluation, we used UCI data-sets, available at the CP4IM repository[1]. For each dataset, we set the minimal support threshold such that the size of $\mathcal{P}$ is approximately $145,000$ frequent patterns. Table 1 shows the data set statistics. Each experiment involves 100 iterations. We compare DISPALE with two state-of-the-art interactive methods, LETSIP [4], an interactive sampling method to learn a logistic function, and

---

[1] https://dtai.cs.kuleuven.be/CP4IM/datasets/.

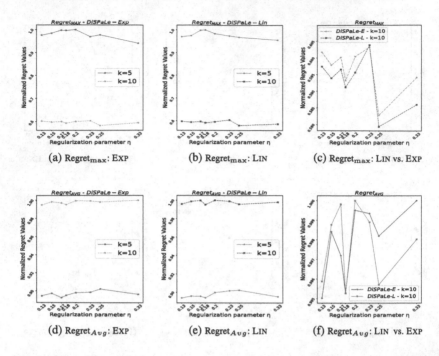

(a) Regret$_{\mathrm{max}}$: EXP   (b) Regret$_{\mathrm{max}}$: LIN   (c) Regret$_{\mathrm{max}}$: LIN vs. EXP

(d) Regret$_{Avg}$: EXP   (e) Regret$_{Avg}$: LIN   (f) Regret$_{Avg}$: LIN vs. EXP

Fig. 2. Effects of the parameter η of DISPALE on $Regret_{max}$, $Regret_{Avg}$ and $\ell = 1$.

an active preference learning to learn a linear ranking function using RANKSVM. We address the following two **research questions**: (i) What effect do DISPALE's parameters have on the quality of learned patterns? (ii) How does DISPALE compares to LETSIPand RANKSVM?

To select the k patterns to show to the user, we use EFLEXICS [6] to draw the k weighted random samples proportional to $\varphi_{logistic}$ as in [4] (see Suppl. Mat. [7] for more details). These patterns are selected according to a TOP(m) strategy, which picks the m highest-quality patterns. The same procedure is also used in RANKSVM. Moreover, to help users to relate the queries to each other, we retain the top ℓ patterns from the previous query and only sample $(k-\ell)$ new patterns. We use the default values suggested by [4] for the parameters in EFLEXICS: $\lambda = 0.001$, $\kappa = 0.9$, $A = 0.1$ and TOP(1).

b) Parameter Settings of DISPALE. We evaluate the effects of different parameter settings on DISPALE: the query size $k \in \{5, 10\}$, the updating rule and the regularization parameter η. We use the following feature combination: $Items$ (I); $Items + Transactions$ (IT); and $Items + Length + Frequency + Transactions$ (ILFT). We consider two settings for ℓ: $\ell = 0$ and $\ell = 1$. Figure 1 shows the evolution of the values of $Regret_{max}$ and $Regret_{Avg}$ for different values of η and for $\ell = 0$. Figures 1a and 1b show that both updating rules (LIN and EXP) ensure the lowest quality regrets with $k = 10$ w.r.t. $Regret_{max}$. This indicates that our approach is able to identify the properties of the target ranking from ordered lists of patterns even when the query size

Table 2. Evaluation of the importance of pattern features and comparison of DISPALE-EXP ($\eta = 0.13$) with LETSIPand RANKSVMfor $k = 10$. Results are aggregated over all datasets. (1): LETSIP, (2): DISPALE-EXP, (3): RANKSVM. Detailed values of the regret for each dataset and results of DISPALE-LIN are given in [7].

	$\ell = 0$						$\ell = 1$					
	Regret: $Regret_{max}$			Regret: $Regret_{Avg}$			Regret: $Regret_{max}$			Regret: $Regret_{Avg}$		
Descriptors	(1)	(2)	(3)	(1)	(2)	(3)	(1)	(2)	(3)	(1)	(2)	(3)
I	**112.137**	114.567	123.130	554.816	**553.050**	582.303	**10.438**	11.438	11.465	**496.918**	499.151	521.634
IT	108.446	**91.528**	101.635	543.556	**492.967**	542.595	10.761	11.465	**9.192**	483.689	**449.444**	491.014
ITLF	106.006	**88.391**	100.162	538.848	**487.537**	540.844	11.275	11.579	**9.601**	490.818	**450.202**	490.649

increases. Additionally, the lowest regret values are obtained with $\eta = 0.13$. Regarding $Regret_{Avg}$ (see Figs. 1d and 1e), $k = 10$ continues to be the better query size and $\eta = 0.13$ gives the lowest regret values. Finally, Figs. 1c and 1f compares the regret values of DISPALE-EXP and DISPALE-LIN for $k = 10$. As we can seen, DISPALE-EXP allows to achieve the best regrets. Figure 2 shows the effect of DISPALE's parameters on regret values for $\ell = 1$. Retaining one highest-ranked pattern from the previous query w.r.t. $Regret_{max}$ does not affect the conclusions drawn previously: $k = 10$ being the better query size. However, we can see the opposite behaviour w.r.t. $Regret_{Avg}$ (see Figs. 2d and 2e): querying 5 patterns allows attaining low regret values. Interestingly, as Fig. 2f shows, DISPALE-EXP outperforms DISPALE-LIN on almost all values of η. Based on these findings, we set $k = 10$ and $\eta = 0.13$ for the next experiments.

c) Evaluating the Importance of Pattern Features. Table 2 compare different combination of feature representations of patterns for two settings of query retention ℓ. As we can see, additional features provide valuable information to learn more accurate pattern rankings, particularly for DISPALE-EXP where the regrets decrease when adding the feature T. However, the importance of features depends on the pattern type and the target measure Φ [5]. For surprising pattern mining, $Length$ is the most likely to be included in the best feature set, because long patterns tend to have higher values of Surprisingness. $Items$ are important as well, because individual item frequencies are directly included in the formula of Surprisingness. $Transactions$ are important because this feature set helps capture interactions between other features, albeit indirectly.

d) Comparing DISPALE with LETSIP and RANKSVM. Table 2 reports the regret values. When considering $Items$ as a feature, LETSIP performs the best. However, selecting queries uniformly at random allows DISPALE-EXP to improve slightly the $Regret_{Avg}$. Moreover, regarding the other features (IT and ITLF), DISPALE-EXP always outperforms LETSIPand RANKSVM. When retaining one highest-ranked pattern ($\ell = 1$), RANKSVMexhibits the lowest $Regret_{max}$ values, while LETSIPand DISPALE-EXP perform very similarly. However, for $Regret_{Avg}$, DISPALE-EXP performs the best. These results indicate that the learned ranks by DISPALE-EXP in the target ranking are more accurate compared to those learned by the alternatives. This confirm the advantage of using discriminants sub-patterns as descriptors.

Figure 3 presents a detailed view of comparison on GERMAN-CREDIT dataset (other results are given in [7]). Curves show the evolution of the regret (cumulative and non cumulative) over 100 iterations of learning for different combination of features. The results confirm again the capacity of DISPALE-EXP to identify frequent patterns with lowest regrets. Figures 3c and 3d compares the performance of the three approaches in terms of CPU-times on two datasets. Overall, learning more complex descriptors from the user-defined pattern ranking does not add significantly to the runtimes of our approach: on most of the data sets considered, DISPALE-EXP and LETSIPbehave very similarly, and the difference is very negligible (see Suppl. Mat [7] for other results). However, RANKSVMrequires much more time to learn a linear ranking function.

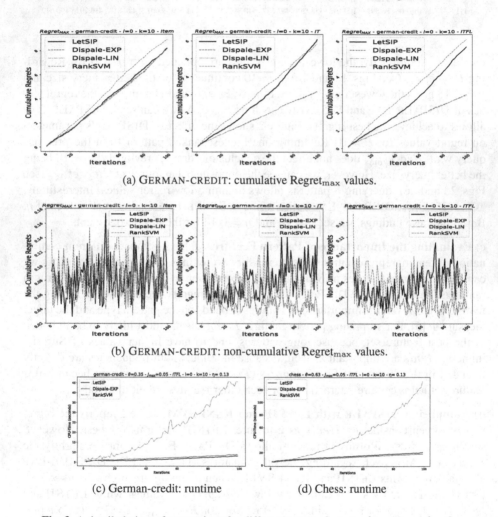

(a) GERMAN-CREDIT: cumulative $Regret_{max}$ values.

(b) GERMAN-CREDIT: non-cumulative $Regret_{max}$ values.

(c) German-credit: runtime   (d) Chess: runtime

Fig. 3. A detailed view of comparison for different pattern features, $k = 10$ and $\ell = 0$.

5   Conclusion

In this paper, we have proposed a new approach to the state-of-the art of interactive pattern mining: instead of using static low-level features that have been pre-defined before the process starts, our approach learns more complex descriptors from the user-defined pattern ranking. These features allow to capture the importance of item interactions, and, as shown experimentally, lead to lower cumulative and individual regret than using low-level features. We have explored two multiplicative updating rules for mapping weights learned for complex features back to their component items, and find that the exponential factor gives better results on most of the data sets we worked with. We have evaluated our proposal only on itemset data so far since the majority of existing work is defined for this kind of data. But the importance of using complex dynamic features can be expected to be even higher when interactively mining complex, i.e. sequential, tree, or graph-structured data. We will explore this direction in future work.

Acknowledgements. A. Hien and S. Loudni were financially support by the ANR project InvolvD (ANR-20-CE23-0023).

References

1. Agrawal, R., Srikant, R.: Fast algorithms for mining association rules in large databases. In: Proceedings of the 20th VLDB, pp. 487–499. Santiago de Chile, Chile (1994)
2. Arora, S., Hazan, E., Kale, S.: The multiplicative weights update method: a meta-algorithm and applications. Theory Comput. **8**(1), 121–164 (2012)
3. Bhuiyan, M., Hasan, M.A.: Interactive knowledge discovery from hidden data through sampling of frequent patterns. Stat. Anal. Data Min. **9**(4), 205–229 (2016)
4. Dzyuba, V., van Leeuwen, M.: Learning what matters - sampling interesting patterns. In: PAKDD 2017, Proceedings, Part I, pp. 534–546 (2017)
5. Dzyuba, V., van Leeuwen, M., Nijssen, S., Raedt, L.D.: Interactive learning of pattern rankings. Int. J. Artif. Intell. Tools **23**(6), 1460026 (2014)
6. Dzyuba, V., van Leeuwen, M., De Raedt, L.: Flexible constrained sampling with guarantees for pattern mining. Data Min. Knowl. Disc. **31**(5), 1266–1293 (2017). https://doi.org/10.1007/s10618-017-0501-6
7. Hien, A., Loudni, S., Aribi, N., Ouali, A., Zimmermann, A.: Code and supplementary material. https://gitlab.com/phdhien/dispale (2023)
8. Joachims, T.: Optimizing search engines using clickthrough data. In: Proceedings of the ACM SIGKDD KDD 2002, pp. 133–142. New York, NY, USA (2002)
9. Kamishima, T., Kazawa, H., Akaho, S.: A survey and empirical comparison of object ranking methods, pp. 181–201 (2010)
10. Morishita, S., Sese, J.: Traversing itemset lattices with statistical metric pruning. In: Proceedings of the Nineteenth ACM SIGACT-SIGMOD-SIGART Symposium, pp. 226–236 (2000)
11. Raedt, L.D., Zimmermann, A.: Constraint-based pattern set mining. In: Proceedings of the 17th SIAM ICDM 2007, Minneapolis, Minnesota, USA, pp. 237–248. SIAM (2007)
12. Rüping, S.: Ranking interesting subgroups. In: Danyluk, A.P., Bottou, L., Littman, M.L. (eds.) Proceedings of ICML 2009, vol. 382, pp. 913–920 (2009)
13. Shalev-Shwartz, S., Tewari, A.: Stochastic methods for l_1-regularized loss minimization. J. Mach. Learn. Res. **12**, 1865–1892 (2011)
14. Tan, P.N., Kumar, V., Srivastava, J.: Selecting the right interestingness measure for association patterns. In: KDD, pp. 32–41 (2002)

A Novel Explainable Link Forecasting Framework for Temporal Knowledge Graphs Using Time-Relaxed Cyclic and Acyclic Rules

Uday Kiran Rage[1]([✉])[iD], Abinash Maharana[2][iD],
and Krishna Reddy Polepalli[2][iD]

[1] The University of Aizu, Aizu-Wakamatsu, Fukushima, Japan
udayrage@u-aizu.ac.jp
[2] International Institute of Information Technology, Hyderabad, Telangana, India
abinash.maharana@research.iiit.ac.in, pkreddy@iiit.ac.in

Abstract. Link forecasting in a temporal Knowledge Graph (tKG) involves predicting a future event from a given set of past events. Most previous studies suffered from reduced performance as they disregarded acyclic rules and enforced a tight constraint that all past events must exist in a strict temporal order. This paper proposes a novel explainable rule-based link forecasting framework by introducing two new concepts, namely *'relaxed temporal cyclic and acyclic random walks'* and *'link-star rules'*. The former concept involves generating rules by performing cyclic and acyclic random walks on a tKG by taking into account the real-world phenomenon that the order of any two events may be ignored if their occurrence time gap is within a threshold value. Link-star rules are a special class of acyclic rules generated based on the natural phenomenon that history repeats itself after a particular time. Link-star rules eliminate the problem of combinatorial rule explosion, thereby making our framework practicable. Experimental results demonstrate that our framework outperforms the state-of-the-art by a substantial margin. The evaluation measures *hits@1* and *mean reciprocal rank* were improved by 45% and 23%, respectively.

Keywords: Knowledge Graphs · Graph Analytics · Forecasting

1 Introduction

A temporal Knowledge Graph (tKG) is a graph in which the nodes correspond to (real-world) entities, and the edges correspond to binary relations between the entities at a particular timestamp. Crucial information that can facilitate end-users to achieve socio-economic development lies hidden in these graphs. *Link forecasting* is an important graph analytical technique that aims to predict a

R. U. Kiran and A. Maharana—Both the authors contributed equally to this work.

© The Author(s), under exclusive license to Springer Nature Switzerland AG 2023
H. Kashima et al. (Eds.): PAKDD 2023, LNAI 13935, pp. 264–275, 2023.
https://doi.org/10.1007/978-3-031-33374-3_21

future event based on a given set of past events drawn from a tKG. Briefly, a link forecasting framework involves the following steps: (i) generating logical rules by performing random walks on a tKG and (ii) identifying suitable candidates for a missing entity of a given query using the generated rules.

Example 1. Figure 1 shows a hypothetical temporal Knowledge Graph containing 5 distinct entities (nodes) and 7 links (temporal edges). A link, say (Robin, visit, France, 05/06), means that a relation 'visit' exists between the subject entity 'Robin' and the object entity 'France' at a timestamp '05/06'. Given a query link ($Fred, visit, ?, 13/06$), the problem of link forecasting is to traverse the tKG shown in Fig. 1, which contains all the past events, generate rules, and use them to identify an appropriate candidate entity to replace ?.

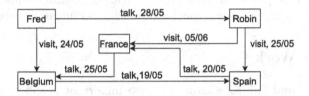

Fig. 1. A temporal Knowledge Graph. The format of the timestamps is DD/MM.

Several frameworks [10,14,15] were described in the literature for link forecasting. Unfortunately, these frameworks completely disregard the temporal information of a link. Many embedding-based frameworks [3–8,16] have exploited temporal information in a graph to perform link forecasting. However, these frameworks lack explainability, which is crucial for developing transparent and interpretable applications. Recently, Liu et al. [9] exploited the concept of 'logical rules' and presented an explainable framework known as TLogic. This framework implicitly assumes that past events follow a strict temporal order and that cyclic rules[1] alone are sufficient (that is, acyclic rules are not needed) to capture all the types of patterns in past events. However, this is seldom the case in the real world. In many applications, the order of two events may not matter if they occurred within a particular time window. Further, both cyclic and acyclic rules can be taken into account for link forecasting purposes. As a result, TLogic misses many useful rules and thus suffers from reduced accuracy.

Example 2. Consider the link (Fred, talk, Robin, 28/05) in Fig. 1. If we enforce a constraint that past events must follow a strict temporal ascending order, we cannot perform any random walk from this link to generate rules. However, if we relax the temporal constraints and permit acyclic walks, we can generate useful rules for link forecasting.

---

[1] Cyclic/acyclic rules represent the rules generated by performing cyclic/acyclic random walks on a tKG, respectively.

With this motivation, we propose a generic rule-based link forecasting framework called Temporally Relaxed Knowledge Graph Miner (TRKG-Miner). The proposed framework employs a new time-relaxation parameter, namely *maximum time gap* (δ), to capture the irregular occurrence order of temporal events. Using δ, TRKG-Miner performs *relaxed temporal random walks* on a tKG and produces cyclic and *link-star* rules. Link-star rules represent a subset of acyclic rules inspired by the star structural pattern [1] in graphs. They facilitate TRKG-Miner to be practicable by controlling rule explosion. Experimental results on various real-world datasets demonstrate that our framework outperforms the state-of-the-art by a very large margin (improved the metrics *hits@1* by 45% and *mean reciprocal rank* by 23%).

The rest of the paper is organized as follows. Section 2 describes related work on link forecasting in tKGs. Section 3 describes the problem statement and Sect. 4 describes the proposed framework. The experimental results are reported in Sect. 5. Finally, in Sect. 6, we conclude and discuss future research.

2   Related Work

Link prediction and link forecasting are two important graph analytical techniques [13]. Link prediction focuses on predicting missing/incomplete links in a tKG for any given set of events, whereas link forecasting aims at predicting links only for future timestamps. In this paper, we focus on link forecasting, which is a very active research problem that lies at the intersection of three prevalent fields: graph theory, machine learning, and data mining.

Several embedding-based frameworks, such as RESCAL [12], TransE [2], DistMult [15], and ComplEx [14], have been described in the literature for predicting links in a static Knowledge Graph. A key limitation of these frameworks is that they disregard the temporal information that may exist between the nodes. When confronted with this problem in real-world applications, researchers initially extended the existing frameworks to accommodate temporal information. For instance, Leblay and Chekol [8] described TTransE by extending TransE to handle tKGs, García-Durán et al. [3] discussed TA-DistMult by extending DistMult, and Lacroix et al. [7] described TNTComplEx by extending ComplEx. It was observed that these extended frameworks do not capture the temporal information of the graphs properly and thus suffer from the performance issues.

CyGNet by Zhu et al. [16], leverages the concept that facts/links often show repetition to describe an embedding-based model that predicts future events and also identifies recurring facts. Jin et al. [6] described a Recurring Neural Network-based embedding framework for link forecasting called RE-Net. Han et al. [5] described an explainable approach for link forecasting using graph neighborhood sampling and attention propagation networks. They were also the first to propose the concept of inverse relation for unrestricted back-and-forth walking on temporal knowledge graphs, to the best of our knowledge. Unfortunately, these embedding-based frameworks lack explainability, which is crucial for developing transparent and interpretable applications.

An advantage of rule-based frameworks over embedding-based ones is that they greatly help in human understanding of both the results and the link prediction process since embeddings can only be treated as a black box. Another advantage is that such frameworks are highly generalizable since we can reuse the rules mined from one dataset to another [9]. AnyBURL [11] samples random walks for generating rules which are then used for link prediction in static KGs. The key limitation of this framework is that it is designed only for static Knowledge Graphs. TLogic [9] extends AnyBURL to accommodate temporal dynamics and is the first explainable framework that successfully does link forecasting on tKGs. Although TLogic is the current best algorithm available for link forecasting in tKGs, it suffers from performance issues as it only considers cyclic rules and enforces a strict temporal constraint on the occurrence order of events. This paper aims to improve the performance of link forecasting by tackling the issues encountered by TLogic.

3   Problem Statement

Let $\mathcal{E} = \{e_1, e_2, \cdots, e_m\}$, $m \geq 1$, be a set of entities. Let $\mathcal{R} = \{r_1, r_2, \cdots, r_n\}$, $n \geq 1$, be a set of relations. Let $\mathcal{T} = \{t_1, t_2, \cdots, t_o\}$, $o \geq 1$, be an ordered set of timestamps. A link (or an edge), denoted as l_i, $i > 0$, is a quadruple (e_{sub}, r, e_{obj}, t), where $e_{sub} \in \mathcal{E}$ corresponds to a subject entity, $r \in \mathcal{R}$ represents a relation, $e_{obj} \in \mathcal{E}$ is an object entity, and $t \in \mathcal{T}$ represents a timestamp. For each link $l_i = (e_{sub}, r, e_{obj}, t)$, there exists an inverse link, denoted as $\hat{l}_i = (e_{obj}, r^{-1}, e_{sub}, t)$, that allows us to perform graph walks along this link in both directions. The relation r^{-1} is called the *inverse relation* of r. A temporal Knowledge Graph, denoted as tKG, is a set of links, i.e., $tKG = \bigcup_{i=1}^{p} l_i$, $p \geq 1$.

Example 3. Let $\mathcal{E} = \{Fred, Robin, France, Belgium, Spain\}$ be the set of entities. Let $\mathcal{R} = \{talk, visit\}$ be the set of relations. Let $\mathcal{T} = \{14/05, \cdots, 05/06\}$ be the set of timestamps. A link, say $l_1 = (Robin, visit, France, 05/06)$, where 'Robin' is a subject entity, 'visit' is a relation, 'France' is an object entity, and '05/06' is a timestamp. For this link, there exists an inverse link $\hat{l}_1 = (France, visit^{-1}, Robin, 05/06)$. A hypothetical temporal Knowledge Graph containing all such links is shown in Fig. 1.

Problem Statement. Given a query containing a future timestamp, say $(e_{sub}, r_x, ?, t_{future})$ or $(?, r_x, e_{obj}, t_{future})$, the goal of link forecasting is to predict a missing (object or subject) entity. The term t_{future} represents a future timestamp that does not exist in the tKG. That is, $t_{future} > t_o$ and $t_{future} \notin \mathcal{T}$.

Example 4. The tKG shown in Fig. 1 records the events that happened from '19/05' to '05/06' in a year. Given a query with an unseen future timestamp, say (Fred, visit, ?, 13/06), the goal of link forecasting is to predict an appropriate candidate entity to replace ?.

4   Proposed Framework: TRKG-Miner

Temporal Relaxation. Since the real world is complex and driven by convenience, events may or may not happen in an exact temporal order. Hence, this paper argues that the order between past events must be considered (respectively, ignored) if there exists (respectively, does not exist) a significant time gap between them. To capture our argument, we introduce a new parameter, called *maximum time gap* (δ). We ignore the temporal occurrence order of two events if their time gap is less than or equal to δ.

Definition 1. *Relaxed Temporal Random Walk:* *A Relaxed Temporal Random Walk[2], denoted as W, is an ordered set of links such that the time gap between any two consecutive links is no more than the user-specified maximum time gap (δ).*
That is, $W = \langle(e_1, r_1, e_2, t_1), (e_2, r_2, e_3, t_2) \ldots (e_n, r_n, e_{n+1}, t_n)\rangle$, $n > 0$, $t_i \geq (t_{i-1} - \delta)$, and $i \in [1, n]$.

Example 5. Consider the following sequence of links in Fig. 1: $\langle(Belgium, visit^{-1}, Fred, 24/05), (Fred, talk, Robin, 28/05), (Robin, visit, Spain, 25/05)\rangle$. The existing frameworks do not consider the above sequence of links as a random walk because the link $(Fred, talk, Robin, 28/05)$ violates the strict temporal constraint that it has to occur before the link $(Robin, visit, Spain, 25/05)$. However, if we relax the temporal constraint that the actual order of events can be ignored if their occurrence is within a particular time gap, say 4 days, i.e., $\delta = 4$, we can consider the above sequence of links as a relaxed temporal random walk.

Definition 2. *Relaxed Temporal Rule:* *Let E_i and T_i represent entity and timestamp variables, respectively. A Relaxed Temporal Rule (See footnote 2) of length $n \in \mathbb{N}$ is defined as $((E_s, r_h, E_o, T_{n+1}) \leftarrow \wedge_i (E_i, r_i, E_{i+1}, T_i))$, with the constraint $T_i \geq T_{i-1} - \delta; i \in [1, n]$.*

Example 6. The following is an example of a Relaxed Temporal Rule extracted from the tKG in Fig. 1: $(E_1, visit, E_3, T_3) \leftarrow (E_1, visit, E_2, T_1) \wedge (E_2, talk, E_3, T_2)$

In a Relaxed Temporal Rule, the left-hand side of the arrow is called the *rule head*, while the right-hand side is called the *rule body*, which is represented by an ordered conjunction of body links (E_i, r_i, E_{i+1}, T_i). All the entities and timestamps are variables, whereas all the relations are fixed. r_h is called as the *head relation*. A rule of this form implies that if the rule body holds along with the given time constraints, then the rule head is true.

Definition 3. *Cyclic Relaxed Temporal Random Walk:* *A relaxed temporal random walk, W, is said to be a Cyclic Relaxed Temporal Random Walk, denoted as CW, if the walk starts and end at the same entity (or node). That is, $CW = \langle(e_1, r_1, e_2, t_1), (e_2, r_2, e_3, t_2) \ldots (e_n, r_n, e_1, t_n)\rangle$*

---

[2] Throughout this paper, we will use 'walks' and 'rules' to refer to relaxed temporal random walks and relaxed temporal rules, respectively.

Example 7. Consider the following walk on the tKG in Fig. 1: $\langle(Robin, visit,$ $Spain, 15/05), (Spain, talk, France, 20/05), (France, visit^{-1}, Robin, 05/06)\rangle$.

This is a cyclic relaxed temporal random walk because it starts and ends with the same entity, i.e., *Robin*.

Definition 4. *Relaxed Temporal Cyclic Rule:* *A Relaxed Temporal Cyclic Rule is a Relaxed Temporal Rule generated from a cyclic walk by generalizing the entities and timestamps with variables. While the inverse of the last link becomes the rule head $(E_1, r_h^{-1}, E_n, T_n)$, the other links are mapped to body atoms, where each link (e_i, r_i, e_{i+1}, t_i) is converted to the body atom (E_i, r_i, E_{i+1}, T_i). That is, the final rule is of the form $(E_1, r_h^{-1}, E_n, T_n) \leftarrow \wedge_{i=1}^l (E_i, r_i, E_{i+1}, T_i)$*

Example 8. Continuing with Example 7, the cyclic relaxed temporal random walk $\langle(Robin, visit, Spain, 15/05), (Spain, talk, France, 20/05), (France,$ $visit^{-1}, Robin, 05/06)\rangle$ can be generalized using variables as $\langle(E_1, visit, E_2, T_1),$ $(E_2, talk, E_3, T_2), (E_3, visit^{-1}, E_1, T_3)$, where E_1, E_2, E_3, T_1, T_2 and T_3 are generalized from 'Robin,' 'Spain,' 'France,' '15/05,' '20/05,' and '05/06' respectively. The relaxed temporal cyclic rule generated from the generalized cyclic relaxed temporal random walk is:

$$(E_1, visit, E_3, T_3) \leftarrow (E_1, visit, E_2, T_1) \wedge (E_2, talk, E_3, T_2) \tag{1}$$

Link-Star Rules. Since cyclic rules alone can be inadequate for efficient link forecasting, there is a need for generating acyclic rules as well. However, the limitation of producing both cyclic and acyclic rules is the combinatorial explosion problem, which involves producing too many rules. To address this problem, we exploit the concept of star structure in graph theory to discover a subset of acyclic rules called Link-star rules, which are formed from acyclic temporal random walks.

Definition 5. *Acyclic Relaxed Temporal Random Walk:* *An Acyclic Relaxed Temporal Random Walk is a Relaxed Temporal Random Walk that does not contain any cycles (repeated entities in the walk).*

Example 5 shows an Acyclic Relaxed Temporal Random Walk of length 3.

Definition 6. *Link-star (L-star) Rule:* *A Link-star (or L-star) Rule is a Relaxed Temporal Rule generated from an acyclic walk of length 3 by generalizing the entities and timestamps with variables. While the second link in the walk becomes the rule head (E_2, r_h, E_3, T_2), the other two links are mapped to body atoms, where each link (e_i, r_i, e_{i+1}, t_i) is converted to the body atom (E_i, r_i, E_{i+1}, T_i). The final rule is of the form $(E_1, r_h, E_2, T_1) \leftarrow (E_0, r_1, E_1, T_0)\wedge$ (E_2, r_3, E_3, T_2)*

Example 9. The following is an L-star rule formed from the acyclic walk given in Example 5.

$$(E_1, talk, E_2, T_1) \leftarrow (E_0, visit^{-1}, E_1, T_0) \wedge (E_2, visit, E_3, T_2) \tag{2}$$

A single link-star rule encompasses several acyclic rules. Consider a crude generalization of the link-star rule given in Eq. 2 as $B \leftarrow A \wedge C$ by representing links as distinct letters. Then $B \leftarrow A$, $B \leftarrow C$, and $B \leftarrow A \wedge C$ are all subsets of this rule. In this way, we are able to prevent the repetition of acyclic rules. Further, TLogic has a strict criteria for forming cyclic rules, where the rule body must also terminate at the rule head, all while following the temporal constraints. It employs a brute-force search to find such cyclic walks. Thus a lot of random walks are wasted as they do not lead to the creation of any rules. In contrast, the search for link-star rules does not contain any such criteria, and thus almost every sampled walk is utilized to generate a rule.

Algorithm 1. Rule Generation

Input: A temporal Knowledge Graph
Parameters: Number of searches $s \in \mathbb{N}$, ACR, δ
Output: Set of Relaxed Temporal rules RT_R
 1: **for** links with relation $r \in \mathcal{R}$ **do**
 2:     **for** i $\in [s]$ **do**   ▷ Repeat with $[ACR * s]$ for L-star rules
 3:         Sample a walk w
 4:         Create rule rt from w
 5:         Compute rule confidence $conf(rt)$
 6:         $RT_R \leftarrow \cup(rt_r, conf(rt))$
 7:     **end for**
 8: **end for**

To evaluate the quality of the generated rules, we consider the conventional confidence measure. The **confidence** of a cyclic rule is the ratio of its *rule groundings* to *body groundings* in the tKG. A body grounding (also known as 'matching' or 'instantiation') for a rule is a tuple of entities and timestamps $(e_1, t_1, e_2, t_2, \cdots, t_n, e_{n+1})$, such that each body link (e_i, r_i, e_{i+1}, t_i) exists in the tKG. For cyclic rules, a body grounding is also a rule grounding if the head link with a future timestamp, i.e., $(e_1, r_h, e_{n+1}, t_{n+1})$; $t_n \leq t_{n+1} + \delta$, also exists in the tKG. We do not use the same measure for L-star rules. Instead, we implicitly mine acyclic rules with high confidence by starting with the head relation while performing acyclic walks. This saves a significant amount of time and computation resources, and also results in better forecasting accuracy.

Example 10. For the rule given by Eq. 1, one possible body grounding is $(Fred, 14/05, Belgium, 19/05, Spain, 20/05, France)$. However, there is no link of the form $(Fred, visit, France, t_x)$; $t_x > 25/05$. Thus it is not a rule grounding.

$(Robin, 15/05, Spain, 20/05, France, 25/05, Belgium)$ is another valid body grounding for that rule. Further, this tKG contains the link $(Robin, visit, France, 05/06)$, and $05/06 > 25/05$. Thus, this body grounding is also a rule grounding.

To allow for flexibility in the number of walks performed during rule generation, we define a parameter which can be controlled by the user. The ratio of acyclic walks to cyclic walks to be performed during rule generation is given by **Acyclic to Cyclic Ratio** or ACR. For example, if the *number of searches* is 100 and $ACR = 0.5$, then 50 acyclic walks and 100 cyclic walks will be performed during rule generation. Our experimentation indicates that increasing ACR results in improved forecasting accuracy but longer runtime. Using the above concepts, we present TRKG-Miner, which consists of algorithms for the following two steps: (i) Rule Generation (ii) Rule Application

Rule Generation. The rule generation process is shown in Algorithm 1. We iterate over all the relations to generate both cyclic and L-star rules. Iterating over all relations is necessary to extract rules for less frequently occurring relations. First, both cyclic and acyclic relaxed temporal random walks are performed on the tKG. During implementation, we always begin sampling walks with the head relation. For cyclic rules, we begin with the last link and work backwards, and for L-star rules we begin with the middle link and then sample the other two links. During walk sampling, since there can be multiple possible next links, priority is given to the links which have a timestamp close to the head link according to the probability distribution given in Eq. 3 (t_u is the current timestamp and $T_c = \{t_{c_1}, ...t_{c_i}, ..., t_{c_s}\}$ is a set of timestamps of the candidates) Next, the entities and timestamps are generalized with variables to create a rule from the performed walk.

$$P(t_c; T_c, t_u) = \frac{exp(-|t_u - t_c|)}{\sum_{T_c} exp(-|t_u - t_{c_i}|)} \tag{3}$$

Rule Application. The algorithm for finding candidates for a given query $(e_{query}, r_{query}, ?, t_{query})$ begins by pruning the rules according to a chosen minimum support ($minsup$). Then, all the rules whose head relation matches the query relation are selected, and body groundings are determined for each rule. From each grounding of the form $(e_0, t_0, e_{query}, t_1, e_2, t_2, e_3)$, e_2 represents a candidate and t_1 represents its candidate timestamp for each L-star rule. For cyclic rules, from each grounding of the form $(e_{query}, t_0, ..., t_i, e_i, ..., t_n, e_n)$, the last entity e_n and timestamp t_n are selected as a candidate and its candidate timestamp.

To compare between the different candidates, a score is calculated for each candidate according to the scoring function given in Algorithm 2 (Line 6, λ and α are experimentally-determined parameters). The reason for considering this scoring function is that recent events are more relevant to the query than older events and that the candidates from rules with a better rule quality metric should receive a higher score. Cyclic rules get lesser priority over acyclic rules because the search criteria for generating cyclic walks is based on loops *of a short length like 2-3*, which does not work well in sparse segments of the tKG. This is not the case with l-star rules, since they do not have any such criteria and they are acyclic by nature. Finally, the top-k candidates ordered by score are returned.

Algorithm 2. Rule Application

Input: $RT_{\mathcal{R}}$, Query $(e_{query}, r_{query}, ?, t_{query})$
Parameters: No. of candidates required (k), $minsup$
Output: Answer candidates A
1: $RT_R \leftarrow \{rt | support(rt) > minsup\}$
2: **for each** rt with relation r_{query} **do**
3:     Find all body groundings for rt
4:     **for each** body grounding **do**
5:         Get candidate a and candidate timestamp t_a
6:         $score(a) \leftarrow \alpha(exp(-\lambda(t_{query} - t_a))) + (1 - \alpha)conf(rt)$
7:     **end for**
8: **end for**
9: Return top-k candidates ordered by score

5   Experiments and Results

In this section, we evaluate the proposed TRKG-Miner[3] against 11 existing frameworks (DistMult [15], ComplEx [14], AnyBURL [10], TTransE [8], TA-DistMult [3], DE-SimplE [4] , TNTComplEx [7], CyGNet [16], RE-Net [6], xERTE [5], and TLogic [9]) on three real-world datasets and show that TRKG-Miner outperforms all of the evaluated frameworks by a very large margin. Integrated Crisis Early Warning System (ICEWS) is the most widely used dataset for link prediction on tKGs. We use the datasets ICEWS14, ICEWS18, and ICEWS0515. They contain information about world events from 2014, 2018, and 2005-2015 respectively. We use the same train-test-validation split as xERTE [5]. The test and validation sets contains timestamps which are relatively in the future from the training set. We use hits@k ($k=\{1,3,10\}$) and Mean Reciprocal Rank as performance metrics. The hits@k metrics measures the fraction of times that the correct entity is present among the top-k returned candidates. For example, $hits@1$ measures the fraction of times when the top-ranked candidate is the correct entity for a given query. MRR is defined as the average of all reciprocal ranks of the correct query answers across all queries, where reciprocal rank is $1/x$ for a rank x. Please note that we have obtained consistent results on all the datasets, but some of them are omitted owing to page constraints.

Table 1 presents the performance results of TRKG-Miner against the existing link forecasting frameworks. It can be observed that TRKG-Miner outperforms all of its competitors on all of the evaluated datasets. Notably, the results of TRKG-Miner for $hits@1$ metrics improve by approximately 45% on average, which means the accuracy of finding the correct result from the first candidate increases. This helps in making accurate predictions in one shot without having to rely upon multiple candidate predictions. There is also a significant improvement in $hits@3$. Consequently, the MRR is also improved by about 23%.

Table 2 records the change in the number of rules and forecasting efficiency when varying ACR during rule generation. We find that increasing the fraction

---
[3] Code available at https://github.com/ab1nash/TRKG-Miner.

Table 1. Results of link forecasting using various approaches. The best results in each category are highlighted in bold while the second best ones are italicised. Note that MRR can be considered as the weighted average of hits@1 (h@1), hits@3 (h@3), and hits@10 (h@10). Parameters: $\delta = 1$, $ACR = 3$ for ICEWS14, $ACR = 1$ for ICEWS18 and ICEWS0515

Dataset	ICEWS14				ICEWS18				ICEWS0515			
Model	MRR	h@1	h@3	h@10	MRR	h@1	h@3	h@10	MRR	h@1	h@3	h@10
DistMult	0.2767	0.1816	0.3115	0.4696	0.1017	0.0452	0.1033	0.2125	0.2873	0.1933	0.3219	0.4754
ComplEx	0.3084	0.2151	0.3448	0.4958	0.2101	0.1187	0.2347	0.3987	0.3169	0.2144	0.3574	0.5204
AnyBURL	0.2967	0.2126	0.3333	0.4673	0.2277	0.1510	0.2544	0.3891	0.3205	0.2372	0.3545	0.5046
TTransE	0.1343	0.0311	0.1732	0.3455	0.0831	0.0192	0.0856	0.2189	0.1571	0.0500	0.1972	0.3802
TA-DistMult	0.2647	0.1709	0.3022	0.4541	0.1675	0.0861	0.1841	0.3359	0.2431	0.1458	0.2792	0.4421
DE-SimplE	0.3267	0.2443	0.3569	0.4911	0.1930	0.1153	0.2186	0.3480	0.3502	0.2591	0.3899	0.5275
TNTComplEx	0.3212	0.2335	0.3603	0.4913	0.2123	0.1328	0.2402	0.3691	0.2754	0.1952	0.3080	0.4286
CyGNet	0.3273	0.2369	0.3631	0.5067	0.2493	0.1590	0.2828	0.4261	0.3497	0.2567	0.3909	0.5294
RE-Net	0.3828	0.2868	0.4134	0.5452	0.2881	0.1905	0.3244	0.4751	0.4297	0.3126	0.4685	0.6347
xERTE	0.4079	0.3270	0.4567	0.5730	0.2931	*0.2103*	0.3351	0.4648	0.4662	*0.3784*	0.5231	0.6392
TLogic	*0.4304*	*0.3356*	*0.4827*	*0.6123*	*0.2982*	0.2054	*0.3395*	*0.4853*	*0.4697*	0.3621	*0.5313*	*0.6743*
TRKG-Miner	**0.5028**	**0.4514**	**0.5189**	**0.6130**	**0.3796**	**0.3262**	**0.3905**	**0.4952**	**0.5856**	**0.5360**	**0.6028**	**0.6912**

Table 2. Model performance on varying ACR (ICEWS14, $\delta = 1$)

ACR	No. of Rules	MRR	h@1	h@3
0	31746	0.4437	0.3626	0.4825
0.25	44609	0.4666	0.3973	0.4968
0.5	53787	0.4805	0.4183	0.5053
1	67713	0.4907	0.4339	0.5113
2	88854	0.4950	0.4402	0.5145
3	104251	0.5028	0.4514	0.5189

Table 3. Model performance on varying δ (ICEWS14, $ACR = 1$)

$delta$	No. of rules	MRR	h@1	h@3
0	63037	0.4403	0.3662	0.4712
1	67713	0.4907	0.4339	0.5113
3	67283	0.4895	0.4313	0.5099
7	65351	0.4867	0.4271	0.5073
15	65279	0.4821	0.4207	0.5043
30	65718	0.4803	0.4181	0.5036
45	66925	0.4844	0.4230	0.5068
90	70627	0.4867	0.4269	0.5084

of L-star rules leads to an increase in prediction accuracy. However, this comes at the increased cost of run-time since more rules are being mined. ACR can be tuned based on the time and computing power available. Another important observation is that when $ACR = 0$, only cyclic rules are mined. This demonstrates how our approach generates a significant number of new useful rules as compared to previous approaches.

Table 3 records the change in the number of rules and forecasting efficiency when varying *maximum time gap* (δ). With an increase in δ, the number of generated rules increases. There is also a significant increase in forecasting efficiency when changing δ from 0 to 1, which indicates the usefulness of introducing temporal relaxation. We can also observe that with subsequent increase in the parameter, there is a decrease in error metrics. This leads to the conclusion that between $\delta = 1$ to $\delta = 90$, a large number of the new rules introduced are noisy. For other datasets, the optimal value of δ can be found by simple parameter tuning. Generally, it is best to keep it as small as possible.

6   Conclusion

We introduce TRKG-Miner, a rule-based link forecasting framework that improves forecasting efficiency by finding additional useful rules over previous approaches. We achieve this by the means of two improvements: (i) Adding time relaxation parameters to better capture the irregular occurrences of events within a time gap. (ii) Introducing and mining Link-star rules based on the principle that past events repeat themselves. Our experimentation indicates that we achieve significantly better results than the past approaches. Future extensions can include link forecasting on datasets with heterogeneous timestamps, a framework for handling time intervals (and not just timestamps), and adding more complex variations of cyclic and acyclic rules.

Acknowledgements. The research of Abinash Maharana and P. Krishna Reddy is supported by Microsoft Research Lab India, and iHub Anubhuti-IIITD Foundation. This research is also funded by JSPS Kakenhi 21K12034.

References

1. CSPM: discovering compressing stars in attributed graphs. Inf. Sci. **611**, 126–158 (2022)
2. Bordes, A., Usunier, N., Garcia-Duran, A., Weston, J., Yakhnenko, O.: Translating embeddings for modeling multi-relational data. Advances in Neural Information Processing Systems 26 (2013)
3. García-Durán, A., Dumančić, S., Niepert, M.: Learning sequence encoders for temporal knowledge graph completion. arXiv preprint arXiv:1809.03202 (2018)
4. Goel, R., Kazemi, S.M., Brubaker, M., Poupart, P.: Diachronic embedding for temporal knowledge graph completion. In: Proceedings of the AAAI Conference on Artificial Intelligence, vol. 34, pp. 3988–3995 (2020)
5. Han, Z., Chen, P., Ma, Y., Tresp, V.: Explainable subgraph reasoning for forecasting on temporal knowledge graphs. In: International Conference on Learning Representations (2020)
6. Jin, W., Qu, M., Jin, X., Ren, X.: Recurrent event network: autoregressive structure inference over temporal knowledge graphs. arXiv preprint arXiv:1904.05530 (2019)
7. Lacroix, T., Usunier, N., Obozinski, G.: Canonical tensor decomposition for knowledge base completion. In: Dy, J., Krause, A. (eds.) Proceedings of the 35th International Conference on Machine Learning. Proceedings of Machine Learning Research, vol. 80, pp. 2863–2872. PMLR (10–15 Jul 2018)
8. Leblay, J., Chekol, M.W.: Deriving validity time in knowledge graph. In: Companion Proceedings of the The Web Conference 2018, pp. 1771–1776 (2018)
9. Liu, Y., Ma, Y., Hildebrandt, M., Joblin, M., Tresp, V.: Tlogic: Temporal logical rules for explainable link forecasting on temporal knowledge graphs. In: AAAI (2022)
10. Meilicke, C., Chekol, M.W., Fink, M., Stuckenschmidt, H.: Reinforced anytime bottom up rule learning for knowledge graph completion. arXiv preprint arXiv:2004.04412 (2020)
11. Meilicke, C., Fink, M., Wang, Y., Ruffinelli, D., Gemulla, R., Stuckenschmidt, H.: Fine-grained evaluation of rule- and embedding-based systems for knowledge graph completion. In: Vrandečić, D., et al. (eds.) ISWC 2018. LNCS, vol. 11136, pp. 3–20. Springer, Cham (2018). https://doi.org/10.1007/978-3-030-00671-6_1
12. Nickel, M., Tresp, V., Kriegel, H.P.: A three-way model for collective learning on multi-relational data. In: ICML (2011)
13. Rossi, A., Barbosa, D., Firmani, D., Matinata, A., Merialdo, P.: Knowledge graph embedding for link prediction: A comparative analysis. ACM Trans. Knowl. Discov. Data (TKDD) **15**(2), 1–49 (2021)
14. Trouillon, T., Welbl, J., Riedel, S., Gaussier, É., Bouchard, G.: Complex embeddings for simple link prediction. In: International Conference on Machine Learning, pp. 2071–2080. PMLR (2016)
15. Yang, B., tau Yih, W., He, X., Gao, J., Deng, L.: Embedding entities and relations for learning and inference in knowledge bases. CoRR abs/1412.6575 (2015)
16. Zhu, C., Chen, M., Fan, C., Cheng, G., Zhang, Y.: Learning from history: Modeling temporal knowledge graphs with sequential copy-generation networks. In: Proceedings of the AAAI Conference on Artificial Intelligence, vol. 35, pp. 4732–4740 (2021)

A Consumer-Good-Type Aware Itemset Placement Framework for Retail Businesses

Raghav Mittal[1]([✉]), Anirban Mondal[1], and P. Krishna Reddy[2]

[1] Ashoka University, Sonipat, India
raghav.mittal@alumni.ashoka.edu.in, anirban.mondal@ashoka.edu.in
[2] IIIT, Hyderabad, India
pkreddy@iiit.ac.in

Abstract. It is a well-established fact that strategic placement of items on the shelves of a retail store significantly impacts the revenue of the retailer. Consumer goods sold in retail stores can be classified into essential and typically low-priced *convenience items*, and non-essential and high-priced *shopping items*. Notably, the lower-priced convenience items are critical to ensuring consumer foot-traffic, thereby also driving the sales of shopping items. Moreover, users typically buy multiple items together (i.e., *itemsets*) to facilitate one-stop shopping. Hence, it becomes a necessity to strategically index and place itemsets that contain both convenience items and shopping items. In this regard, we propose a consumer-good-type aware and revenue-conscious itemset indexing scheme for *efficiently* retrieving high-revenue itemsets containing both convenience and shopping items. Moreover, we propose an itemset placement scheme, which exploits our indexing scheme, for improving retailer revenue. Our performance study with two real datasets shows that our framework is indeed effective in improving retailer revenue w.r.t. a reference scheme.

Keywords: Pattern Mining · Utility Mining · Retail · Indexing ·
Itemset Placement · Convenience Items · Shopping Items

1 Introduction

Retail stores provide consumers with easy access to items, and try to improve their revenue through increased sales. Typical retail stores comprise multiple shelves (racks), which contain slots for placement of items. Retail slots can either be premium or non-premium. Premium slots are those with high product visibility/accessibility e.g., slots near the eye/shoulder level of users and impulse-buy slots near checkout counters; all other slots are non-premium. Items placed in premium slots have higher probability of sales than items placed in non-premium slots. Moreover, customers typically prefer to buy a set of items (i.e., *itemsets* [1]) as opposed to individual items for the convenience of *one-stop shopping*. Therefore, there is an opportunity for the retailer to improve its revenue through placement of itemsets in premium retail slots [2,5,6,10,16–18,20].

H. Kashima et al. (Eds.): PAKDD 2023, LNAI 13935, pp. 276–288, 2023.
https://doi.org/10.1007/978-3-031-33374-3_22

Consumer goods can be classified into three broad categories, namely *convenience items*, *shopping items* and *specialty items* [4,11,15]. Convenience items are essential and low-priced daily-use products with frequent consumer demand e.g., newspapers, dairy products, cigarettes and medicines. They reflect the *needs* of consumers. In contrast, shopping items are higher-priced and non-essential products with infrequent consumer demand e.g., televisions, furniture and refrigerators. They reflect the *wants* of consumers. Specialty goods, such as rare works of art, are purchased by connoisseurs typically at auction houses, and are generally never sold at retail stores. Since our focus is on retail stores, we consider placement *only* for convenience and shopping items in the premium slots.

Given that convenience items are essential products that consumers *frequently need*, they cater to a loyal consumer base, thereby *significantly driving consumer foot-traffic*; this also drives the sales of shopping items. Convenience items tend to attract consumers to the retail store and once a consumer is already in that store, she can purchase itemsets containing the more expensive shopping items as well. Hence, if retailers greedily placed *only* itemsets containing the higher-priced shopping items in the premium slots, they would fail to capture the *importance of convenience items in attracting foot-traffic*. This would lead to some of the convenience items not being placed as premium slots are limited in number. This would reduce foot-traffic, thereby significantly degrading retailer revenue.

As a case in point, in the past decade, SuperFresh [7] (a supermarket brand of Key Food Stores) decided to halt sales of convenience items (e.g., low-priced grocery products) to increase its focus on selling higher-priced shopping items. However, they started losing foot-traffic precipitously and soon became bankrupt [7]. *This motivates the need to place at least a minimum number min_{c_i} of instances of each convenience item c_i to ensure adequate foot-traffic for driving sales.* We refer to this as the minimum placement criteria (MPC). Here, min_{c_i} is application-dependent and should be decided by the retailer. In essence, the retailer needs to strategically place itemsets in *a consumer-good-type aware manner* i.e., the placed itemsets should contain both convenience and shopping items.

Existing research efforts can be broadly categorized into three types, namely (a) approaches that consider consumer-good-types [4,11,14,15] (b) approaches for mining high-utility itemsets (HUIs) [3,8,9,12,21,22] and (c) approaches for retail itemset placement [5,16–20]. While the research in (a) considers consumer-good-types, it considers only the placement of individual items as opposed to itemsets. The research in (b) and (c) focuses on identifying and placing HUIs respectively, but fails to consider consumer-good-types. Given that existing HUI mining approaches typically apply a utility (revenue) threshold for pruning away the low-utility itemsets, they often tend to prioritize the placement of itemsets containing expensive shopping items as compared to the low-priced convenience items. However, they fail to capture how the pruned-away low-revenue itemsets containing convenience items also play a significant role in driving foot-traffic by attracting consumers to the store. In essence, absence of itemsets containing convenience items could significantly degrade retailer revenue due to loss of foot-traffic. Notably, *none* of the existing works address the issue of consumer-good-type aware placement of HUIs in retail stores.

This work addresses the problem of incorporating consumer-good-type awareness during itemset placement in premium retail slots for improving retailer revenue. Our proposed problem framework requires as input a database D of user purchase transactions and a set H of HUIs extracted from D using any existing HUI mining algorithm [8,22]. Each item is associated with a price, a frequency of sales and a consumer-good-type (i.e., convenience versus shopping items). Further, each convenience item has a minimum placement criterion (MPC). Given T_S premium slots in a retail store, the problem is to place HUIs in these slots to improve retailer revenue such that MPC for each convenience item is satisfied.

To address this problem, we need: (a) to *efficiently* retrieve itemsets containing convenience items to satisfy their MPC and (b) to place itemsets in the limited number of premium slots in a consumer-good-type aware manner. To address (a), we propose the **C**onsumer-good-type aware and **R**evenue-conscious **I**temset indexing **S**cheme (CRIS). CRIS comprises N_{con} hash buckets, one for each convenience item. Each hash bucket is associated with a linked list of the top-k high-revenue itemsets containing the convenience item corresponding to that hash bucket. Observe how this enables CRIS to quickly retrieve itemsets containing any given convenience item to satisfy MPC.

To address (b), we propose the **C**onsumer-good-type aware and **R**evenue-conscious **I**tem**S**et **P**lacement Scheme (CRISP). Initially, CRISP *greedily* places the top-revenue itemsets in the slots until all slots are exhausted. Then CRISP keeps progressively replacing the already placed low-revenue itemsets (starting from the placed itemset with the lowest revenue) with itemsets containing convenience items. These itemsets are obtained from the CRIS index. This process continues until MPC of all convenience items is satisfied. Notably, CRISP does not replace the placed itemsets containing convenience items to prevent inefficiencies. Observe how CRISP satisfies MPC by replacing only the low-revenue placed itemsets. Our key contributions are three-fold:

1. We introduce the problem of consumer-good-type aware itemset placement in retail stores for improving retailer revenue.
2. We propose the CRIS index for efficiently retrieving HUIs in a consumer-good-type aware manner. We further propose the CRISP itemset placement scheme, which exploits CRIS, for improving retailer revenue.
3. We conduct a performance study with two real datasets to demonstrate that our framework is indeed effective in improving retailer revenue w.r.t. a reference scheme.

To our knowledge, this is the first work to address itemset placement in retail stores in a consumer-good-type aware manner. The remainder of this paper is organized as follows. Section 2 discusses related works, while Sect. 3 describes the proposed problem framework. Section 4 presents our proposed CRIS index and our proposed CRISP itemset placement scheme. Section 5 reports our performance study. Finally, we conclude in Sect. 6 with directions for future work.

2 Related Work

In this section, we discuss (a) approaches concerning consumer-good types (b) utility mining approaches (c) itemset placement approaches for retail stores.

The work in [11] has made the distinction between convenience goods, shopping goods and specialty goods. Moreover, the work in [4] discussed how such consumer goods classification can be integrated with retail strategy. The work in [15] suggested that convenience goods and shopping goods should follow different modes of advertising. The work in [14] aimed at demand forecasting for new convenience goods by using a stochastic evolutionary adoption model.

The Utility Pattern Growth algorithm [22] exploits the *Utility Pattern Tree*, which maintains information about HUIs for pruning purposes. Moreover, the work in [21] discussed an approach for mining closed HUIs for both dense and sparse datasets. The work in [9] proposed the LHUI-Miner and PHUI-Miner algorithms, which consider that itemset utilities may vary w.r.t. time. Further, the work in [12] finds HUIs by using two upper-bounds, namely the tight maximum average utility upper-bound and the maximum remaining average utility upper-bound, in conjunction with a list-based data structure. The work in [3] proposed an approximate HUI mining approach for scenarios with noisy data. Additionally, the work in [13] proposed a high average utility pattern mining approach, which uses the HAUP-List data structure for compactly storing information about patterns, in the context of dynamic databases.

Moreover, efforts have been made to improve HUI placement approaches based on different item sizes [5], slot premiumness [20], sale urgency and product expiry [16,18], diversification of products [17], and item inventory [19]. Notably, *none* of the existing approaches consider the consumer-good type of the items.

3 Proposed Framework of the Problem

Consider a set D of user purchase transactions over a set Υ of items, where all transactions in D contain unique and non-repeating items. Each item i in Υ occupies only a single retail slot, and is associated with price ρ_i, frequency of sales σ_i, consumer-good-type μ_i, and minimum placement criterion (MPC), quantified by θ_i. Recall that the minimum placement criteria (MPC) pertains to the need for placing at least a minimum number of instances of each convenience item to ensure adequate foot-traffic. We set $\mu = 1$ for all convenience items and $\mu = 0$ for all shopping items. The value of θ is application-dependent and can be decided by the retailer based on domain knowledge. Shopping items are not associated with any MPC. Hence, $\theta = 0$ for all shopping items.

Given that all transactions in D comprise unique items, we simply count the number of instances of a given item i in D to compute its frequency of sales σ_i. Moreover, in this paper, we consider the net revenue of an item as a measure of utility of the item and the average net revenue of an itemset as a measure of utility of the itemset based on the approach proposed in [5]. We shall now explain these terms and present the proposed problem framework.

We compute the **net revenue** NR_i **of an item** i as the product of its frequency of sales σ_i and its price ρ_i. Hence, $NR_i = \sigma_i * \rho_i$.

For computing the frequency of sales of an itemset, we determine the number of occurrences of that itemset across all transactions in D. We compute the **net revenue** NR_z **of an itemset** z as the product of its frequency of sales σ_z and its price ρ_z. Here, the value of ρ_z is computed as the sum of the prices of all the individual items in itemset z. Hence, $NR_z = \sigma_z * \sum_{i \in z} \rho_i$. Observe that the definition of the net revenue of an itemset is biased towards larger itemsets as they are likely to have higher NR albeit at the cost of occupying more slots. To address this bias, we introduce the notion of the **average net revenue** (ANR) of an itemset. We compute ANR_z of an itemset z as the net revenue of z divided by the total number of the items in z. Hence, $ANR_z = (\sigma_z * \sum_{i \in z} \rho_i) \ / \ |z|$.

Problem Statement: Consider a database D of user purchase transactions over items in set Υ. Each item i in Υ has price ρ_i, frequency of sales σ_i, a consumer-good-type μ_i and a minimum placement criterion (MPC). Given a set H of HUIs extracted from D (using any existing HUI mining approach [8,22]) and T_S premium slots, the problem is to place HUIs in these slots to improve retailer revenue such that MPC for each convenience item is satisfied.

4   Proposed Itemset Placement Framework

This section discusses our proposed framework, which includes (a) our proposed CRIS index and (b) our proposed CRISP itemset placement scheme, which exploits the CRIS index, for improving retailer revenue.

Basic Idea: Placement of itemsets in premium slots of a retail store significantly impacts retailer revenue. High-utility itemsets (HUIs) can be extracted for populating the slots using existing HUI mining approaches. However, these approaches are oblivious to *consumer-good-types*. Given that they typically apply a utility (revenue) threshold for pruning away the low-utility itemsets, they often tend to prioritize the placement of itemsets containing expensive shopping items as compared to the low-priced convenience items.

It is well known that convenience items act as the key driver of foot-traffic [7], which in turn influences the sales of both shopping and convenience items. Hence, if we were to greedily place HUIs based only on their average net revenue (ANR), there is a possibility of some of the convenience items not being placed. This would reduce foot-traffic, thereby degrading retailer revenue. In essence, *at least a minimum number of instances* need to be placed for each of the convenience items to ensure adequate foot-traffic for driving sales. As discussed previously in Sect. 1, this is the minimum placement criteria (MPC). To satisfy MPC, we propose the CRIS index for *efficient* retrieval of HUIs associated with each of the convenience items. Our CRISP itemset placement scheme exploits CRIS for placing itemsets in a consumer-good-type aware manner to satisfy MPC. Figure 1 presents an overview of our proposed framework.

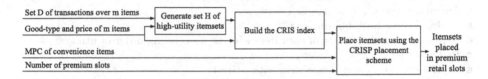

Fig. 1. Schematic diagram of our proposed itemset placement framework

Description of the CRIS Index: Given N_{con} convenience items, CRIS comprises N_{con} hash buckets, one for each convenience item. Each hash bucket i has an entry of the form $\{i, NR_i, ptr_i\}$, where i is the unique identifier of a given convenience item, NR_i is the net revenue of the item and ptr_i is the pointer to the linked list of HUIs containing item i. Each entry of the linked lists is of the form $\{z, ANR_z\}$, where z refers to an itemset and ANR_z is the average net revenue of z. The entries in the linked lists are sorted in descending order of ANR. Further, each linked list stores only the top-k HUIs containing the corresponding convenience item. Here, the value of k is application-dependent. For *efficiently* retrieving HUIs containing the ith convenience item, CRIS can directly traverse to the ith hash bucket and traverse the corresponding linked list to retrieve the relevant HUIs, until MPC is satisfied.

Notably, the hash buckets in CRIS are sorted in descending order of NR of the convenience items for prioritizing the placement of convenience items with higher NR. This helps in further improving the retailer revenue. Figure 2(b) depicts an illustrative example of CRIS with five convenience items, namely C_1 to C_5. Observe how the linked list corresponding to each convenience item comprises itemsets containing that convenience item. Further, observe that the linked lists of HUIs are sorted in descending order of ANR and the hash buckets corresponding to the convenience items are sorted in descending order of NR.

Description of CRISP Itemset Placement Scheme: Initially, CRISP *greedily* places the input HUIs from set H in the slots starting from the highest-revenue HUI until all slots have been exhausted. Then it checks MPC w.r.t. each convenience item. For each convenience item not satisfying MPC, it traverses the

---

Algorithm 1: Itemset placement framework

Input: D: set of transactions; Υ: list of m tuples of the form $< i, util, \mu >$ (i, $util$, and μ denote the item ID, utility, and good-type of i respectively); I: list of tuples of form $< i, \theta >$ (θ is the value of MPC of i); T_S: number of slots

Output: Items placed in T_S slots

Variables: H, L: list of tuples of the form $< z, util >$ (z is a set, $util$ is a real value), C: list of tuples of the form $< i, cnt >$ (i is an item ID, cnt is an int.)

1 Using D, compute set H of HUIs using any HUI mining algorithm

Populating the CRIS index

2 From H, extract itemsets with at least one convenience item and store in L

3 Sort L in the descending order of itemset utility values

4 **foreach** entry $< i, util, \mu >$ in Υ with $\mu = 1$

5   Initialize a hash bucket for i

6 Arrange the hash buckets of each item i in the descending order of utility

7 **foreach** itemset $z \in L$; **foreach** item $i \in z$

8   **if** item i is a convenience item with $\mu(i)=1$

9   Insert z into the hash bucket of i

Placing the itemsets in the premium slots

10 Sort H and place HUIs in T_S slots in the descending order of utility

11 Compute the count cnt of each convenience item i in the HUIs placed in T_S slots and insert $< i, cnt >$ in C

12 **foreach** hash bucket i /*starting with the highest-utility item i*/

13   **while** $\theta(i) > cnt(i)$ /*$\theta(i)$, $cnt(i)$ are retrieved from I and C respectively*/

14   Replace the lowest-utility itemset (not containing items from C) in T_S slots, with highest-utility itemset z containing i from i^{th} hash bucket

15   **foreach** item $j \in z$ **if** item j is a convenience item **then** $cnt_j + = 1$

16   Remove z from the ith hash bucket

CRIS index to retrieve a high-revenue HUI z containing that convenience item. Next, it replaces the lowest-revenue placed HUI (not containing any convenience item(s)) in the retail slots with z. This process is repeated until MPC is satisfied for all convenience items. CRISP avoids replacing the already placed HUIs containing convenience items as the convenience items contained in those HUIs would need to be re-examined, thereby causing inefficiencies.

Algorithm 1 contains the pseudocode for our proposed framework.

Illustrative Example: Figure 2 depicts an illustrative example of our proposed framework. Inputs to our proposed framework include convenience items (C_1 to C_5) and shopping items (S_1 to S_{10}), along with their corresponding attributes. For simplicity, we assign all convenience items with equal values of θ. Observe how our approach computes NR and ANR for the inputs HUIs in H, based on our discussion in Sect. 3. Figure 2(b) depicts an example of the CRIS index. Observe how for each convenience item, CRIS maintains a linked list of HUIs sorted based on ANR. For instance, observe how convenience item C_3 is associ-

Shopping items: S_1 - S_{10}, Convenience items: C_1 - C_5, ρ: Price,
σ: Frequency of sales, NR: Net Revenue, ANR: Average Net Revenue,
θ : Placement requirement

Shopping items	S_1	S_2	S_3	S_4	S_5	S_6	S_7	S_8	S_9	S_{10}
Price (ρ)	4	3	4	2	6	5	8	9	4	7
Frequency of sales (σ)	9	11	10	12	11	11	9	8	8	10
Net revenue (NR)	36	33	40	24	66	55	72	72	32	70
Placement requirement (θ)	-	-	-	-	-	-	-	-	-	-

Convenience items	C_1	C_2	C_3	C_4	C_5
Price (ρ)	7	2	6	1	3
Frequency of sales (σ)	16	15	14	17	19
Net Revenue (NR)	112	30	84	17	57
Placement requirement (θ)	2	2	2	2	2

Input itemsets	ρ	σ	NR	ANR
S_1, S_3, S_4, S_6	15	5	75	18.75
C_1, S_3	11	5	55	27.5
C_4, S_1, S_3	9	5	45	15
S_5, C_3	12	5	60	30
S_2, S_4, S_5	11	8	88	29.33
C_4, S_1	5	7	35	17.5
S_1, S_3, S_9, C_4	15	5	75	18.75
S_2, S_5	9	8	72	36
S_6, C_2, C_4, C_5	11	5	55	13.8
S_1, S_3, C_3	14	6	84	28
S_2, S_9, C_2, S_1	15	4	60	15
S_1, S_3, S_5	14	6	84	28
S_1, S_3, S_4, S_5	16	6	96	24
S_2, C_5, S_5	12	8	96	32
C_1, C_3	13	2	26	13
S_2, S_3	7	5	35	17.5

(a) Inputs to our proposed itemset placement framework

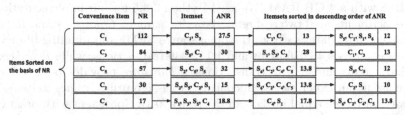

(b) Illustrative example of the CRIS index

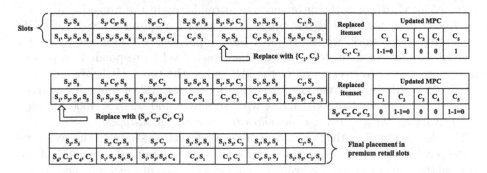

(c) Example of CRISP itemset placement Scheme

Fig. 2. Illustrative example of our proposed itemset placement framework

ated with a linked list containing itemsets $\{S_5, C_3\}$, $\{S_1, S_3, C_3\}$, and $\{C_1, C_3\}$, that are sorted in descending order of *ANR*.

Figure 2(c) depicts an illustrative example of CRISP. Initially, CRISP greed-ily places HUIs in descending order of ANR until all slots are exhausted. To address MPC for convenience items, CRISP exploits the CRIS index to place HUIs containing convenience items. Starting with the convenience item with highest NR (i.e., item C_1), CRISP retrieves the top-1 HUI containing C_1 from CRIS. Since the top-1 HUI (i.e., $\{C_1, S_3\}$) has already been placed, CRISP retrieves the top-2 itemset containing C_1 to satisfy its MPC. Notice how CRISP replaces the lowest-revenue HUI not containing any convenience item(s) to sat-isfy MPC for C_1. Next, since MPC for C_3 has already been satisfied, CRISP progresses to place HUIs for C_5, which is the next convenience item in order of NR. Observe how satisfying MPC for C_5 also satisfies MPC for C_2. Thus, CRISP is able to satisfy MPC for all convenience items.

5 Performance Evaluation

This section reports the results of our performance study. We conducted our experiments on an Intel(R) Pentium(R) 2.20 GHz processor running on Ubuntu 20.04.1 LTS with a 4 GB RAM. We used Python 3.8.5 for our implementation.

Our experiments used two real retail datasets, namely, *Chainstore* and *Retail*. We extracted these datasets from the open-source SPMF data-mining library[1]. *Chainstore* comprises 46,086 items and 1,112,949 transactions, while *Retail* con-tains 16,470 items and 88,162 transactions. *Chainstore* provides price values, which we use in our experiments. In contrast, to assign utility values to items in *Retail*, we created ten equal buckets in the range [0,1]. Consistent with practice, we assigned items with lower sales to high-priced buckets and vice-versa. Next, we assigned a random price to every item corresponding to its range. We set the value of MPC (θ) for each convenience item randomly in the range [1, 5].

Recall from Sect. 4 that our proposed approach requires a set H of HUIs as input. In this work, we use the kUI index approach [17] to generate set H. The kUI index is a multi-level index, where each level corresponds to a partic-ular itemset size. At every level of the kUI index, the top-λ HUIs are stored in descending order of revenue for facilitating quick retrieval of a queried itemset size. We implemented the kUI index with six levels and set $\lambda = 4000$ for each level. *Notably, the kUI index is oblivious to consumer-good-types.*

Table 1 summarizes the parameters of our performance study. In this work, we divide each dataset into training and test sets, which contain 70% and 30% of the transactions respectively. We evaluate the performance of both schemes on the test set. Our performance metrics include total revenue TR, number N_C of con-venience items, whose MPC is completely satisfied by the placement scheme, and execution time ET required for identification and placement of HUIs. Note that

---

[1] http://www.philippe-fournier-viger.com/spmf/datasets.

in the test phase, we iterate through transactions in the test set and add (item-set) prices to TR only if the items belonging to the transaction had been placed as itemsets during the training phase. To our knowledge, existing approaches do not consider consumer-good-types for itemset placement. Hence, for mean-ingful comparison, we propose a reference scheme, designated as **R**andomized high-utility **I**temset **P**lacement scheme (RIP). RIP works as follows. First, RIP segregates the input HUIs into clusters based on their size i.e., HUIs of size k are inserted in the kth cluster. Next, RIP *randomly* selects a cluster and then ran-domly chooses an HUI from that cluster, and places it. This process is repeated until all slots are populated. We set k=6 in this work. Notably, RIP places HUIs of different sizes, thereby making it an efficient approach for improving retailer revenue. Since RIP is oblivious to consumer-good-types, it tends to ignore the placement of convenience items.

Table 1. Performance study parameters

Parameter	Default	Variations
Total no. of premium slots (T_S) (10^3)	15	5, 10, 20, 25
No. of convenience items (N_{con})	120	40, 80, 160, 200

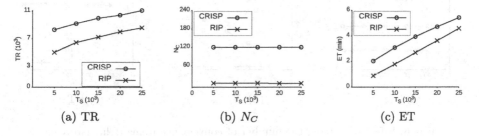

(a) TR   (b) N_C (c) ET

Fig. 3. Effect of varying the total number of premium slots (Retail)

Effect of Variations in the Number of Premium Slots: The results in Fig. 3 depict the effect of varying the total number T_S of premium slots for *Retail*. As T_S increases, TR increases for both schemes as more slots imply that more HUIs can be placed, thereby leading to higher revenue. CRISP outper-forms RIP in terms of TR and N_C as it places HUIs in a consumer-good-type aware manner, thus including HUIs containing both convenience and shopping items. RIP is oblivious to consumer-good-types, hence it tends to ignore HUIs containing convenience items. Notably, as T_S increases, N_C remains constant for CRISP because CRISP already satisfies MPC for all convenience items. Hence, increasing T_S does not further increase N_C. N_C remains comparable for RIP with increase in T_S as it places fewer HUIs containing convenience items.

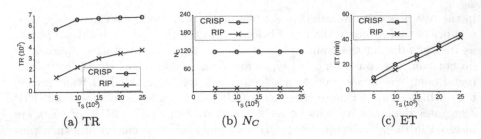

Fig. 4. Effect of varying the total number of premium slots (Chainstore)

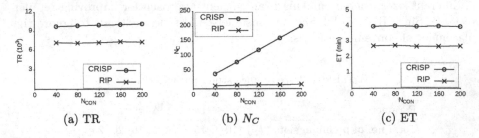

Fig. 5. Effect of varying the number of convenience items (Retail)

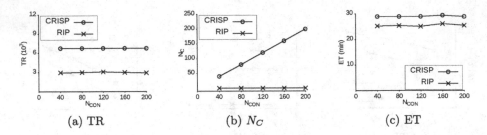

Fig. 6. Effect of varying the number of convenience items (Chainstore)

The results in Fig. 3(c) indicate that ET increases for both CRISP and RIP with increase in T_S since more HUIs need to be examined for populating a larger number of slots. CRISP incurs higher ET than RIP since it meticulously places HUIs in a consumer-good-type aware manner. This is a small price to pay for improved retailer revenue. The results in Fig. 4 for *Chainstore* exhibit comparable trends; the differences are due to varying dataset sizes.

Effect of Variations in the Number of Convenience Items: The results in Fig. 5 depict the effect of varying the number N_{con} of convenience items for *Retail*. As N_{con} increases, TR remains comparable for both CRISP and RIP. This occurs as the number of HUIs placed does not change with variations in N_{con}. CRISP outperforms RIP in terms of TR as per the rationale provided for the results in Fig. 3(a). As N_{con} increases, CRISP exhibits considerably increased N_C since it satisfies MPC for a larger number of convenience items as there are more convenience items in the input. In contrast, N_C remains comparable for RIP with increase in N_{con} since RIP is oblivious to consumer-good-types.

The results in Fig. 5(c) depict that RIP incurs lower ET than CRISP as per the rationale provided for results for Fig. 3(c). Further, RIP incurs comparable ET across variations in N_{con} as it is oblivious to consumer-good-types. Notably, the results in Fig. 6 for *Chainstore* exhibit comparable trends.

6   Conclusion

Strategic placement of items in a retail store significantly impacts retailer revenue. Retail consumer goods can be classified into lower-priced and frequently purchased *convenience items*, and higher-priced and less frequently purchased *shopping items*. Lower-priced convenience items ensure consumer foot-traffic and drive the sales of shopping items [7]. Further, since users typically buy itemsets instead of individual items, we have introduced the problem of consumer-good-type aware itemset placement in retail stores. Our proposed framework comprises the CRIS index for efficiently retrieving HUIs in a consumer-good-type aware manner and the CRISP itemset placement scheme, which exploits CRIS for improving retailer revenue. Our performance study with two real datasets shows that our framework is indeed effective in improving retailer revenue w.r.t. a reference scheme. In the near future, we plan to investigate the performance of our proposed approach by considering the impact of placement of distinct consumer-good-types on the consumer footfall in a retail store.

References

1. Agrawal, R., Srikant, R.: Fast algorithms for mining association rules. In: Proceedings of the VLDB, vol. 1215, pp. 487–499 (1994)
2. Ahn, K.I.: Effective product assignment based on association rule mining in retail. Expert Syst. Appl. **39**, 12551–12556 (2012)
3. Baek, Y., et al.: Approximate high utility itemset mining in noisy environments. Knowl.-Based Syst. **212**, 106596 (2021)
4. Bucklin, L.P.: Retail strategy and the classification of consumer goods. J. Mark. **27**(1), 50–55 (1963)
5. Chaudhary, P., Mondal, A., Reddy, P.K.: An improved scheme for determining top-revenue itemsets for placement in retail businesses. Int. J. Data Sci. Anal. **10**(4), 359–375 (2020). https://doi.org/10.1007/s41060-020-00221-5
6. Chen, M., Lin, C.: A data mining approach to product assortment and shelf space allocation. Expert Syst. Appl. **32**, 976–986 (2007)
7. Fisher, M., Vaidyanathan, R.: Which products should you stock? a new approach to assortment planning turns an art into a science. Harvard Business Review (2012)
8. Fournier-Viger, P., Lin, J.C.-W., Wu, C.-W., Tseng, V.S., Faghihi, U.: Mining minimal high-utility itemsets. In: Hartmann, S., Ma, H. (eds.) DEXA 2016. LNCS, vol. 9827, pp. 88–101. Springer, Cham (2016). https://doi.org/10.1007/978-3-319-44403-1_6
9. Fournier-Viger, P., Zhang, Y., Lin, J.C.W., Fujita, H., Koh, Y.S.: Mining local and peak high utility itemsets. Inf. Sci. **481**, 344–367 (2019)

10. Hansen, P., Heinsbroek, H.: Product selection and space allocation in supermarkets. Eur. J. Oper. Res. **3**, 474–484 (1979)
11. Holton, R.H.: The distinction between convenience goods, shopping goods, and specialty goods. J. Mark. **23**(1), 53–56 (1958)
12. Kim, H., et al.: Efficient list based mining of high average utility patterns with maximum average pruning strategies. Inf. Sci. **543**, 85–105 (2021)
13. Kim, J., Yun, U., Yoon, E., Lin, J.C.W., Fournier-Viger, P.: One scan based high average-utility pattern mining in static and dynamic databases. Futur. Gener. Comput. Syst. **111**, 143–158 (2020)
14. Massy, W.F.: Forecasting the demand for new convenience products. J. Mark. Res. **6**(4), 405–412 (1969)
15. Meisel, J.B.: Demand and supply determinants of advertising intensity among convenience goods. Southern Econ. J. 233–243 (1979)
16. Mittal, R., Mondal, A., Chaudhary, P., Reddy, P.K.: An urgency-aware and revenue-based itemset placement framework for retail stores. In: Strauss, C., Kotsis, G., Tjoa, A.M., Khalil, I. (eds.) DEXA 2021. LNCS, vol. 12924, pp. 51–57. Springer, Cham (2021). https://doi.org/10.1007/978-3-030-86475-0_5
17. Mondal, A., Mittal, R., Chaudhary, P., Reddy, P.K.: A framework for itemset placement with diversification for retail businesses. Appl. Intell. 1–19 (2022)
18. Mondal, A., Mittal, R., Khandelwal, V., Chaudhary, P., Reddy, P.K.: PEAR: a product expiry-aware and revenue-conscious itemset placement scheme. In: Proceedings of the DSAA, pp. 1–10. IEEE (2021)
19. Mondal, A., Mittal, R., Saurabh, S., Chaudhary, P., Reddy, P.K.: An inventory-aware and revenue-based itemset placement framework for retail stores. Expert Syst. Appl. **216**, 119404 (2023)
20. Mondal, A., Saurabh, S., Chaudhary, P., Mittal, R., Reddy, P.K.: A retail itemset placement framework based on premiumness of slots and utility mining. IEEE Access **9**, 155207–155223 (2021)
21. Nguyen, L.T., et al.: An efficient method for mining high utility closed itemsets. Inf. Sci. **495**, 78–99 (2019)
22. Tseng, V.S., Wu, C., Shie, B., Yu, P.S.: UP-Growth: an efficient algorithm for high utility itemset mining. In: Proceedings of the ACM SIGKDD, pp. 253–262. ACM (2010)

Deep Learning

M-EBM: Towards Understanding the Manifolds of Energy-Based Models

Xiulong Yang[iD] and Shihao Ji[✉]

Georgia State University, Atlanta, USA
{xyang22,sji}@gsu.edu

Abstract. Energy-based models (EBMs) exhibit a variety of desirable properties in predictive tasks, such as generality, simplicity and compositionality. However, training EBMs on high-dimensional datasets remains unstable and expensive. In this paper, we present a Manifold EBM (M-EBM) to boost the overall performance of unconditional EBM and Joint Energy-based Model (JEM). Despite its simplicity, M-EBM significantly improves unconditional EBMs in training stability and speed on a host of benchmark datasets, such as CIFAR10, CIFAR100, CelebA-HQ, and ImageNet 32×32. Once class labels are available, label-incorporated M-EBM (M-JEM) further surpasses M-EBM in image generation quality with an over 40% FID improvement, while enjoying improved accuracy. The code can be found in https://github.com/sndnyang/mebm.

Keywords: Generative Model · Energy-based Model · Joint Energy-based Model

1 Introduction

Energy-Based Models (EBMs) are an class of probabilistic models, which are widely applicable in image generation, out of distribution detection, adversarial robustness, and hybrid discriminative-generative modeling [3–7,16,20,21]. However, training EBMs on high-dimensional datasets remains very challenging. Most of the works utilize the Markov Chain Monte Carlo (MCMC) sampling [19] to generate samples from the model distribution represented by an EBM. Specifically, they require K-step Langevin Dynamics sampling [19] to generate samples from the model distribution in every iteration, which can be extremely expensive when using a large number of sampling steps, or highly unstable with a small number of steps. The trade-off between the training time and stability prevents the MCMC sampling based EBMs from scaling to large-scale datasets.

Recently, there are a flurry of works on improving EBMs. The most recent studies [3,5] on the MCMC-based approach focus on improving the generation quality and stability. However, they still resort to a long sampling chain and requires expensive training. Another branch of works [6,20] augment the EBM with a regularized generator in a GAN-style training to improve the stability and speed, sacrificing the desired property of learning a single object. Moreover,

H. Kashima et al. (Eds.): PAKDD 2023, LNAI 13935, pp. 291–302, 2023.
https://doi.org/10.1007/978-3-031-33374-3_23

Fig. 1. Generated samples of CelebA-HQ 128 × 128 from our M-EBM.

JEM [7] proposes an elegant framework to reinterpret the modern CNN classifier as an EBM and achieves impressive performances in image classification and generation simultaneously. However, it also suffers from the divergence issue of the MCMC-based sampling, and its generative performance falls behind state-of-the-art EBMs. Tackling the limitations of JEM, JEM++ [21] introduces a variety of training procedures and architecture features to improve JEM in terms of accuracy, speed and stability altogether. Furthermore, JEM++ demonstrates a trade-off between classification accuracy and image quality, but it still cannot improve image generation quality notably.

In this paper, we introduce simple yet effective training techniques to improve unconditional EBM and JEM in terms of image generation quality, training stability and speed altogether. First, the informative initialization introduced in JEM++ dramatically improves the training stability and reduces the required MCMC sampling steps. However, it's not scalable for high-resolution and large-scale datasets. Hence, we introduce a simplified informative initialization that is suitable for unconditional EBM and JEM for high-resolution images and a large number of classes (e.g., 128 × 128 CelebA-HQ and 1000-class ImageNet 32 × 32 datasets). We name our models as Manifold EBM (M-EBM) and Manifold JEM (M-JEM) respectively. Second, we find the L_2 regularization of the energy magnitude does not work with the energy function utilized in JEM. To enable L_2 regularization and improve the training stability, we augment the standard softmax classifier with a new energy head, which is then L_2 regularized. Despite the simplicity, these techniques allow us to reduce the number of MCMC sampling steps of EBM dramatically, while retaining or sometimes improving classification accuracy of prior state-of-the-art EBMs.

Our main contributions are summarised as follows:

1. We simplify the informative initialization in JEM++ for the SGLD chain, which stabilizes and accelerates the training of unconditional EBM and JEM, while being scalable for high-resolution and large-scale datasets.
2. Adding an L_2-regularized energy head on top of a CNN feature extractor to represent an energy function stabilizes the training of JEM. Then we train M-JEM using two mini-batches: one with data augmentation for classification, and the other one without data augmentation for maximum likelihood estimation of EBMs.
3. We conduct extensive experiments on four benchmark datasets. M-EBM matches or outperforms prior state-of-the-art unconditional EBMs, while significantly improves training stability and reduces the number of sampling

steps. Moreover, M-JEM improves JEM's training stability and speed, image generation quality, and classification accuracy altogether, while outperforming M-EBM in image generation quality.

2   Background

Energy-based Models (EBMs) [13] utilizes the idea that any probability density $p_\theta(x)$ can be expressed as

$$p_\theta(x) = \frac{\exp\left(-E_\theta(x)\right)}{Z(\theta)}, \tag{1}$$

where $E_\theta(x)$ is named the energy function that maps each input $x \in \mathcal{X}$ to a scalar, and $Z(\theta) = \int_x \exp\left(-E_\theta(x)\right) dx$ is the normalizing constant w.r.t x (also known as the partition function). Ideally, an energy function should assign low energy values to samples drawn from data distribution and high values otherwise.

The key challenge of EBM training is estimating the intractable partition function $Z(\theta)$, and the maximum likelihood estimation of parameters θ is not straightforward. A number of sampling-based approaches have been proposed to approximate the partition function effectively. Specifically, the derivative of the log-likelihood of $x \in \mathcal{X}$ w.r.t. θ can be expressed as

$$\frac{\partial \log p_\theta(x)}{\partial \theta} = \mathbb{E}_{p_\theta(x')}\left[\frac{\partial E_\theta(x')}{\partial \theta}\right] - \mathbb{E}_{p_d(x)}\left[\frac{\partial E_\theta(x)}{\partial \theta}\right], \tag{2}$$

where the first expectation is over the model density $p_\theta(x')$, which is challenging due to the intractable $Z(\theta)$.

To estimate it efficiently, MCMC and Gibbs sampling [10] have been proposed. Moreover, to speed up the sampling, recently Stochastic Gradient Langevin Dynamics (SGLD) [19] is employed to train EBMs [4,7,16]. Specifically, to sample from $p_\theta(x)$, the SGLD follows

$$x^0 \sim p_0(x), \qquad x^{t+1} = x^t - \frac{\alpha}{2}\frac{\partial E_\theta(x^t)}{\partial x^t} + \alpha\epsilon^t, \ \ \epsilon^t \sim \mathcal{N}(0,1), \tag{3}$$

where $p_0(x)$ is typically a uniform distribution over $[-1,1]$, whose samples are refined via a noisy gradient decent with step-size α over a sampling chain.

Prior works [4,7,15,16] have investigated the effect of hyper-parameters in SGLD sampling and showed that the SGLD-based approaches suffer from poor stability and prolonged computation of sampling at every iteration. Nijkamp et al. [15] find that it's desirable to generate samples from the SGLD chain after it converges. The convergence requires the step-size α to decay with a polynomial schedule and infinite sampling steps, which is impractical. Therefore, Short-Run and Long-Run MCMC samplings are utilized for EBM training. Moreover, most works [3,4,7] use a constant step-size α during sampling and approximate the samples with a sampler that runs only for a finite number of steps, which is still computationally very expensive. Another recent work [5] combines the

SGLD-based approach with diffusion models [11] under a framework of conditional EBMs. They achieve state-of-the-art image generation quality and obtain a faithful energy potential.

Joint Energy-based Models (JEM) [7] demonstrates that standard softmax-based classifiers can be trained as EBMs. Given an input $x \in R^D$, a classifier of parameters θ maps the input to a vector of C real-valued numbers (known as logits): $f_\theta(x)[y], \forall y \in [1, \cdots, C]$, where C is the number of classes. Then the softmax function is employed to convert the logits into a categorical distribution: $p_\theta(y|x) = e^{f_\theta(x)[y]} / \sum_{y'} e^{f_\theta(x)[y']}$. The authors reuse the logits to define an energy function for the joint density: $p_\theta(x, y) = e^{f_\theta(x)[y]} / Z(\theta)$. Then a marginal density of x can be achieved by marginalizing out y as: $p_\theta(x) = \sum_y p_\theta(x, y) = \sum_y e^{f_\theta(x)[y]} / Z(\theta)$. As a result, the corresponding energy function of x is defined as

$$E_\theta(x) = -\log \sum_y e^{f_\theta(x)[y]} = -\text{LSE}(f_\theta(x)), \qquad (4)$$

where LSE($\cdot$) denotes the Log-Sum-Exp function. The advantage of this LSE energy function is that an additional degree of freedom in the scale of the logit vector now can model the data distribution.

To optimize the model parameter θ, JEM maximizes the logarithm of joint density function $p_\theta(x, y)$:

$$\log p_\theta(x, y) = \log p_\theta(y|x) + \log p_\theta(x), \qquad (5)$$

where the first term is the cross-entropy objective for classification, and the second term is the maximum likelihood learning of EBM as shown in Eq. 2. We can also interpret the second term as an unsupervised regularization on the model parameters θ.

3   Manifold EBM

3.1   Informative Initialization and M-EBM

As shown in Eq. 3, the SGLD sampling starts from an initial distribution $p_0(x)$. To train the EBM as a generative model, Short-Run MCMC sampling [16] utilizes an MCMC sampler that starts from a random noise distribution such as a uniform distribution. A concurrent work IGEBM [4] proposes an initialization approach with a sample replay buffer in which they store past generated samples and draw samples from either replay buffer or uniform random noise to initialize the Langevin dynamics procedure. This is also the sampling approach adopted by [7,25]. Furthermore, JEM++ [21] introduces an informative initialization with the replay buffer by using a Gaussian mixture distribution estimated from the training images, which significantly reduces the number of sampling steps required by SGLD while improving its training stability.

However, the per-class covariance matrices of the Gaussian mixture distribution utilized by JEM++ can be huge for high-resolution image datasets with a large number of classes. Hence, we estimate a single Gaussian distribution from the whole training dataset. That is, we estimate the initial sampling distribution as

$$p_0(\boldsymbol{x}) = \mathcal{N}(\boldsymbol{\mu}, \boldsymbol{\Sigma}) \tag{6}$$

$$\text{with} \quad \boldsymbol{\mu} = \mathbb{E}_{\boldsymbol{x}\sim\mathcal{D}}[\boldsymbol{x}], \quad \boldsymbol{\Sigma} = \mathbb{E}_{\boldsymbol{x}\sim\mathcal{D}}\left[(\boldsymbol{x} - \boldsymbol{\mu})(\boldsymbol{x} - \boldsymbol{\mu})^{\top}\right],$$

where $\mathcal{D}$ denotes the whole training set. The visualization of the estimated centers and samples from $p_0(\boldsymbol{x})$ of different datasets are provided in the appendix. Since only one Gaussian distribution is estimated from the whole training set, we can apply it for unconditional datasets such as CelebA, and reduce the memory and space required for the large covariance matrices[1]. Although $\boldsymbol{\mu}$ and $\boldsymbol{\Sigma}$ can be well estimated with sufficient samples, they still lead to a biased initialization with higher variance, compared to the Gaussian mixture initialization utilized in JEM++. But our empirical study shows that our simplified initialization won't deteriorate the performance and is comparable to bias-reduced Gaussian mixture initialization.

Since the manifold of $\boldsymbol{x}_0$ from our informative initialization is much closer to the real data manifold than that of uniform initialization, this informative initialization reduces the required sampling steps (and thus accelerates training), and also improves training stability as we will demonstrate in the experiments. We therefore call the EBM with this simplified informative initialization as M-EBM throughout this work.

3.2   Injected Noise in M-EBM

Existing work [16] studied the effect of injected noise on training stability via smoothing p_{data} with additive Gaussian noises $\boldsymbol{x} \leftarrow \boldsymbol{x} + \boldsymbol{\epsilon}, \boldsymbol{\epsilon} \sim \mathcal{N}(0, \sigma^2 I)$. Their results demonstrated that the fidelity of the examples in terms of IS and FID improves, when lowering σ^2. And they depict the tradeoff between the sampling steps K and the level of injected noise, indicating the training time and the stability. After it, several following methods [3,5,22] successfully remove the injected noise and achieve better image quality. However, they require a very large $K \geq 30$ to stabilize the training. Thanks to the informative initialization, it not only allows us to significantly reduce K, but also removes the injected noise to improve the image quality while keeping high stability. As shown in Fig. 4(b), the manifolds of real data and $\boldsymbol{x}_0$ sampled from informative initialization are very close, even mixing together when M-EBM is trained without energy regularization. Hence, we suppose the gradients $\nabla_{\boldsymbol{x}} E(\boldsymbol{x})$ are defined (almost) everywhere in such manifolds and thus can reduce the perturbation with noise which is originally explained in NCSN [18].

---

[1] One covariance matrix of CIFAR10 has $(3 \times 32 \times 32)^2 \approx 9.4M$ parameters and uses 37.6MB memory. A dataset with C classes will take $37.6 \times C$ MB.

4 Manifold JEM

4.1 Injected Noise in M-JEM

As discussed in previous section, the injected noise smoothing p_{data} would hurt the generative performance of EBMs. For JEM and JEM++, we suppose it would also decrease the classification accuracy. Hence, it's critical to remove the injected noise and gain benefits in terms of classification accuracy and generation quality. We use N to denote the noise-adding operation. Then the actual objective of JEM is

$$\log p_\theta(N(\boldsymbol{x}), y) = \log p_\theta(y|N(\boldsymbol{x})) + \log p_\theta(N(\boldsymbol{x})). \tag{7}$$

Interestingly, we find that if we only remove the injected noise, the training is not stable. However, if we further disable the data augmentation when learning maximum likelihood $\log p_\theta(\boldsymbol{x})$, it becomes even more stable than JEM++ and enjoys improved accuracy and better sampling quality. Following the observation, we train our M-JEM using two mini-batches: one with data augmentation for classification, and the other one without data augmentation for maximum likelihood estimation of EBMs.

4.2 Energy Function Regularization in M-JEM

IGEBM [4] finds that constraining the Lipschitz constant of the energy network can ease the instability issue in Langevin dynamics. Hence, they weakly L_2 regularize energy magnitudes for both positive and negative samples to the contrastive divergence as:

$$\mathcal{L} = \frac{1}{B} \sum_{i=1}^{B} \left(E_i^+ - E_i^- + \alpha(E_i^{+2} + E_i^{-2}) \right), \tag{8}$$

where $E^+ = E_\theta(\boldsymbol{x}^+)$ with $\boldsymbol{x}^+$ sampled from the data distribution p_d, and $E^- = E_\theta(\boldsymbol{x}^-)$ with $\boldsymbol{x}^-$ sampled from the model distribution $p_\theta(\boldsymbol{x})$. The effect of L_2 regularization on EBMs can be viewed as Fig. 4(b). However, since L_2 regularization would force the vector of logits $f_\theta(\boldsymbol{x})$ to be uniform, while maximizing $p_\theta(y|\boldsymbol{x})$ boosts $f_\theta(\boldsymbol{x})[y]$. Hence, the L_2 regularization is incompatible with Eq. 4 and cannot be directly applied to vanilla JEM.

To incorporate L_2 regularization to JEM, we propose to augment the standard CNN softmax classifier with an extra fully connected layer, called Energy Head, as shown in Fig. 2(a). Then the L_2 regularization is applied on the energy head (instead of the LSE classification head) to improve the training stability.

We provide the pseudo-code for M-EBM/JEM as in Algorithm 1, which follows the framework of IGEBM [4] and JEM [7].

5 Experiments

In this section, we first evaluate the generative performance of M-EBM on multiple datasets, including CIFAR10, CIFAR100, CelebA-HQ 128×128 and ImageNet 32×32. Then, we investigate the efficacy of the M-JEM on CIFAR10

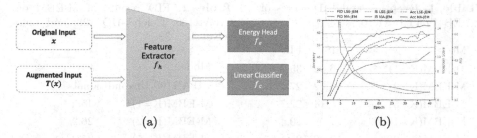

(a)   (b)

Fig. 2. a) The architecture of M-JEM. An energy head f_e is augmented for energy magnitude regularization and two mini-batches are used for the training of classifier and the maximum likelihood estimate of EBM, respectively. b) Comparison between M-JEM and LSE-JEM on CIFAR100.

Algorithm 1. M-EBM/JEM Training: Given network f_θ, SGLD step-size α, SGLD noise σ, replay buffer B, SGLD steps K, reinitialization frequency ρ

1: **while** not converged **do**
2:     Sample x^+ and y from dataset
3:     Sample $\widehat{x}_0 \sim B$ with probability $1 - \rho$, else $\widehat{x}_0 \sim p_0(x)$ as Eq. 6
4:     **for** $t \in [1, 2, \ldots, K]$ **do**
5:         $\widehat{x}_t = \widehat{x}_{t-1} - \alpha \cdot \frac{\partial E(\widehat{x}_{t-1})}{\partial \widehat{x}_{t-1}} + \sigma \cdot \mathcal{N}(0, I)$
6:     **end for**
7:     $x^- = \text{StopGrad}(\widehat{x}_K)$
8:     $L_{\text{gen}}(\theta) = E(x^+) - E(x^-) + \alpha \left(E(x^+)^2 + E(x^-)^2 \right)$ as Eq. 8.
9:     $L(\theta) = L_{\text{gen}}(\theta)$ for M-EBM
10:    $L(\theta) = L_{\text{clf}}(\theta) + L_{\text{gen}}(\theta)$ with $L_{\text{clf}}(\theta) = \text{xent}(f_\theta(x), y)$ for M-JEM
11:    Calculate gradient $\frac{\partial L(\theta)}{\partial \theta}$ to update θ
12:    Add x^- to B
13: **end while**

---

and CIFAR100. Finally, the study and visualization of the differences between trained EBM and JEM are provided to analyze their generative capability.

Our code is largely built on top of JEM [7][2]. For a fair comparison with JEM, we update each model with 390 iterations in 1 epoch. Empirically, we find a batch size of 128 for $p_\theta(y|x)$ achieves the best classification accuracy on CIFAR10, while we use 64, the same batch size as in JEM, for $p_\theta(x)$. We train our models on ImageNet 32×32 for 50 epochs and other datasets for 150 epochs at most. All our experiments are performed with PyTorch on Nvidia GPUs. For CIFAR10 and CIFAR100, we train the backbone Wide-ResNet 28-10 [23] on a single GPU. Due to limited computational resources, we use Wide-ResNet 28-2 for ImageNet 32×32 on a single GPU, and Wide-ResNet 28-5 for CelebA-HQ 128×128 on 2 GPUs. Due to page limitations, we show a detailed comparison of the training speed of different methods in the appendix.

---

[2] https://github.com/wgrathwohl/JEM.

Table 1. Inception and FID scores of M-EBM on CIFAR10.

Model	IS ↑	FID ↓
M-EBM(K=1)*	6.02	35.7
M-EBM(K=2)	6.72	27.1
M-EBM(K=5)	7.14	22.7
M-EBM(K=10)	7.08	20.4
M-EBM(K=20)	7.20	21.1
Explicit EBM(Unconditional)		
ShortRun(K=100) [16]	6.72	32.1
IGEBM(K=60) [4]	6.78	38.2
f-EBM(K=60) [22]	8.61	30.8
CF-EBM(K=50) [24]	–	16.7
KL-EBM(K=40) [3]	7.85	25.1
DiffuRecov(K=30) [5]	8.31	9.58
Regularized Generator		
GEBM [1]	–	23.02
VAEBM(K=6) [20]	8.43	12.19
Other		
SNGAN [14]	8.59	21.7
NCSN [18]	8.91	25.3
StyleGAN2-ADA [12]	**9.74**	**2.92**
DDPM [11]	9.46	3.17

* M-EBM diverges with $K = 1$, and we report the best FID before diverging.

Table 2. FID results of M-EBM on CIFAR100, CelebA-HQ 128, and ImageNet 32 × 32.

Model	FID ↓
CIFAR100 Unconditional	
M-EBM(K=1)*	45.5
M-EBM(K=2)	26.2
M-EBM(K=5)	27.2
M-EBM(K=10)	26.9
SNGAN [14]	22.4
CelebA-HQ 128 Unconditional	
M-EBM(K=5)*	57.76
M-EBM(K=10)	39.87
KL-EBM(K=40) [3]	28.78
SNGAN [14]	24.36
ImageNet 32 × 32 Unconditional	
M-EBM(K=2)	54.52
M-EBM(K=5)	52.71
IGEBM(K=60) [4]	62.23
KL-EBM(K=40) [3]	32.48

* Our models diverge during training with given K, and we report the best FID before diverging.

5.1 M-EBM

We first evaluate the performance of M-EBM on CIFAR10, CIFAR100, CelebA-HQ 128 and ImageNet 32 × 32. We utilize the Inception Score (IS) [17] and Fréchet Inception Distance (FID) [9] to evaluate the quality of generated images.

The results are reported in Table 1 and 2, respectively. It can be observed that our method consistently surpasses existing methods in terms of sampling steps by a significant margin. On CIFAR10, M-EBM outperforms many EBM approaches and SNGAN in terms of FID, while the performance is slightly worse than SNGAN on CIFAR100. Some EBM approaches show better performance, such VAEBM, CF-EBM and DiffuRecov. However, they require an extra pre-trained generator, or special architecture, or much larger sampling steps, while M-EBM can train on a classical architecture as the backbone with least K. On ImageNet 32 × 32, we note that M-EBM with $K = 2$ is incredibly stable and achieves FID 54.52 within 30 epochs and outperforms IGEBM. In addition, increasing sampling steps K further doesn't have an obvious improvement. Finally, on CelebA-HQ, M-EBM is worse than baseline methods as we find it is less stable and requires more sampling steps due to the high resolution of CelebA-HQ. Nevertheless, our method builds a new solid baseline on different large-scale benchmarks for further investigations of EBM training in these more challenging tasks. Samples generated by M-EBMs for CIFAR10, CIFAR100, and CelebA-HQ are shown in Fig. 3 and Fig. 1, respectively. The generated samples of ImageNet 32×32 can be found in the appendix.

5.2   M-JEM

We train M-JEM on two benchmark datasets: CIFAR10 and CIFAR100, and compare its performance to the state-of-the-art hybrid models and some representative generative models. Table 3 and 4 report results on CIFAR10 and CIFAR100, respectively. As we can see, M-JEM improves JEM's image generation quality, stability, speed, and accuracy by a notable margin. It also boosts the IS and FID scores over M-EBM. Compared with JEM++, FID of M-JEM drops dramatically since we exclude the noise, and the notable gain of accuracy when $K = 5$ indicates M-JEM($K = 5$) is much more stable than JEM++($K = 5$). On CIFAR100, IS and FID scores are not commonly reported by state-of-the-art hybrid models, such as JEM [7], VERA [6], and JEM++ [21]. Hence, our work builds a baseline for hybrid modeling on CIFAR100 with decent classification accuracy and image generation quality for future investigations. Images generated by M-JEM for CIFAR10 and CIFAR100 are can be found in Fig. 3.

5.3   Analysis

Is Energy Head Better Than LSE? To evaluate the effect of the energy head, we conduct an experiment comparing M-JEM (with energy head) and LSE-JEM (without energy head) on CIFAR100. Figure 2(b) shows that M-JEM achieves much higher classification accuracy, comparable FID but a lower Inception Score than LSE-JEM. However, we empirically find LSE-JEM is less stable than M-JEM after 40 epochs which leads us to analyze the manifolds learned by different models.

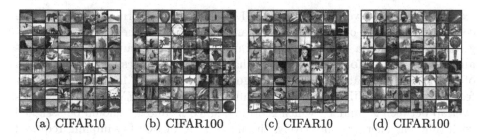

(a) CIFAR10   (b) CIFAR100   (c) CIFAR10   (d) CIFAR100

Fig. 3. M-EBM (a, b) and M-JEM (c, d) generated samples of CIFAR10 and CIFAR100.

Manifold Analysis. To facilitate better understanding of different approaches, we utilize the t-SNE visualization for manifold analysis shown in Fig. 4(a) and 4(b). To have a fair comparison, we pick fixed samples from CIFAR10 as x^+, initialize samples from $p_0(x)$ as x^0, and randomly select samples from the replay buffer of each pre-trained models as x^-. Given the inputs from x^+, x^- and x^0, we collect the outputs of the penultimate layer as features and apply the t-SNE technique to generate the visualization. For Fig. 4(a), three CIFAR10-trained M-EBM, M-JEM, and LSE-JEM with $K = 10$ are involved. We further conduct the comparison between M-EBMs($K = 5$) with and without energy L_2 regularization in Fig. 4(b).

Table 3. Hybrid Modeling Results on CIFAR10.

Model	Acc % ↑	IS ↑	FID ↓
M-JEM(K=1)*	78.4	7.91	29.8
M-JEM(K=2)*	86.5	8.64	19.3
M-JEM(K=5)	93.1	8.71	12.1
M-JEM(K=10)	93.8	8.52	**11.5**
M-JEM(K=20)	**94.2**	8.72	12.2
Single Hybrid Model			
Residual Flow [2]	70.3	3.60	46.4
IGEBM(K=60) [4]	49.1	8.30	37.9
JEM(K=20)⁺ [7]	92.9	8.76	38.4
JEM++(M=5)⁺ [21]	91.1	7.81	37.9
JEM++(M=10)	93.5	8.29	37.1
JEM++(M=20)	94.1	8.11	38.0
JEAT [25]	85.2	**8.80**	38.2
EBM + Generator			
VERA(α=100)	93.2	8.11	30.5
VERA(α=1) [6]	76.1	8.00	27.5
softmax	95.8	-	-

* We report the best performance before the diverging of training.
⁺ They suffer from high instability and regularly diverge.

Table 4. Hybrid Modeling Results on CIFAR100.

Model	Acc % ↑	IS ↑	FID ↓
Softmax	78.9	–	–
SNGAN(Cond)	–	9.30	15.6
BigGAN(Cond)	–	11.0	11.73
JEM(K=20)*	70.4	10.32	51.7
JEM(K=30)*	72.8	10.84	34.2
JEM++(K=5)*	72.0	8.19	37.7
JEM++(K=10)*	74.5	10.23	32.9
VERA(α=100)*	69.3	8.14	28.2
VERA(α=1)*	48.7	7.97	26.6
M-JEM(K=1)⁺	46.5	8.71	26.2
M-JEM(K=2)⁺	63.5	11.22	15.1
M-JEM(K=5)	73.5	**11.95**	13.5
M-JEM(K=10)	**75.1**	11.72	**12.7**

* No official IS and FID scores are reported.
⁺ We report the best FID before diverging.

As we can observe in Fig. 4(a), M-JEM with label information forms more compact manifolds of x^+ and x^- than M-EBM. In other words, M-JEM-generated samples x^- match the distribution of real data and have lower variance and less **manifold intrusion** [8] than M-EBM. It gives us an explanation of why label information can improve generation quality. Moreover, the latent feature space of M-JEM is better formulated than LSE-JEM, and there's less overlap between x^0 and x^+ which is desired since x^0 and x^+ should be assigned with different energies. Intuitively, the number K of SGLD sampling required for stable training is correlated to the distance between manifolds of x^0 and x^+. However, in Figs. 4(a) and 4(b), we can also observe that x^0 and x^+ are roughly mixed together from M-EBM without regularization and LSE-JEM. Hence, it's interesting to reconsider the distance and the training instability when x^0 and x^+ are somewhat mixing together. We leave the exploration of this phenomenon as an existing direction for future.

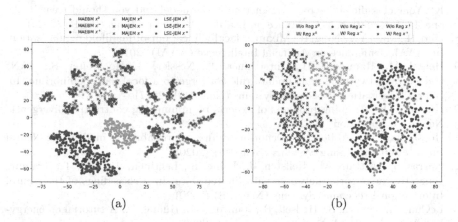

Fig. 4. t-SNE visualization of the latent feature spaces learned by different models trained on CIFAR10. We use different colors to represent (x^0 initial samples, x^- samples in replay buffer, x^+ real data), and different shapes($\circ, \times, \star$) to indicate different EBMs.

6   Conclusion

In this paper, we propose simple yet effective training techniques to improve the image generation quality, training speed, and stability of unconditional EBM and JEM altogether. The experimental results demonstrate that our models surpass prior hybrid models, achieve comparable performance on unconditional EBMs and enable us to scale the MCMC-based EBM learning to high-resolution large-scale image datasets, such as CelebA-HQ 128×128 and ImageNet 32×32 with the least MCMC sampling steps, making EBM training more practical for a research lab in academia to afford and explore.

References

1. Arbel, M., Zhou, L., Gretton, A.: Generalized energy based models. In: International Conference on Learning Representations (ICLR) (2021)
2. Chen, R.T., Behrmann, J., Duvenaud, D., Jacobsen, J.H.: Residual flows for invertible generative modeling. In: Neural Information Processing Systems (2019)
3. Du, Y., Li, S., Tenenbaum, J., Mordatch, I.: Improved Contrastive Divergence Training of Energy Based Models. In: International Conference on Machine Learning (ICML) (2021)
4. Du, Y., Mordatch, I.: Implicit generation and generalization in energy-based models. In: Neural Information Processing Systems (NeurIPS) (2019)
5. Gao, R., Song, Y., Poole, B., Wu, Y.N., Kingma, D.P.: Learning energy-based models by diffusion recovery likelihood. In: International Conference on Learning Representations (ICLR) (2021)

6. Grathwohl, W., Kelly, J., Hashemi, M., Norouzi, M., Swersky, K., Duvenaud, D.: No MCMC for me: amortized sampling for fast and stable training of energy-based models. In: International Conference on Learning Representations (ICLR) (2021)
7. Grathwohl, W., Wang, K.C., Jacobsen, J.H., Duvenaud, D., Norouzi, M., Swersky, K.: Your classifier is secretly an energy based model and you should treat it like one. In: International Conference on Learning Representations (ICLR) (2020)
8. Guo, H., Mao, Y., Zhang, R.: Mixup as locally linear out-of-manifold regularization. In: AAAI Conference on Artificial Intelligence (AAAI) (2019)
9. Heusel, M., Ramsauer, H., Unterthiner, T., Nessler, B., Hochreiter, S.: GANs trained by a two time-scale update rule converge to a local nash equilibrium. In: Neural Information Processing Systems (NeurIPS) (2017)
10. Hinton, G.E.: Training products of experts by minimizing contrastive divergence. Neural Computation (2002)
11. Ho, J., Jain, A., Abbeel, P.: Denoising diffusion probabilistic models. In: Neural Information Processing Systems (NeurIPS) (2020)
12. Karras, T., Aittala, M., Hellsten, J., Laine, S., Lehtinen, J., Aila, T.: Training generative adversarial networks with limited data. In: Proceedings of the Neural Information Processing Systems (NeurIPS) (2020)
13. LeCun, Y., Chopra, S., Hadsell, R., Ranzato, M., Huang, F.: A tutorial on energy-based learning. Predicting structured data (2006)
14. Miyato, T., Kataoka, T., Koyama, M., Yoshida, Y.: Spectral normalization for generative adversarial networks. In: International Conference on Learning Representations (ICLR) (2018)
15. Nijkamp, E., Hill, M., Han, T., Zhu, S.C., Wu, Y.N.: On the anatomy of MCMC-based maximum likelihood learning of energy-based models. In: AAAI Conference on Artificial Intelligence (AAAI) (2019)
16. Nijkamp, E., Zhu, S.C., Wu, Y.N.: Learning non-convergent short-run MCMC toward energy-based model. In: Neural Information Processing Systems (NeurIPS) (2019)
17. Salimans, T., Goodfellow, I., Zaremba, W., Cheung, V., Radford, A., Chen, X.: Improved techniques for training gans. In: Neural Information Processing Systems (NeurIPS) (2016)
18. Song, Y., Ermon, S.: Generative modeling by estimating gradients of the data distribution. In: Neural Information Processing Systems (NeurIPS) (2019)
19. Welling, M., Teh, Y.W.: Bayesian learning via stochastic gradient langevin dynamics. In: International Conference on Machine Learning (ICML) (2011)
20. Xiao, Z., Kreis, K., Kautz, J., Vahdat, A.: VAEBM: a symbiosis between variational autoencoders and Energy-based Models. In: International Conference on Learning Representations (ICLR) (2021)
21. Yang, X., Ji, S.: JEM++: improved techniques for training JEM. In: International Conference on Computer Vision (ICCV) (2021)
22. Yu, L., Song, Y., Song, J., Ermon, S.: Training deep energy-based models with f-divergence minimization. In: International Conference on Machine Learning (ICML) (2020)
23. Zagoruyko, S., Komodakis, N.: Wide residual networks. In: BMVC (2016)
24. Zhao, Y., Xie, J., Li, P.: Learning energy-based generative models via coarse-to-fine expanding and sampling. In: International Conference on Learning Representations (ICLR) (2021)
25. Zhu, Y., Ma, J., Sun, J., Chen, Z., Jiang, R., Li, Z.: Towards understanding the generative capability of adversarially robust classifiers. In: IEEE International Conference on Computer Vision (ICCV) (2021)

Small Temperature is All You Need for Differentiable Architecture Search

Jiuling Zhang[1,2] and Zhiming Ding[1,2(✉)]

[1] University of Chinese Academy of Sciences, Beijing 100049, China
zhangjiuling19@mails.ucas.ac.cn, zhiming@iscas.ac.cn
[2] Institute of Software, Chinese Academy of Sciences, Beijing 100190, China

Abstract. Differentiable architecture search (DARTS) yields highly efficient gradient-based neural architecture search (NAS) by relaxing the discrete operation selection to optimize continuous architecture parameters that maps NAS from the discrete optimization to a continuous problem. DARTS then remaps the relaxed supernet back to the discrete space by one-off post-search pruning to obtain the final architecture (finalnet). Some emerging works argue that this remap is inherently prone to mismatch the network between training and evaluation which leads to performance discrepancy and even model collapse in extreme cases. We propose to close the gap between the relaxed supernet in training and the pruned finalnet in evaluation through utilizing small temperature to sparsify the continuous distribution in the training phase. To this end, we first formulate sparse-noisy softmax to get around gradient saturation. We then propose an exponential temperature schedule to better control the outbound distribution and elaborate an entropy-based adaptive scheme to finally achieve the enhancement. We conduct extensive experiments to verify the efficiency and efficacy of our method.

Keywords: Deep learning architecture · Neural architecture search

1 Introduction

DARTS abstracts the search space as a cell-based directed acyclic graph composed by V nodes $H = \{h_1, h_2, ..., h_V\}$ and compound edges $C = \{c_{1,2}, ..., c_{V-1,V}\}$. Every node represents feature maps and each edge subsumes all operation candidates to express the transformations between nodes. Compound edge $c_{u,v}$ connects node u to v and associates three attributes: candidate operation set $O_c = \{o_c^1, o_c^2, ..., o_c^M\}$, corresponding operation parameter set $A_c = \{a_c^1, a_c^2, ..., a_c^M\}$, probability distribution of the parameters $\beta_c = \text{softmax}(A_c)$. Every intermediate node is densely-connected to all its predecessor through an edge $h_v = c_{u,v}(h_u)$ and weighted product sum $c_{u,v}(h_u) = \langle \beta_c, O_c(h_u) \rangle$ for $u < v$. Generally, a unified set of operation candidates $O = \{o^1, o^2, ..., o^M\}$ is defined for all edges in the space. The network that encodes all architectural candidates is termed supernet. In the training phase, DARTS first divides data into training and validation sets and then formulates a bilevel objective depicted in Eq. (1) to alternately optimize architecture parameter a and operation weight ω on the validation set and training set respectively. We refer to [13] for more details of DARTS.

$$\min_{a} L_{val}(\omega^*(a), a) \;\; \text{s.t.} \;\; \omega^* = \arg\min_{\omega} L_{train}(\omega, a) \tag{1}$$

Henceforth, we abbreviate operation weight ω as weights and architecture parameter $a \in A$ as parameters. After the training phase, DARTS selects the operation associated with the largest entry (probability component after normalized by softmax) through a post-search pruning depicted in Eq. (2) to obtain the final architecture (finalnet) for evaluation. In sum, the supernet is trained in a multi-path manner while the finalnet is evaluated in a single-path manner [14].

$$o_c^i \in O \text{ for } c \in C \text{ where } i = \arg\max_{i \in \{1,...,M\}} \beta_c, \; \beta_c = \text{softmax}(A_c) \tag{2}$$

where the operation o_c^i is selected on edge c due to the largest entry β_c^i in β_c. However, the parameter-value-based post-search pruning depicted in Eq. (2) inevitably risks mismatching the architecture between supernet and finalnet which ultimately leads to performance discrepancy between training and evaluation in DARTS. Gradient-based NAS methods are deemed to favor architectures that are easier to be trained [15]. More skip connections can obviously help the network to converge faster which is considered as an unfair advantage in an exclusive competition [7,8]. This unfairness leads to an abnormal preference for the skip connection in some cases during optimization. The architecture mismatch exacerbates this issue since the one-off post-pruning generally discards all operations except the dominant one $(O \setminus o^{\arg\max_i(\beta)})$. In extreme cases, the pruning removes all operations except the skip connection which causes collapse or the catastrophic failure of DARTS [1,5,10,12].

Both Fair-DARTS [8] and GAEA [11] pointed out that the sparse parameters are crucial to alleviating the architecture mismatch. In particular, GAEA emphasized that obtaining sparse final architecture parameters is critical for good performance, both for the mixture relaxation, where it alleviates the effect of overfitting, and for the stochastic relaxation, where it reduces noise when sampling architectures. In DARTS, *the sparse parameters refer to the sparse distribution β (low entropy) after the parameters normalized by softmax*, i.e. $\beta = \text{softmax}(A)$. Sparse parameters intrinsically reduce the gap between multi-path supernet and single-path finalnet as shown in Fig. 1. [17] proposed to combine single-path and multi-path space by a Sparse Group Lasso constraint but impose non-trivial additional time consumption due to it's harder to converge. Temperature coefficient is widely used to control the sparseness and smoothness of the (gumbel) softmax output. In this paper, we propose to achieve sparse training straightforwardly by employing a small temperature on softmax. Our contributions can be summarized:

- We propose sparse-noisy softmax (sn-softmax) to alleviate the gradient saturation which causes premature convergence of the training of parameters while utilizing small temperature in the training phase;

 – We propose exponential temperature schedule (ETS) and entropy-based
   adaptive scheme to maintain and better control the sparsity of β that inherently narrows the gap between the relaxed supernet in training and the pruned
   finalnet in evaluation;
 – We carry out extensive evaluations on multiple spaces and datasets and conduct further ablations to show the effect of different hyperparameter choices.

2   Methodology

Utilizing small temperature to sparsify β in training is non-trivial for DARTS
because the gradient saturation will impede the propagation when softmax converges. Likewise, the operation weights will not be updated either when the
operation output are weighted by a zero entry in β. If that happens in the middle training of DARTS, supernet converges prematurely. In this section, we start
with formulating sparse-noisy softmax to alleviate the gradient saturation when
the outbound β converges. After that, we propose an exponential temperature
schedule (ETS) to better control the temperature t to smooth the swing of the
outbound β in training. We provide an entropy-based dynamic decay (EDD) to
finally realize a flexible and robust enhancement for DARTS.

2.1   Sparse-Noisy Softmax

Softmax normalizes the input vector $A = \{a^1, ..., a^M\}$ to a probability distribution $\beta = \{\beta^1, ..., \beta^M\}$ depicted in Eq. (3).

$$\text{softmax}(\frac{A}{t}) = \beta_t \text{ where } \beta_t^i = \frac{\exp(a^i/t)}{\sum_{j=1}^{M} \exp(a^j/t)} \tag{3}$$

where t is the temperature coefficient and M is the total entries of the input of
softmax in DARTS. The smaller the temperature t, the sharper the outbound
β is and the closer the distribution compared to post-search one-hot argmax
pruning in Eq. (2). The derivative of softmax can be gotten through:

$$\frac{\partial \beta_t^i}{\partial a^i} = \begin{cases} \frac{\beta_t^i(1-\beta_t^i)}{t} & \text{for } i = j \\ \frac{-\beta_t^i\beta_t^j}{t} & \text{for } i \neq j \end{cases} \tag{4}$$

by which we can get the Jacobian matrix as:

$$\frac{\partial \beta_t}{\partial A} = \frac{1}{t} \begin{bmatrix} \beta_t^1 - [\beta_t^1]^2 & -\beta_t^1\beta_t^2 & -\beta_t^1\beta_t^3 & \cdots & -\beta_t^1\beta_t^M \\ -\beta_t^2\beta_t^1 & \beta_t^2 - [\beta_t^2]^2 & -\beta_t^2\beta_t^3 & \cdots & -\beta_t^2\beta_t^M \\ \vdots & \vdots & \vdots & \cdots & \vdots \\ -\beta_t^M\beta_t^1 & -\beta_t^M\beta_t^2 & -\beta_t^M\beta_t^3 & \cdots & \beta_t^M - [\beta_t^M]^2 \end{bmatrix} \tag{5}$$

Let $\mathcal{A}_t$ denotes the rightmost matrix in Eq. (5) and $\mathcal{A}_t/t$ is the Jacobian with the
temperature t. Let l indicates the loss actually used, then the backpropagation
to a can be written as:

$$\frac{\partial \beta_t}{\partial A} \frac{\partial l}{\partial \beta_t} = \frac{\mathcal{A}_t}{t} \frac{\partial l}{\partial \beta_t} \tag{6}$$

By leveraging a smaller temperature t to get a sparser β_t, only one single entry in β_t approaches one ($\beta_t^i \to 1$) and the others are thereby close to zero ($\beta_t^{j \neq i} \to 0$), Jacobian matrix is then overall trapped in zero $\lim\limits_{\substack{\beta_t^i \to 1 \\ \beta_t^{j \neq i} \to 0}} \partial \beta_t / \partial A = 0$ and the gradient saturates. When softmax is saturated, back propagation depicted in Eq. (6) stops propagating gradients through softmax to parameters thus the parameters stop updating. To deal with this problem, ideally, we hope that our softmax outputs a sparse β in feedforward as the middle pane of Fig. 1, but the backpropagation is smoother (not saturated). To this end, we propose sparse-noisy softmax (sn-softmax), summarized in Algorithm 1, to approximate the ideal case by *combining different temperature for feedforward and backpropagation* respectively. Our goal is to keep the Jacobian matrix not zero even softmax converges $\lim\limits_{\substack{\beta_t^i \to 1 \\ \beta_t^{j \neq i} \to 0}} \partial \beta_t / \partial A \neq 0$, which is different to the previous research that injects noise to postpone convergence [2].

Set the forward-pass temperature as t, we formulate the Jacobian matrix of sn-softmax as:

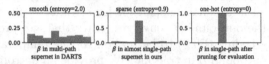

Fig. 1. Multi-path, sparse and single-path.

$$\frac{\partial \beta_t}{\partial A} = \frac{\mathcal{A}_t}{t} + \frac{\mathcal{A}_{st}}{st} \quad \text{for } s > 1 \tag{7}$$

where $\mathcal{A}_{st}/st$ is the Jacobian of β_{st}. The backpropagation from the loss l w.r.t A can be written as:

$$\frac{\partial \beta_t}{\partial A} \frac{\partial l}{\partial \beta_t} = (\frac{\mathcal{A}_t}{t} + \frac{\mathcal{A}_{st}}{st}) \frac{\partial l}{\partial \beta_t} \tag{8}$$

Since the equal sign can not be guaranteed:

$$sign(\frac{\mathcal{A}_t}{t} \frac{\partial l}{\partial \beta_t}) \neq sign(\frac{\mathcal{A}_{st}}{st} \frac{\partial l}{\partial \beta_t}) \quad \text{for } s > 1 \tag{9}$$

Sn-softmax is equivalent to adding noise into the backpropagation by which the Jacobian matrix is not absolute zero after the forward term $\mathcal{A}_t/t$ converges. s acts as a scaling factor in Eq. (8) to tune the noise intensity. A smaller s brings stronger gradient noise, but is easier to saturate either as the training progresses. We can also keep st as a constant and determine s accordingly and dynamically and thereby get out of the saturation in the whole training phase.

In practice, sn-softmax leverages a small temperature value t for feedforward to obtain a sparse output β_t while setting $s \gg 1$ (generally $s > 50$) thus get $1/st \ll 1/t$. When the feedforward does not converge ($\mathcal{A}_t \neq 0$), $\mathcal{A}_t/t$ plays a leading role in Eq. (7) because of $1/st \ll 1/t$ and $\mathcal{A}_{st}/st \ll \mathcal{A}_t/t$. After the feedforward convergence, the gradient saturates ($\mathcal{A}_t = 0$, $\mathcal{A}_t/t = 0$). Since the scaling factor $s \gg 1$, β_{st} is much smoother than β_t which makes $\mathcal{A}_{st}$ less convergent ($\mathcal{A}_{st} \neq 0$). As a result, $\mathcal{A}_{st}/st$ supersedes the first term $\mathcal{A}_t/t$ in the Jacobian matrix and comes into play in the backpropagation of sn-softmax depicted in

Eq. (8). This way, when sn-softmax does not converge, the normal term repels the noise term and dominate the update direction due to $\mathcal{A}_{st}/st \ll \mathcal{A}_t/t$ but allows sn-softmax to have a chance to escape the premature convergence after it's saturated. We schematically visualize the backpropagation dynamic of sn-softmax in terms of the gradient norm on different values of s in Fig. 2A.

Sn-softmax only needs *feed-forward and backpropagation once like softmax* to get $\partial l/\partial\beta_t$. Then it calculates the smoother Jacobian based on temperature st through Eq. (5). The calculation of another Jacobian and the final addition depicted in Eq. (7) only takes negligible additional budget in the whole training.

Algorithm 1. Sparse-noisy (sn) softmax

Input: logits $A = \{a^1, ..., a^M\}$, feedforward temperature t, scaling factor s.

Output: normalized distribution β_t

Feedforward: softmax$(\frac{A}{t}) = \beta_t$ for $\beta_t^i = \frac{\exp(a^i/t)}{\sum_j^M \exp(a^j/t)}$

Backpropagation: $\frac{\partial\beta_t}{\partial A} = \frac{1}{t}\mathcal{A}_t + \frac{1}{st}\mathcal{A}_{st}$ for $s > 1$

2.2 Exponential Temperature Schedule

In general, the temperature t is gradually decayed to drive β converge to a sparse solution. As shown in Eq. (3), t is in the exponential term and nonlinearly affects the transformation of $a \to \beta$ in softmax. Linearly scheduling t swings β nonlinearly and results in that the early temperature decay has little effect on the β while decay in the later stage of training has too much impact on the sparsity and causes the training converges precipitately (see LTS in Fig. 2B and LPCD-DARTS in Fig. 4A). Therefore, the naive linear temperature schedule is inappropriate for the training of DARTS. In this section, we focus on temperature t so let $t_a^{\exp} = \exp(a/t)$ refers to the exponential function with e as the base and t, a as the exponent where we add the superscript "exp" in $t_a^{\exp}$ to emphasize the value after the exponential transformation $\exp(a/t)$. We then apply variable substitution in softmax as:

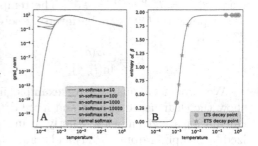

Fig. 2. Visualize the empirical analysis of sn-softmax and ETS. Tests are based on the order of magnitude of a initialized in DARTS. (A). By setting $s \gg 1$, gradients of sn-softmax is consistent with normal softmax before saturated. After saturated, we can get different backpropagation dynamics by setting different s. (B). For equidistant decay, ETS yields much smoother swings than the linear counterpart in terms of the entropy of outbound distribution β.

$$\beta^i = \frac{t_{a^i}^{\exp}}{\sum_{j=1}^M t_{a^j}^{\exp}} \quad \text{where} \quad t_{a^*}^{\exp} = \exp(\frac{a^*}{t}) \tag{10}$$

where the transformation from $t_a^{\exp}$ to β^i is linear. We can then linearly schedule $t_a^{\exp}$ instead of t to better control the sparsity of β. We call this design of scheduling $t_a^{\exp}$ after the exponent substitution as exponential temperature schedule (ETS).

To calculate $t_a^{\exp}$, we first approximate a by the order of its expectation $E(a)$, $a \in A$ and get the temperature $t_{E(a)}^{\exp}$. From now on, we omit $E(a)$ in the subscript of $t_{E(a)}^{\exp}$ since it's fixed in this analysis. We then specify the initial temperature t_0 and the decay target t_N and get $t_0^{\exp} = e^{E(a)/t_0}$ and $t_N^{\exp} = e^{E(a)/t_N}$ for the start and target in the exponential space (after $\exp(a/t)$ transformation) respectively. Since t act as the denominator in the exponential term $\exp(a/t)$ of softmax, the temperature decay from t_0 to t_N corresponds to an ascent in the exponential space from $t_0^{\exp}$ to $t_N^{\exp}$. For the equidistant temperature decay, the decay strength $d^{\exp}$ indicates the variation amplitude within $[t_0^{\exp}, t_N^{\exp}]$ which can be calculated statically via $d^{\exp} = \frac{t_N^{\exp}-t_0^{\exp}}{N}$ where N is the number of decay points. All decay points together with $t_0^{\exp}$ form a list $L^{\exp}$ as

$$L^{\exp} = [t_n^{\exp}|t_n^{\exp} = t_0^{\exp} + nd^{\exp}, \ d^{\exp} = \frac{t_N^{\exp} - t_0^{\exp}}{N}, n = 0, 1, ..., N] \quad (11)$$

where $t_0^{\exp} = t_0^{\exp}$, $t_N^{\exp} = t_N^{\exp}$. The temperature value corresponding to each decay point in $L^{\exp}$ can be inversely solved by Eq. (12) where $t_n > 0$ when $t_n^{\exp} > 1$.

$$t_n = \frac{E(a)}{\ln(t_n^{\exp})} \quad \text{for } t_n^{\exp} \in L^{\exp}, \ n = 0, 1, ..., N \quad (12)$$

We provide a quantitative example in Fig. 3 to show the clear usage of ETS and more details on the calculation are given below. Firstly, as in A are initialized by sampling from $N(0, 1)$ and scaling the samples by $1e-3$ in DARTS. When $a \leq 0$, the variation of t has little effect on the value of $t_a^{\exp}$, so we only consider $E(a)$ for $a > 0$. The scaled expectation of a can be get through $E(a) = (1e-3)E(N(0,1))=(1e-3)\sqrt{2/\pi}/2 \approx 4e-4$ for $a > 0$. By specifying the initial temperature $t_0 = 1$ and the decay target $t_N = 1e-3$, we can then get $t_0^{\exp}= \exp(E(a)/t_0)=e^{4e-4} \approx 1$ and $t_N^{\exp} = \exp(E(a)/t_N)=e^{4e-1} \approx 1.492$ in the exponential space respectively. The temperature decay from t_0 to t_N $(1 \rightarrow 1e\text{-}3)$ thus corresponds to an ascent from $t_0^{\exp}$ to $t_N^{\exp}$ $(1 \rightarrow 1.492)$ in the exponential space. For

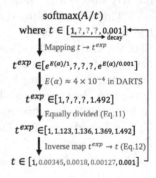

softmax(A/t)

where $t \in [1, ?, ?, ?, 0.001]$ ← decay

↓ Mapping $t \rightarrow t^{exp}$

$t^{exp} \in [e^{E(\alpha)/1}, ?, ?, ?, e^{E(\alpha)/0.001}]$

↓ $E(\alpha) \approx 4 \times 10^{-4}$ in DARTS

$t^{exp} \in [1, ?, ?, ?, 1.492]$

↓ Equally divided (Eq. 11)

$t^{exp} \in [1, 1.123, 1.136, 1.369, 1.492]$

↓ Inverse map $t^{exp} \rightarrow t$ (Eq. 12)

$t \in [1, 0.00345, 0.0018, 0.00127, 0.001]$

Fig. 3. Diagram of the quantitative example.

the equidistant temperature decay, we preset $N = 4$ and $t_4^{\exp} = t_N^{\exp} = 1.492$, then $[1, 1.492]$ is equidistantly divided into 4 segments by the other 3 decay points. We get the decay strength accordingly as $d^{\exp} = \frac{1.492-1}{4} = 0.123$ and determine the remaining three decay points as $1.123, 1.246, 1.369$ and form $L^{\exp} = [1, 1.123, 1.246, 1.369, 1.492]$ by Eq. (11). After that, we can get the temperatures w.r.t $t_{1.123}^{\exp}$ as $t_{1.123} = 4e\text{-}4/\ln(1.123) \approx 0.00345$ by Eq. (12). Similarly, the temperature of the other 3 decay points are $t_{1.246} = 0.0018$, $t_{1.369} = 0.00127$, $t_{1.492} = 0.001$ from which we can clearly find that the temperature decay scheduled equidistantly in the exponential space $(1, 0.00345, 0.0018, 0.00127, 0.001)$ leads to a radical difference from the typical linear temperature scheduling LTS $(1, 0.75, 0.5, 0.25, 0.001)$. We visualize the effect of the differences between ETS and LTS in terms of the entropy of β in a more sensible way in Fig. 2B.

For linear temperature schedule, we always need to preset t_0, t_N, N and calculate the decay strength statically to *prevent temperature from decreasing to less than* 0. In contrast, Eq. (12) is always greater than 0 for $t^{\exp} > 1$ which yields added flexibility for the design of ETS-based training scheme. We can calculate $d^{\exp}$ and $L^{\exp}$ after we specified $[t_0^{\exp}, t_N^{\exp}]$ and N. Alternatively, we can first determine $t_0^{\exp}$, $d^{\exp}$ and get $t_N^{\exp}$ or N by $t_N^{\exp} = t_0^{\exp} + Nd^{\exp}$, so that we can still build $L^{\exp}$ through Eq. (11).

2.3 Entropy-Based Adaptive Scheme

If operation output $o_c^i(h)$ on edge c is zero-weighted by as $\beta_c^i o_c^i(h)$ for $\beta_c^i \in \beta_c$ and $\beta_c^i = 0$, both forward and backward paths of the operation o_c^i are blocked so that the operation weights ω within o_c^i cannot obtain effective gradients. Employing a small temperature at the beginning of training will lead to exaggerated swings of β and finally biases the search result. For an appropriate scheme, presetting fixed t_0, t_N, $d^{\exp}$ and updating $E(x)$ and $L^{\exp}$

Algorithm 2. EDD sparse training scheme

Input: s for sn-softmax, λ for EDD, $t_0 = 1$ (mostly)
Get the expectation of parameter $E(a)$
Get $t_0^{\exp} = e^{E(a)/t_0}$
while training epoch k **do**
    Training a and ω by Algorithm 1 in [13] with $t^{(k)}$, s.
    Update $[d^{\exp}]^{(k)}$ by Eq.(14).
    Update $t^{(k)}$ via $[d^{\exp}]^{(k)}$ by Eq.(15).
end while
---

every epoch is *cumbersome and inflexible*. Empirically, we also observe that the same architecture under different initialization exhibits various optimization dynamics. Some lead to strong convergence that the parameters of supernet converge quickly under mild temperature decay. In other cases, the search is indecisive among two or three operations. In sum, a fixed decay strength $d^{\exp}$ is *not robust in practice*.

Being equipped with the additional design freedom supported by ETS, we further propose **entropy-based dynamic decay (EDD)** to adaptively determine $d^{\exp}$ in terms of both the sparsity of β and the current epoch k by which *we need only to tune one single hyperparameter λ to control the training process*. We first define the expectation of entropies of βs as $E(H(\beta_c))$, $c \in C$ in Eq. (13) to represent the sparsity of βs over all compound edges in the cell space.

$$E(H(\beta_c)) = \frac{-\sum_c^{|C|} \sum_i^M \beta_c^i \log \beta_c^i}{|C|} \quad \text{where } \beta_c = \text{softmax}(\Lambda_c), \ c \in C \quad (13)$$

Table 1. Evaluation results on NB201&C10. "clip" refers to the gradient clip on a to alleviate the effect of the noisy none (zero) operation which has been identified as unsearchable in DARTS [13].

Search dataset	Method	Search (seconds)	CIFAR-10		CIFAR-100		ImageNet-16-120	
			validation	test	validation	test	validation	test
C10	DARTS-V2 [13]	22323	39.77±0.00	54.30±0.00	15.03±0.00	15.61±0.00	16.43±0.00	16.32±0.00
	DARTS-V1 [13]	7253	39.77±0.00	54.30±0.00	15.03±0.00	15.61±0.00	16.43±0.00	16.32±0.00
	GDAS [9]	19720	90.00±0.21	93.51±0.13	71.14±0.27	70.61±0.26	41.70±1.26	41.84±0.90
	ETS-GDAS	19755	90.57±0.35	93.75±0.22	71.45±0.64	71.39±0.66	42.96±0.93	42.92±0.87
	DrNAS [4]	7544	90.15±0.10	93.74±0.03	70.82±0.27	71.07±0.08	40.76±0.05	41.37±0.17
	GAEA-Bilevel [11]	8280	82.80±1.01	84.64±1.00	55.24±1.47	55.35±1.72	27.72±1.35	26.40±0.85
	GAEA-ERM [11]	14464	84.59±0.00	86.59±0.00	58.12±0.00	58.43±0.00	29.54±0.00	28.19±0.00
			(91.50±0.06)	(94.34±0.06)	(73.12±0.26)	(73.11±0.06)	(45.71±0.28)	(46.38±0.18)
	GibbsNAS [19]	-	90.02±0.60	92.72±0.60	68.88±1.43	69.20±1.44	42.31±1.69	42.08±1.95
	DARTS- [7]	-	91.03±0.44	93.80±0.40	71.36±1.51	71.53±1.51	44.87±1.46	45.12±0.82
	SurgeNAS [14]	-	90.2	93.7	71.2	71.6	44.5	45.2
	EDD-DARTS	7392	90.95±0.44	**93.80±0.33**	**71.44±1.21**	71.42±1.25	**45.14±0.78**	45.12±0.56
	EDD-DARTS (clip)	7400	**91.12±0.19**	**94.05±0.11**	**72.59±0.62**	**72.43±0.64**	**45.89±0.41**	**45.80±0.29**

Table 2. Experimental results on S1~S4&C100 and S1~S4&SVHN.

Dataset	Space	DARTS	PC-DARTS	DARTS-ES	R-DARTS (DP/L2)	SDARTS (RS/ADV)	DARTS+PT (unfixed/fixed)	EDD-DARTS
C100	S1	29.46	24.69	28.37	25.93/24.25	23.51/22.33	24.48/24.40	**22.27**
	S2	26.05	22.48	23.25	22.30/22.44	22.28/**20.56**	23.16/23.30	21.73
	S3	28.90	21.69	23.73	22.36/23.99	21.09/**21.08**	22.03/21.94	**21.08**
	S4	22.85	21.50	21.26	22.18/21.94	21.46/21.25	20.80/**20.66**	**20.66**
SVHN	S1	4.58	2.47	2.72	2.55/4.79	2.35/2.29	2.62/2.39	**2.23**
	S2	3.53	2.42	2.60	2.52/2.51	2.39/2.35	2.53/2.32	**2.30**
	S3	3.41	2.41	2.50	2.49/2.48	2.36/2.40	2.42/**2.32**	**2.32**
	S4	3.05	2.43	2.51	2.61/2.50	2.46/2.42	2.42/**2.39**	2.45

where M operation candidates and $|C|$ compound edges in search space. β_c^i denotes the ith entry in the β_c that used to weight the feature maps from operation o^i on edge c . The design principles can be summarized as follows:

- $d^{\exp}$ is stronger when $E(H(\beta))$ is higher (smoother β in softmax);
- Gradually increase $d^{\exp}$ w.r.t the training epoch k.

Based on the above design principles, We update $d^{\exp}$ according to Eq. (14) for epoch k.

$$[d^{\exp}]^{(k)} = \lambda(1 - \rho)E(H(\beta)) + \rho[d^{\exp}]^{(k-1)} \tag{14}$$

where we set $[d^{\exp}]^{(0)} = 0$ in practice and keep an exponentially moving average with a momentum $\rho \equiv 0.5$ to avoid oscillations. λ is the hyperparameter to determine the influence of $E(H(\beta))$ on $d^{\exp}$. Since $d^{\exp}$ is adaptively decided through Eq. (14), $L^{\exp}$ cannot be calculated in advance. After kth update of $d^{\exp}$, we can get temperature $t^{(k)}$ accordingly by Eq. (15)

$$t^{(k)} = \frac{E(a)}{\ln(t_0^{\exp} + k[d^{\exp}]^{(k)})} \tag{15}$$

results in that $d^{\exp}$ is proportional to the training epoch k. EDD is summarised in Algorithm 2 where λ is the only hyperparameter for the scheme.

3   Evaluations

We evaluate EDD, namely EDD-DARTS, on CIFAR-10 (C10), CIFAR-100 (C100), ImageNet-1k (IN-1k), SVHN and multiple spaces: NAS-BENCH-201 (NB201), DARTS space (DS), S1~S4 [1].

Evaluations on NB201: NB201 supports three datasets (C10, C100, ImageNet-16 [6]) and has a unified cell-based search space with 15,625 architectures. We refer their paper [10] for more details of the search space. All our baselines come from recent top venues. Experiments of DrNAS and GAEA are both based on the released codes. We provide extra results for GAEA-ERM in parentheses by excluding the none operation since it's particularly fragile for"none" in NB201. The experimental results shown in Table 1 demonstrates that the our enhancement effectively eliminates the performance collapse of DARTS. Remarkably, EDD-DARTS claims superior searching on both C10 and C100 on the standard space (include none) of NB201.

Evaluations on DS: We employ the same search recipe as on NB201. Our evaluation is based on the source code released by DrNAS. We keep the hyperparameter settings unchanged except for replacing the cell genotypes. As [4,11], we repeat the search 3 times under different seeds, evaluate each result independently and report the mean accuracies and standard deviations in Table 3.

Transfer to ImageNet-1k: As a common practice, we transfer the most prominent architecture on C10 to ImageNet-1k (IN1K) for additional performance evaluation. As shown in Table 3 and Table 4, the EDD itself is already very competitive, and the combination of the EDD with partial channel trick [18] can deliver a clear new art scores with the lowest budget on both C10 and ImageNet-1k.

Table 3. Search and evaluate on DS&C10.

Method	Error (%)	Params (M)	GPU days
PC-DARTS [18]	2.57±0.07	3.6	0.1
GAEA+PC-DARTS [11]	2.50±0.06	3.7	0.1
DARTS+PT [16]	2.61±0.08	3.0	0.8
SDARTS-RS+PT [16]	2.54±0.10	3.3	0.8
SGAS+PT [16]	2.56±0.10	3.9	0.29
DrNAS [4]	2.54±0.03	4.0	0.4
DARTS- [7]	2.59±0.08	3.5	0.4
GibbsNAS [19]	2.53±0.02	4.1	0.5
SparseNAS [17]	2.69±0.03	4.2	1
β-DARTS [20]	2.53±0.08	3.7	0.4
EDD-DARTS	2.52±0.10	3.6	0.4
EDD-PC-DARTS	**2.47±0.06**	4.2	**0.1**

Comparison with Regularization: We evaluate EDD on four specially designed search spaces S1~S4 by [1] which are particularly challenging for DARTS-based methods. Unregularized DARTS is always prone to make a wrong choice of non-parametric operations on these spaces. We refer to [1] for more details of S1~S4. We

Table 4. Transfer to evaluate on IN1K.

Method	Top-1	Top-5	Params (M)	GPU days
PC-DARTS [18]	25.1	7.8	5.3	0.1
GAEA+PC-DARTS [11]	24.3	7.3	5.6	0.1
GibbsNAS [19]	24.6	-	5.1	0.5
DrNAS [4]	24.2	7.3	5.2	3.9[*]
DARTS- [7]	24.8	7	4.9	4.5[*]
SparseNAS [17]	24.6	7.6	5.7	-
β-DARTS [20]	24.2	7.1	5.4	0.4
EDD-DARTS	24.6	7.4	5.0	0.4
EDD-PC-DARTS	**24.0**	7.2	5.6	**0.1**

[*] Search on ImagetNet.

evaluate our method against two strong baselines Smooth DARTS [3] and DARTS+PT [16] on this benchmark on C10, C100 and SVHN respectively. As illustrated in Table 2, EDD-DARTS performs well under all datasets and search spaces and outperforms baselines in aggregate.

4 Further Experiments, Analyses and Conclusion

To understand what makes EDD effective, we conduct ablation studies along two axes, s and λ, on both NB201 and DS. We come up with and elaborate another ETS-based baseline to validate our claim of superior robustness of EDD.

Table 5. Ablate s and λ on NB201&C10.

Search dataset	Softmax	s	λ	CIFAR-10		CIFAR-100		ImageNet-16-120	
				valiation	test	valiation	test	valiation	test
C10	Normal softmax	-	0.06	90.56±0.75	93.29±0.70	70.03±1.48	70.21±1.54	43.59±2.05	43.33±1.80
			0.12	90.74±0.31	93.60±0.29	70.46±0.84	70.45±0.96	44.17±0.92	44.34±1.23
			0.24	89.93±1.44	92.82±1.16	69.38±2.05	69.60±1.49	43.01±2.16	43.07±2.14
	sn-softmax	$st \equiv 1$	0.06	**90.95±0.44**	**93.80±0.33**	**71.44±1.21**	**71.42±1.25**	**45.14±0.78**	**45.12±0.56**
			0.12	90.57±0.58	93.38±0.52	70.06±1.46	69.86±1.42	43.87±1.92	43.89±2.33
			0.24	90.06±0.81	92.83±0.52	69.07±1.35	69.26±1.29	42.28±2.07	42.43±1.83
		$s = 100$	0.06	90.68±0.71	93.44±0.54	70.45±1.69	70.43±1.61	43.98±1.85	44.08±2.17
			0.12	**91.02±0.30**	**93.75±0.31**	**71.36±0.96**	**71.33±1.21**	44.73±0.64	44.95±0.62
			0.24	90.54±0.74	93.22±0.58	69.90±1.60	70.19±1.88	43.43±1.82	43.49±1.68

Ablations on NB201: We ablate the impact of three configurations of softmax and λ on NB201. The experimental results on C10 are shown in Table 5. For the searching on C10, sn-softmax is helpful where both the highest and second-highest accuracies in the experiments of EDD-DARTS come from the results equipped with sn-softmax ($st \equiv 1$ or $s = 100$).

Ablations on DS: We ablate EDD-DARTS further on DS and the results are shown in Table 6. We find that a larger λ brings stronger temperature decay, EDD-DARTS tends to

Table 6. Ablate s and λ on DS&C10.

Search dataset	λ	normal softmax Test Error (%)	sn-softmax $st \equiv 1$ Test Error (%)	sn-softmax $s = 100$ Test Error (%)
	0.06	2.60 ± 0.10	2.55 ± 0.10	**2.52 ± 0.10**
C10	0.12	2.63 ± 0.13	2.56 ± 0.10	2.54 ± 0.09
	0.24	2.65 ± 0.13	2.59 ± 0.12	2.55 ± 0.10

find higher capacity architectures some of which are tricky to be trained properly but sn-softmax alleviates this trend and ensures that EDD keeps delivering efficient results.

Robustness Validations: To validate the robustness of our adaptive scheme EDD, we elaborate another scheme i.e. periodic cyclic decay (PCD) which excluding EDD and determine L^{exp} statically before training as the additional tailored baseline. We first *finetune the hyperparameters of both EDD and PCD on NB201&C10 and then transfer the settings and recipes exactly to all other search spaces and datasets.* we finetune PCD even marginally surpass EDD in Table 7. On the contrary, by transferring the configuration of NB201&C10, EDD starkly

Table 7. Finetune (FT) on NB201&C10 and transfer to other datasets and search spaces to evaluate the robustness.

Search space	Search dataset	Evaluation dataset	PCD-DARTS validation	PCD-DARTS test	EDD-DARTS validation	EDD-DARTS test
NB201	C10 (FT)	C10	91.06±0.49	93.78±0.43	90.95±0.44	93.80±0.33
		C100	71.61±1.30	72.00±1.48	71.44±1.21	71.42±1.25
		IN-16	45.17±0.84	45.63±0.86	45.14±0.78	45.12±0.56
	C100	C10	90.04±1.35	92.97±1.78	90.27±0.81	93.44±0.49
		C100	69.97±1.82	70.02±2.00	70.46±1.52	70.57±1.40
		IN-16	42.34±2.16	42.79±2.01	42.43±1.80	42.88±1.81
			Error (%)	Param (M)	Error (%)	Param (M)
DS	C10	C10	2.57±0.09	3.0	2.55±0.10	3.6
		IN-1k	26.5	4.2	24.6	5.1
			Error (%)		Error (%)	
S1		C10	2.75		2.77	
		C100	22.35		22.27	
		SVHN	2.27		2.23	
S2		C10	2.56		2.54	
		C100	21.44		21.73	
		SVHN	2.33		2.30	
S3		C10	2.50		2.49	
		C100	21.05		21.08	
		SVHN	2.32		2.32	
S4		C10	2.95		2.61	
		C100	21.48		20.66	
		SVHN	2.44		2.45	

outperforms PCD on most other cases especially on NB201&C100, DS&C10 and DS&IN-1k. This results underpin the robustness virtue of EDD over less flexible decay scheme PCD. Further investigation identify that PCD is more brittle than EDD to the warmup epoch and preseted decay strength on NB201 and DS respectively. We also note that simply increase these two values can recover the performance.

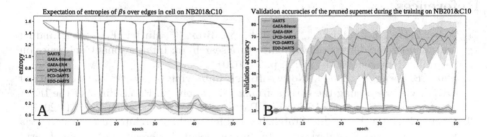

Fig. 4. (A). Trajectories of the expectation of entropies of βs during training on NB201&C10. (B). Discretized accuracies on validation set in training on NB201&C10.

Analyses: According to [11], the entropy of β and the discretized accuracy of the pruned finalnet on validation set are two main measurements for evaluating the impact of the sparsity on method. We illustrate the dynamics of the expectation of entropies of βs over edges in space during the training on NB201&C10 in Fig. 4A. in which the distribution entropy of EDD are much lower than other baselines. This validates our proposal of employing small temperature to sparsify β in training. To better characterize the effect of the sparse β on alleviating the mismatch, we also illustrate the corresponding discretized validation accuracies during the training on NB201&C10 in Fig. 4B. Both DARTS and GEAE in Fig. 4B are basically stuck at the random discretized accuracies which is obviously due to the insufficient sparsity of parameters shown in Fig. 4A. In contrast, EDD-DARTS, shown by the brown line in Fig. 4B, maintains an appropriate sparsity of β and steadily improve the discretized accuracies throughout the training. We observe the similar phenomenon on DARTS space. The drift of $E(a)$ can be seen in Fig. 4A in which the three decay cycles of PCD exhibit slightly different dynamics of β for the same temperature sequence $L^{\exp}$. This is the downside of PCD since it determines the whole $L^{\exp}$ beforehand and fixes it during training. In contrast, EDD adaptively gets $t^{\exp}$ directly based on the sparsity depicted in Eq. (15), thereby finds the appropriate t timely to compensate that drift of the expectation.

5   Conclusion

In this paper, we focus on sparsifying the β via utilizing and scheduling small temperature in DARTS. We first propose sn-softmax to alleviate the gradient saturation of the premature convergence. Next, we propose ETS to better control the sparsity of β and we elaborate an entropy-based adaptive scheme EDD to finally deliver the effective enhancement in DARTS.

Acknowledgment. National Key R&D Program of China (2022YFF0503900).

References

1. Arber Zela, T.E., Saikia, T., Marrakchi, Y., Brox, T., Hutter, F.: Understanding and robustifying differentiable architecture search. In: International Conference on Learning Representations, vol. 3, p. 7 (2020)
2. Chen, B., Deng, W., Du, J.: Noisy softmax: Improving the generalization ability of DCNN via postponing the early softmax saturation. In: Proceedings of the IEEE Conference on Computer Vision and Pattern Recognition, pp. 5372–5381 (2017)
3. Chen, X., Hsieh, C.J.: Stabilizing differentiable architecture search via perturbation-based regularization. In: International Conference on Machine Learning, pp. 1554–1565. PMLR (2020)
4. Chen, X., Wang, R., Cheng, M., Tang, X., Hsieh, C.J.: DrNAS: Dirichlet neural architecture search. In: International Conference on Learning Representations (2021). https://openreview.net/forum?id=9FWas6YbmB3
5. Cheng, X., et al.: Hierarchical neural architecture search for deep stereo matching. Adv. Neural. Inf. Process. Syst. **33**, 22158–22169 (2020)
6. Chrabaszcz, P., Loshchilov, I., Hutter, F.: A downsampled variant of imagenet as an alternative to the CIFAR datasets. arXiv preprint arXiv:1707.08819 (2017)
7. Chu, X., Wang, X., Zhang, B., Lu, S., Wei, X., Yan, J.: Darts-: Robustly stepping out of performance collapse without indicators. In: International Conference on Learning Representations (2021)
8. Chu, X., Zhou, T., Zhang, B., Li, J.: Fair DARTS: eliminating unfair advantages in differentiable architecture search. In: Vedaldi, A., Bischof, H., Brox, T., Frahm, J.-M. (eds.) ECCV 2020. LNCS, vol. 12360, pp. 465–480. Springer, Cham (2020). https://doi.org/10.1007/978-3-030-58555-6_28

9. Dong, X., Yang, Y.: Searching for a robust neural architecture in four gpu hours. In: Proceedings of the IEEE/CVF Conference on Computer Vision and Pattern Recognition. pp. 1761–1770 (2019)

10. Dong, X., Yang, Y.: NAS-bench-201: extending the scope of reproducible neural architecture search. In: International Conference on Learning Representations (2020)

11. Li, L., Khodak, M., Balcan, M.F., Talwalkar, A.: Geometry-aware gradient algorithms for neural architecture search. In: International Conference on Learning Representations (2021). https://openreview.net/forum?id=MuSYkd1hxRP

12. Liang, H., et al.: Darts+: improved differentiable architecture search with early stopping. arXiv preprint arXiv:1909.06035 (2019)

13. Liu, H., Simonyan, K., Yang, Y.: Darts: differentiable architecture search. In: International Conference on Learning Representations (2019)

14. Luo, X., Liu, D., Kong, H., Huai, S., Chen, H., Liu, W.: SurgeNAS: a comprehensive surgery on hardware-aware differentiable neural architecture search. IEEE Transactions on Computers (2022)

15. Shu, Y., Wang, W., Cai, S.: Understanding architectures learnt by cell-based neural architecture search. In: International Conference on Learning Representations (2019)

16. Wang, R., Cheng, M., Chen, X., Tang, X., Hsieh, C.J.: Rethinking architecture selection in differentiable NAS. In: International Conference on Learning Representations (2021)

17. Wu, Y., Liu, A., Huang, Z., Zhang, S., Van Gool, L.: Neural architecture search as sparse supernet. In: Proceedings of the AAAI Conference on Artificial Intelligence, vol. 35, pp. 10379–10387 (2021)

18. Xu, Y., et al.: PC-DARTS: Partial channel connections for memory-efficient architecture search. In: International Conference on Learning Representations (2020)

19. Xue, C., Wang, X., Yan, J., Hu, Y., Yang, X., Sun, K.: Rethinking bi-level optimization in neural architecture search: A gibbs sampling perspective. In: Proceedings of the AAAI Conference on Artificial Intelligence, vol. 35, pp. 10551–10559 (2021)

20. Ye, P., Li, B., Li, Y., Chen, T., Fan, J., Ouyang, W.: β-darts: Beta-decay regularization for differentiable architecture search. In: Proceedings of the IEEE/CVF Conference on Computer Vision and Pattern Recognition, pp. 10874–10883 (2022)

Document-Level Relation Extraction
with Cross-sentence Reasoning Graph

Hongfei Liu[1], Zhao Kang[1(✉)], Lizong Zhang[1], Ling Tian[1], and Fujun Hua[2]

[1] University of Electronic Science and Technology of China, Chengdu, China
202022081525@std.uestc.edu.cn,
{zkang,l.zhang,lingtian}@uestc.edu.cn
[2] Research and Development Center TROY Information Technology Co. Ltd., Chengdu, China
huafj@troy.cn

Abstract. Relation extraction (RE) has recently moved from the sentence-level to document-level, which requires aggregating document information and using entities and mentions for reasoning. Existing works put entity nodes and mention nodes with similar representations in a document-level graph, whose complex edges may incur redundant information. Furthermore, existing studies only focus on entity-level reasoning paths without considering global interactions among entities cross-sentence. To these ends, we propose a novel document-level RE model with a **GR**aph information **A**ggregation and **C**ross-sentence **R**easoning network (GRACR). Specifically, a simplified document-level graph is constructed to model the semantic information of all mentions and sentences in a document, and an entity-level graph is designed to explore relations of long-distance cross-sentence entity pairs. Experimental results show that GRACR achieves excellent performance on two public datasets of document-level RE. It is especially effective in extracting potential relations of cross-sentence entity pairs. Our code is available at https://github.com/UESTC-LHF/GRACR.

Keywords: Deep learning · Relation extraction · Document-level RE

1 Introduction

Relation extraction (RE) is to identify the semantic relation between a pair of named entities in text. Document-level RE requires the model to extract relations from the document and faces some intractable challenges. Firstly, a document contains multiple sentences, thus relation extraction task needs to deal with more rich and complex semantic information. Secondly, subject and object entities in the same triple may appear in different sentences, and some entities have aliase, which are often named entity mentions. Hence, the information utilized by document-level RE may not come from a single sentence. Thirdly, there may be interactions among different triples. Extracting the relation between two entities from different triples requires reasoning with contextual features. Figure 1 shows an example from DocRED dataset [21]. It is easy to predict intra-sentence relations because the subject and object appear in the same sentence. However, it has a problem in identifying the inter-sentence relation between "Swedish"

H. Kashima et al. (Eds.): PAKDD 2023, LNAI 13935, pp. 316–328, 2023.
https://doi.org/10.1007/978-3-031-33374-3_25

Johan Gottlieb Gahn
[S1] Johan Gottlieb Gahn (19 August 1745 – 8 December 1818) was a Swedish chemist and metallurgist who Discovered manganese In 1774.··· [S7] In 1784, Gahn was elected a member of the Royal Swedish Academy of Sciences. [S8] He also made a managerial career in Swedish mining.
Triples:
<Johan Gottlieb Gahn, country of citizenship, Swedish>
<Johan Gottlieb Gahn, member of, Royal Swedish Academy>
<Royal Swedish Academy, country, Swedish>

Fig. 1. An example of document-level RE excerpted from DocRED dataset.

and "Royal Swedish Academy", whose mentions are distributed across different sentences and there exists long-distance dependencies.

[21] proposed DocRED dataset, which contains large-scale human-annotated documents, to promote the development of sentence-level RE to document-level RE. In order to make full use of the complex semantic information of documents, recent works design document-level graph and propose models based on graph neural networks (GNN) [4]. [1] proposed an edge-oriented model that constructs a document-level graph with different types of nodes and edges to obtain a global representation for relation classification. [12] defined the document-level graph as a latent variable and induced it based on structured attention to improve the performance of document-level RE models by optimizing the structure of document-level graph. [17] proposed a model that learns global representations of entities through a document-level graph, and learns local representations of entities based on their contexts. However, these models simply average the embeddings of mentions to obtain entity embeddings and feed them into classifiers to obtain relation labels. Entity and mention nodes share a similar embedding if certain entity has only one mention. Therefore, putting them in the same graph will introduce redundant information and reduce discrimination.

To address above issues, we propose a novel GNN-based document-level RE model with two graphs constructed by semantic information from the document. Our key idea is to build document-level graph and entity-level graph to fully exploit the semantic information of documents and reason about relations between entity pairs across sentences. Specifically, we solve two problems:

First, how to integrate rich semantic information of a document to obtain entity representations? We construct a document-level graph to integrate complex semantic information, which is a heterogeneous graph containing mention nodes and sentence nodes. Representations of mention nodes and sentence nodes are computed by the pre-trained language model BERT [3]. The built document-level graph is input into the R-GCNs [13], a relational graph neural network, to make nodes contain the information of their neighbor nodes. Then, representations of entities are obtained by performing logsumexp pooling operation on representations of mention nodes. In previous methods, representations of entity nodes are obtained from representations of mention nodes. Hence putting them in the same graph will introduce redundant information and reduce discriminability. Unlike previous document-level graph construction, our document-level graph contains only sentence nodes and mention nodes to avoid redundant information caused by repeated node representations.

Second, how to use connections between entities for reasoning? In this paper, we exploit connections between entities and propose an entity-level graph for reasoning.

The entity-level graph is built by the positional connections between sentences and entities to make full use of cross-sentence information. It connects long-distance cross-sentence entity pairs. Through the learning of GNN, each entity node can aggregate the information of its most relevant entity nodes, which is beneficial to discover potential relations of long-distance cross-sentence entity pairs.

In summary, we propose a novel model called GRACR for document-level RE. Our main contributions are as follows:

- We propose a simplified document-level graph to integrate rich semantic information. The graph contains sentence nodes and mention nodes but not entity nodes, which avoids introducing redundant information caused by repeated node representations.
- We propose an entity-level graph for reasoning to discover potential relations of long-distance cross-sentence entity pairs. An attention mechanism is applied to fuse document embedding, aggregation, and inference information to extract relations of entity pairs.
- We conduct experiments on two public document-level relation extraction datasets. Experimental results demonstrate that our model outperforms many state-of-the-art methods.

2   Related Work

The research on document-level RE has a long history. The document-level graph provides more features for entity pairs. The relevance between entities can be captured through graph learning using GNN [10]. For example, [2] utilized GNN to aggregate the neighborhood information of text graph nodes for text classification. Following this, [1] constructed a document-level graph with heterogeneous nodes and proposed an edge-oriented model to obtain a global representation. [7] characterized the interaction between sentences and entity pairs to improve inter-sentence reasoning. [25] introduced context of entity pairs as edges between entity nodes to model semantic interactions among multiple entities. [24] constructed a dual-tier heterogeneous graph to encode the inherent structure of document and reason multi-hop relations of entities. [17] learned global representations of entities through a document-level graph, and learned local representations based on their contexts. [12] defined the document-level graph as a latent variable to improve the performance of RE models by optimizing the structure of the document-level graph. [23] proposed a double graph-based graph aggregation and inference network (GAIN). Different from GAIN, our entity-level graph is a heterogeneous graph and we use R-GCNs to enable interactions between entity nodes to discover potential relations of long-distance cross-sentence entity pairs. [18] constructed a document-level graph with rhetorical structure theory and used evidence to reasoning. [14] constructed the input documents as heterogeneous graphs and utilized Graph Transformer Networks to generate semantic paths.

Unlike above document-level graph construction methods, our document-level graph contains only sentence nodes and mention nodes to avoid introducing redundant information. Moreover, previous works don't directly deal with cross-sentence entity pairs. Although entities in different sentences are indirectly connected in the graph, e.g., the minimum distance between entities across sentences is 3 and the information needs

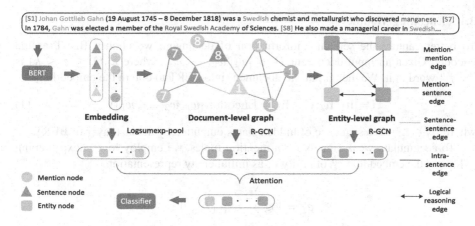

Fig. 2. Architecture of our proposed model.

to pass through two different nodes when interacting in GLRE [17]. We directly connect cross-sentence entity pairs with potential relations through bridge entities to shorten the distance of information transmission, which reduces the introduction of noise.

In addition, there are some works that try to use pre-trained models directly instead of introducing graph structures. [16] applied a hierarchical inference method to aggregate the inference information of different granularity. [22] captured the coreferential relations in context by a pre-training task. [9] proposed a mention-based reasoning network to capture local and global contextual information. [20] used mention dependencies to construct structured self-attention mechanism. [26] proposed adaptive thresholding and localized context pooling to solve the multi-label and multi-entity problems. These models take advantage of the multi-head attention of Transformer instead of GNN to aggregate information.

However, these studies focused on the local entity representation, which overlooks the interaction between entities distributed in different sentences [11]. To discover potential relations of long-distance cross-sentence entity pairs, we introduce an entity-level graph built by the positional connections between sentences and entities for reasoning.

3   Methodology

In this section, we describe our proposed GRACR model that constructs a document-level graph and an entity-level graph to improve document-level RE. As shown in Fig. 2, GRACR mainly consists of 4 modules: encoding module, document-level graph aggregation module, entity-level graph reasoning module, and classification module. First, in encoding module, we use a pre-trained language model such as BERT [3] to encode the document. Next, in document-level graph aggregation module, we construct a heterogeneous graph containing mention nodes and sentence nodes to integrate rich semantic information of a document. Then, in entity-level graph reasoning module, we also propose a graph for reasoning to discover potential relations of long-distance and cross-sentence entity pairs. Finally, in classification module, we merge the context information of relation representations obtained by self-attention [15] to make final relation prediction.

3.1 Encoding Module

To better capture the semantic information of document, we choose BERT as the encoder. Given an input document $D = [w_1, w_2, \ldots, w_k]$, where $w_j (1 \leq j \leq k)$ is the jth word in it. We then input the document into BERT to obtain the embeddings:

$$\mathbf{H} = [\mathbf{h}_1, \mathbf{h}_2, \ldots, \mathbf{h}_k] = \text{Encoder}([w_1, w_2, \ldots, w_k]) \tag{1}$$

where $\mathbf{h}_j \in \mathbb{R}^{d_w}$ is a sequence of hidden states outputted by the last layer of BERT.

To accumulate weak signals from mention tuples, we employ logsumexp pooling [5] to get the embedding e_i^h of entity $\mathbf{e}_i$ as initial entity representation.

$$e_i^h = \log \sum_{j=1}^{N_{\mathbf{e}_i}} \exp \left(h_{\mathbf{m}_j^i} \right) \tag{2}$$

where $\mathbf{m}_j^i$ is the mention $\mathbf{m}_j$ of entity $\mathbf{e}_i$, $h_{\mathbf{m}_j^i}$ is the embedding of $\mathbf{m}_j^i$, $N_{\mathbf{e}_i}$ is the number of mentions of entity $\mathbf{e}_i$ in D.

As shown in Eq. (2), the logsumexp pooling generates an embedding for each entity by accumulating the embeddings of its all mentions across the whole document.

3.2 Document-Level Graph Aggregation Module

To integrate rich semantic information of a document to obtain entity representations, we construct a document-level graph (*Dlg*) based on $\mathbf{H}$.

Dlg has two different kinds of nodes:

Sentence nodes, which represent sentences in D. The representation of a sentence node s_i is obtained by averaging the representations of contained words. We concatenate a node type representation $\mathbf{t}_s \in \mathbb{R}^{d_t}$ to differentiate node types. Therefore, the representations of s_i is $\mathbf{h}_{s_i} = \left[\text{avg}_{w_j \in s_i} (\mathbf{h}_j) ; \mathbf{t}_s \right]$, where $[;]$ is the concatenation operator.

Mention nodes, which represent mentions in D. The representation of a mention node m_i is achieved by averaging the representations of words that make up the mention. We concatenate a node type representation $\mathbf{t}_m \in \mathbb{R}^{d_t}$. Similar to sentence nodes, the representation of m_i is $\mathbf{h}_{m_i} = \left[\text{avg}_{w_j \in m_i} (\mathbf{h}_j) ; \mathbf{t}_m \right]$.

There are three types of edges in *Dlg*:

- Mention-mention edge. To exploit the co-occurrence dependence between mention pairs, we create a mention-mention edge. Mention nodes of two different entities are connected by mention-mention edges if their mentions co-occur in the same sentence.
- Mention-sentence edge. Mention-sentence edge is created to better capture the context information of mention. Mention node and sentence node are connected by mention-sentence edges if the mention appears in the sentence.
- Sentence-sentence edge. All sentence nodes are connected by sentence-sentence edges to eliminate the effect of sentences sequence in the document and facilitate inter-sentence interactions.

Then, we use an L-layer stacked R-GCNs [13] to learn the document-level graph. R-GCNs can better model heterogeneous graph that has various types of edges than GCN. Specifically, its node forward-pass update for the $(l+1)^{(th)}$ layer is defined as follows:

$$\mathbf{n}_i^{l+1} = \sigma \left(\mathbf{W}_0^l \mathbf{n}_i^l + \sum_{x \in X} \sum_{j \in N_i^x} \frac{1}{|N_i^x|} \mathbf{W}_x^l \mathbf{n}_j^l \right) \tag{3}$$

where $\sigma(\cdot)$ means the activation function, N_i^x denotes the set of neighbors of node i linked with edge x, and X denotes the set of edge types. $\mathbf{W}_x^l, \mathbf{W}_0^l \in \mathbb{R}^{d_n \times d_n}$ are trainable parameter matrices and d_n is the dimension of node representation.

We use the representations of mention nodes after graph convolution to compute the preliminary representation of entity node e_i by logsumexp pooling as e_i^{pre}, which incorporates the semantic information of e_i throughout the whole document. However, the information of the whole document inevitably introduce noise. We employ attention mechanism to fuse the initial embedding information and semantic information of entities to reduce noise. Specifically, we define the entity representation e_i^{Dlg} as follows:

$$e_i^{Dlg} = \text{softmax} \left(\frac{e_i^{pre} \mathbf{W}_i^{e^{pre}} \left(e_i^h \mathbf{W}_i^{e^h} \right)^T}{\sqrt{d_{e_i^h}}} \right) e_i^h \mathbf{W}_i^{e^h} \tag{4}$$

and

$$e_i^{pre} = \log \sum_{j=1}^{N_{e_i}} \exp \left(n_{m_j^i} \right) \tag{5}$$

where $\mathbf{W}_i^{e^{pre}}$ and $\mathbf{W}_i^{e^h} \in \mathbb{R}^{d_n \times d_n}$ are trainable parameter matrices. $n_{m_j^i}$ is mention semantic representations after graph convolution. $d_{e_i^h}$ is the dimension of e_i^h.

3.3 Entity-Level Graph Reasoning Module

To discover potential relations of long-distance cross-sentence entity pairs, we introduce an entity-level graph (*Elg*) reasoning module. *Elg* contains only one kind of node:

Entity node, which represents entities in D. The representation of an entity node e_i is obtained from document-level graph defined by Eq. (5). We concatenate a node type representation $\mathbf{t}_e \subset \mathbb{R}^{t_e}$. The representations of e_i is $\mathbf{h}_{e_i} = [e_i^{pre}; \mathbf{t}_e]$.

There are two kinds of edges in *Elg*:

- Intra-sentence edge. Two different entities are connected by an intra-sentence edge if their mentions co-occur in the same sentence. For example, *Elg* uses an intra-sentence edge to connect entity nodes e_i and e_j if there is a path $PI_{i,j}$ denoted as $\mathbf{m}_i^{\mathbf{s}_1} \to \mathbf{s}_1 \to \mathbf{m}_j^{\mathbf{s}_1}$. $\mathbf{m}_i^{\mathbf{s}_1}$ and $\mathbf{m}_j^{\mathbf{s}_1}$ are mentions of an entity pair <e_i, e_j> and they appear in sentence $\mathbf{s}_1$. "$\to$" denotes one reasoning step on the reasoning path from entity node e_i to e_j.
- Logical reasoning edge. If the mention of entity $\mathbf{e}_k$ has co-occurrence dependencies with mentions of other two entities in different sentences, we suppose that $\mathbf{e}_k$ can be used as a bridge between entities. Two entities distributed in different sentences are connected by a logical reasoning edge if a bridge entity connects them. There is a

logical reasoning path $PL_{i,j}$ denoted as $\mathbf{m}_i^{s_1} \rightarrow s_1 \rightarrow \mathbf{m}_k^{s_1} \rightarrow \mathbf{m}_k^{s_2} \rightarrow s_2 \rightarrow \mathbf{m}_j^{s_2}$, and we apply a logical reasoning edge to connect entity nodes e_i and e_j.

Similar to Dlg, we apply an L-layer stacked R-GCNs to convolute the entity-level graph to get the reasoned representation of entity e_i^{Elg}. In order to better integrate the information of entities, we employ the attention mechanism to fuse the aggregated information, the reasoned information, and the initial information of entity to form the final representation of entity.

$$e_i^{rep} = \text{softmax} \left(\frac{e_i^{Dlg} \mathbf{W}_i^{e_i^{Dlg}} \left(e_i^{Elg} \mathbf{W}_i^{e_i^{Elg}} \right)^T}{\sqrt{d_{e_i^{Elg}}}} \right) e_i^h \mathbf{W}_i^{e_i^h} \qquad (6)$$

where $\mathbf{W}_i^{e_i^{Dlg}}$ and $\mathbf{W}_i^{e_i^{Elg}} \in \mathbb{R}^{d_n \times d_n}$ are trainable parameter matrices. $d_{e_i^{Elg}}$ is the dimension of e_i^{Elg}.

3.4 Classification Module

To classify the target relation r for an entity pair $<e_m, e_n>$, we concatenate entity final representations and relative distance representations to represent one entity pair:

$$\hat{e}_m = [e_m^{rep}; s_{mn}], \hat{e}_n = [e_n^{rep}; s_{nm}] \qquad (7)$$

where s_{mn} denotes the embedding of relative distance from the first mention of e_m to that of e_n in the document. s_{nm} is similarly defined.

Then, we concatenate the representations of $\hat{e}_m$, $\hat{e}_n$ to form the target relation representation $\mathbf{o}_r = [\hat{e}_m; \hat{e}_n]$.

Furthermore, following [17], we employ self-attention [15] to capture context relation representations, which can help us exploit the topic information of the document:

$$\mathbf{o}_c = \sum_{i=1}^{p} \theta_i \mathbf{o}_i = \sum_{i=1}^{p} \frac{\exp\left(\mathbf{o}_i \mathbf{W} \mathbf{o}_r^T\right)}{\sum_{j=1}^{p} \exp\left(\mathbf{o}_j \mathbf{W} \mathbf{o}_r^T\right)} \mathbf{o}_i \qquad (8)$$

where $\mathbf{W} \in \mathbb{R}^{d_r \times d_r}$ is a trainable parameter matrix, d_r is the dimension of target relation representations. $\mathbf{o}_i$ is the relation representation of the ith entity pair. θ_i is the attention weight for $\mathbf{o}_i$. p is the number of entity pairs.

Finally, we use a feed-forward neural network (FFNN) on the target relation representation $\mathbf{o}_r$ and the context relation representation $\mathbf{o}_c$ for prediction. What's more, we transform the multi-classification problem into multiple binary classification problems, since an entity pair may have different relations. The predicted probability distribution of r over the set R of all relations is defined as follows:

$$y_r = \text{sigmoid}\left(\text{FFNN}\left([\mathbf{o}_r; \mathbf{o}_c]\right)\right) \qquad (9)$$

where $y_r \in \{0, 1\}$.

We define the loss function as follows:

$$L = -\sum_{r \in R} \left(y_r^* \log\left(y_r\right) + (1 - y_r^*) \log\left(1 - y_r\right) \right) \qquad (10)$$

where $y_r^* \in \{0, 1\}$ denotes the true label of r.

4   Experiments and Results

Table 1. Statistics of the datasets.

Statistics	DocRED	CDR
# Train	3053	500
# Dev	1000	500
# Test	1000	500
# Relations	97	2
Avg.# Ents per Doc.	19.5	7.6

Table 2. Results on the development and test set of DocRED. Some results are quoted from respective paper.

Model		Dev		Test	
		Ign F_1	F_1	Ign F_1	F_1
Sequence-based	CNN [21]	41.58	43.45	40.33	42.26
	LSTM [21]	48.44	50.68	47.71	50.07
	BiLSTM [21]	48.87	50.94	48.78	51.06
	Context-aware [21]	48.94	51.09	48.40	50.70
Transformer-based	BERT [19]	–	54.16	–	53.20
	HIN [16]	54.29	56.31	53.70	55.60
	CorefBERT [22]	55.32	57.51	54.54	56.96
	SSAN [20]	57.03	59.19	55.84	58.16
Graph-based	EoG [1]	45.94	52.15	49.48	51.82
	GEDA [7]	54.52	56.16	53.71	55.74
	GCGCN [25]	55.43	57.35	54.53	56.67
	GLRE [17]	–	–	55.40	57.40
	DISCO [18]	55.91	57.78	55.01	55.70
Ours	GRACR	**57.85**	**59.73**	**56.47**	**58.54**

Table 3. Results on CDR.

Model	F_1	intra-F_1	inter-F_1
LSR [12]	64.8	68.9	53.1
DHG [24]	65.9	70.1	54.6
HGNN [14]	64.4	69.2	51.2
MRN [9]	65.9	70.4	54.2
GRACR	**68.8**	**73.9**	**55.8**

Table 4. Ablation study on the development set of DocRED.

Model	Ign F_1	F_1
GRACR	**57.85**	**59.73**
w/o both module	57.33	59.16
w/o reasoning module	57.44	59.30
w/o aggregation module	57.61	59.57
w/o reasoning edge	57.52	59.48
w/o intra-sentence edge	57.51	59.46
w/ previous Dlg	57.13	58.97

Table 5. Intra-F_1 and inter-F_1 results on DocRED.

Model	intra-F_1	inter-F_1
CNN [21]	51.87	37.58
LSTM [21]	56.57	41.47
BiLSTM [21]	57.05	43.39
Context-aware [21]	56.74	42.26
GEDA [7]	61.85	49.46
LSR [12]	65.26	52.05
GRACR	**65.88**	**52.49**

4.1 Dataset

We evaluate our model on DocRED and CDR dataset. The dataset statistics are shown in Table 1. The DocRED dataset [21], a large-scale human-annotated dataset constructed from Wikipedia, has 3,053 documents, 132,275 entities, and 56,354 relation facts in total. DocRED covers a wide variety of relations related to science, art, time, personal life, etc. The Chemical-Disease Relations (CDR) dataset [8] is a human-annotated dataset, which is built for the BioCreative V challenge. CDR contains 1,500 PubMed abstracts about chemical and disease with 3,116 relational facts.

4.2 Experiment Settings and Evaluation Metrics

To implement our model, we choose uncased BERT-base [3] as the encoder on DocRED and set the embedding dimension to 768. For CDR dataset, we pick up BioBERT-Base v1.1 [6], which re-trained the BERT-base-cased model on biomedical corpora.

All hyper-parameters are tuned based on the development set. Other parameters in the network are all obtained by random orthogonal initialization [17] and updated during training.

For a fair comparison, we follow the same experimental settings from previous works. We apply F_1 and Ign F_1 as the evaluation metrics on DocRED. F_1 scores can be obtained by calculation through an online interface. Furthermore, Ign F_1 means that the F_1 score ignores the relational facts shared by the training and development/test sets. We compare our model with three categories of models. Sequence-based models use neural architectures such as CNN and bidirectional LSTM as encoder to acquire embeddings of entities. Graph-based models construct document graphs and use GNN to learn graph structures and implement inference. Instead of using document graph, transformer-based models adopt pre-trained language models to extract relation.

For CDR dataset, we use training subset to train the model. Depending on whether relation between two entities occur within one sentence or not, F_1 can be further split into intra-F_1 and inter-F_1 to evaluate the model's performance on intra-sentence relations and inter-sentence relations. To make a comprehensive comparison, we also measure the corresponding F_1, intra-F_1 and inter-F_1 scores on development set.

4.3 Main Results

Results on DocRED. As shown in Table 2, our model outperforms all baseline methods on both development and test sets. Compared with graph-based models, both F_1 and Ign F_1 of our model are significantly improved. Compared to GLRE, which is the most relevant approach to our method, the performance improves 1.07% for F_1 and 1.14% for Ign F_1 on test set. Furthermore, compared to Transformer-based model SSAN, our method improves by 0.54% for F_1 and 0.84% for Ign F_1 on development set. With respect to sequence-based methods, the improvement is considerable.

Results on CDR. Table 3 depicts the comparisons with state-of-the-art models on CDR. Compared to MRN [9], the performance of our model approximately improves about 2.9% for F_1, and 3.9% for intra-F_1 and 1.6% for inter-F_1. DHG and MRN produce similar results. In summary, these results demonstrate that our method is effective in extracting both intra-sentence relations and inter-sentence relations.

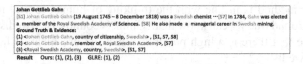

Fig. 3. Case study on the DocRED development set. Entities are colored accordingly.

4.4 Ablation Study

We conduct a thorough ablation study to investigate the effectiveness of two key modules in our method: an aggregation module and an reasoning module. From Table 4, we can observe that all components contribute to model performance.

(1) When the reasoning module is removed, the performance of our model on the DocRED development set for Ign F_1 and F_1 scores drops by 0.41% and 0.43%, respectively. Furthermore, we analyze the role of each edge in the reasoning module. F_1 drops by 0.23% or 0.25% when we remove intra-sentence edge or logical reasoning edge. Likewise, removing the aggregation module results in 0.24% and 0.16% drops in Ign F_1 and F_1. This phenomenon verifies the effectiveness of the aggregation module and the reasoning module.

(2) A larger drop occurs when two modules are removed. The F_1 score dropped from 59.73% to 59.16% and the Ign F_1 score dropped from 57.85% to 57.33%. This study validates that all modules work together can handle RE task more effective.

(3) When we apply the document-level graph with entity nodes and more complex edge types like GLRE, the F_1 score dropped from 59.73% to 58.97% and the Ign F_1 score dropped from 57.85% to 57.13%. This result suggests that document-level graph containing complex and repetitive node information and edges can lead to information redundancy and degrade model performance.

4.5 Intra- and Inter-sentence Relation Extraction

In this subsection, we further analyze both intra- and inter-sentence RE performance on DocRED. The experimental results are listed in Table 5, from which we can find that GRACR outperforms the compared models in terms of intra- and inter-F_1. For example, our model obtains 0.62% intra-F_1 and 0.44% inter-F_1 gain on DocRED. The improvements suggest that GRACR not only considers intra-sentence relations, but also handles long-distance inter-sentence relations well.

4.6 Case Study

As shown in Fig. 3, GRACR infers the relations of <Swedish, Royal Swedish Academy of Sciences> based on the information of $S1$ and $S7$. "Swedish" and "Royal Swedish Academy of Sciences" distributed in different sentences are connected by entity-level graph because they appear in the same sentence with "Johan Gottlieb Gahn". Entity-level graph connects them together to facilitate reasoning about their relations. More importantly, our method is in line with the thinking of human logical reasoning. For example, from ground true we can know that "Gahn"'s country is "Swedish". Therefore, we can speculate that there is a high possibility that the organization he joined has a relation with "Swedish".

5   Conclusion

In this paper, we propose GRACR, a graph information aggregation and logical cross-sentence reasoning network, to better cope with document-level RE. GRACR applies a document-level graph and attention mechanism to model the semantic information of all mentions and sentences in a document. It also constructs an entity-level graph to utilize the interaction among different entities to reason the relations. Finally, it uses an attention mechanism to fuse document embedding, aggregation, and inference information to help identify relations. Experimental results show that our model achieves excellent performance on DocRED and CDR.

Acknowledgements. This work was supported by the National Natural Science Foundation of China (Nos. 62276053, 62271125) and the Sichuan Science and Technology Program (No. 22ZDYF3621).

References

1. Christopoulou, F., Miwa, M., Ananiadou, S.: Connecting the dots: document-level neural relation extraction with edge-oriented graphs. In: Proceedings of the 2019 Conference on Empirical Methods in Natural Language Processing and the 9th International Joint Conference on Natural Language Processing (EMNLP-IJCNLP), pp. 4925–4936 (2019)
2. Dai, Y., Shou, L., Gong, M., Xia, X., Kang, Z., Xu, Z., Jiang, D.: Graph fusion network for text classification. Knowl.-Based Syst. **236**, 107659 (2022)

3. Devlin, J., Chang, M.W., Lee, K., Toutanova, K.: Bert: pre-training of deep bidirectional transformers for language understanding. In: Proceedings of the 2019 Conference of the North American Chapter of the Association for Computational Linguistics, pp. 4171–4186 (2019)

4. Fang, R., Wen, L., Kang, Z., Liu, J.: Structure-preserving graph representation learning. In: ICDM (2022)

5. Jia, R., Wong, C., Poon, H.: Document-level n-ary relation extraction with multiscale representation learning. In: Proceedings of the 2019 Conference of the North American Chapter of the Association for Computational Linguistics: Human Language Technologies, Volume 1 (Long and Short Papers), pp. 3693–3704 (2019)

6. Lee, J., et al.: Biobert: a pre-trained biomedical language representation model for biomedical text mining. Bioinformatics **36**(4), 1234–1240 (2020)

7. Li, B., Ye, W., Sheng, Z., Xie, R., Xi, X., Zhang, S.: Graph enhanced dual attention network for document-level relation extraction. In: Proceedings of the 28th International Conference on Computational Linguistics, pp. 1551–1560 (2020)

8. Li, J., Sun, Y., Johnson, R.J., Sciaky, D., Wei, C.H., Leaman, R., Davis, A.P., Mattingly, C.J., Wiegers, T.C., Lu, Z.: Biocreative v cdr task corpus: a resource for chemical disease relation extraction. Database **1**, 10 (2016)

9. Li, J., Xu, K., Li, F., Fei, H., Ren, Y., Ji, D.: MRN: a locally and globally mention-based reasoning network for document-level relation extraction. In: Findings of the Association for Computational Linguistics: ACL-IJCNLP 2021, pp. 1359–1370 (2021)

10. Liu, L., Kang, Z., Ruan, J., He, X.: Multilayer graph contrastive clustering network. Inf. Sci. **613**, 256–267 (2022)

11. Luoma, J., Pyysalo, S.: Exploring cross-sentence contexts for named entity recognition with Bert. In: Proceedings of the 28th International Conference on Computational Linguistics, pp. 904–914 (2020)

12. Nan, G., Guo, Z., Sekulić, I., Lu, W.: Reasoning with latent structure refinement for document-level relation extraction. In: Proceedings of the 58th Annual Meeting of the Association for Computational Linguistics, pp. 1546–1557 (2020)

13. Schlichtkrull, M., Kipf, T.N., Bloem, P., van den Berg, R., Titov, I., Welling, M.: Modeling relational data with graph convolutional networks. In: Gangemi, A., et al. (eds.) ESWC 2018. LNCS, vol. 10843, pp. 593–607. Springer, Cham (2018). https://doi.org/10.1007/978-3-319-93417-4_38

14. Shi, Y., Xiao, Y., Quan, P., Lei, M., Niu, L.: Document-level relation extraction via graph transformer networks and temporal convolutional networks. Pattern Recogn. Lett. **149**, 150–156 (2021)

15. Sorokin, D., Gurevych, I.: Context-aware representations for knowledge base relation extraction. In: Proceedings of the 2017 Conference on Empirical Methods in Natural Language Processing, pp. 1784–1789 (2017)

16. Tang, H., et al.: HIN: hierarchical inference network for document-level relation extraction. In: Lauw, H.W., Wong, R.C.-W., Ntoulas, A., Lim, E.-P., Ng, S.-K., Pan, S.J. (eds.) PAKDD 2020. LNCS (LNAI), vol. 12084, pp. 197–209. Springer, Cham (2020). https://doi.org/10.1007/978-3-030-47426-3_16

17. Wang, D., Hu, W., Cao, E., Sun, W.: Global-to-local neural networks for document-level relation extraction. In: Proceedings of the 2020 Conference on Empirical Methods in Natural Language Processing (EMNLP), pp. 3711–3721 (2020)

18. Wang, H., Qin, K., Lu, G., Yin, J., Zakari, R.Y., Owusu, J.W.: Document-level relation extraction using evidence reasoning on RST-graph. Knowl.-Based Syst. **228**, 107274 (2021)

19. Wang, H., Focke, C., Sylvester, R., Mishra, N., Wang, W.: Fine-tune Bert for docred with two-step process. arXiv preprint arXiv:1909.11898 (2019)

20. Xu, B., Wang, Q., Lyu, Y., Zhu, Y., Mao, Z.: Entity structure within and throughout: Modeling mention dependencies for document-level relation extraction. In: Proceedings of the AAAI Conference on Artificial Intelligence, vol. 35, pp. 14149–14157 (2021)
21. Yao, Y., et al.: Docred: A large-scale document-level relation extraction dataset. In: Proceedings of the 57th Annual Meeting of the Association for Computational Linguistics, pp. 764–777 (2019)
22. Ye, D., et al.: Coreferential reasoning learning for language representation. In: Proceedings of the 2020 Conference on Empirical Methods in Natural Language Processing (EMNLP), pp. 7170–7186 (2020)
23. Zeng, S., Xu, R., Chang, B., Li, L.: Double graph based reasoning for document-level relation extraction. In: Proceedings of the 2020 Conference on Empirical Methods in Natural Language Processing (EMNLP), pp. 1630–1640 (2020)
24. Zhang, Z., et al.: Document-level relation extraction with dual-tier heterogeneous graph. In: Proceedings of the 28th International Conference on Computational Linguistics, pp. 1630–1641 (2020)
25. Zhou, H., Xu, Y., Yao, W., Liu, Z., Lang, C., Jiang, H.: Global context-enhanced graph convolutional networks for document-level relation extraction. In: Proceedings of the 28th International Conference on Computational Linguistics, pp. 5259–5270 (2020)
26. Zhou, W., Huang, K., Ma, T., Huang, J.: Document-level relation extraction with adaptive thresholding and localized context pooling. In: Proceedings of the AAAI Conference on Artificial Intelligence, vol. 35, pp. 14612–14620 (2021)

Weight Prediction Boosts the Convergence of AdamW

Lei Guan(✉) (iD)

Department of Mathematics, National University of Defense Technology,
Changsha, China
guanleimath@163.com

Abstract. In this paper, we introduce weight prediction into the AdamW optimizer to boost its convergence when training the deep neural network (DNN) models. In particular, ahead of each mini-batch training, we predict the future weights according to the update rule of AdamW and then apply the predicted future weights to do both forward pass and backward propagation. In this way, the AdamW optimizer always utilizes the gradients w.r.t. the future weights instead of current weights to update the DNN parameters, making the AdamW optimizer achieve better convergence. Our proposal is simple and straightforward to implement but effective in boosting the convergence of DNN training. We performed extensive experimental evaluations on image classification and language modeling tasks to verify the effectiveness of our proposal. The experimental results validate that our proposal can boost the convergence of AdamW and achieve better accuracy than AdamW when training the DNN models.

Keywords: Deep learning · Weight prediction · Convergence · AdamW

1 Introduction

The optimization of deep neural network models is to find the optimal parameters using an optimizer which has a decisive influence on the convergence of the models and thus directly affects the total training time. Adaptive gradient methods, such as RMSprop [20], AdaGrad [3], Adam [7] and AdamW [11], are currently of core practical importance in deep learning training as they are able to attain rapid training of modern deep learning models. Particularly, AdamW [11], also known as Adam with decoupled weight decay, has been used as a default optimizer for training various DNN models [1,10,11,20,21]. The major advantage of AdamW lies in that it improves the generalization performance of Adam [7] and thus works as effectively as SGD with momentum [18] on image classification tasks.

As with other popular gradient-based optimization methods, when using AdamW as an optimizer, each iteration of DNN training, *i.e.*, a mini-batch training, generally consists of one forward pass and one backward propagation, where the gradients w.r.t. all the parameters (also known as weights) are computed during the backward propagation. The generated gradients are then used

© The Author(s), under exclusive license to Springer Nature Switzerland AG 2023
H. Kashima et al. (Eds.): PAKDD 2023, LNAI 13935, pp. 329–340, 2023.
https://doi.org/10.1007/978-3-031-33374-3_26

by the AdamW optimizer to calculate the update values for all parameters, which are finally applied to updating the weights. The remarkable features of using AdamW to update parameters include: 1) the updates of weights are continuous; 2) each mini-batch uses the currently available weights to do both forward pass and backward propagation.

Motivated by the fact that DNN weights are updated in a continuous manner and the update values calculated by the AdamW should reflect the "correct" direction for updating the weights, we introduce weight prediction [2,4] into the DNN training to further boost the convergence of AdamW. Concretely, ahead of each mini-batch training, we first perform weight prediction according to the currently available weights and the update rule of AdamW. Following that, we use the predicted future weights instead of current weights to perform both forward pass and backward propagation. Finally, the AdamW optimizer utilizes the gradients w.r.t. the predicted weights to update the DNN parameters. We experiment with two typical machine learning tasks, including image classification and language modeling. The experimental results demonstrate that our proposal outperforms AdamW in terms of convergence and accuracy. For instance, when training four convolution neural network (CNN) models on CIFAR-10 dataset, our proposal yields an average accuracy improvement of 0.47% (up to 0.74%) over AdamW. When training LSTMs on Penn TreeBank dataset, our proposal achieves 5.52 less perplexity than AdamW on average.

The contributions of this paper can be summarized as follows:

(1) We, for the first time, construct the mathematical relationship between currently available weights and future weights after several continuous updates when using AdamW as an optimizer.
(2) We devise an effective way to incorporate weight prediction into AdamW. To the best of our knowledge, this is the first time that uses weight prediction strategy to boost the convergence of AdamW. The proposed weight prediction strategy is believed to be well suited for other popular optimization methods such as RMSprop [20], AdaGrad [3], Adam [7], et al.
(3) We conducted extensive experimental evaluations to validate the effectiveness of our proposal, which demonstrates that our proposal is able to boost the convergence of AdamW when training the DNN models.

2   Related Work

When using the gradient-based optimization methods to train DNN models, the differences in optimization methods lie in that the ways using gradients to update model parameters are different. Generally, the commonly used first-order gradient methods can be categorized into two groups: the accelerated stochastic gradient descent (SGD) family [15,16,18] and adaptive gradient methods [7,23, 24].

Adaptive gradient methods, also known as adaptive learning methods, have been heavily studied in prior research and widely used in deep learning training. Very different from the SGD methods (e.g., Momentum SGD [18]), which use a unified learning rate for all parameters, adaptive gradient methods compute a specific learning rate for each individual parameter [24]. In 2011, Duchi et al. [3] proposed the AdaGrad, which can dynamically adjust the learning rate according to the history gradients from previous iterations and utilize the quadratic sum of all previous gradients to update the model parameters. Zeiler [23] proposed AdaDelta, seeking to alleviate the continual decay of the learning rate of Ada-Grad. AdaDelta does not require manual tuning of a learning rate and is robust to noisy gradient information. Tieleman and Hinton [20] refined AdaGrad and proposed RMSprop. The same as AdaGrad, RMSprop adjusts the learning rate via element-wise computation and then updates the variables. One remarkable feature of RMSprop is that it can avoid decaying the learning rate too quickly. In order to combine the advantages of both AdaGrad and RMSprop, Kingma and Ba [7] proposed another famous adaptive gradient method, Adam, which has become an extremely important choice for deep learning training. Loshchilov and Hutter [11] found that the major factor of the poor generalization of Adam is due to that L_2 regularization for it is not as effective as for its competitor, the Momentum SGD. They thus proposed decoupled weight decay regularization for Adam, which is also known as AdamW. The experimental results demonstrate that AdamW substantially improves the generalization performance of Adam and illustrates competitive performance as Momentum SGD [18] when tackling image classification tasks. To simultaneously achieve fast convergence and good generalization, Zhuang et al. [24] proposed another adaptive gradient method called AdaBelief, which adapts the stepsize according to the "belief" in the current gradient direction. Other adaptive gradient methods include AdaBound [12], RAdam [9], Yogi [22], et al. It is worth noting that all these adaptive gradient methods share a common feature: weight updates are continuous and each mini-batch training always uses currently available weights to perform both forward pass and backward propagation.

Weight prediction was previously used to overcome the weight inconsistency issue in the asynchronous pipeline parallelism. Chen et al. [2] used the smoothed gradient to replace the true gradient in order to predict future weights when using Momentum SGD [18] as the optimizer. Guan et al. [4] proposed using the update values of Adam [7] to make weight predictions. Yet, both approaches use weight prediction to ensure the weight consistency of pipeline training rather than considering the impact of weight prediction on the optimizers themselves.

3   Methods

Ahead of any t-th $(t \geq 1)$ iteration, we assume that the current available DNN weights are $\boldsymbol{\theta}_{t-1}$. Given the initial learning rate $\gamma \in \mathbb{R}$, momentum factor $\beta_1 \in \mathbb{R}$ and $\beta_2 \in \mathbb{R}$, and weight decay value $\lambda \in \mathbb{R}$, we reformulate the update of

AdamW [11] as

$$\boldsymbol{\theta}_t = (1 - \gamma\lambda)\boldsymbol{\theta}_{t-1} - \frac{\gamma\hat{\mathbf{m}}_t}{\sqrt{\hat{\mathbf{v}}_t} + \epsilon},$$

$$\text{s.t.} \begin{cases} \mathbf{g}_t = \nabla_\theta f_t(\boldsymbol{\theta}_{t-1}), \\ \mathbf{m}_t = \beta_1 \cdot \mathbf{m}_{t-1} + (1 - \beta_1) \cdot \mathbf{g}_t, \\ \mathbf{v}_t = \beta_2 \cdot \mathbf{v}_{t-1} + (1 - \beta_2) \cdot \mathbf{g}_t^2, \\ \hat{\mathbf{m}}_t = \frac{\mathbf{m}_t}{1-\beta_1^t}, \\ \hat{\mathbf{v}}_t = \frac{\mathbf{v}_t}{1-\beta_2^t}. \end{cases} \tag{1}$$

In (1), $\mathbf{m}_t$ and $\mathbf{v}_t$ refer to the first and second moment vector respectively, ϵ is the smoothing term which can prevent division by zero.

Letting $\boldsymbol{\theta}_0$ denote the initial weights of a DNN model, then in the following s times of continuous mini-batch training, the DNN weights are updated via

$$\boldsymbol{\theta}_1 = (1 - \gamma\lambda)\boldsymbol{\theta}_0 - \frac{\gamma\hat{\mathbf{m}}_1}{\sqrt{\hat{\mathbf{v}}_1} + \epsilon},$$

$$\boldsymbol{\theta}_2 = (1 - \gamma\lambda)\boldsymbol{\theta}_1 - \frac{\gamma\hat{\mathbf{m}}_2}{\sqrt{\hat{\mathbf{v}}_2} + \epsilon}, \tag{2}$$

$$\cdots$$

$$\boldsymbol{\theta}_s = (1 - \gamma\lambda)\boldsymbol{\theta}_{s-1} - \frac{\gamma\hat{\mathbf{m}}_s}{\sqrt{\hat{\mathbf{v}}_s} + \epsilon},$$

where for any $i \in \{1, 2, \cdots, s\}$, we have

$$\begin{cases} \mathbf{g}_i = \nabla_\theta f_i(\boldsymbol{\theta}_{i-1}), \\ \mathbf{m}_i = \beta_1 \cdot \mathbf{m}_{i-1} + (1 - \beta_1) \cdot \mathbf{g}_i, \\ \mathbf{v}_i = \beta_2 \cdot \mathbf{v}_{i-1} + (1 - \beta_2) \cdot \mathbf{g}_i^2, \\ \hat{\mathbf{m}}_i = \frac{\mathbf{m}_i}{1-\beta_1^i}, \\ \hat{\mathbf{v}}_i = \frac{\mathbf{v}_i}{1-\beta_2^i}. \end{cases} \tag{3}$$

When summing up all weight update equations in (2), we have

$$\boldsymbol{\theta}_s = \boldsymbol{\theta}_0 - \gamma\lambda \sum_{i=0}^{s-1} \boldsymbol{\theta}_i - \sum_{i=1}^{s} \frac{\gamma\hat{\mathbf{m}}_i}{\sqrt{\hat{\mathbf{v}}_i} + \epsilon},$$

$$\text{s.t.} \begin{cases} \mathbf{g}_i = \nabla_\theta f_i(\boldsymbol{\theta}_{i-1}), \\ \mathbf{m}_i = \beta_1 \cdot \mathbf{m}_{i-1} + (1 - \beta_1) \cdot \mathbf{g}_i, \\ \mathbf{v}_i = \beta_2 \cdot \mathbf{v}_{i-1} + (1 - \beta_2) \cdot \mathbf{g}_i^2, \\ \hat{\mathbf{m}}_i = \frac{\mathbf{m}_i}{1-\beta_1^i}, \\ \hat{\mathbf{v}}_i = \frac{\mathbf{v}_i}{1-\beta_2^i}. \end{cases} \tag{4}$$

It is well known that the weight decay value λ is generally set to an extremely small value (e.g., $5e^{-4}$), and the learning rate γ is commonly set to a value smaller than 1 (e.g., 0.01). Consequently, $\gamma\lambda$ is pretty close to zero, and thus, the second term of the right hand of (4) can be neglected. This, therefore, generates the following equation:

$$\boldsymbol{\theta}_s \approx \boldsymbol{\theta}_0 - \sum_{i=1}^{s} \frac{\gamma \hat{\mathbf{m}}_i}{\sqrt{\hat{\mathbf{v}}_i} + \epsilon}. \tag{5}$$

(5) illustrates that given the initial weights $\boldsymbol{\theta}_0$, the weights after s times of continuous updates can be approximately calculated. Correspondingly, given $\boldsymbol{\theta}_t$, the weights after s times of continuous updates can be approximately calculated via

$$\boldsymbol{\theta}_{t+s} \approx \boldsymbol{\theta}_t - \sum_{i=t+1}^{t+s} \frac{\gamma \hat{\mathbf{m}}_i}{\sqrt{\hat{\mathbf{v}}_i} + \epsilon}. \tag{6}$$

From (6), we see that given the initial weights $\boldsymbol{\theta}_t$, $\boldsymbol{\theta}_{t+s}$ can be approximately calculated by letting $\boldsymbol{\theta}_t$ subtract the sum of s continuous relative variation of the weights. Note that the relative increments of the weights in each iteration should reflect the trend of the weight updates in each iteration. In (6), $\frac{\gamma \hat{\mathbf{m}}_i}{\sqrt{\hat{\mathbf{v}}_i} + \epsilon}$ should reflect the "correct" direction for updating the weights $\boldsymbol{\theta}_t$ as it is calculated by the AdamW, and the weights are updated in a continuous manner and along the way of inertia directions.

We can therefore replace $\sum_{i=t+1}^{t+s} \frac{\gamma \hat{\mathbf{m}}_i}{\sqrt{\hat{\mathbf{v}}_i} + \epsilon}$ in (6) with $s \frac{\gamma \hat{\mathbf{m}}_{t+1}}{\sqrt{\hat{\mathbf{v}}_{t+1}} + \epsilon}$ in an effort to approximately predict $\boldsymbol{\theta}_{t+s}$ for the case when only $\boldsymbol{\theta}_t$, $\mathbf{g}_t$ and the weight prediction steps s are available. Note that at any t-th iteration, the gradients of stochastic objective, i.e., $\mathbf{g}_t = \nabla_t(\boldsymbol{\theta}_{t-1})$, can be calculated when the backward propagation is completed. Letting $\hat{\boldsymbol{\theta}}_{t+s}$ denote the approximately predicted weights for $\boldsymbol{\theta}_{t+s}$, we can construct the mathematical relationship between $\boldsymbol{\theta}_t$ and $\hat{\boldsymbol{\theta}}_{t+s}$ as

$$\hat{\boldsymbol{\theta}}_{t+s} = \boldsymbol{\theta}_t - s \frac{\gamma \hat{\mathbf{m}}_{t+1}}{\sqrt{\hat{\mathbf{v}}_{t+1}} + \epsilon},$$

$$\text{s.t.} \begin{cases} \mathbf{g}_t = \nabla_{\boldsymbol{\theta}} f_t(\boldsymbol{\theta}_{t-1}), \\ \mathbf{m}_t = \beta_1 \cdot \mathbf{m}_{t-1} + (1 - \beta_1) \cdot \mathbf{g}_t, \\ \mathbf{v}_t = \beta_2 \cdot \mathbf{v}_{t-1} + (1 - \beta_2) \cdot \mathbf{g}_t^2, \\ \hat{\mathbf{m}}_t = \frac{\mathbf{m}_t}{1 - \beta_1^t}, \\ \hat{\mathbf{v}}_t = \frac{\mathbf{v}_t}{1 - \beta_2^t}. \end{cases} \tag{7}$$

In the following, we showcase how to incorporate weight prediction into the DNN training when using AdamW [11] as an optimizer. Algorithm 1 illustrates the detailed information. The weight prediction step s and other hyperparameters are required ahead of the DNN training. At each iteration, the current available weights $\boldsymbol{\theta}_t$ should be cached before the forward pass starts (Line 4). Then weight prediction are performed using (7) and the predicted weights $\hat{\boldsymbol{\theta}}_{t+s}$ is generated (Line 5). Following that, the predicted weights $\hat{\boldsymbol{\theta}}_{t+s}$ are used to do both forward pass and backward propagation (Lines 6 and 7). Finally, the cached weights $\boldsymbol{\theta}_t$ is recovered and updated using the AdamW optimizer (Lines 8 and 9).

Algorithm 1. Weight prediction for AdamW

Require: Weight prediction step s, other hyper-parameters such as γ, β_1, β_2, γ, ϵ.
1: Initialize or load DNN weights $\boldsymbol{\theta}_0$.
2: $t \leftarrow 1$.
3: **while** stopping criterion is not met **do**
4:    Cache the current weights $\boldsymbol{\theta}_t$.
5:    Calculate $\hat{\boldsymbol{\theta}}_{t+s}$ using (7).
6:    Do forward pass with $\hat{\boldsymbol{\theta}}_{t+s}$.
7:    Do backward propagation with $\hat{\boldsymbol{\theta}}_{t+s}$.
8:    Recover the cached weights $\boldsymbol{\theta}_t$.
9:    Update the weights $\boldsymbol{\theta}_t$ using the AdamW optimizer.
10:    $t \leftarrow t + 1$.
11: **end while**

---

4 Experiments

4.1 Experiment Settings

In this section, we mainly compare our proposal with AdamW [11]. We evaluated our proposal with three different weight prediction steps (*i.e.*, $s = 1$, $s = 2$, and $s = 3$), which were respectively denoted as Ours-S1, Ours-S2, and Ours-S3 for convenience purposes. We conducted experimental evaluations on two different machine learning tasks: image classification on the CIFAR-10 [8] dataset with four CNN models and language modeling on Penn TreeBank [14] dataset with two LSTM [13] models. All the experiments were conducted on a multi-GPU platform which is equipped with four NVIDIA Tesla P100 GPUs, each with 16 GB of memory size. The CPU on the platform is Intel Xeon E5-2680 with 128 GB DDR4-2400 off-chip main memory.

The CIFAR-10 dataset totally includes 60k 32×32 images, 50k images for training, and 10k images for validation. The Penn TreeBank dataset consists of 929k training words, 73k validation words as well as 82k test words. For image classification, the used CNN models are VGG-11 [17], ResNet-34 [5], DenseNet-121 [6], and Inception-V3 [19]. For language modeling, we trained the LSTM models with two sizes: 1-layer LSTM and 2-layer LSTM. Each layer was configured with 650 units and was applied 50% dropout on the non-recurrent connections.

We trained all CNN models for 120 epochs with a mini-batch size of 128. The learning rate was initialized as $1e^{-4}$, and divided by ten at the 90th epoch. For training 1-layer and 2-layer LSTM models, we set the size of each mini-batch to 20. We trained both LSTM models for 100 epochs with an initial learning rate of 0.01 and decreased the learning rate by a factor of 10 at the 60th and 80th epochs. For AdamW [11] and our proposal, we always evaluated them with the default parameters, *i.e.*, $\beta_1 = 0.9$, $\beta_2 = 0.999$, and $\epsilon = 10^{-8}$. The weight decay for both approaches was set to $\lambda = 5e^{-4}$.

4.2   CNNs on CIFAR-10

In this section, we report the experimental results when training four CNN models on the CIFAR-10 dataset. Table 1 summarizes the maximum validation top-1 accuracy and Table 2 presents the minimum validation loss. Figure 1 depicts the learning curves of validation accuracy vs. epochs for training CNNs using AdamW, Ours-S1, Ours-S2, and Ours-S3, respectively. The learning curves about validation loss vs. epochs are shown in Fig. 2.

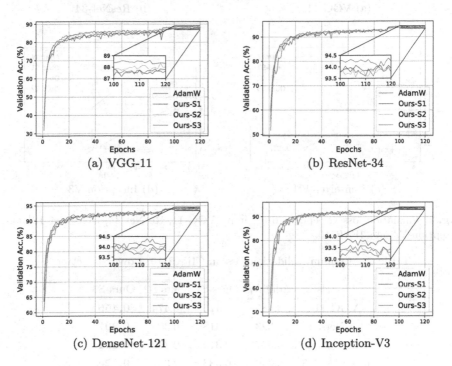

(a) VGG-11   (b) ResNet-34

(c) DenseNet-121   (d) Inception-V3

Fig. 1. Validation accuracy vs. epochs of training VGG-11, ResNet-34, DenseNet-121 and Inception-V3 on CIFAR-10.

Table 1. Maximum validation top-1 accuracy on CIFAR-10. **Higher** is better.

Models	AdamW	Ours-S1	Ours-S2	Ours-S3
VGG-11	87.85%	87.83%	88.34%	**88.59%**
ResNet-34	94.03%	94.06%	93.95%	**94.37%**
DenseNet-121	93.97%	94.13%	94.04%	**94.39%**
Inception-V3	93.53%	**93.90%**	93.60%	93.61%

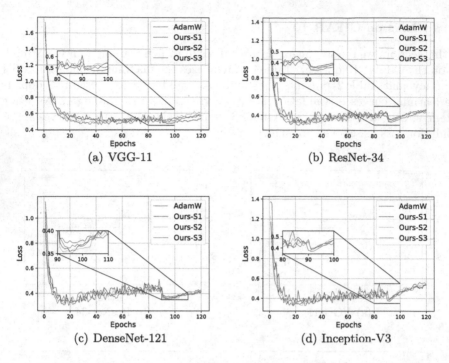

Fig. 2. Validation loss vs. epochs of training VGG-11, ResNet-34, DenseNet-121 and Inception-V3 on CIFAR-10.

Table 2. Minimum validation loss on CIFAR-10. **Lower** is better.

Models	AdamW	Ours-S1	Ours-S2	Ours-S3
VGG-11	0.485	0.475	0.474	**0.456**
ResNet-34	0.323	0.318	0.305	**0.297**
DenseNet-121	0.327	0.319	0.319	**0.302**
Inception-V3	0.346	0.333	0.331	**0.324**

Based on the observation of Table 1 and Fig. 1, we can immediately reach the following conclusions. First, Fig. 1 shows that the learning curves of our proposal with different weight prediction steps match well with that of AdamW but converge faster than that of AdamW, especially at the beginning of training epochs. The learning curves in Figs. 1(a), 1(b), 1(b), and 1(d) also illustrate that our proposal generally attains higher validation accuracy than AdamW at the end of the training. Second, Table 1 shows that our proposal outperforms AdamW on all evaluated CNN models in terms of the obtained maximum validation accuracy. In particular, our proposal achieves consistently higher validation top-1 accuracy than AdamW. Compared to AdamW, our proposal achieves 0.74%, 0.34%, 0.42%, and 0.37% for training VGG-11, ResNet-34, DenseNet-121, and Inception-V3, respectively. On average, our proposal yields 0.47% (up to 0.74%)

top-1 accuracy improvement over AdamW. Third, comparing the experimental results of Ours-S1, Ours-S2, and Ours-S3, we can see that our proposal with different weight prediction steps consistently gets good results, which demonstrates that the performance of our proposal is independent of the settings of the weight prediction step. Particularly, the experimental results show that Ours-S3 works the best for VGG-11, ResNet-34, and DenseNet-121, while Ours-S1 works the best for Inception-V3. Similar conclusions can be drawn from the observation of Table 2 and Fig. 2. Our proposal consistently obtains less validation loss than AdamW which again verifies that weight prediction can boost the convergence of AdamW when training DNN models.

4.3 LSTMs on Penn TreeBank

In this section, we report the experimental results when training 1-layer and 2-layer LSTM models on the Penn TreeBank dataset [14]. Figures 3 and 4 depict the learning curves. Table 3 presents the obtained minimum perplexity (lower is better), and Table 4 summarizes the obtained minimum validation loss (lower is better).

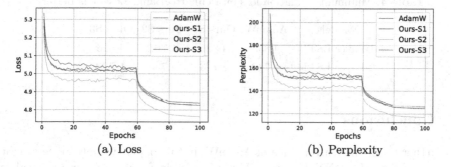

(a) Loss   (b) Perplexity

Fig. 3. Training 1-layer LSTM on Penn TreeBank. Left: Loss vs. epochs; Right: Perplexity vs. epochs.

We can draw the following conclusions from the experiment results. First, as shown in Table 3, for both 1-layer and 2-layer LSTM models, our proposal achieves lower perplexity and validation loss than AdamW, validating the fast convergence and good accuracy of our proposal. Second, for 1-layer LSTM, our proposal with $s = 2$ yields 9.22 less perplexity than AdamW. For 2-layer LSTM, our proposal with $s = 3$ yields 2.02 less perplexity than AdamW. On average, our proposal achieves 5.52 less perplexity than AdamW. Second, similar conclusions can be drawn based on the observation of the loss vs. epochs learning curves in Figs. 3(a) and 4(a) and Table 4. Our proposal consistently achieves less validation loss than AdamW, again validating that weight prediction can boost the convergence of AdamW.

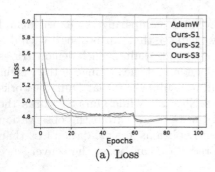 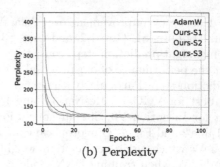

(a) Loss　　　　　　　　　　　　　(b) Perplexity

Fig. 4. Training 2-layer LSTM on Penn TreeBank. Left: Loss vs. epochs; Right: Perplexity vs. epochs.

Table 3. Minimum perplexity on Penn TreeBank. **Lower** is better.

Models	AdamW	Ours-S1	Ours-S2	Ours-S3
1-Layer LSTM	126.21	124.75	**116.99**	124.71
2-Layer LSTM	114.68	114.80	113.99	**112.64**

Table 4. Minimum validation loss on Penn TreeBank. **Lower** is better.

Models	AdamW	Ours-S1	Ours-S2	Ours-S3
1-Layer LSTM	4.838	4.826	**4.762**	4.826
2-Layer LSTM	4.742	4.743	4.736	**4.724**

5 Conclusions

To further boost the convergence of AdamW, in this paper, we introduce weight prediction into the DNN training. The remarkable feature of our proposal is that we perform both forward pass and backward propagation using the future weights which are predicted according to the update of AdamW. In particular, we construct the mathematical relationship between current weights and future weights and devise an effective way to incorporate weight prediction into DNN training. Our proposal is easy to implement and works well in boosting the convergence of DNN training. The experimental results on image classification and language modeling tasks verify the effectiveness of our proposal.

The weight prediction should also work well for other adaptive optimization methods such as RMSprop [20], AdaGrad [3], and Adam [7] et al. when training the DNN models. For future work, we would like to apply weight prediction to those popular optimization methods.

References

1. Bai, Y., Mei, J., Yuille, A.L., Xie, C.: Are transformers more robust than CNNs? In: Advances in Neural Information Processing Systems, vol. 34, pp. 26831–26843 (2021)
2. Chen, C.C., Yang, C.L., Cheng, H.Y.: Efficient and robust parallel DNN training through model parallelism on multi-GPU platform. arXiv preprint arXiv:1809.02839 (2018)
3. Duchi, J., Hazan, E., Singer, Y.: Adaptive subgradient methods for online learning and stochastic optimization. J. Mach. Learn. Res. **12**(7) (2011)
4. Guan, L., Yin, W., Li, D., Lu, X.: Xpipe: efficient pipeline model parallelism for multi-GPU DNN training. arXiv preprint arXiv:1911.04610 (2019)
5. He, K., Zhang, X., Ren, S., Sun, J.: Deep residual learning for image recognition. In: Proceedings of the IEEE Conference on Computer Vision and Pattern Recognition, pp. 770–778 (2016)
6. Huang, G., Liu, Z., Van Der Maaten, L., Weinberger, K.Q.: Densely connected convolutional networks. In: Proceedings of the IEEE Conference on Computer Vision and Pattern Recognition, pp. 4700–4708 (2017)
7. Kingma, D.P., Ba, J.: Adam: a method for stochastic optimization. arXiv preprint arXiv:1412.6980 (2014)
8. Krizhevsky, A., Hinton, G., et al.: Learning multiple layers of features from tiny images (2009)
9. Liu, L., et al.: On the variance of the adaptive learning rate and beyond. arXiv preprint arXiv:1908.03265 (2019)
10. Liu, Z., et al.: Swin transformer: hierarchical vision transformer using shifted windows. In: Proceedings of the IEEE/CVF International Conference on Computer Vision, pp. 10012–10022 (2021)
11. Loshchilov, I., Hutter, F.: Decoupled weight decay regularization. arXiv preprint arXiv:1711.05101 (2017)
12. Luo, L., Xiong, Y., Liu, Y., Sun, X.: Adaptive gradient methods with dynamic bound of learning rate. arXiv preprint arXiv:1902.09843 (2019)
13. Ma, X., Tao, Z., Wang, Y., Yu, H., Wang, Y.: Long short-term memory neural network for traffic speed prediction using remote microwave sensor data. Transport. Res. Part C: Emerg. Technol. **54**, 187–197 (2015)
14. Marcinkiewicz, M.A.: Building a large annotated corpus of English: the PENN treebank. Using Large Corpora **273** (1994)
15. Nesterov, Y.: A method of solving a convex programming problem with convergence rate mathcal {O}(1/k {2}). In: Sov. Math. Dokl. vol. 27
16. Polyak, B.T.: Some methods of speeding up the convergence of iteration methods. USSR Comput. Math. Math. Phys. **4**(5), 1–17 (1964)
17. Simonyan, K., Zisserman, A.: Very deep convolutional networks for large-scale image recognition. arXiv preprint arXiv:1409.1556 (2014)
18. Sutskever, I., Martens, J., Dahl, G., Hinton, G.: On the importance of initialization and momentum in deep learning. In: International Conference on Machine Learning, pp. 1139–1147. PMLR (2013)
19. Szegedy, C., Vanhoucke, V., Ioffe, S., Shlens, J., Wojna, Z.: Rethinking the inception architecture for computer vision. In: Proceedings of the IEEE Conference on Computer Vision and Pattern Recognition, pp. 2818–2826 (2016)
20. Tieleman, T., Hinton, G.: Lecture 6.5-rmsprop, coursera: neural networks for machine learning. Technical report 6, University of Toronto (2012)

21. Wang, P., et al.: Scaled ReLU matters for training vision transformers. In: Proceedings of the AAAI Conference on Artificial Intelligence, vol. 36, pp. 2495–2503 (2022)
22. Zaheer, M., Reddi, S., Sachan, D., Kale, S., Kumar, S.: Adaptive methods for nonconvex optimization. In: Advances in Neural Information Processing Systems, vol. 31, pp. 9815–9825 (2018)
23. Zeiler, M.D.: Adadelta: an adaptive learning rate method. arXiv preprint arXiv:1212.5701 (2012)
24. Zhuang, J., et al.: Adabelief optimizer: adapting stepsizes by the belief in observed gradients. Adv. Neural. Inf. Process. Syst. **33**, 18795–18806 (2020)

Model-Agnostic Reachability Analysis on Deep Neural Networks

Chi Zhang[1], Wenjie Ruan[1]([✉]), Fu Wang[1], Peipei Xu[2], Geyong Min[1], and Xiaowei Huang[2]

[1] Department of Computer Science, University of Exeter, Exeter, UK
{cz338,w.ruan,fw377,g.min}@exeter.ac.uk
[2] Department of Computer Science, University of Liverpool, Liverpool, UK
{peipei.xu,xiaowei.huang}@liverpool.ac.uk

Abstract. Verification plays an essential role in the formal analysis of safety-critical systems. Most current verification methods have specific requirements when working on Deep Neural Networks (DNNs). They either target one particular network category, e.g., Feedforward Neural Networks (FNNs), or networks with specific activation functions, e.g., ReLU. In this paper, we develop a model-agnostic verification framework, called DeepAgn, and show that it can be applied to FNNs, Recurrent Neural Networks (RNNs), or a mixture of both. Under the assumption of Lipschitz continuity, DeepAgn analyses the reachability of DNNs based on a novel optimisation scheme with a global convergence guarantee. It does not require access to the network's internal structures, such as layers and parameters. Through reachability analysis, DeepAgn can tackle several well-known robustness problems, including computing the maximum safe radius for a given input, and generating the ground-truth adversarial example. We also empirically demonstrate DeepAgn's superior capability and efficiency in handling a broader class of deep neural networks, including both FNNs and RNNs with very deep layers and millions of neurons, than other state-of-the-art verification approaches. Our tool is available at https://github.com/TrustAI/DeepAgn

Keywords: Verification · Deep Learning · Model-agnostic · Reachability

1 Introduction

DNNs, or systems with neural network components, are widely applied in many applications such as image processing, speech recognition, and medical diagnosis [10]. However, DNNs are vulnerable to adversarial examples [12,25,33]. It is vital to analyse the safety and robustness of DNNs before deploying them in practice, particularly in safety-critical applications.

Supplementary Information The online version contains supplementary material available at https://doi.org/10.1007/978-3-031-33374-3_27.

The research on evaluating the robustness of DNNs mainly falls into two categories: falsification-based and verification-based approaches. While falsification approaches (e.g. adversarial attacks) [11] can effectively find adversarial examples, they cannot provide theoretical guarantees. Verification techniques, on the other hand, can rigorously prove the robustness of deep learning systems with guarantees [12,13,18,19,24]. Some researchers propose to reduce the safety verification problems to constraint satisfaction problems that can be tackled by constraint solvers such as Mixed-Integer Linear Programming (MILP) [1], Boolean Satisfiability (SAT) [20], or Satisfiability Modulo Theories (SMT) [15]. Another popular technique is to apply search algorithms [13] or Monte Carlo tree search [32] over discretised vector spaces on the inputs of DNNs. To improve the efficiency, these methods can also be combined with a heuristic searching strategy to search for a counter-example or an activation pattern that satisfies certain constraints, such as SHERLOCK [4] and Reluplex [15]. Nevertheless, the study subjects of these verification methods are restricted. They either target specific layers (e.g., fully-connected or convolutional layers), have restrictions on activation functions (e.g., ReLU activation only), or are only workable on a specific neural network structure (e.g., feedforward neural networks). Particularly, in comparison to FNNs, verification on RNNs is still in its infancy, with only a handful of representative works available, including [14,16,34]. The adoption of [16] requires short input sequences, and [14,34] can result in irresolvable over-approximation error.

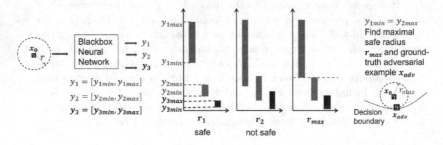

Fig. 1. Illustration of DeepAgn working on a black-box three-output neural network. *In reachability problem, given a set of inputs (quantified by a predefined L_p-norm ball) and a well-trained black-box neural network, DeepAgn can calculate the output range, namely, the minimal and maximum output confidence of each label (i.e., $[y_{1min}, y_{1max}]$, $[y_{2min}, y_{2max}]$, and $[y_{3min}, y_{3max}]$). For the safety verification problem, we can use a binary search upon the reachability to find the maximum safe radius r_{max} where the confidence intervals of the original label y_1 and target label y_2 meet.*

This paper proposes a novel model-agnostic solution for safety verification on both feedforward and recurrent neural networks without suffering from the above weaknesses. Figure 1 outlines the working principle of DeepAgn, demonstrating its safety evaluation process and the calculation of the maximum safety radius. To the best of our knowledge, DeepAgn is one of the pioneering attempts on *model-agnostic* verification that can work on both modern feedforward and recurrent neural networks under a unified framework. DeepAgn can deal with DNNs with very deep layers, a large number of neurons, and any type of activation function, via a black-box manner (without access to the internal structures/parameters of the network). Our contributions are summarised below:

- To theoretically justify the applicability of DeepAgn, we prove that recurrent neural networks are also Lipschitz continuous for bounded inputs.

Table 1. Comparison with other verification techniques from different aspects

	Guarantees	Core Techniques	Neural Network Types	Model Agnostic	Exact Computation	Model Access
Reluplex [15]	Deterministic	SMT+LP	ReLu-based FNNs	✗	✓	Model parameters
Planet [5]	Deterministic	SAT+LP	ReLU-based FNNs	✗	✓	Model parameters
AI2 [6]	Upper bound	Abstract Interpretation	ReLU-based FNNs	✗	✗	Model Parameters
ConDual [31]	Upper bound	Convex relaxation	ReLU-based FNNs	✗	✗	Model parameters
DeepGO [24]	Converging bound	Lipschitz Optimisation	FNNs with Lipschitz continuous layers (ReLU, Sigmoid, Tanh, etc.)	✗	✓	Confidence values
FastLip [29]	Upper bound	Lipschitz estimation	ReLU-based FNNs	✗	✗	Model parameters
DeepGame [32]	Approximated converging bound	Search based	ReLU/Tanh/Sigmod based FNNs	✗	✓	Confidence values
POPQORN [16]	Upper bound	Unrolling	RNNs, LSTMs, GRUs	✗	✗	Model parameters
RnnVerify [14]	Upper bound	Invariant Inference	RNNs	✗	✗	Model parameters
VERRNN [34]	Upper bound	Unrolling+MILP	RNNs	✗	✗	Model parameters
DeepAgn	Converging bound	Lipschitz Optimisation	FNNs (CNNs), RNNs, Hybrid networks with Lipschitz continuous layers	✓	✓	Confidence values

- We develop an efficient method for reachability analysis on DNNs. We demonstrate that this *generic* and *unified* model-agnostic verification framework can work on FNNs, RNNs, and a hybrid of both. DeepAgn is an anytime algorithm, i.e., it can return both intermediate lower and upper bounds that are gradually, but strictly, improved as the computation proceeds; and it has provable guarantees, i.e., both the bounds can converge to the optimal value within an arbitrarily small error with provable guarantees.
- Our experiments demonstrate that DeepAgn outperforms the state-of-the-art verification tools in terms of both accuracy and efficiency when dealing with complex, large and hybrid deep learning models.

2 Related Work

Adversarial Attacks. Attacks apply heuristic search algorithms to find adversarial examples. Attacking methods are mainly guided by the forward or cost gradient of the target DNNs. Major approaches include L-BFGS [25], FGSM [11], Carlini & Wagner attack [2], Universal Adversarial Attack [36], etc. Adversarial attacks for FNNs can be applied to cultivate adversarial examples for RNNs with proper adjustments. The concepts of adversarial example and adversarial sequence for RNNs are introduced in [22], in which they concrete adversarial examples for Long Short Term Memory (LSTM) networks. Based on the C&W attack [2], attacks are implemented against DeepSpeech in [3]. The method in [9] is the first approach to analyse and perturb the raw waveform of audio directly.

Verification on DNNs. The recent advances of DNN verification include a layer-by-layer exhaustive search approach [13], methods using constraint solvers [15,23], global optimisation approaches [24,27], and the abstract interpretation approach [6,17]. The properties studied include robustness [13,15,35], or reachability [24], i.e., whether a given output is possible from properties expressible with SMT constraints, or a given output is reachable from a given subspace of inputs. Verification approaches aim to not only find adversarial examples but also provide guarantees on the results obtained. However, efficient verification on large-scale deep neural networks is still an *open problem*. Constraint-based approaches such as Reluplex can only work with a neural network with a few hundred hidden neurons [15,23]. Exhaustive search suffers from the state-space explosion problem [13], although it can be partially alleviated by Monte Carlo tree search [32]. Moreover, the work [1] considers determining whether an output value of a DNN is reachable from a given input subspace. It proposes a MILP-based solution. SHERLORCK [4] studies the range of output values from a given input subspace. This method interleaves local search (based on gradient descent) with global search (based on reduction to MILP). Both approaches can only work with small neural networks.

The research on RNN verification is still relatively new and limited compared with verification on FNNs. Approaches in [16,26,34] start with unrolling RNNs and then use the equivalent FNNs for further analysis. POPQORN [16] is an algorithm to quantify the robustness of RNNs, in which upper and lower planes are introduced to bound the non-linear parts of the estimated neural networks. The authors in [14] introduce invariant inference and over-approximation, transferring the RNN to a simple FNN model, demonstrating better scalability. However, the search for a proper invariant form is not straightforward. In Table 1, we compare DeepAgn with other safety verification works from *six* aspects. Deep-Agn is the *only* model-agnostic verification tool that can verify hybrid networks consisting of both RNN and FNN structures. DeepAgn only requires access to the confidence values of the target model, enabling the verification in a *black-box* manner. Its precision can reach an arbitrarily small (pre-defined) error with a global convergence *guarantee*.

3   Preliminaries

Let $o : [0,1]^m \rightarrow \mathbb{R}$ be a generic function that is Lipschitz continuous. The generic term o is cascaded with the Softmax layer of the neural network for statistically evaluating the outputs of the network. Our problem is to find its upper and lower bounds given the set X' of inputs to the network.

Definition 1 (Generic Reachability of Neural Networks). *Let* $X' \subseteq [0,1]^n$ *be an input subspace and* $f : \mathbb{R}^n \rightarrow \mathbb{R}^m$ *is a neural network. The generic reachability of neural networks is defined as the reachable set* $R(o, X', \epsilon) = [l, u]$ *of network* f *over the generic term* o *under an error tolerance* $\epsilon \geq 0$ *such that*

$$\inf_{x' \in X'} o(f(x')) - \epsilon \leq l \leq \inf_{x' \in X'} o(f(x')) + \epsilon$$

$$\sup_{x' \in X'} o(f(x')) - \epsilon \leq u \leq \sup_{x' \in X'} o(f(x')) + \epsilon \tag{1}$$

We write $u(o, X', \epsilon) = u$ and $l(o, X', \epsilon) = l$ for the upper and lower bound respectively. Then the reachability diameter is $D(o, X', \epsilon) = u(o, X', \epsilon) - l(o, X', \epsilon)$ Assuming these notations, we may write $D(o, X', \epsilon; f)$ if we need to explicitly refer to the network f.

Definition 2 (Safety of Neural Network). *A network f is safe with respect to an input x_0 and an input subspace $X' \subseteq [0, 1]^n$ with $x_0 \in X'$, if*

$$\forall x' \in X' : \arg\max_j c_j(x') = \arg\max_j c_j(x_0) \tag{2}$$

where $c_j(x_0) = f(x_0)_j$ returns N's confidence in classifying x_0 as label j.

Definition 3 (Verified Safe Radius). *Given a neural network $f : \mathbb{R}^n \to \mathbb{R}^m$ and an input sample x_0, a verifier V returns a verified safe radius r_v regarding the safety of neural network. For input x' with $\|x' - x_0\| \leq r_v$, the verifier guarantees that $\arg\max_j c_j(x') = \arg\max_j c_j(x)$. For $\|x' - x_0\| > r$, the verifier either confirms $\arg\max_j c_j(x') \neq \arg\max_j c_j(x)$ or provides an unclear answer.*

Verified safe radius is important merit for robustness analysis, which is adopted by many verification tools such as CLEVER [30] and POPQORN [16]. Verification tools can further determine the safety of the neural network by comparing the verified safe radius r_v and the perturbation radius. A neural network f is determined safe by verifier V with respect to input x_0, if $\|x' - x_0\| \leq r_v$. In Fig. 2, the verification tool V_2 with higher verified $r_2 > r_1$ radius have a higher evaluation accuracy. The sample x_2 is misjudged as unsafe by V_2.

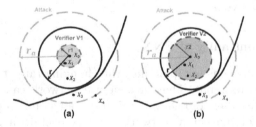

(a)   (b)

Fig. 2. Verification of samples x_1, x_2, x_3, x_4 by different verifiers (a) Verifier V1 (b) Verifier V2, with verified safe radius $r_1 < r_2$. *According to V1, x_1 is safe since $\|x_1 - x_0\| < r_1$. x_2, x_3 and x_4 are determined as unsafe since $\|x_2 - x_0\| > r_1, \|x_3 - x_0\| > r_1, \|x_4 - x_0\| > r_1$. According to V2, x_1, x_2 are safe since $\|x_1 - x_0\| < r_2, \|x_2 - x_0\| < r_2$. x_3, x_4 are determined as unsafe since $\|x_3 - x_0\| > r_2, \|x_4 - x_0\| > r_2$. x_4 is generated by the attack method. Adversarial example x_3 still exists in the attack radius.*

Definition 4 (Maximum Radius of a Safe Norm Ball). *Given a neural network* $f : \mathbb{R}^{m \times n} \to \mathbb{R}^s$, *an distance metric* $\| \cdot \|_D$, *an input* $x_0 \in \mathbb{R}^{m \times n}$, *a norm ball* $B(f, x_0, \| \cdot \|_D, r)$ *is a subspace of* $[a, b]^{m \times n}$ *such that* $B(f, x_0, \| \cdot \|_D, r) = \{x' | \|x' - x_0\|_D \leq r\}$. *When* f *is safe in* $B(f, x_0, \| \cdot \|_D, r)$ *and not safe in any input subspace* $B(f, x_0, \| \cdot \|_D, r')$ *with* $r' > r$, *we call* r *here the maximum radius of a safe norm ball.*

Definition 5 (Successful Attack on Inputs). *Given a neural network* f *and input* x_0, *a* α-*bounded attack* A_α *create input sets* $X' = \{x', \|x' - x_0\| \leq \alpha\}$. A_α *is a successful attack, if an* $x_a \in X'$ *exists, where* $\arg\max_j c_j(x_a) \neq \arg\max_j c_j(x_0)$. *We call* $r_a = \|x_a - x_0\| \leq \alpha$ *the perturbation radius of a successful attack.*

Ideally, the verification solution should provide the maximum radius r of a safe norm ball as the verified safe radius, i.e., the black circle in Fig. 2. However, most sound verifiers can only calculate a lower bound of the maximum safe radius, i.e., a radius that is smaller than r, such as r_1 and r_2. Distinguishing from baseline methods, DeepAgn can estimate the maximum safe radius.

4 Lipschitz Analysis on Neural Networks

This section will theoretically prove that most neural networks, including recurrent neural networks, are Lipschitz continuous. We first introduce the definition of Lipschitz continuity.

Definition 6 (Lipschitz Continuity [21]). *Given two metric spaces* (X, d_X) *and* (Y, d_Y), *where* d_X *and* d_Y *are the metrics on the sets* X *and* Y *respectively, a function* $f : X \to Y$ *is called* Lipschitz continuous *if there exists a real constant* $K \geq 0$ *such that, for all* $x_1, x_2 \in X$: $d_Y(f(x_1), f(x_2)) \leq K d_X(x_1, x_2)$. K *is called the* Lipschitz constant *of* f. *The smallest* K *is called the* Best Lipschitz constant, *denoted as* K_{best}.

4.1 Lipschitz Continuity of FNN

Intuitively, a Lipschitz constant quantifies the changing rate of a function's output with respect to its input. Thus, if a neural network can be proved to be Lipschitz continuous, then Lipschitz continuity can potentially be utilized to bound the output of the neural network with respect to a given input perturbation. The authors in [24,25] demonstrated that deep neural networks with convolutional, max-pooling layer and fully-connected layers with ReLU, Sigmoid activation function, Hyperbolic Tangent, and Softmax activation functions are Lipschitz continuous. According to the chain rule, the composition of Lipschitz continuous functions is still Lipschitz continuous. Thus we can conclude that a majority of deep feedforward neural networks are Lipschitz continuous.

4.2 Lipschitz Analysis on Recurrent Neural Networks

In this paper, we further prove that any recurrent neural network with finite input is Lipschitz continuous. Different from FNNs, RNNs contain feedback loops for processing sequential data, which can be unfolded into FNNs by eliminating loops [10].

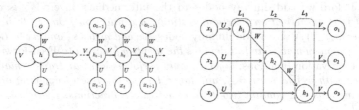

Fig. 3. (a) Unfolded recurrent neural network; (b) A feedforward neural network by unfolding a RNN with input length 3: *The layers are denoted by L_i for $1 \leq i \leq 3$. The node h_2 is located in the middle of layer L_2, taking x_2 and h_1. The rest of the L_2 are obtained by simply copying the information in L_1.*

Figure 3 illustrates such a process, by fixing the input size and direct unrolling the RNNs, we can eliminate the loops and build an equivalent feed-forward neural network. The FNN however contains structures that do not appear in regular FNNs. They are time-delays between nodes in Fig. 4 (a) and different activation functions in the same layer, see Fig. 4 (c).

For the time delay situation, we add dummy nodes to intermediary layers. These dummy nodes use the identity matrix for weight and use the identity function as an activation function, as illustrated in Fig. 4 (b). After the modification, the intermediary layer is equivalent to a regular FNN layer.

The time delay between nodes occurs even by simple structure RNNs, such as in Fig. 3 (a), while the same layer with different activation functions appears only by unfolding complex RNNs. Figure 4 (c) demonstrates the layer with different activation functions after unrolling. When the appeared different activation functions are Lipschitz continuous, the layer is Lipschitz continuous based on the sub-multiplicative property in matrix norms. See **Appendix-A** for detailed proof.

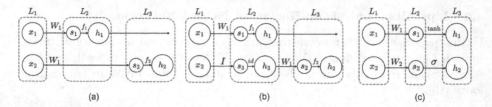

(a)   (b)   (c)

Fig. 4. (a) Before we add dummy nodes to the intermediary layer: W_i *is a weight matrix, and* f_i *is an activation function. Initially, a connection from* x_2 *to* h_2 *crosses over the layer* L_2. (b) After we add dummy nodes: *we add nodes* s_3 *and* h_3 *to* L_2, *where I denotes the identity matrix and id denotes the identity function.*(c)Feed-forward layer with distinct activation functions: *layer* L_2 *only performs a linear transformation (i.e. multiplication with weight matrices), and layer* L_3 *has the role of applying non-linear activation functions that contain two distinct activation functions: Hyperbolic Tangent and Sigmoid.*

5 Reachability Analysis with Provable Guarantees

5.1 Verification via Lipschitz Optimization

In the Lipschitz optimization [8] we asymptotically approach the global minimum. Practically, we execute a finite number of iterations by using an error tolerance ϵ to control the termination. As shown in Fig. 5 (a), we first generate two straight lines with slope K and $-K$, concreting a cross point Z_0. Since Z_0 is the minimal value of the generated piecewise-linear lower bound function (blue lines), we use the projected W_1 for the next iteration. In Fig. 5 (b), new W and Z points are generated. In i-th iteration, the minimal value of W is the upper bound u_i, and the minimal value of Z is the lower bound l_i. Our approach constructs a sequence of lower and upper bounds, terminates the iteration whenever $|u_i - l_i| \leq \epsilon$,

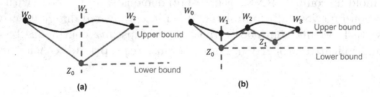

(a)   (b)

Fig. 5. A lower-bound function designed via Lipschitz constant

For the multi-dimensional optimization problem, we decompose it into a sequence of nested one-dimensional subproblems [7]. Then the minima of those one-dimensional minimization subproblems are back-propagated into the original dimension, and the final global minimum is obtained with $\min_{x \in [a_i, b_i]^n}$

$w(x) = \min_{x_1 \in [a_1, b_1]} \dots \min_{x_n \in [a_n, b_n]} w(x_1, \dots, x_n)$. We define that for $1 \le k \le n - 1$, $\phi_k(x_1, \dots, x_k) = \min_{x_{k+1} \in [a_{k+1}, b_{k+1}]} \phi_{k+1}(x_1, \dots, x_k, x_{k+1})$ and for $k = n$, $\phi_n(x_1, \dots, x_n) = w(x_1, \dots, x_n)$. Thus we can conclude that $\min_{x \in [a_i, b_i]^n} w(x) = \min_{x_1 \in [a_1, b_1]} \phi_1(x_1)$ which is actually a one-dimensional optimization problem.

We design a practical approach to dynamically update the current Lipschitz constant according to the previous iteration: $K = \eta \max_{j=1,\dots,i-1} \left\| \frac{w(y_j) - w(y_{j-1})}{y_j - y_{j-1}} \right\|$ where $\eta > 1$, so that $\lim_{i \to \infty} K > K_{best}$. We use the Lipschitz optimisation to find the minimum and maximum function values of the neural network. With binary search, we further estimate the maximum safe radius for target attack.

5.2 Global Convergence Analysis

We first analyse the convergence for a one-dimensional case. In the one dimensional case convergence exists under two conditions: $\lim_{i \to \infty} l_i = \min_{x \in [a,b]} w(x)$; $\lim_{i \to \infty} (u_i - l_i) = 0$. It can be easily proved since the lower bound sequence $\mathcal{L}_i$ is strictly monotonically increasing and bounded from above by $\min_{x \in [a,b]} w(x)$.

We use mathematical induction to prove convergence for the multi-dimension case. The convergence conditions of the inductive step: if, for all $x \in \mathbb{R}^k$, $\lim_{i \to \infty} l_i = \inf_{x \in [a,b]^k} w(x)$ and $\lim_{i \to \infty} (u_i - l_i) = 0$ are satisfied, then, for all $x \in \mathbb{R}^{k+1}$, $\lim_{i \to \infty} l_i = \inf_{x \in [a,b]^{k+1}} w(x)$ and $\lim_{i \to \infty} (u_i - l_i) = 0$ hold.

Proof. (sketch) By the nested optimisation scheme, we have $\min_{x \in [a_i, b_i]^{k+1}} w(x) = \min_{x \in [a,b]} \Phi(x)$, $\Phi(x) = \min_{y \in [a_i, b_i]^k} w(x, y)$. Since $\min_{y \in [a_i, b_i]^k} w(x, y)$ is bounded by an interval error ϵ_y, assuming $\Phi^*(x)$ is the accurate global minimum, then we have $\Phi^*(x) - \epsilon_y \le \Phi(x) \le \Phi^*(x) + \epsilon_y$ $\Phi(x)$ is not accurate but bounded by $|\Phi(x) - \Phi^*(x)| \le \epsilon_y, \forall x \in [a, b]$, where $\Phi^*(x)$ is the accurate function evaluation.

For the inaccurate evaluation case, we assume $\Phi_{min} = \min_{x \in [a,b]} \Phi(x)$, and its lower and bound sequences are, respectively, $\{l_0, \dots, l_i\}$ and $\{u_0, \dots, u_i\}$. The termination criteria for both cases are $|u_i^* - l_i^*| \le \epsilon_x$ and $|u_i - l_i| \le \epsilon_x$, and ϕ^* represents the ideal global minimum. Then we have $\phi^* - \epsilon_x \le l_i$. Assuming that $l_i^* \in [x_k, x_{k+1}]$ and x_k, x_{k+1} are adjacent evaluation points, then due to the fact that $l_i^* = \inf_{x \in [a,b]} H(x; \mathcal{Y}_i)$ and the search scheme, we have $\phi^* - l_i \le \epsilon_y + \epsilon_x$. Similarly, we can get $\phi^* + \epsilon_x \ge u_i^* = \inf_{y \in \mathcal{Y}_i} \Phi^*(y) \ge u_i - \epsilon_y$ so $u_i - \phi^* \le \epsilon_x + \epsilon_y$. By $\phi^* - l_i \le \epsilon_y + \epsilon_x$ and the termination criteria $u_i - l_i \le \epsilon_x$, we have $l_i - \epsilon_y \le \phi^* \le u_i + \epsilon_y$, *i.e.*, the accurate global minimum is also bounded. See more theoretical analysis of the global convergence in **Appendix-B**.

6 Experiments

6.1 Performance Comparison with State-of-the-art Methods

In this section, we compare DeepAgn with baseline methods. Their performance in feedforward neural networks and more details of the technique are demonstrated in **Appendix-C**. Here, we mainly focus on the verification of RNN. We choose POPQORN [16] as the baseline method since it can solve RNN verification problems analogously, i.e., calculating safe input bounds for given samples. Both methods were run on a PC with an i7-4770 CPU and 24 GB RAM. Table 2 demonstrates the verified safe radius of baseline r_b, DeepAgn r, and the radius of CW attack r_a. We fixed the number of hidden neurons and manipulated the input lengths in Table 3 to compare the average safe radius and the time costs. It can be seen that increasing the input length does not dramatically increase the time consumption of DeepAgn because it is independent of the models' architectures. As in Table 4, we fixed the input length and employed RNNs and LSTMs with different numbers of hidden neurons.

Table 2. Average radius of attack and standard deviations ($\cdot/\cdot$) of Attack, DeepAgn, and POPQORN on MNIST

Model	Attack (r_a)	**DeepAgn (r)**	POPQORN (r_b)
rnn 7_64	0.8427/0.3723	**0.5227/0.2157**	0.0198/0.014
rnn 4_32	0.8424/0.4641	**0.6189/0.3231**	0.0182/0.0201
lstm 4_32	0.6211/ 0.4329	**0.3223/0.3563**	0.0081/0.0052
lstm 7_64	0.7126/ 0.3987	**0.4023/0.2112**	0.0194/0.0165

Table 3. Average safe radius and time cost of DeepAgn and POPQORN on NN verification with different frame lengths

Models	DeepAgn		POPQORN	
	safe radius	time	safe radius	time
rnn 4_64	0.1336	253.64 s	0.0328	1.31 s
rnn 14_64	0.3248	228.23 s	0.2344	11.73 s
rnn 28_64	0.3551	285.35 s	nan	nan
rnn 56_64	0.4369	314.1 s	nan	nan
lstm 4_64	0.3195	250.99 s	0.0004	307.93 s
lstm 14_64	0.3883	382 s	0.0123	400.83 s
lstm 28_64	0.6469	512.45 s	0.0296	532.47 s
lstm 56_64	0.6344	491.64 s	0.0309	557.22 s

Table 4. Average safe radius and time cost of DeepAgn and POPQORN on NN verification with different hidden neurons

Models	DeepAgn		POPQORN	
	safe radius	time	safe radius	time
rnn 7_16	0.5580	117.82 s	0.2038	2.14 s
rnn 7_32	0.2371	175.92 s	0.1340	2.44 s
rnn 7_128	0.6633	240.59 s	0.1052	4.25 s
rnn 7_256	0.6656	187.83 s	0.2038	1.89 s
lstm 7_16	0.3789	175.11 s	0.0007	243.60 s
lstm 7_32	0.3461	189.51 s	0.0015	256.77 s
lstm 7_128	0.3625	256.50 s	0.0050	375.85 s

6.2 Ablation Study

In this section, we present an empirical analysis of the Lipschitz constant K and the number of perturbed pixels, which both affect the precision of the results and the cost of time. As shown in Fig. 6, DeepAgn with $K = 0.1$ gives a false safe radius, indicating that 0.1 is not a suitable choice. When the Lipschitz constant is larger than the minimal Lipschitz constant ($K \geq 1$), DeepAgn can always provide the exact maximum safe radius. However, with larger K, we need more iterations to achieve the convergence condition when solving the optimisation problem. As for the number of perturbed pixels, we treat an n-pixel perturbation as an n-dimensional optimisation problem. Therefore, when the number of pixels increases, the evaluation time grows exponentially.

6.3   Case Study 1

In this experiment, we use our method to verify a deep neural network in an audio classification task. The evaluated model is a deep CNN and is trained under the PyTorch framework. The data set is adopted from [28], where each one-second raw audio is transformed into a sequence input with 8000 frames and classified into 35 categories. We perturb the input value of the frame $(1000, 2000..., 7000)$ and verify the network of different perturbation radii. For the deep CNN case, the baseline method has a lower verification accuracy, while DeepAgn can still provide the output ranges and the maximal safe radius. Figure 7 shows the boundary of the radio waveform with perturbation $\theta = 0.1$, $\theta = 0.2$ and $\theta = 0.3$. Their differences from the original audio are imperceptible to human ears. We performed a binary search and found the exact maximum safe radius $r = 0.1591$.

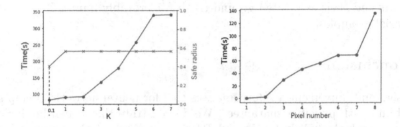

Fig. 6. Time cost and safe radius with different K and perturbed pixel numbers

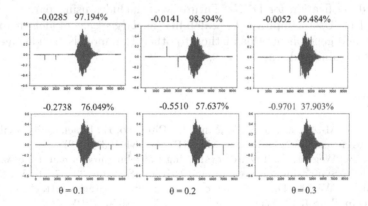

Fig. 7. Input on lower and upper bounds of different perturbations. The first value indicates the logit output and the second shows the confidence value. For perturbation $\theta = 0.1$ and $\theta = 0.2$, the network remains safe. It is not safe for perturbation $\theta = 0.3$.

6.4   Case Study 2

In this case study, we verify a hybrid neural network CRNN that contains convolutional layers and LSTM layers with CTC loss. The network converts characters

(a)   (b)   (c)

Fig. 8. We perturb six pixels of the image to generate the ground-truth adversarial examples(a) $\theta = 0.638$, the first letter recognized as "K" with 48.174% confidence, as "L" with 48.396%, the word recognized as "IKEVIN"; (b) $\theta = 0.0631$, confidence for fourth letter recognized as "R" 49.10%, "B" 49.12%, recognized as "CHEBPIN" (c) $\theta = 0.464$, the third letter as "L" 25.96%, as "I" 25.97%, recognized as "JUIES".

from scanned documents into digital forms. As far as we know, there is no existing verification tool that can deal with this complex hybrid network. However, DeepAgn can analyze the output range of this CRNN and compute the maximum safe radius of a given input. In Fig. 8, we present the maximum safe radius of the inputs and their associated ground-truth (or provably minimally-distorted) adversarial examples.

7   Conclusion

We design and implement a safety analysis tool for neural networks, computing reachability with provable guarantees. We demonstrate that it can be deployed in any network, including FNNs and RNNs regardless of the complex structure or activation function, as long as the network is Lipschitz continuous. We envision that DeepAgn marks an important step towards practical and provably-guaranteed verification for DNNs. Future work includes using parallel computation and GPUs to improve its scalability on large-scale models trained on ImageNet, and generalising this method to other deep models such as deep reinforcement learning and transformers.

References

1. Akintunde, M., Lomuscio, A., Maganti, L., Pirovano, E.: Reachability analysis for neural agent-environment systems. In: KR, pp. 184–193 (2018)
2. Carlini, N., Wagner, D.: Towards evaluating the robustness of neural networks. In: 2017 IEEE Symposium on Security and Privacy (SP), pp. 39–57. IEEE (2017)
3. Carlini, N., Wagner, D.: Audio adversarial examples: targeted attacks on speech-to-text. In: 2018 IEEE Security and Privacy Workshops (SPW) (2018)
4. Dutta, S., Jha, S., Sanakaranarayanan, S., Tiwari, A.: Output range analysis for deep neural networks. arXiv preprint arXiv:1709.09130 (2017)
5. Ehlers, R.: Formal verification of piece-wise linear feed-forward neural networks. In: International Symposium on Automated Technology for Verification and Analysis (2017)
6. Gehr, T., Mirman, M., Drachsler-Cohen, D., Tsankov, P., Chaudhuri, S., Vechev, M.: Ai2: safety and robustness certification of neural networks with abstract interpretation. In: 2018 IEEE Symposium on Security and Privacy (SP), pp. 3–18. IEEE (2018)

7. Gergel, V., Grishagin, V., Gergel, A.: Adaptive nested optimization scheme for multidimensional global search. J. Glob. Optim. **66**(1), 35–51 (2016)
8. Goldstein, A.: Optimization of lipschitz continuous functions. Math. Program. (1977)
9. Gong, Y., Poellabauer, C.: Crafting adversarial examples for speech paralinguistics applications. arXiv preprint arXiv:1711.03280 (2017)
10. Goodfellow, I., Bengio, Y., Courville, A.: Deep Learning. MIT Press, Cambridge (2016)
11. Goodfellow, I.J., Shlens, J., Szegedy, C.: Explaining and harnessing adversarial examples. arXiv preprint arXiv:1412.6572 (2014)
12. Huang, X., Kroening, D., Ruan, W., et al.: A survey of safety and trustworthiness of deep neural networks: verification, testing, adversarial attack and defence, and interpretability. Comput. Sci. Rev. **37** (2020)
13. Huang, X., Kwiatkowska, M., Wang, S., Wu, M.: Safety verification of deep neural networks. In: International Conference on Computer Aided Verification (2017)
14. Jacoby, Y., Barrett, C., Katz, G.: Verifying recurrent neural networks using invariant inference. arXiv preprint arXiv:2004.02462 (2020)
15. Katz, G., Barrett, C., et al.: Reluplex: an efficient SMT solver for verifying deep neural networks. In: International Conference on Computer Aided Verification (2017)
16. Ko, C.Y., Lyu, Z., Weng, T.W., et al.: Popqorn: quantifying robustness of recurrent neural networks. arXiv preprint arXiv:1905.07387 (2019)
17. Mirman, M., Gehr, T., Vechev, M.: Differentiable abstract interpretation for provably robust neural networks. In: ICML (2018)
18. Mu, R., Ruan, W., Marcolino, L.S., Ni, Q.: 3Dverifier: efficient robustness verification for 3D point cloud models. Mach. Learn. 1–28 (2022)
19. Mu, R., Ruan, W., Marcolino, L.S., Jin, G., Ni, Q.: Certified policy smoothing for cooperative multi-agent reinforcement learning. In: Proceedings of the AAAI Conference on Artificial Intelligence (AAAI'23) (2023)
20. Narodytska, N., Kasiviswanathan, S., Ryzhyk, L., Sagiv, M., Walsh, T.: Verifying properties of binarized deep neural networks. In: Proceedings of the AAAI Conference on Artificial Intelligence, vol. 32 (2018)
21. O'Searcoid, M.: Metric Spaces. Springer, London (2006). https://doi.org/10.1007/978-1-84628-627-8
22. Papernot, N., McDaniel, P., Swami, A., Harang, R.: Crafting adversarial input sequences for recurrent neural networks. In: MILCOM 2016–2016 IEEE Military Communications Conference, pp. 49–54. IEEE (2016)
23. Pulina, L., Tacchella, A.: An abstraction-refinement approach to verification of artificial neural networks. In: International Conference on Computer Aided Verification (2010)
24. Ruan, W., Huang, X., Kwiatkowska, M.: Reachability analysis of deep neural networks with provable guarantees. In: Proceedings of the 27th International Joint Conference on Artificial Intelligence, pp. 2651–2659 (2018)
25. Szegedy, C., et al.: Intriguing properties of neural networks. arXiv preprint arXiv:1312.6199 (2013)
26. Vengertsev, D., Sherman, E.: Recurrent neural network properties and their verification with Monte Carlo techniques. In: SafeAI@AAAI (2020)
27. Wang, F., Xu, P., Ruan, W., Huang, X.: Towards verifying the geometric robustness of large-scale neural networks. In: Proceedings of the AAAI Conference on Artificial Intelligence (AAAI'23) (2023)

28. Warden, P.: Speech commands: a dataset for limited-vocabulary speech recognition. arXiv preprint arXiv:1804.03209 (2018)
29. Weng, L., Zhang, H., Chen, H., et al.: Towards fast computation of certified robustness for ReLU networks. In: ICML (2018)
30. Weng, T.W., et al.: Evaluating the robustness of neural networks: an extreme value theory approach. arXiv preprint arXiv:1801.10578 (2018)
31. Wong, E., Kolter, Z.: Provable defenses against adversarial examples via the convex outer adversarial polytope. In: ICML (2018)
32. Wu, M., Wicker, M., Ruan, W., Huang, X., Kwiatkowska, M.: A game-based approximate verification of deep neural networks with provable guarantees. Theor. Comput. Sci. **807**, 298–329 (2020)
33. Yin, X., Ruan, W., Fieldsend, J.: Dimba: discretely masked black-box attack in single object tracking. Mach. Learn. 1–19 (2022)
34. Zhang, H., Shinn, M., Gupta, A., Gurfinkel, A., Le, N., Narodytska, N.: Verification of recurrent neural networks for cognitive tasks via reachability analysis. In: ECAI 2020, pp. 1690–1697. IOS Press (2020)
35. Zhang, T., Ruan, W., Fieldsend, J.E.: Proa: a probabilistic robustness assessment against functional perturbations. In: Joint European Conference on Machine Learning and Knowledge Discovery in Databases (ECML/PKDD'22) (2022)
36. Zhang, Y., Ruan, W., Wang, F., Huang, X.: Generalizing universal adversarial perturbations for deep neural networks. Mach. Learn. (2023)

Adaptive Bi-nonlinear Neural Networks Based on Complex Numbers with Weights Constrained Along the Unit Circle

Felip Guimerà Cuevas[1(✉)], Thomy Phan[2], and Helmut Schmid[3]

[1] BMW Group Munich, Munich, Germany
felip.guimera-cuevas@bmw.de
[2] Institute for Informatics at LMU Munich, Munich, Germany
thomy.phan@ifi.lmu.de
[3] Center for Information and Language Processing at LMU Munich,
Munich, Germany
schmid@cis.lmu.de

Abstract. Traditional real-valued neural networks can suppress neural inputs by setting the weights to zero or overshadow other inputs by using extreme weight values. Large network weights are undesirable because they may cause network instability and lead to exploding gradients. To penalize such large weights, adequate regularization is typically required. This work presents a feed-forward and convolutional layer architecture that constrains weights along the unit circle such that neural connections can never be eliminated or suppressed by weights, ensuring that no incoming information is lost by dying neurons. The neural network's decision boundaries are redefined by expressing model weights as angles of phase rotations and layer inputs as amplitude modulations, with trainable weights always remaining within a fixed range. The approach can be quickly and readily integrated into existing layers while preserving the model architecture of the original network. The classification performance was tested and assessed on basic computer vision data sets using ShuffleNetv2, ResNet18, and GoogLeNet at high learning rates.

Keywords: Deep Learning · Complex numbers · Neural architecture

1 Introduction

Deep learning is one of the most common and promising approaches to solving many machine learning tasks [14,20]. Conventional Deep Neural Networks (DNN) use real-valued (RV) connection weights to compute the output values of activation functions to fire neurons. Each output of a DNN's activation function can be thought of as a representation of a decision boundary; determined by the weighting of all input connections under (optional) consideration of bias terms. The purpose of an activation function is to introduce non-linearity between successive layers and respectively in the decision boundaries; otherwise, a DNN would collapse back to an equivalent of a single-layer perceptron [29]. It is intended that through training neural networks (NN), generalizable decision

© The Author(s), under exclusive license to Springer Nature Switzerland AG 2023
H. Kashima et al. (Eds.): PAKDD 2023, LNAI 13935, pp. 355–366, 2023.
https://doi.org/10.1007/978-3-031-33374-3_28

boundaries are learned. The more representative they are, the better their final performance on new and previously unseen data is. Linear decision boundaries have limitations. The notorious XOR problem *cannot* be solved by a single-layer perceptron with no activation function since XOR is not linearly separable, but it can be solved by a NN with at least one hidden layer and non-linear activation functions. Decision boundaries are fundamental for DNNs; their importance cannot be overstated.

This work proposes and evaluates a new technique for computing decision boundaries that draws motivation from complex numbers. It integrates complex numbers into a bi-nonlinear NN, resulting in an architecture that retains incoming information by preventing the weights from eliminating or suppressing the inputs. The method first uses weights as *angles* to represent artificial complex unit numbers, i.e. complex numbers z with $||z|| = 1$; then scales these unit numbers accordingly by the signed numeric magnitude of the input connections to the neurons and aggregates them by summation; finally, it applies a non-linear transformation on that sum to obtain the final output value. The model is designed for RV inputs and outputs. It does not affect the compatibility e.g. with RV loss functions and can be used on the same problem types as RV networks. It forms a novel alternative approach to Complex-Valued Neural Networks (CVNNs), which differs notably from prior work - in particular from Multi-Valued Neurons (MVNs) [2] and models where RV inputs are encoded into phases of complex numbers of unity magnitude and multiplied by complex-valued (CV) weights [10] - by the following key points: (1) Setting weights to zero does *not* cancel out inputs; (2) Complex back-propagation is *not* necessary; (3) Neural inputs do *not* require re-mapping from $\mathbb{R} \to \mathbb{C}$; (4) The weight matrix is RV. Furthermore, compared to other CVNN methods for RV inputs, this work's approach has several advantages: There is *no* explicit mapping of inputs from real to complex numbers, e.g. to scale down the domain of each input feature to phases of a unit circle. Therefore, *no* domain knowledge of the upper and lower bounds of the input features is required. The layers' outputs are also RV, which avoids having to repeatedly re-map values when stacking multiple layers sequentially. Further, neural connections can never be suppressed or eliminated by setting the weights to zero because weights purely denote phase rotations and correspond to angles of complex unit numbers. The weights can also never become too large since there are always equivalent weights in the range of $[0, 2\pi)$, counteracting exploding gradients [27]. Conventional weight regularization techniques are therefore unnecessary. Lastly, the proposed CV-based layers can be easily and directly incorporated into existing models without modifying the actual architecture.

2   Background

The Wirtinger calculus [9] can be applied to enable differentiability, with the idea being to rewrite a complex function $f(z)$ as a function of two variables, namely z and its complex conjugate $\overline{z}$, denoted as $f(z, \overline{z})$. It uses the property that the real and imaginary components of z can be re-written as $Re(z) = \frac{z + \overline{z}}{2}$ and $Im(z) = \frac{z - \overline{z}}{2i}$, on which this work's approach is motivated.

3   Related Work

The idea of CVNNs themselves is not new [11,13,15,22], but has often been shadowed by the success of the classic RV DNNs [23] and by the increased complexity of performing CV back-propagation for gradient descent - which, however, is no longer a concern with today's deep learning frameworks. One difficulty of designing CVNN architectures for RV inputs (or outputs) is how real numbers are mapped to complex numbers and vice versa, and how activation functions should be defined or handled. If learning occurs through gradient descent, the entire forward-call of the network must be *complex*-differentiable and thus *holomorphic*, which is the case if and only if it fulfills the Cauchy-Riemann (CR) equations and has continuous first partial derivatives [1]. The first pioneering methods for CV back-propagation algorithms (e.g. for multi-layer CVNNs) were proposed in the early 1990s s [12,21]. Yet, most loss functions of interest are RV or non-holomorphic. According to the CR conditions, an RV holomorphic function must be constant, so typical transformations on complex numbers z like $||z||, arg(z), \bar{z}, Re(z)$ and $Im(z)$ are not holomorphic. Liouville's theorem further states that any bounded function that is holomorphic at every point must be constant. Therefore, complex activation functions cannot be bounded and complex differentiable at once.

The architecture behind a CVNN is not always entirely consistent in the literature and is often characterized by the CV activation function. Approaches with non-holomorphic complex activation functions are called split activation functions, categorized as *Type A* if the real and imaginary parts are bounded, or *Type B* if only the magnitude is bounded but the phase is retained [19]. Although networks are typically considered CVNNs if they handle CV information via CV variables and parameters [16,17], a less stringent constraint may be imposed to designate CVNNs if merely weights are CV. The weight matrix of a CVNN in the complex domain typically denotes a rotation matrix. The concept of a discrete MVN was already proposed in 1992 [8] as a neural element based on the mathematical model of multi-valued threshold logic for complex numbers [18]. Since then, there have been several advancements in MVNs, in addition to approaches integrating them into DNNs [5–7,25], and also concerning back-propagation [4]. Initially, the activation function of an MVN was merely dependent on the phase of a weighted sum, involving inputs and outputs on the unit circle and CV weights. Learning was reduced along movement on the unit circle by applying a mapping from the complex domain onto it (falling so into category Type B). Discrete MVNs can also be extended to continuous-valued inputs. The use of periodic activation functions has also been suggested to increase the functionality of a single MVN [3]. In 2004, a Type A activation function of the form $f_C(z) = f_{\mathbb{R}}(x) + i f_{\mathbb{R}}(y)$ was analyzed [26], with $f_{\mathbb{R}}$ being the sigmoid function. It was shown that the decision boundaries of such a single CV neuron are formed by two orthogonal intersecting hypersurfaces; and the boundaries of a CVNN whose weights, thresholds, values, inputs, and outputs are all complex numbers approach orthogonality as the network's inputs increase. It was claimed that the orthogonality of the decision boundaries improved generalization and that learning with complex back-propagation occurred on average faster than with RV back-propagation. Another CVNN approach [10] maps RV inputs to complex numbers of unit size using a linear transformation and multiplies them by CV weights. The activation function then takes the weighted sum of inputs and re-maps the resulting complex values to real numbers. CVNN architectures were also suggested for convolutional layers [30] along with various activation functions, and techniques for batch normalization and complex weight initialization.

4 Methods

4.1 Feed-Forward Layer

A CV-motivated adaptive bi-nonlinear model architecture similar to an RV-NN is proposed. It consists of k layers with m_k neurons each; $k, m_k \in \mathbb{N}^+$. However, a neuron's output is no longer a weighted sum of real numbers, but a more intricate mathematical operation. The weights are still RV and *not* complex since angles in radians represent complex unit numbers. Using $e^{i\theta} = e^{i(\theta+2\pi)}$ and $\overline{e^{i\theta}} = e^{-i\theta}$ for $\theta \in \mathbb{R}$, a weight matrix is initialized by randomly sampling radians and then deriving the CV matrix from it. The entire search space of the CVNN's weight matrices can effectively be limited to $[0, 2\pi)$; or $[-\pi, \pi)$. Since the former breaks the symmetry around zero, the latter is advocated. By using a modulo operation, weights may be explicitly limited within the range. Because the maximum weight value is constrained, large network weight changes (e.g. induced by exploding gradients) are mitigated. Furthermore, zero weights no longer erase inputs, but indicate a zero-angle, preserving the input information.

Let $1 \leq k \leq n$ refer to the kth layer in a network with n layers and $\mathbf{h}_k$ the RV state of the layer k of size m_k. $\mathbf{h}_0$ is the network input of size m_0. Given $m_k, m_0 \geq 1$, let $W_k \in \mathbb{R}^{m_k \times m_{k-1}}$ denote the (angular) weight matrix and $\sin(W_k), \cos(W_k)$ denote the point-wise applications of the sine and cosine to all values in W_k. By trivial decomposition over the real and imaginary parts, an equivalence for a polar vs. rectangular expression can be obtained:

$$\mathbf{z}_k := e^{iW_k}\mathbf{h}_{k-1} = \left[\cos(W_k)\mathbf{h}_{k-1}\right] + i\left[\sin(W_k)\mathbf{h}_{k-1}\right] \tag{1}$$

Because the real and imaginary parts are independent and uncorrelated during the matrix operation, an activation function is defined to introduce **nonlinear** dependencies: the product of the real and imaginary parts divided by the magnitude of the complex number; dropping the imaginary number i. Therefore, let $\mathbf{x}_k := \cos(W_k)\mathbf{h}_{k-1} \in \mathbb{R}^{m_k}$ and $\mathbf{y}_k := \sin(W_k)\mathbf{h}_{k-1} \in \mathbb{R}^{m_k}$. State $\mathbf{h}_k$ then is:

$$h_k^{[j]} = \alpha_k^{[j]}\left(\frac{x_k^{[j]}y_k^{[j]}}{\sqrt{\left(x_k^{[j]}\right)^2 + \left(y_k^{[j]}\right)^2}}\right) + \beta_k^{[j]} : \forall_{1 \leq j \leq m_k} \tag{2}$$

where $h_k^{[j]}$ is the jth component of $\mathbf{h}_k$ for $1 \leq j \leq m_k$, and $\boldsymbol{\alpha}_k, \boldsymbol{\beta}_k \in \mathbb{R}^{m_k}$ trainable vector variables. Thus, "bi-nonlinear" refers to using $\mathbf{x}_k, \mathbf{y}_k$ non-linearly to express $\mathbf{h}_k$. The activation function of the CV model is plotted in Fig. 1.

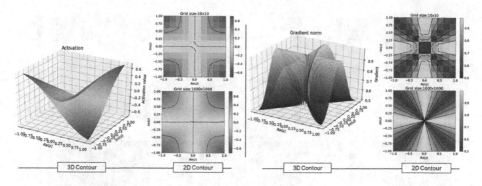

Fig. 1. A 3D and 2D visualization of the activation function (left figure) and its corresponding gradient norm surface (right figure). The 2D visualization's loss space is shown for two different grid sizes (i.e. thresholds). Point symmetry exists around the origin and decision boundaries intersect orthogonally along the coordinate axes.

Let $\theta := arg[z_k^{[j]}]$ denote the angle of the complex argument function and let $r := ||z_k^{[j]}||_2$ be the complex norm, such that $z_k^{[j]} \equiv re^{i\theta}$ is the polar form of the complex number. Respectively, w.l.o.g. $r \in \mathbb{R}^+$ is the magnitude (due to $re^{i(\pi+\theta)} = -re^{i\theta}$), and $\theta \in [-\pi, \pi]$ is the angle of z in radians. Using $Re(z_k^{[j]}) = \frac{z_k^{[j]} + \overline{z}_k^{[j]}}{2} = x_k^{[j]}$ and $Im(z_k^{[j]}) = \frac{z_k^{[j]} - \overline{z}}{i2} = y_k^{[j]}$, it holds $x_k^{[j]} y_k^{[j]} = \frac{(z_k^{[j]})^2 - (\overline{z_k}^{[j]})^2}{i4} = \frac{r^2 \sin(2\theta)}{2}$. Inserted into Eq. 2 yields the closed-form expression of layer k: $h_k^{[j]} = \frac{r}{2} \sin(2\theta)$. The magnitude $r = ||z_k^{[j]}||$ may increase the more neurons the layer k has (Eq. 1), i.e. the larger the size of the input m_{k-1} is. Thus, it is sensible to include a normalization factor $\epsilon > 0$ (e.g. $\epsilon := \sqrt{m_{k-1}}$) to reduce the magnitude relative to the dimension of the input vector. The equation resembles the general formula of a *sinusoid* $y(t) = a \cdot \sin(2\pi ft + \varphi)$, where a is an amplitude, f an ordinal frequency, and φ a phase in radians. $2\pi f$ is referred to as the angular velocity. Drawing an analogy, the amplitude would correspond to $a \triangleq \frac{1}{2\epsilon}$, and the term $\pi ft \triangleq \theta$. If a new trainable vector variable is introduced as the angular velocity $\omega := 2\pi f$ (i.e. rate of change), and $\hat{\alpha} := \frac{\alpha}{2\epsilon}$, then Eq. 2 is equivalent to:

$$h_k^{[j]} \triangleq ||z_k^{[j]}|| \sin\left(\omega_k^{[j]} arg[z_k^{[j]}]\right)\hat{\alpha}_k^{[j]} + \beta_k^{[j]} \tag{3}$$

An advantage of Eq. 3 over Eq. 2 is the directly adjustable (i.e. trainable) angular velocity. Thus, Eq. 3 will be used as the transformation function for the network. It should be noted that one might falsely assume that Eq. 3 precludes zero-division. This is not true because the *arg* function can still produce a zero-division (when all inputs are zero; in which case h_k be the zero vector). And, while it is possible to represent a CV weight using a single RV, using complex unit numbers in this manner requires a trade-off in which two matrix multiplications are required, one for the real part and one for the imaginary part; but this is a common concern when using CVs instead of RVs. The effect of using different values for ω is illustrated in Fig. 2 below:

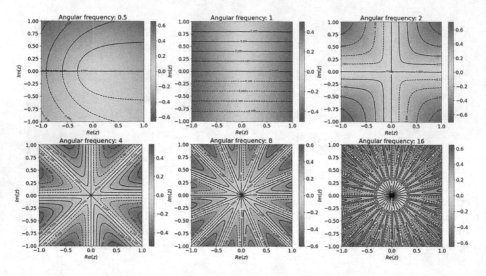

Fig. 2. A visual illustration of using different angular velocities ω on the decision boundaries. As ω increases, the space becomes more divided.

4.2 Convolutional Layer

Let f, g be discrete functions. Then, CV cross-correlation is defined as $(f \star g)[n] := \sum_{u=\infty}^{\infty} \overline{f[u]}g[u + n]$ [28], where $\star$ denotes a valid 2D cross-correlation operator, $\overline{f[u]}$ the complex conjugate and n the displacement (alias *lag*). If g is a real-function, then $(f \star g)[n] \triangleq \sum_{u=-\infty}^{\infty} Re(f[u])g[u + n] - iIm(f[u])g[u + n]$ $= \sum_{u=-\infty}^{\infty} Re(f[u])g[u+n] - i\sum_{u=-\infty}^{\infty} Im(f[u])g[u+n]$, which can be interpreted as running and aggregating two separate convolutions over the same input. The output of the CV convolutional layer is derived analogously to Eq. 3. However, for a convolutional layer at layer k with $c_k^{(in)}$ input channels and a kernel size of $a_k \times b_k$, the magnitude scales differently and setting $\epsilon := \sqrt{a_k b_k c_k^{(in)}}$ smooths out the use of multiple channels and larger kernel sizes.

5   Experimental Methods and Limitations

Models were trained using the same hyper-parameters; results averaged across six runs. Assertive performance was defined as the greatest achieved train and test accuracy per run. CVNNs were constructed by simply substituting the RV convolutional and feed-forward layers with the suggested CV layers. As an optimizer, *AdamW* [24] was utilized. The trainable angular velocity (Eq. 3) was initialized as $\omega := 2$ for feed-forward layers and $\omega := 1$ for convolutional layers. The images were normalized to a fixed size of 64×64 and a single gray color channel. The batch size was 1024; the default learning rate was set to a relatively high value of 0.1. Both models have fundamentally different forward-call architectures. Thus, this work was limited to choosing hyper-parameters, especially the learning rate, to demonstrate the effectiveness of the proposed technique rather than claiming superiority over RV DNNs, which would otherwise give the CVNN an unfair advantage. The models for minimal networks against point classification tasks (Sect. 6.2) were trained for 50k iterations each.

6   Analytical and Experimental Results

6.1   XOR-Problem

Given the XOR problem with two input variables $v_1, v_2 \in \{0, 1\}$, complex connection unit-weights z_1, z_2, output labels $y \in \{0, 1\}$, and scaling and bias terms α, β for the complex forward-call (Eq. 3), it holds $\text{XOR}(v_1, v_2) := \frac{\alpha}{2} \| v_1 z_1 + v_2 z_2 \| \sin \left(2 \cdot arg[v_1 z_1 + v_2 z_2]\right)$, where arg is the argument function. W.l.o.g. radian-valued weights $\theta \in [0, 2\pi)$ are required, such that XOR is solved. Letting $z_1 := e^{i\pi/4}$, $z_2 := -z_1$, $\alpha = 2$ and $\beta = 0$ provides a solution:

$$\text{XOR}(v_1, v_2) = \| z_1(v_1 - v_2) \| \sin \left(2 \cdot arg[z_1(v_1 - v_2)]\right)$$
$$\overset{(a)}{=} |v_1 - v_2| \sin(\pi/2) = \begin{cases} 0 & \text{if } v_1 = v_2 \\ 1 & \text{otherwise} \end{cases} \tag{4}$$

where (a) uses $(v_1 - v_2) \in \{0, 1\}$ and $\|e^{i\theta}\| = 1$. Then, $\sin(\pi/2) = 1$ yields the elegant direct solution. Since the complex weights have a magnitude of one, the scaling factor α is important for producing a value $\text{XOR}(v_1, v_2) \in \{0, 1\}$. Other values of θ and α also provide a solution if classification using threshold values is allowed, i.e.: $y = 0$ if $\text{XOR}(v_1, v_2) < t$, otherwise $y = 1$, with t being the threshold. This is useful for continuous classification with $v_1, v_2 \in [0, 1]$. A threshold can also be introduced e.g. by adding a fixed bias term to the output.

6.2   Minimal Networks and Expressive Power

A 2D point classification task is given, where points belong to the same class if they share the same color. The inputs are the (standardized) 2D coordinates. The models are networks with minimal layer topologies to better analyze neural expressiveness. The output is a single scalar. Because of the learnable angular velocity and magnitude scaling, the CVNN neuron has slightly more weights. To avoid unfair comparisons, two analyses were performed. (1) Comparing neurons with different amounts of weights but the same number of neurons; (2) comparing networks with an equal number of weights but different numbers of neurons.

Equal Number of Trainable Weights. Given is a CVNN with two inputs and a single output. The hidden layer size be n. Thus, the total number of trainable weights (including biases) is $6n + 3 \triangleq [2n + (n + n + n)] + [1n + (1 + 1 + 1)]$. Likewise, for the RV-NN it is: $4k + 1 \triangleq [2k + k] + [1k + 1]$; k be the number of its hidden neurons. For the number of weights to be equal, $6n + 3 = 4k + 1$ must hold, and so $k = \frac{1}{2}(3n + 1)$. Given $n \leftarrow 3, 5$, this respectively corresponds to $k \leftarrow 5, 8$. The results are depicted in Fig. 3. Overall, the CV model outperformed its RV counterpart given the same number of weights.

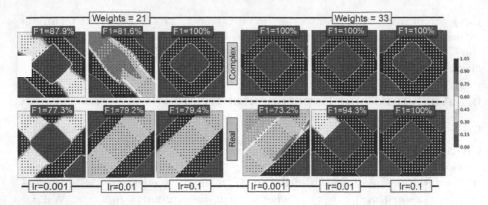

Fig. 3. Decision boundaries of minimal NNs with an equal number of weights are compared for a 2D point fitting task. The yellow and blue colors indicate the labels. (Color figure online)

Equal Number of Neurons. The decision boundaries for the point classification task when both models use the same number of neurons are shown in Fig. 4.

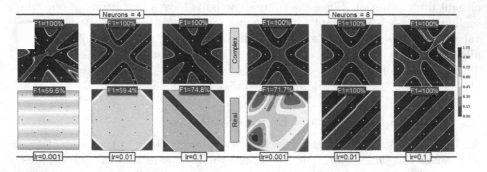

Fig. 4. Decision boundaries of minimal NNs with an equal number of neurons are compared for a 2D point fitting task. The yellow and blue colors indicate the labels. (Color figure online)

The CVNN consistently achieved an F1-score (a score based on precision and recall) of approximately 100% for all learning rates. The RV network struggled with eight hidden neurons (depending on the learning rate) and mostly failed with four neurons. (F1 ≈ 60%). Still, with eight neurons at a learning rate of 0.01, the RV network also achieved perfect classification; finding "linear" diagonal boundaries. In contrast, the CVNN returned "curved" decision boundaries.

6.3   Classification

To analyze the classification performance of the CVNN architecture, three popular larger network architectures, namely ShuffleNetv2, ResNet18, and

GoogLeNet, were utilized on basic computer vision data sets. This allowed us to evaluate CVNN in CNNs and NNs within a common model. The experimental results showed that the CVNNs were successful on all data sets and outperformed the RV equivalent model counterparts when trained on high learning rates; with CIFAR-10 exhibiting the most notable performance difference. The individual results of the classification scores for each model architecture when trained on a high learning rate are shown in Fig. 5 below:

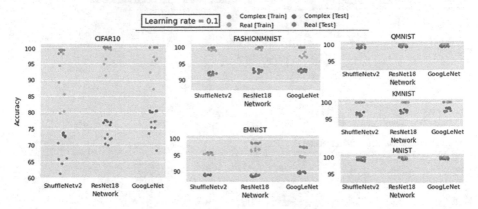

Fig. 5. Scatterplot classification results from independent runs on various data sets, using models ShuffleNetv2, ResNet18, and GoogLeNet. The CVNN was created by replacing neurons, keeping the original network architecture unchanged. Datasets analyzed were CIFAR10, FashionMNIST, EMNIST, QKMNIST, KMNIST, and MNIST.

Overall, the CVNNs and the RV network models achieved similar results using a high learning rate, but the CVNN performed better overall, particularly on CIFAR10. The results indicate that the CVNN outperforms the RV model at higher learning rates. Moreover, all three models (ShuffleNetv2, ResNet18, and GoogLeNet) produced comparable outcomes across all data sets. Table 1 presents the mean classification scores for all models combined (from Fig. 5).

Table 1. Average classification results of all models on different data sets

| Data set | Test | | Train | |
	Complex	Real	Complex	Real
MNIST	**99.44%** (± 0.16)	99.31% (± 0.19)	99.99% (± 0.002)	99.99% (± 0.009)
EMNIST	88.96% (± 0.43)	**89.16%** (± 0.51)	**97.20%** (± 1.25)	95.39% (± 0.95)
QMNIST	**99.33%** (± 0.14)	99.1% (± 0.17)	99.99% (± 0.001)	99.99% (± 0.009)
FMNIST	**92.79%** (± 0.51)	92.21% (± 0.44)	**99.92%** (± 0.08)	98.71% (± 1.02)
KMNIST	**97.49%** (± 0.60)	96.71% (± 0.58)	99.99% (± 0.001)	99.99% (± 0.004)
CIFAR10	**76.84%** (± 3.03)	70.18% (± 5.15)	**99.60%** (± 0.600)	91.86% (± 7.16)

Fake-Data. The ability to fit random noise was further examined on randomly produced pixel images. The labels were arbitrary but balanced. Because there is no correlation between the instances of randomly generated data, attempting to make the model "generalizable" to new (random) data makes no sense. As a result, there was no test set. The results are depicted in Fig. 6 below:

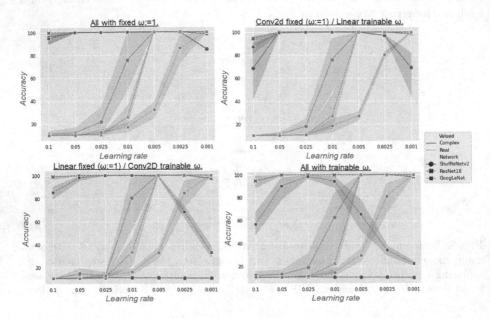

Fig. 6. The classification accuracy for random images is shown at different learning rates and combinations of fixed and learnable angular velocities.

Particularly at higher learning rates, the CVNN appeared to fit random pixel images better than its counterpart. On uncorrelated random data, a fixed angular velocity ω performed better; a trainable ω for convolutional layers was detrimental here. The RV network, however, performed well at low learning rates.

7   Summary and Discussion

This work presented an adaptive bi-nonlinear layer architecture that learns with model weights constrained along the unit circle, so neural connections cannot be eliminated by setting the weights to zero, and other inputs cannot be overshadowed by extreme connection weights. The weights of the CV model apply phase rotations; amplitude scaling is determined by the input. The neural inputs and outputs remain real-valued. The model's layer topology is analogous to that of traditional NNs, with a comparable amount of trainable parameters (the CVNN has more; equal without additional bias and angular weights). The model is trained using traditional gradient descent and back-propagation techniques and is well effective at high learning rates; successfully completing a variety of objective tasks on different architectures, and outperforming the real-valued network in terms of expressiveness. The forward-pass, however, requires twice as many matrix multiplications, one for the real- and one for the imaginary parts.

References

1. Ahlfors, L.V.: Complex analysis: an introduction to the theory of analytic functions of one complex variable. New York, London **177** (1953)
2. Aizenberg, I.: Complex-Valued Neural Networks with Multi-valued Neurons, vol. 353. Springer, Heidelberg (2011). https://doi.org/10.1007/978-3-642-20353-4
3. Aizenberg, I.: The multi-valued neuron. In: Aizenberg, I. (ed.) Complex-Valued Neural Networks with Multi-valued Neurons. SCI, vol. 353, pp. 55–94. Springer, Heidelberg (2011). https://doi.org/10.1007/978-3-642-20353-4_2
4. Aizenberg, I., Moraga, C.: Multilayer feedforward neural network based on multi-valued neurons (MLMVN) and a backpropagation learning algorithm. Soft. Comput. **11**(2), 169–183 (2007)
5. Aizenberg, I., Moraga, C., Paliy, D.: A feedforward neural network based on multi-valued neurons. In: Reusch, B. (eds.) Computational Intelligence, Theory and Applications. ASC, vol. 33, pp. 599–612. Springer, Heidelberg (2005). https://doi.org/10.1007/3-540-31182-3_55
6. Aizenberg, I., Paliy, D., Astola, J.T.: Multilayer neural network based on multi-valued neurons and the blur identification problem. In: The 2006 IEEE International Joint Conference on Neural Network Proceedings, pp. 473–480. IEEE (2006)
7. Aizenberg, I., Sheremetov, L., Villa-Vargas, L., Martinez-Muñoz, J.: Multilayer neural network with multi-valued neurons in time series forecasting of oil production. Neurocomputing **175**, 980–989 (2016)
8. Aizenberg, N.N., Aizenberg, I.N.: Cnn based on multi-valued neuron as a model of associative memory for grey scale images. In: CNNA'92 Proceedings Second International Workshop on Cellular Neural Networks and Their Applications, pp. 36–41. IEEE (1992)
9. Amin, M.F., Amin, M.I., Al-Nuaimi, A.Y.H., Murase, K.: Wirtinger calculus based gradient descent and Levenberg-Marquardt learning algorithms in complex-valued neural networks. In: Lu, B.-L., Zhang, L., Kwok, J. (eds.) ICONIP 2011. LNCS, vol. 7062, pp. 550–559. Springer, Heidelberg (2011). https://doi.org/10.1007/978-3-642-24955-6_66
10. Amin, M.F., Murase, K.: Single-layered complex-valued neural network for real-valued classification problems. Neurocomputing **72**(4–6), 945–955 (2009)
11. Bassey, J., Qian, L., Li, X.: A survey of complex-valued neural networks. arXiv preprint arXiv:2101.12249 (2021)
12. Benvenuto, N., Piazza, F.: On the complex backpropagation algorithm. IEEE Trans. Sig. Process. **40**(4), 967–969 (1992)
13. Clarke, T.L.: Generalization of neural networks to the complex plane. In: 1990 IJCNN International Joint Conference on Neural Networks, pp. 435–440. IEEE (1990)
14. Goodfellow, I., Bengio, Y., Courville, A., Bengio, Y.: Deep Learning, vol. 1. MIT Press, Cambridge (2016)
15. Hirose, A.: Dynamics of fully complex-valued neural networks. Electron. Lett. **28**(16), 1492–1494 (1992)

16. Hirose, A.: Complex-Valued Neural Networks: Theories and Applications, vol. 5. World Scientific, Singapore (2003)

17. Hirose, A.: Complex-Valued Neural Networks, vol. 400. Springer, Heidelberg (2012). https://doi.org/10.1007/978-3-642-27632-3

18. Hirose, A.: Complex-Valued Neural Networks: Advances and Applications, vol. 18. Wiley, Hoboken (2013)

19. Kuroe, Y., Yoshid, M., Mori, T.: On activation functions for complex-valued neural networks—existence of energy functions—. In: Kaynak, O., Alpaydin, E., Oja, E., Xu, L. (eds.) ICANN/ICONIP -2003. LNCS, vol. 2714, pp. 985–992. Springer, Heidelberg (2003). https://doi.org/10.1007/3-540-44989-2_117

20. LeCun, Y., Bengio, Y., Hinton, G.: Deep learning. Nature **521**(7553), 436–444 (2015)

21. Leung, H., Haykin, S.: The complex backpropagation algorithm. IEEE Trans. Sig. Process. **39**(9), 2101–2104 (1991)

22. Little, G.R., Gustafson, S.C., Senn, R.A.: Generalization of the backpropagation neural network learning algorithm to permit complex weights. Appl. Opt. **29**(11), 1591–1592 (1990)

23. Liu, W., Wang, Z., Liu, X., Zeng, N., Liu, Y., Alsaadi, F.E.: A survey of deep neural network architectures and their applications. Neurocomputing **234**, 11–26 (2017)

24. Loshchilov, I., Hutter, F.: Decoupled weight decay regularization. arXiv preprint arXiv:1711.05101 (2017)

25. Lupea, V.M.: Multi-valued neuron with a periodic activation function-new learning strategy. In 2012 IEEE 8th International Conference on Intelligent Computer Communication and Processing, pp. 79–82. IEEE (2012)

26. Nitta, T.: Orthogonality of decision boundaries in complex-valued neural networks. Neural Comput. **16**(1), 73–97 (2004)

27. Philipp, G., Song, D., Carbonell, J.G.: The exploding gradient problem demystified-definition, prevalence, impact, origin, tradeoffs, and solutions. arXiv preprint arXiv:1712.05577 (2017)

28. Rabiner, L.R., Gold, B.: Theory and Application of Digital Signal Processing. Prentice-Hall, Englewood Cliffs (1975)

29. Rosenblatt, F.: The perceptron, a perceiving and recognizing automaton Project Para. Cornell Aeronautical Laboratory (1957)

30. Trabelsi, C., et al.: Deep complex networks. arXiv preprint arXiv:1705.09792 (2017)

CopyCAT: Masking Strategy Conscious Augmented Text for Machine Generated Text Detection

Chien-Liang Liu$^{(\boxtimes)}$ and Hung-Yu Kao

Department of Computer Science and Information Engineering, National Cheng Kung University, Tainan, Taiwan
nail1021734@gmail.com, hykao@mail.ncku.edu.tw

Abstract. Recent developments in natural language generation have made it possible to generate fluent articles automatically. If it is maliciously used to mislead the public, it may cause potential social risks. To avoid these risks, building automatic discriminators for detecting machine-generated text is required. However, in real-world situations, it is hard for humans to identify the machine-generated text, which causes the collection of machine-generated text to be difficult, and discriminators can only be trained on insufficient data. Also, it's hard to generate synthetic machine data ourselves because we are unable to know the masking strategy of collected machine-generated text in real-world situations. In this paper, we found that even if there is a small amount of training data, the saliency score computed by the trained discriminator can reveal the masking strategy of the machine-generated text in the training set. Based on this observation, we propose a data augmentation method, CopyCAT. CopyCAT can mimic the masking strategy of the collected machine data by the information revealed by the saliency score. Our experiments show that the discriminator trained with our augmented data can have up to 10% accuracy gain.

Keywords: Machine-generated text detection · Data augmentation

1 Introduction

Natural language generation technologies have been widely used in various fields. However, it may be used to generate mass articles for inappropriate purposes, such as generating fake news [17] or fake reviews [2]. We need an automatic discriminator for detecting machine-generated text to avoid these risks.

Past research used sufficient data to fine-tune a pre-trained model as the discriminator and achieved a great performance [5,6,13]. However, past research focuses on detecting articles generated by the left-to-right language model [6,13, 17], which generates articles using a given prefix. In recent years, the text-infilling language model [4,19] has become popular. Text-infilling language models can predict appropriate text to infill the mask tokens in the input prompt. While

H. Kashima et al. (Eds.): PAKDD 2023, LNAI 13935, pp. 367–379, 2023.
https://doi.org/10.1007/978-3-031-33374-3_29

generating, text-infilling language models can apply different masking strategies, which may lead to the generated article having different characteristics. Unlike the left-to-right language model, which can only mask tokens from the end of the article, the text-infilling language model can flexibly choose the position and masking length of each mask, and the masking strategy can be more diverse. Most past research focused on detecting articles generated by the left-to-right language model and using sufficient data to train the discriminator. But in real-world situations, it's hard to collect sufficient data from the same adversary, and the adversary does not necessarily use the left-to-right language model for generating. They may use the text-infilling language model for generating, which is not explored by past research.

For building a powerful discriminator, it is inevitable to use a large amount of data for training. According to past research, the more training data, the better the trained discriminator performs [7]. However, because machine-generated articles can easily deceive humans [2,6,17], it is impossible to label the data by humans. This makes it difficult to collect large amounts of machine data generated by the adversary in real-world situations. Even if we want to generate more machine data by language modeling technologies, it is also difficult because we cannot know the masking strategy of the machine-generated text that the adversary used. If the synthetic machine data is generated by an arbitrary masking strategy, the performance of the discriminator may get worse in detecting articles generated by the adversary.

A masking strategy similar to the adversary is needed to generate synthetic machine data that can enhance the discriminator's performance. Our preliminary analysis found that the trained discriminator uses different features to make predictions depending on the masking strategy of machine data in the training set. The saliency score computed by the trained discriminator can reveal information about the masking strategy of machine data. Based on this observation, we propose the CopyCAT method. CopyCAT can generate synthetic machine data depending on the masking strategy of the adversary, and the generated synthetic machine data can effectively improve the discriminator.

2   Related Work

2.1   Detecting Machine-Generated Articles in Real-World Settings

Our paper adopts the framework proposed by Zellers et al. [17]. They divide this task into two roles, adversary and verifier. To meet the realistic settings, the verifier will have some restrictions.

Adversary: The adversary's objective is to generate articles that humans and the verifier cannot detect.

Verifier: The verifier's objective is to detect articles generated by the adversary. To meet realistic settings, the verifier can only obtain a few machine-generated articles, but human-written articles are unlimited. In addition, the verifier cannot know the adversary's masking strategy.

2.2 Left-to-Right Language Modeling

Left-to-right next token prediction is an unsupervised training method for the left-to-right language model. During training, a previous text is given, and the model learns to predict the next token depending on the previous text. Thereby the model can learn to predict the subsequent text of the given previous text. Currently, many released pre-trained language models are trained with this training method [8, 11, 17].

The left-to-right language model can generate articles by conditional and unconditional generation. The model generates the text following a given prompt when using conditional generation. Conditional generation can control the topic of generated articles by the given prompt. When using unconditional generation, we only give a BOS token, and the model will generate the whole article. Unconditional generation cannot control the generated article into a specific topic. We use GPT2 small to generate machine articles for the experiment.

2.3 Text-Infilling

Text-infilling has become popular in recent years [4, 16, 19]. The text-infilling task aims to infill appropriate text at the masked position. Donahue et al. [4] proposed a training method that enables unidirectional architectures to learn the text-infilling task. This training method solves the problem of using the bidirectional architecture in the text-infilling task. In addition, the trained text-infilling language model can control the generated result of each mask by specifying different mask tokens. We train a text-infilling language model (ILM) using the training method proposed by Donahue et al. on the OpenWebText [1] dataset. The trained ILM will be used to generate machine articles for the experiment.

2.4 Saliency Score

The saliency score is used to explain that the model makes predictions based on which input feature. A basic way to get the saliency score is to compute the gradient of the output logits with respect to the input [12]. The magnitude of the gradient to each input feature is the saliency score of the feature. We use the same method as that of Li et al. [9]. Suppose an article x contains n tokens $x = \{t_1, t_2, ..., t_n\}$ and is convert into word embeddings $E = \{e_1, e_2, ..., e_n\}$ by the embedding layer. We obtain the saliency score of each e_i in E with respect to the predicted score of class c by the following equation, where $S_c(E)$ is the output logits of class c, and w_i is the gradient of e_i.

$$w_i = \frac{\partial S_c(E)}{\partial e_i} \tag{1}$$

$$\text{Saliency score}_i = \|w_i\|_1 \tag{2}$$

3   Preliminary Analysis

In this section, we observe whether the discriminators trained on different training sets will focus on different features when making prediction.

3.1   Datasets

We have created 6 different datasets, and the machine data in each dataset was generated using a single masking strategy, each dataset uses a different masking strategy from the others. During generating, the nucleus sampling with a p value randomly selected from 0.8 to 1.0 was used as the decoding strategy, and the masking strategy of each dataset is as follows.

GPT2(C) 15%, 25%, 35%: GPT2 [11] with conditional generation was used to generate the machine data of these datasets. During generation, articles in the OpenWebText dataset were truncated to 255 (15%), 225 (25%), and 195 (35%) tokens as prompts, and the generator generated tokens up to the maximum sequence length (300 tokens).

GPT2(U): The gpt2-output-dataset[1] released by OpenAI is used, which is generated by GPT2 with unconditional generation.

ILM 7-gram, 20-gram: The trained ILM is used to generate the machine data of these datasets. Before generating, we fine-tune the ILM on the masked data containing only 7-gram or 20-gram masks so that ILM can learn how to generate with this masking strategy.

Each dataset contains 250,000 machine-generated and 250,000 human-written articles for training. To avoid human data causing bias, the same set of human articles is used in each dataset.

3.2   Experiment

We individually fine-tune the pre-trained Bert [3] on the above 6 datasets. To know whether different discriminators focus on different features to make the prediction, we use 3,500 human-written articles in the validation set to compute the saliency score of each token to the machine class.

Figure 1a counts the number of tokens with a saliency score higher than the average saliency score of the data in the 3,500 articles. Figure 1a shows that each discriminator focuses on different features. Discriminators trained on GPT2(C) datasets focus on the later position of articles to make predictions. Since only the BOS token is given while generating the GPT2(U) dataset, the repetition rate at the beginning of each article is relatively higher. The discriminator trained on the GPT2(U) dataset tends to make predictions based on the repetition tokens at the beginning of articles. Discriminators trained on the ILM dataset does not focus on specific positions because the mask positions are randomly

---

[1] https://github.com/openai/gpt-2-output-dataset.

Table 1. We count the character level continuous length of the top 90 tokens with the highest saliency score in each data. This table shows the occurrence times of each length interval. Since the tokenizer of the generator and discriminator is different, we compute the character level's continuous length for a fair comparison.

Length	7-gram	20-gram
0-30	74015	**78479**
30-60	**9550**	8596
60-90	**3081**	2919
90-120	1152	**1242**
120-150	405	**568**
150-180	172	**258**
180-210	52	**124**

Table 2. This table shows the number of times each token interrupted the continuous mask, i.e., the token whose adjacent tokens have a saliency score higher than the average but not themselves.

Token	Interrupt times
the	4704
,	4652
.	4392
of	3324
to	3321
a	2236
and	1928
in	1764
-	1722
##s	818

chosen. To verify whether the two ILM discriminators rely on different features for prediction, we take out the top 90 tokens with the largest saliency score and compare whether there is a difference across the continuous length.

Table 1 shows that the tokens with a high saliency score are more likely to be connected when we use the discriminator trained on the 20-gram dataset to compute the saliency score. Saliency scores reveal the difference between these two masking strategies.

Because it is hard to obtain sufficient data in real-world situations, in Fig. 1b, we investigate the influence on saliency scores when the amount of training data changes. The subset of the GPT2(C) 35% dataset is used in this experiment. Each discriminator in Fig. 1b only trains one epoch. Figure 1b shows that as the amount of training data increases, saliency scores can better reflect the masking strategy of machine data. Even if there is only a small amount of data, the saliency score can reveal the masking strategy of machine data to a certain extent.

4 Methodology

Experiments in Sect. 3 show that the discriminator trained by different training data focuses on different features to make predictions. Using the saliency score computed by the trained discriminator, the masking strategy of the adversary can be revealed to a certain extent. Based on this observation, we propose the CopyCAT method. CopyCAT can use the saliency score computed by the discriminator to mimic the masking strategy of the adversary. The overview of

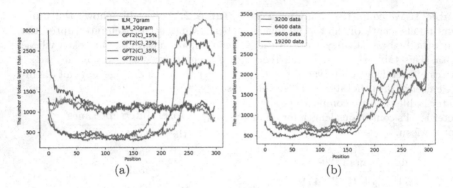

Fig. 1. The x-axis is the position of the token in the article, and the y-axis indicates the number of saliency scores above average.

CopyCAT is shown in Fig. 2. In this section, we introduce CopyCAT in three parts, the training method of the CopyCAT generator, the way to create the masked dataset, and the post-processing methods for the masked dataset.

4.1 Training CopyCAT Generator

Since the masking strategy that the adversary used is unknown. We need a flexible training method that can mimic various masking strategies for training the CopyCAT generator. We don't use the left-to-right training method because the masking strategy of the left-to-right language model has some restrictions. It can't mimic the masking strategy that uses the subsequent text to generate the previous text. Instead, we use the text-infilling training method to train the CopyCAT generator. The text-infilling language model can refer to the context during generation, and we can mimic various masking strategies by adjusting the mask level and position.

4.2 Creating Masked Dataset

The masked dataset should have a similar masking strategy to the adversary. Therefore, adjusting the masking strategy based on the collected machine data is necessary. Combined with the experiment results of Sect. 3, which shows that the information of masking strategy can be revealed by the saliency score of tokens in the input article. To create the masked dataset, we will mask tokens with a high saliency score and select the tokens to be masked by the following three steps.

Due to the varying ranges of saliency scores in each data, using a fixed threshold to determine which token should be masked is difficult. Therefore, we normalize the computed saliency scores to z-scores first. This normalization allows us to use a fixed threshold to select which token should be masked. Secondly, we use the sigmoid function to convert the saliency scores to values between 0 and

1. When performing the sigmoid function, we increase the slope of the sigmoid function so that the gap between the high and low saliency score will be larger, and the token with a saliency score higher than average can be more obvious. In the third step, we specify a value between 0 and 1 as the threshold and mask the tokens with a saliency score higher than the threshold.

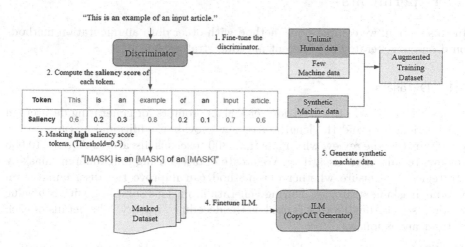

Fig. 2. Overview of CopyCAT

4.3   Post-processing Methods

When creating the masked dataset, we found that some tokens with high frequency tend to have low saliency, resulting in the continuous mask tokens will be interrupted by these tokens. We count the times each token interrupted the continuous mask tokens and show the top 10 in Table 2. Since these tokens usually have little or no meaning, they can't provide additional information during the generation of synthetic machine data. Keeping these tokens in the prompt may limit the generated article and make the data generated by the CopyCAT generator worse. In this section, we test three post-processing methods to reduce the cases where these tokens interrupt the continuous mask.

Masking Interruption Tokens: When the adjacency tokens are masked, the middle token will also be masked.

Gaussian Blur: Use Gaussian blur to smooth the computed saliency scores. When the token with a low saliency score is between tokens with a high saliency score, the saliency score of the token will be raised. In addition, when the token with a high saliency score is between tokens with a low saliency score, the saliency score of this token will also be reduced. This makes the mask positions more continuous.

Masking Low Saliency Tokens: This post-processing method needs to compute the average saliency score of each token first and take the $n\%$ token with the lowest average saliency score as the target token to be masked. The target token will be masked when there are only target tokens between two mask tokens.

5   Experiments

In this section, we compare our method with other data augmentation methods on datasets generated by different masking strategies.

5.1   Datasets

We use the review in the Yelp dataset [18] as prompts for generating datasets in this section. To avoid the length of review providing additional features to the discriminator, the review which less than 300 tokens is discarded. About 110,000 reviews remain after discarding. We create datasets using four different masking strategies to measure whether our method can improve the discriminator on various masking strategies. During generation, nucleus sampling with a p value random select from 0.8 to 1 is used as the decoding strategy. The details of each dataset are as follows.

GPT2(C): GPT2 with conditional generation was used to generate the machine data of this dataset. During generation, we truncate the reviews in the YELP dataset to 195 (35%) tokens as prompts and let the generator generate up to 300 tokens.

GPT2(U): GPT2 with unconditional generation was used to generate the machine data of this dataset. During generation, we use only a BOS token as the prompt and let the generator generate up to 300 tokens.

ILM(Mix): The trained ILM is used to generate the machine data of this dataset. During generation, we randomly mask the sentence, n-gram token, and single token in reviews as prompts. The average ratio of generated tokens to the whole review is about 20%.

ILM(S): The trained ILM is used to generate the machine data of this dataset. During generation, we only use the sentence-level mask to mask the sentences in reviews as prompts. The average ratio of generated tokens to the whole review is about 40%.

Each generator is fine-tuned with 50,000 YELP data to make the generated reviews have a similar style to the YELP dataset before generating the machine data. The remaining 70,000 data are used to create the dataset of the discriminator. In real-world situations, the machine data from the same adversary is hard to collect, so we only use 1,000 machine data in the training set. To balance the amount of data between human and machine labels in the training set, we treat the remaining human data as extra data. This meets the realistic setting that we mentioned in Sect. 2.

Table 3. Dataset statistics.

	Training data	Validation data	Test data	Extra data
Human data	1,000	1,000	5,000	48,039
Machine data	1,000	1,000	5,000	0
Total	2,000	2,000	10,000	48,039

5.2 Data Augmentation Methods

In this section, we introduce the details of baselines and CopyCAT methods.

Conditional Bert (CBert) [15]: We fine-tune a class-conditional Bert (CBert) on the dataset containing extra human and machine data in the training set. The fine-tuned CBert model is used to increase the amount of machine training data to 50,000. The extra human data is added to the human training data to maintain the balance between human and machine data in the training set.

Easy Data Augmentation (EDA) [14]: The amount of machine data is increased to 50,000 by EDA, and the extra human data is added to the human training data.

GPT2 Data Augmentation: We fine-tune GPT2 on the extra human data and then used the fine-tuned GPT2 to generate synthetic machine data. During generation, the prompt is randomly truncated into 150 to 240 tokens. In our experiment, we use half of the extra human data as prompts to generate synthetic machine data and add the other half into human training data to maintain the balance between the two classes.

Random masked ILM: We fine-tune ILM on the extra human data and used the fine-tuned ILM to generate synthetic machine data. Mask positions in the prompt are selected randomly. Similar to the GPT2 Data Augmentation method, we used half of the extra human data as prompts to generate synthetic machine data and added the other half into human training data.

CopyCAT: CopyCAT is similar to the random masked ILM method. The only difference between CopyCAT and random masked ILM is that the mask positions of CopyCAT in the prompt are selected using the method we described in Sect. 4.

The token mask ratio of each data augmentation method except EDA is approximately 35% of the entire article.

5.3 Experiment Results

The experiment results show that task-agnostic data augmentation methods (CBert and EDA) degenerate the discriminators. We thought discriminators have a lower accuracy because the prompts used to generate synthetic data are created by the machine data in the training set, which cause the synthetic

Table 4. This table shows the accuracy(%) of discriminators with different data augmentation methods on each dataset.

Method	Test set				
	GPT2(C)	**ILM(Mix)**	**GPT2(U)**	**ILM(S)**	**Avg**
Training set only	65.3	53.3	**81.1**	57.9	64.4
EDA	53.5	50.7	80.0	51.3	58.9
CBert	54.9	50.7	80.5	50.9	59.3
GPT2 Data Augmentation	**73.9**	53.8	75.2	57.3	65.1
Random masked ILM	59.2	**64.5**	78.6	62.0	66.0
CopyCAT (Origin)	66.2	61.5	74.8	64.5	66.8
CopyCAT (Mask Interruption Tokens)	68.9	62.8	77.0	64.6	68.3
CopyCAT (Gaussian Blur)	71.4	60.8	75.6	65.4	68.3
CopyCAT (Mask Low Saliency Tokens)	70.2	63.3	79.0	**67.8**	**70.1**

data hard to create new linguistic patterns [10], and certain useful features in the origin machine data may be faded when generating synthetic data.

Discriminators achieve better accuracy than task-agnostic data augmentation methods when using GPT2 or Random masked ILM to generate synthetic data. Furthermore, the choice of masking strategy has a significant impact on the performance of discriminators. The synthetic data generated by the GPT2 model does not perform well on ILM datasets and vice versa.

The CopyCAT (Origin) method can improve discriminators' performance on each dataset. However, using the post-processing method to mask the token that interrupts continuous mask tokens can improve the discriminator performance further. The experiment results show that even if we only use a simple post-processing method (Masking Interruption Tokens) to deal with the low saliency tokens, the performance of discriminators improves significantly on each dataset. CopyCAT (Gaussian Blur) improves the performance on datasets with more continuous machine-generated tokens, such as GPT2(C) and ILM(S), but performed worse on the ILM(Mix) dataset, possibly because the ILM(Mix) dataset uses token-level masks to mask the prompt when generating machine data. Gaussian blurring makes the single token hard to be masked, leading to a greater difference between the masking strategy of synthetic machine data and the original machine data, resulting in worse performance. CopyCAT (Masking Low Saliency Tokens) has the best average performance, but it takes time to compute the average saliency score of each token before masking data.

5.4   Selection of Mask Threshold

None of the methods in Table 4 could improve the performance of the discriminator in the GPT2(U) dataset. In this section, we reduced the amount of extra human data to find the best masking threshold for each dataset, and re-run the experiments using whole extra human data on the GPT2(U) dataset. Using the suitable threshold and CopyCAT (Masking Low Saliency Tokens) method, we

Table 5. This table tests the most suitable threshold of each dataset by using the dataset which only contains 5,000 extra human data. The values in this table are the validation accuracies(%) of each threshold.

Dataset	Threshold					
	0.05	0.1	0.2	0.3	0.4	0.5
GPT2(C)	64.0	66.0	67.3	66.4	**68.1**	64.3
ILM(Mix)	55.8	58.3	**59.7**	53.1	56.8	53.0
GPT2(U)	**81.6**	79.1	80.3	76.5	76.1	75.7
ILM(S)	64.1	64.0	**64.8**	62.0	62.6	62.2

improved the accuracy to 83.0% in the GPT2(U) dataset. In addition to the GPT2(U) dataset, we also re-run experiments on other datasets and achieved accuracies of 70.0%, 65.4%, and 66.5% on the GPT2(C), ILM(Mix), and ILM(S) datasets, respectively. The average is about 71.2%. The experimental results show that choosing a suitable masking threshold can further improve discriminators' performance.

6 Conclusion

In this paper, we proposed the CopyCAT method and enhanced the performance of the discriminator regardless of the masking strategy used to generate the datasets. However, we assume the adversary only uses a single masking strategy to mask prompts. When the adversary uses multiple masking strategies to generate articles, CopyCAT may not work well. Besides, the choice of the threshold will significantly affect the CopyCAT method. How to effectively select the threshold and deal with multiple masking strategies are problems that need to be solved.

References

1. Gokaslan, A., Cohen, V., Pavlick, E., Tellex, S.: OpenWebText Corpus (2019)
2. Adelani, D.I., Mai, H., Fang, F., Nguyen, H.H., Yamagishi, J., Echizen, I.: Generating sentiment-preserving fake online reviews using neural language models and their human- and machine-based detection. CoRR abs/1907.09177 (2019). http://arxiv.org/abs/1907.09177
3. Devlin, J., Chang, M.W., Lee, K., Toutanova, K.: BERT: Pre-training of deep bidirectional transformers for language understanding. In: Proceedings of the 2019 Conference of the North American Chapter of the Association for Computational Linguistics: Human Language Technologies, Volume 1 (Long and Short Papers), pp. 4171–4186. Association for Computational Linguistics, Minneapolis, Minnesota (2019). https://doi.org/10.18653/v1/N19-1423. https://aclanthology.org/N19-1423

4. Donahue, C., Lee, M., Liang, P.: Enabling language models to fill in the blanks. In: Proceedings of the 58th Annual Meeting of the Association for Computational Linguistics, pp. 2492–2501. Association for Computational Linguistics, Online (2020). https://doi.org/10.18653/v1/2020.acl-main.225. https://aclanthology.org/2020.acl-main.225

5. Fagni, T., Falchi, F., Gambini, M., Martella, A., Tesconi, M.: TweepFake: about detecting deepfake tweets. CoRR abs/2008.00036 (2020). https://arxiv.org/abs/2008.00036

6. Ippolito, D., Duckworth, D., Callison-Burch, C., Eck, D.: Automatic detection of generated text is easiest when humans are fooled. In: Proceedings of the 58th Annual Meeting of the Association for Computational Linguistics, pp. 1808–1822. Association for Computational Linguistics, Online (Jul 2020). https://doi.org/10.18653/v1/2020.acl-main.164. https://aclanthology.org/2020.acl-main.164

7. Jawahar, G., Abdul-Mageed, M., Lakshmanan, V.S., L.: Automatic detection of machine generated text: a critical survey. In: Proceedings of the 28th International Conference on Computational Linguistics, pp. 2296–2309. International Committee on Computational Linguistics, Barcelona, Spain (Online) (2020). https://doi.org/10.18653/v1/2020.coling-main.208. https://aclanthology.org/2020.coling-main.208

8. Keskar, N.S., McCann, B., Varshney, L., Xiong, C., Socher, R.: CTRL - a conditional transformer language model for controllable generation. arXiv preprint arXiv:1909.05858 (2019)

9. Li, J., Chen, X., Hovy, E., Jurafsky, D.: Visualizing and understanding neural models in nlp. In: Proceedings of the 2016 Conference of the North American Chapter of the Association for Computational Linguistics: Human Language Technologies, pp. 681–691 (2016)

10. Longpre, S., Wang, Y., DuBois, C.: How effective is task-agnostic data augmentation for pretrained transformers? In: Findings of the Association for Computational Linguistics: EMNLP 2020, pp. 4401–4411. Association for Computational Linguistics, Online (2020). https://doi.org/10.18653/v1/2020.findings-emnlp.394. https://aclanthology.org/2020.findings-emnlp.394

11. Radford, A., Wu, J., Child, R., Luan, D., Amodei, D., Sutskever, I.: Language models are unsupervised multitask learners (2019)

12. Simonyan, K., Vedaldi, A., Zisserman, A.: Deep inside convolutional networks: Visualising image classification models and saliency maps. arXiv preprint arXiv:1312.6034 (2013)

13. Solaiman, I., et al.: Release strategies and the social impacts of language models (2019)

14. Wei, J., Zou, K.: EDA: Easy data augmentation techniques for boosting performance on text classification tasks. In: Proceedings of the 2019 Conference on Empirical Methods in Natural Language Processing and the 9th International Joint Conference on Natural Language Processing (EMNLP-IJCNLP), pp. 6382–6388. Association for Computational Linguistics, Hong Kong, China (2019). https://doi.org/10.18653/v1/D19-1670. https://aclanthology.org/D19-1670

15. Wu, X., Lv, S., Zang, L., Han, J., Hu, S.: Conditional BERT contextual augmentation. In: Rodrigues, J., et al. (eds.) ICCS 2019. LNCS, vol. 11539, pp. 84–95. Springer, Cham (2019). https://doi.org/10.1007/978-3-030-22747-0_7

16. Xue, Q., Takiguchi, T., Ariki, Y.: Building a knowledge-based dialogue system with text infilling. In: Proceedings of the 23rd Annual Meeting of the Special Interest Group on Discourse and Dialogue, pp. 237–243. Association for Computational Linguistics, Edinburgh, UK (2022). https://aclanthology.org/2022.sigdial-1.25

17. Zellers, R., et al.: Defending against neural fake news. In: Wallach, H., Larochelle, H., Beygelzimer, A., d'Alché-Buc, F., Fox, E., Garnett, R. (eds.) Advances in Neural Information Processing Systems, vol. 32. Curran Associates, Inc. (2019). https://proceedings.neurips.cc/paper/2019/file/3e9f0fc9b2f89e043bc6233994dfcf76-Paper.pdf
18. Zhang, X., Zhao, J., LeCun, Y.: Character-level convolutional networks for text classification. Advances in neural information processing systems 28 (2015)
19. Zhu, W., Hu, Z., Xing, E.: Text infilling. arXiv preprint arXiv:1901.00158 (2019)

Federated Learning Under Statistical Heterogeneity on Riemannian Manifolds

Adnan Ahmad[1]([✉])[iD], Wei Luo[1][iD], and Antonio Robles-Kelly[1,2][iD]

[1] School of Information Technology, Deakin University, Geelong, VIC 3220, Australia
{ahmadad,wei.luo,antonio.robles-kelly}@deakin.edu.au
[2] Defence Science and Technology Group, Edinburgh, SA 5111, Australia

Abstract. Federated learning (FL) is a collaborative machine learning paradigm in which clients with limited data collaborate to train a single "best" global model based on consensus. One major challenge facing FL is the statistical heterogeneity among the data for each of the local clients. Clients trained with non-IID or imbalanced data whose models are aggregated using averaging schemes such as FedAvg may result in a biased global model with a slow training convergence. To address this challenge, we propose a novel and robust aggregation scheme, FedMan, which assigns each client a weighting factor based on its statistical consistency with other clients. Such statistical consistency is measured on a Riemannian manifold spanned by the covariance of the local client output logits. We demonstrate the superior performance of FedMAN over several FL baselines (FedAvg, FedProx, and Fedcurv) as applied to various benchmark datasets (MNIST, Fashion-MNIST, and CIFAR-10) under a wide variety of degrees of statistical heterogeneity.

1 Introduction

Conventional deep learning approaches need an enormous amount of training data centrally to learn Machine Learning (ML) tasks. In many applications, data exist in silos belonging to different owners. Collecting such data centrally is often infeasible due to privacy and communication concerns. Federated Learning (FL) [12] provides a means for clients with limited data to collaborate in order to train a single global model.

Typical FL training involves multiple communication rounds between a central server and multiple clients. At each round, a central server shares the global model parameters with a (random) subset of the clients. In these methods, the global model is often obtained via aggregation schemes whereby the selected clients initialize their respective local models with the global model parameters, perform a specified number of stochastic gradient descent (SGD) steps on their local data and communicate their respective models (or parameter updates) back to the server. The central server finally aggregate the local updates to update the global model.

In these techniques, the manner in which the central server aggregates local updates is an important consideration that affects how fast the global model

H. Kashima et al. (Eds.): PAKDD 2023, LNAI 13935, pp. 380–392, 2023.
https://doi.org/10.1007/978-3-031-33374-3_30

converges and how biased it may be. Most existing FL aggregation schemes are variants of FedAvg [12], where local parameters are aggregated using an arithmetic mean, weighted by the number of training examples in each client. With FedAvg, FL has shown its superiority in the homogeneous data scenario where all participants carry independent and identically distributed (IID) data. Since each local model is initialized with the same parameters and trained on homogeneously distributed data, every local model follows the same loss landscape trajectory towards its minimum. As a result, all local models converge consistently. However, for data that is not independent and non-identically distributed (non-IID), statistical heterogeneity induces each local model to follow different loss landscape trajectories, causing local models to drift away from each other's optimum. As a result, approaches such as FedAvg, often exhibit catastrophic forgetting in the global model. This problem is also known as "client drift" [7].

Several approaches, such as FedProx [11] and SCAFFOLD [9], have been proposed to mitigate the client drift problem. These methods employ proximal terms in the local objective function of the clients while keeping the standard weighted average approach for model aggregation on the server. The main idea of modifying the local loss function is to penalize local models that tend to deviate from the global model optimum. Note that this treatment may result in slower convergence [14,17]. Here, however, we do not resort to rectifying the local model parameters but rather explicitly use weighting factors for the aggregation operation that reflect how each client model may contribute to a global model.

Thus, we proposed a new aggregation scheme based on the degree of agreement or "proximity" of the client model with the global one. The challenges here are twofold. Firstly, since the model is initialised randomly on both, the clients and the central server, instead of having a natural reference to start with, we require a client model to be consistent with other client models. Secondly, we need to reliably measure the agreement or similarity between models. An option in this regard can be to view the N parameters in a model as a vector in $\mathbb{R}^N$ and to compute the Euclidean distance between the parameter vectors of different models. This is, however, not a robust solution since many model parameters in a deep learning model contain redundant information. Moreover, such an approach may be prone to the curse of dimensionality due to the large number of parameters often found in deep learning models.

As a result, here we opt to employ a metric on a Riemannian manifold spanned by the covariance matrix of the output logits. Here we propose an aggregation scheme, which we name *FedMan*. FedMan utilizes the covariance of the output logits to compare each of the local models with the global one making use of the geodesic distance on a Riemannian manifold. This follows from the notion that the covariance of the model logits belongs to a Riemannian manifold of symmetric and positive-definite (SPD) matrices. This Riemannian treatment of the problem has several desirable advantages. First, the covariance matrix is closely related to the Fisher Information [1], which provides a natural metric reflecting the statistical heterogeneity among clients. Next, this opens up the possibility of using the mathematical machinery associated with Affine Invariant

Riemannian Metrics (AIRMs) and log-Euclidean metrics to perform operations in their native spaces.

Our main contributions are summarized as follows:

- We explicitly address the client divergence problem in the aggregation phase and propose a robust aggregation method based on inter-client consistency.
- We present an approach that leverages the relationship between statistical heterogeneity with the covariance of the output logits in local models, introducing a novel approach based on SPD matrix manifolds.
- Guided by a canonical Riemannian metric on the SPD manifold spanned by the logit covariance matrices, we propose a way to measure local model similarity based on the geodesic distance making use of the Fréchet mean.
- We demonstrate the superior performance of our proposed aggregation scheme by conducting an extensive number of experiments on various image classification tasks.

2   Problem Formulation and Related Work

Consider a set of K clients collaborating in an FL process to learn a global task. The objective at the central server storing the global model with weights w is hence given by:

$$\min_{w \in \mathbb{R}^d} f(w) \quad \text{where} \quad f(w) = \sum_{k=1}^{K} \frac{N_k}{N} f(w_k) \tag{1}$$

$f(.)$ is the loss function, w_k are the weights of the kth local model at the corresponding client which has access to N_k training samples and N is the total amount of training data distributed across all participating clients.

Recall that, in a standard federated averaging approach, such as FedAvg, each participant locally performs a specified number of SGD steps and communicates local weights to the server. The estimates of local parameters at the client indexed k are given by:

$$w_k = w_k - \eta \nabla f(w_k) \tag{2}$$

where, as usual, η is the learning rate. Once the server receives the updates from the clients, it aggregates local updates using the weighted average method given by:

$$w^{t+1} = \sum_{k=1}^{K} \frac{N_k}{N} f(w_k) \tag{3}$$

where we have indexed the central server weights to iteration number t.

Note that recent years have seen a lot of research on the convergence of FedAvg under statistical heterogeneity. Hsu et al. [6] observed that the performance of FedAvg declines with an increase in the degree of statistical heterogeneity in the client's data distributions. Zhao et al. [18] measure the performance degradation rate with the earth mover distance (EMD) between the data distributions of the clients and propose a data-sharing approach to decrease the degree of statistical heterogeneity between the data distributions. Despite its effectiveness, this data-sharing approach is impractical as it conflicts with FL's primary objective of respecting privacy constraints.

Several studies [9–11,16] modify the local loss function of clients to mitigate "client drift". These introduce regularization terms in Eq. 2 via additional hyperparameters that need to be tuned for each task. For instance, FedProx [11] introduces a proximal term in the local loss function to penalize the updates that deviate from the global objective. Similarly, FedDANE [10] incorporate a gradient correction term in the local function to control divergence. SCAFFOLD [9] employs a variance reduction technique in local objectives to control gradient dissimilarity. FedCurv [16] employs the diagonal of the Fisher information matrix to penalize only those parameters that are important to the global model.

Note that the aforementioned approaches address the "client drift" problem via modification in the client's local objective and adopt a standard weighted averaging approach (Eq. 3) to aggregate local updates at the server. Furthermore, these approaches can often be much more expensive than FedAvg in computational terms since they place extra burdens on local clients. FedCurv [16], for example, requires clients to compute the diagonal of the fisher information matrix and communicate it to the server along with local parameter updates. This is important since clients in FL are usually considered resource-constrained devices. It is worth noting in passing that this is an added advantage of our approach, which avoids additional computation on the clients and only assigns extra work to the global server.

3   Riemannian Geometry of SPD Matrices

Before we go on, we require some formalism. Thus, in this section, we cover the fundamental concepts of Riemannian geometry used throughout the remainder of the paper.

Let $M(n)$ be the space of $n \times n$ real matrices and $S(n) = \{V \in M(n) : V^T = V\}$ be the subspace of all symmetric matrices. The subspace $\mathcal{W}(n) = \{\xi \in S(n) : g^T \xi g > 0, \forall g \in \mathbb{R}^n \setminus \{\mathbf{0}\}\}$ contains all the *symmetric positive-definite* (SPD) matrices. SPD matrices are diagonalizable into strictly positive eigenvalues whereby the space of SPD matrices $\mathcal{W}(n)$ forms a differentiable Riemannian manifold $\mathcal{M}$. The derivatives at point ξ lie in the tangent space $\mathcal{T}_\xi$ at point ξ. Recall that each tangent space on the manifold $\mathcal{M}$ has an inner product $\langle, \rangle_\xi$ that varies smoothly from point to point throughout the manifold [3]. This local inner product defines an affine-invariant metric which is a Riemannian one and is invariant under $GL(n)$, a set of $n \times n$ invertible matrices in $M(n)$.

Moreover, an affine-invariant metric is a natural metric in differential geometry and satisfies the following properties:

1. $\delta(\xi_1, \xi_2) = \delta(\xi_2, \xi_1)$
2. $\delta(\xi_1^{-1}, \xi_2^{-1}) = \delta(\xi_1, \xi_2)$

3. $\delta(P^T\xi_1 P, P^T\xi_2 P) = \delta(\xi_1, \xi_2)$ $\forall P \in GL(n)$

where the third property above makes the space of SPD matrices invariant to a projection. This, in turn, allows simplifying complex spaces by projecting data into less complex spaces. The metric is given by:

$$\langle V_1, V_2 \rangle_\xi = \langle \xi V_1 \xi^{-1}, \xi V_2 \xi^{-1} \rangle \tag{4}$$

Further, since the tangent space $S(n)$ is linear and symmetric [2], Eq. (4) can be written as:

$$\langle V_1, V_2 \rangle_\xi = \mathrm{Tr}(V_1 \xi^{-1} V_2 \xi^{-1}) \tag{5}$$

Here we will also employ the exponential and logarithmic maps of the space of SPD matrices above. These project the points on the tangent space $\mathcal{T}_\xi$ at any reference point ξ. Note that an SPD matrix ξ has a logarithmic $log(\xi)$ and a symmetric matrix V has an exponential $\exp(V)$, such that an inverse connection exists between the two assignments [3]. The Logarithmic map Log_ξ projects any point ξ_1 from the manifold to the tangent space at a point ξ, where the operations can be performed in a locally linear space. A logarithmic map is given by:

$$Log_\xi(\xi_1) = (V_1) = \xi^{1/2} \log\left(\xi^{-1/2}\xi_1\xi^{-1/2}\right)\xi^{1/2} \tag{6}$$

Similarly, the exponential map Exp_ξ projects any point V_1 from the tangent space to the manifold $\mathcal{M}$. The exponential map is given as follows:

$$Exp_\xi(V_1) = \xi_1 = \xi^{1/2} \exp\left(\xi^{-1/2}V_1\xi^{-1/2}\right)\xi^{1/2} \tag{7}$$

4   Federated Learning on Riemannian Manifold

The representation of the parameter space of neural networks on an SPD manifold is not a straightforward task. Here we make use of the logits of the neural network to represent its weights w on the manifold $\mathcal{M}$. Recall that for a feed-forward neural network parameterized by w with input $\hat{x}$, logits are defined as:

$$y = z(\hat{x}^{o-1}) = \sigma\left(w^o\hat{x}^{o-1} + b^o\right) \tag{8}$$

where $\sigma(\cdot)$ denotes the soft max function and we have adopted the o supra-index to represent the output layer. The mean and covariance of the logits can be defined as $\mu = [y]$ and $C = [(y - \mu)(y - \mu)^T]$, respectively. Since the covariance matrix C is indeed a family of SPD matrices we can employ the properties in the previous section to use the geodesic distance on the manifold $\mathcal{M}$ so as to perform the aggregation process.

To this end, the global server computes the covariance matrix $\xi_k = C_k$ for each $k \in K$ making use of a small amount of unlabeled data $\hat{D} = (\hat{x})$ which can be easily procured centrally. This covariance matrix can then be used to obtain a geodesic distance on the manifold. Viewed in this manner, the point ξ_k represents the local data distribution of the client k on the manifold, such that

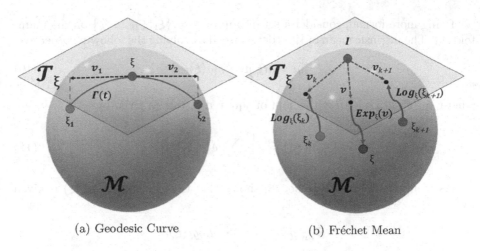

(a) Geodesic Curve   (b) Fréchet Mean

Fig. 1. Figure 1a presents a geodesic curve Γ on Riemannian manifold $\mathcal{M}$. $\mathcal{T}_\xi$ is a tangent space at point ξ on which v_1 and v_2 are the projections of points ξ_1 and ξ_2, respectively. Figure 1b illustrates the procedure to obtain Fréchet Mean of points ξ_k and ξ_{k+1} on the manifold. We start from the identity matrix I and iteratively obtain Fréchet mean ξ on the manifold.

the geodesic is the shortest path that joins two points on the manifold. Since geodesics are analogous to a straight line in Euclidean geometry, our method is consistent with the arithmetic mean for iid data giving rise to a diagonal covariance matrix.

To illustrate this treatment, in Fig. 1a(a), we illustrate how a geodesic $\Gamma(t)$ with arclength $0 \leq t \leq 1$ defines a series of points that joins ξ_1 to ξ_2. In this manner, the initial point of the geodesic is $\Gamma(0)$ at point ξ_1 and its final point is $\Gamma(1)$ at ξ_2. The minimum distance between two points on the manifold is the Riemannian distance that can be computed by computing the arc length on the curve Γ. A Riemannian geodesic distance induced by the natural Riemannian metric in Eq. (4) is given by

$$\delta(\xi_1, \xi_2) = \left\| \log(\xi_1^{-1/2} \xi_2 \xi_1^{-1/2}) \right\|^2 \tag{9}$$

To employ the geodesic distance as a metric to estimate the similarity between local clients, we need to define a reliable reference point on the manifold. This is so that we can calculate the geodesic distance between the reference point and the local covariance matrices. To obtain this reference point, we also need to consider the need to be agnostic regarding the local data distributions of the clients. Interestingly, Fréchet [4] defines the mean of the points on the manifold, and Pennec [15] proposes a gradient decent algorithm to approximate such a mean. This is important since we can use the Fréchet mean of the local covariance matrices as a potential reference point on the manifold so as to use the geodesic distance accordingly.

Being more formal, consider a set of points $W = \{\boldsymbol{\xi}_1, \boldsymbol{\xi}_2, ..., \boldsymbol{\xi}_k\}$ on the manifold $\mathcal{M}$. The intrinsic mean of W is determined by solving the following objective:

$$\mu = \arg\min_{\boldsymbol{\xi} \in \mathcal{M}} \rho_W(\boldsymbol{\xi}) \tag{10}$$

where $\rho_W(\boldsymbol{\xi})$ in Eq. (10) is the sum of squared distances given by:

$$\rho_W(\boldsymbol{\xi}) = \frac{1}{2K} \sum_{i=1}^{K} \delta(\mu, \boldsymbol{\xi}_i)^2 \tag{11}$$

As mentioned above, Karcher [8] shows that the gradient of $\rho_W(\boldsymbol{\xi})$ is given by:

$$\nabla_{\rho_W(\boldsymbol{\xi})} = -\frac{1}{K} \sum_{i=1}^{K} Log_{\boldsymbol{\xi}}(\boldsymbol{\xi}_i) \tag{12}$$

Thus allowing for the gradient descent step sequence presented in Algorithm (1) presented in [15] to be used to solve the minimization problem in Eq. 10. The algorithm starts with $\mu = I$ as the initial point and iteratively finds a final point $\mu = \boldsymbol{\xi}$ such that the sum of the squared distances in Eq. (11) is minimized. A geometrical interpretation of the procedure is shown in Fig. 1(b). Note that, here, these points correspond to the covariance matrices.

Using Eq. (9), we can calculate the geodesic distance between the Fréchet mean and the local covariance updates. Ideally, the geodesic distance $\delta(\boldsymbol{\xi}, \boldsymbol{\xi}_k)$ between $\boldsymbol{\xi}$ and $\boldsymbol{\xi}_k$ is expected to be zero in an ideal IID data setting. This is due to the fact that local updates $\boldsymbol{w}_k$ come from the same data distribution and their covariance matrices $\boldsymbol{\xi}_k$ and hence reside on the same point on the manifold. On the other hand, in non-IID settings, the geodesic distance varies from client to client depending on the degree of heterogeneity in the local data distribution of clients. Intuitively, the local covariance matrices that originated from clients with less heterogeneous data should be far from the covariance matrices that come from clients with extremely heterogeneous data distributions. Therefore, the reliability of each local update can be easily scored based on its distance from the Fréchet mean. Thus, by making use of geodesic distance, contribution

---

Algorithm 1: Fréchet Mean

Data: $\boldsymbol{\xi}_1, \boldsymbol{\xi}_2, ..., \boldsymbol{\xi}_k$
Result: $\mu = \boldsymbol{\xi}$
1 $\mu_0 = I$
2 **do**
3 $\quad \vert \quad v_i = \frac{1}{K} \sum_{k=1}^{K} Log_{\mu_i}(\boldsymbol{\xi}_k)$
4 $\quad \vert \quad \mu_{i+1} = Exp_{\mu_i}(v_i)$
5 **while** $\|v_i\| > \epsilon$;

---

weights of local updates can be calculated as:

$$\tau_k = \frac{\delta(\boldsymbol{\xi}, \boldsymbol{\xi}_k)}{\sum_{l=1}^{K} \delta(\boldsymbol{\xi}, \boldsymbol{\xi}_l)} \tag{13}$$

Equation (13) reflects the notion that the greater the geodesic distance between $\boldsymbol{\xi}_k$ and the Fréchet mean, the more reliable the local update $\boldsymbol{w}_k$ is expect to be. As a result, our proposed method uses the equation above as a substitute for Eq. (3). This yields:

$$\boldsymbol{w}^{t+1} = \sum_{k=1}^{K} \tau_k f(\boldsymbol{w}_k) \tag{14}$$

In Algorithm 2, we show the pseudocode for our approach, which we have named FedMAN. The pseudo-code takes the initial global model, unlabelled data and the number of communication rounds as the server input. For the sake of consistency, we used the same notation and equations as those employed throughout the section. In our implementation, we use Geomstats [13], which is an open-source Python package for computation on manifolds.

Algorithm 2: FedMAN

---

1 **Server Input:** Initial global mdoel $\boldsymbol{w}^t$, unlabelled data $\hat{D} = (\hat{x})$, R: number of communication rounds
2 **Output:** Final global model $\boldsymbol{w}^{t+1}$
3 Let $t = 0$
4 **for** $r = 1$ **to** R **do**
5   Communicate $\boldsymbol{w}^t$ to all clients
6   **for** *For each client* $k \in K$ *in parallel* **do**
7   **Initialize** local model $\boldsymbol{w}_k \leftarrow \boldsymbol{w}^t$
8   **Client Training:** $\boldsymbol{w}_k = \boldsymbol{w}_k - \eta \nabla f(\boldsymbol{w}_k)$
9   **Communicate** updated $\boldsymbol{w}_k$ to the central server.
10   **end**
11   Server computes cov. matrices : $\boldsymbol{\xi}_1, \boldsymbol{\xi}_2, ..., \boldsymbol{\xi}_K \in \mathcal{W}(n)$.
12   Server computes Fréchet mean $\boldsymbol{\xi}$ using algorithm 1
13   **for** $k = 1$ **to** K **do**
14   Server compute Geodesic distance: $D_k = \delta(\boldsymbol{\xi}, \boldsymbol{\xi}_k)$
15   **end**
16   **for** $k = 1$ **to** K **do**
17   Server computes contribution weights: $\tau_k = \frac{D_k}{\sum_{l=1}^{K} D_l}$
18   **end**
19   Server aggregates weights: $\boldsymbol{w}^{t+1} = \sum_{k=1}^{K} \tau_k \boldsymbol{w}_k$
20 **end**
21 **return** $\boldsymbol{w}^{t+1}$

---

5 Experiments

For our experiments, we consider three publicly available image classification datasets. These are the MNIST, Fashion-MNIST and CIFAR-10. The MNIST is a digit classification dataset consisting of 60,000 training and 10,000 test samples of grayscale images each with a size of (28×28). The Fashion-MNIST is a garment classification dataset with the same characteristics as MNIST. CIFAR-10 is a 10 class classification dataset consisting of 50,000 training and 10,000 test samples of RGB images, each having a size of $(3 \times 32 \times 32)$. For all datasets, we compare our method with three baselines. These are FedAvg [12], FedProx [11] and Fedcurv [16]. The selection of these baselines stems from the fact that FedAvg is a widely used baseline that employs a standard aggregation algorithm for FL, whereas the other two are approaches that explicitly address the "client drift" problem under non-IID settings.

We adopt the experimental settings used in [12]. For all our experiments, we distribute training data among 10 clients and assume all clients remain active in each communication round. We employ multi-layer perceptrons (MLPs) for MNIST and Fashion-MNIST datasets. These MLPs consist of 2 hidden layers with 200 units, each with ReLu activations. For CIFAR-10 we train a convolutional neural network (CNN) which consists of three convolutional layers with 3×3 kernels (channel sizes 32,64, and 128) followed by two fully connected layers. We tune the hyper-parameters for all models via cross-validation and set to 5 the number of training epochs between communication rounds with a learning rate of 0.01. For training, we use a stochastic gradient descent (SGD) optimizer with 0.001 weight decay and server momentum of 0.9 at each local client for the local updates. Since FedProx and FedCurv have additional hyper-parameters which are set to 0.01 and 1.0, respectively. To compute local covariance matrices for our FedMAN aggregation approach we assume a small amount of data on the server. To this end, we split the test dataset into two parts. 30% of test data is used for producing local covariance matrices while the remaining test data is used for evaluation. It is worth noting in passing that this is not an overly restrictive condition since this data can be easily obtained in practice by making use of a GAN [5] or communicating the covariances together with the local models at each communication round.

Following [6], we use a Dirichlet distribution to simulate the non-IID setting. In this manner, for each class label, we make a random draw from the Dirichlet distribution $\text{Dir}(\alpha)$. The resulting multinomial distribution is used to determine how many training examples each client is allocated for that particular class label. The degree of heterogeneity depends on the value of α, i.e. a small α value simulates extremely heterogeneous settings and vice versa.

In Table 1 we present the results obtained in our experiments. We repeat each experiment 10 times and report the $mean \pm std$ of the test data accuracy for the global model obtained after 50 communication rounds. FedMAN outperforms the baselines in all classification tasks. Further, note that the difference between the performance of FedAvg and FedMan increases with an increase in the degree of heterogeneity. This is due to the fact that at higher values of α, local points

Table 1. *Mean ± std* of the test accuracy (%) in percentage for the three data sets under consideration when FedMan and the alternatives as used on non-IID data with different α values. The absolute best performance is in bold font.

Dataset	α	FedAvg	FedProx	FedCurv	FedMan
MNIST	0.1	94.52 ± 0.952	94.60 ± 1.041	94.56 ± 0.889	$\mathbf{95.17 \pm 0.479}$
	0.05	90.77 ± 3.851	90.82 ± 3.823	90.81 ± 3.671	$\mathbf{92.50 \pm 2.789}$
	0.01	69.63 ± 5.289	71.30 ± 7.934	67.75 ± 8.108	$\mathbf{72.79 \pm 3.814}$
FashionMNIST	0.1	82.32 ± 1.639	82.16 ± 1.680	82.07 ± 1.869	$\mathbf{83.06 \pm 1.603}$
	0.05	77.74 ± 5.624	78.47 ± 5.263	78.68 ± 5.174	$\mathbf{78.88 \pm 5.955}$
	0.01	58.33 ± 5.094	62.22 ± 1.929	63.75 ± 0.913	$\mathbf{65.25 \pm 1.807}$
CIFAR10	0.1	68.55 ± 0.247	69.0 ± 0.203	68.49 ± 0.254	$\mathbf{69.13 \pm 0.556}$
	0.05	62.47 ± 4.898	62.27 ± 4.202	63.30 ± 4.842	$\mathbf{63.84 \pm 3.827}$
	0.01	46.25 ± 7.230	38.75 ± 7.687	47.68 ± 6.748	$\mathbf{49.32 \pm 5.997}$

on the manifold tend to be closer and their geodesic distances to their Frechet mean are smaller. On the other hand, a small value of α pushes local points away from each other in such a way that their distances to the Frechet mean reflect that they are further apart from each other on the manifold. Finally, in Figs. 2-4 we show the test data accuracy and loss curves as a function of communication round for $\alpha = \{0.1, 0.05, 0.01\}$. These are consistent with our observations made earlier, where FedMAN consistently outperforms alternatives throughout the training process.

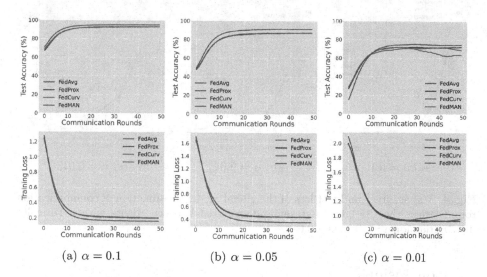

(a) $\alpha = 0.1$ (b) $\alpha = 0.05$ (c) $\alpha = 0.01$

Fig. 2. Test accuracy and test loss of the global model as a function of communication rounds when the MNIST dataset is distributed between clients with a Dirichlet distribution for $\alpha = \{0.1, 0.05, 0.01\}$.

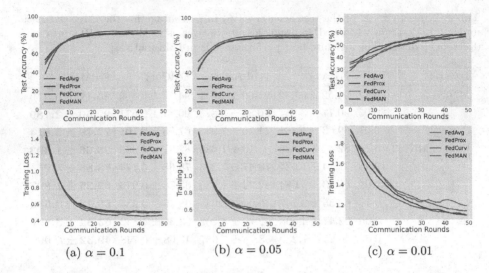

Fig. 3. Test accuracy and test loss of the global model as a function of communication rounds when the Fashion-MNIST dataset is distributed between clients with a Dirichlet distribution for $\alpha = \{0.1, 0.05, 0.01\}$.

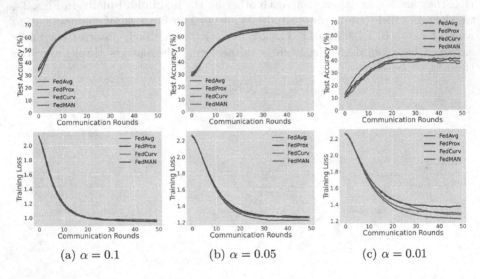

Fig. 4. Test accuracy and test loss of the global model as a function of communication rounds when the CIFAR-10 dataset is distributed between clients with a Dirichlet distribution for $\alpha = \{0.1, 0.05, 0.01\}$.

6   Conclusions

We proposed a novel approach to address statistical heterogeneity in FL. Fed-MAN uses the covariance matrices of the local model output logits and follows

Riemannian Geometry principles to produce aggregation coefficients for each participating model. Unlike existing approaches such as FedCurv, our method does not change the local loss function. Moreover, by leveraging the relationship between statistical heterogeneity and the covariance of the client model output logits it yields a more robust model aggregate. We demonstrate the superior performance of FedMAN on three publicly available datasets for image classification tasks, where our approach consistently outperforms the alternatives.

References

1. Amari, S.: Information Geometry and Its Applications. AMS, vol. 194. Springer, Tokyo (2016). https://doi.org/10.1007/978-4-431-55978-8
2. Fletcher, P.T., Joshi, S.: Principal geodesic analysis on symmetric spaces: statistics of diffusion tensors. In: Sonka, M., Kakadiaris, I.A., Kybic, J. (eds.) CVAMIA/MMBIA -2004. LNCS, vol. 3117, pp. 87–98. Springer, Heidelberg (2004). https://doi.org/10.1007/978-3-540-27816-0_8
3. Förstner, W., Moonen, B.: A metric for covariance matrices (2003)
4. Fréchet, M.: Les éléments aléatoires de nature quelconque dans un espace distancié. In: Annales de l'institut Henri Poincaré, vol. 10, pp. 215–310 (1948)
5. Goodfellow, I.J., Pouget-Abadie, J., et al.: Generative adversarial nets. In: NIPS (2014)
6. Hsu, T., Qi, H., Brown, M.: Measuring the effects of non-identical data distribution for federated visual classification. ArXiv abs/1909.06335 (2019)
7. Kairouz, P., McMahan, H.B., Avent, B., et. al.: Advances and open problems in federated learning. Found. Trends Mach. Learn. **14**, 1–210 (2021)
8. Karcher, H.: Riemannian center of mass and mollifier smoothing. Commun. Pure Appl. Math. **30**(5), 509–541 (1977)
9. Karimireddy, S.P., Kale, S., Mohri, M., et al.: SCAFFOLD: Stochastic controlled averaging for federated learning. In: ICML, pp. 5132–5143 (13–18 Jul 2020)
10. Li, T., Sahu, A.K., et al.: FedDANE: a federated newton-type method. In: 2019 53rd Asilomar Conference on Signals, Systems, and Computers, pp. 1227–1231 (2019)
11. Li, T., Sahu, A.K., et al.: Federated optimization in heterogeneous networks. In: Dhillon, I., Papailiopoulos, D., Sze, V. (eds.) Proceedings of Machine Learning and Systems, vol. 2, pp. 429–450 (2020)
12. McMahan, H.B., Moore, E., et al.: Communication-efficient learning of deep networks from decentralized data. In: AISTATS (2017)
13. Miolane, N., Guigui, N., et. al.: Geomstats: a python package for Riemannian geometry in machine learning. J. Mach. Learn. Res. **21**(223), 1–9 (2020). http://jmlr.org/papers/v21/19-027.html
14. Mitra, A., Jaafar, R.H., Pappas, G.J., Hassani, H.: Linear convergence in federated learning: Tackling client heterogeneity and sparse gradients. In: Neural Information Processing Systems (2021)
15. Pennec, X.: Probabilities and statistics on Riemannian manifolds: basic tools for geometric measurements. In: Proceedings of Nonlinear Signal and Image Processing (NSIP1999), vol. 1, pp. 194–198 (1999)
16. Shoham, N., et al.: Overcoming forgetting in federated learning on non-IID data. In: FL-NeurIPS (2019). arXiv:1910.07796

17. Wang, J., Liu, Q., Liang, H., Joshi, G., Poor, H.V.: Tackling the objective inconsistency problem in heterogeneous federated optimization. Adv. Neural. Inf. Process. Syst. **33**, 7611–7623 (2020)
18. Zhao, Y., Li, M., Lai, L., et al.: Federated learning with non-IID data. ArXiv abs/1806.00582 (2018)

Enhanced Topic Modeling with Multi-modal Representation Learning

Duoyi Zhang[ID], Yue Wang[✉][ID], Md Abul Bashar[ID], and Richi Nayak[ID]

Centre for Data Science, School of Computer Science,
Queensland University of Technology, Brisbane, Queensland 4000, Australia
{duoyi.zhang,y355.wang}@hdr.qut.edu.au, {m1.bashar,r.nayak}@qut.edu.au

Abstract. Existing topic modelling methods primarily use text features to discover topics without considering other data modalities such as images. The recent advances in multi-modal representation learning show that the multi-modality features are useful to enhance the semantic information within the text data for downstream tasks. This paper proposes a novel Neural Topic Model framework in a multi-modal setting where visual and textual information are utilized to derive text-based topic models. The framework includes a **G**ated **D**ata **F**usion module to learn the textual-specific visual representations for generating contextualized multi-modality features. These features are then mapped into a joint latent space by using a **N**eural **T**opic **M**odel to learn topic distributions. Experiments on diverse datasets show that the proposed framework improves topic quality significantly.

Keywords: Deep Learning · Neural Topic Model · Multi-modal Representation Learning

1 Introduction

The growing prominence of user-generated content in Web 3.0 has resulted in a massive explosion of text and image data. Topic modelling methods have been commonly used to understand text data and capture meaningful word patterns [7]. Traditionally, topic models are restricted to solely utilize text data [23]. However, text data often appears with other supplementary data, most commonly with image data. Text-image paired data is commonly seen on social media platforms, review platforms and search engines, and it provides enriched semantic information for human cognitive perception and communication [3]. The meanings of words can be explicitly tied to visual perception [3,18]. A movie poster, for example, is often found to be more precise and expressive than a thousand-word textual description.

In the last few years, a new research area of multi-modal representation learning has emerged [3,21]. This type of method enriches the semantic representation of text data by including other data modalities such as image features.

D. Zhang and Y. Wang—Equal contribution.

H. Kashima et al. (Eds.): PAKDD 2023, LNAI 13935, pp. 393–404, 2023.
https://doi.org/10.1007/978-3-031-33374-3_31

Multi-modal representation learning is reported to improve the performance of subsequent machine learning tasks like classification and clustering [3,21].

In this paper, we propose to combine topic modelling with multi-modal learning and conjecture that images can provide valuable complementary information for topic modelling. Topic modelling with multi-modal data remains a challenging task and received little attention [3]. (1) One primary challenge is how to learn effective multi-modal features in an unsupervised setting. Complexities arise because information from text and image modalities may not be equally important and informative, and text and images may contain irrelevant or inconsistent information. This results in the Feature Mismatch problem [21]. Consequently, simply combining (e.g. addition or concatenation) text-image features may not contribute to a meaningful outcome [21]. (2) Another challenge is incorporating context from multi-modalities into topic models. Moreover, traditional topic models such as Latent Dirichlet Allocation (LDA) [7] are purely based on the Bag-of-words representation [5] which fail to consider word order and context information.

To address the above-mentioned challenges, we propose a novel neural topic model framework in a multi-modal setting where visual and textual information is utilized to generate text-based topics. The **G**ated **D**ata **F**usion **N**eural **T**opic **M**odel (GDF-NTM) consists of two modules: data fusion module and topic modelling module. GDF-NTM exploits the useful interactions between text and image modality and is trained jointly in the same latent space of both text and image features. To verify the effectiveness of our framework, we conduct extensive experiments on three diverse datasets. Empirical analysis shows a significant improvement in topic quality in comparison to the state-of-the-art models. More specifically, the contributions of this paper are listed below:

- The novel data fusion module uses both the text and image as input and learns a joint feature space through the GDF Module through a **G**ating **M**echanism. The gating mechanism exploits the inter-modalities interaction and only learns informative features from the image data. To the best of our knowledge, this is the first work that uses a gating mechanism for topic modelling with multi-modality features.
- We deploy a classic **V**ariational **A**uto **E**ncoder (VAE) based neural topic model architecture to allow the fused features to be projected into a shared latent space and infer the topic words distribution from the latent space.
- We prepare and make a new multi-modal topical dataset, consisting of 5110 text-image pairs from Twitter[1].

2   Related Work

Topic Models in the Age of Deep Neural Networks. The most popular topic modelling method, namely LDA [7], models three important concepts: word (w), documents (d) and topics (z). LDA assumes the observed words in

---

[1]   https://github.com/Duoyi1/GDF-NTM.

each document (i.e. a tweet) are generated by a mixture of corpus-wide K topics. Documents are modelled as mixtures of shared topics. Traditional LDA-based topic models mainly rely on the bag-of-words (BoW) representation of data and are criticized for ignoring the word order and semantic information of documents [23]. To solve this problem, a stream of recent works combines Neural Topic Models with contextualized representation generated from word embedding or pre-trained language models [5,9,22]. The embedding enhanced topic models are reported to produce improved performance (i.e. topic interpretability and coherence) [23]. For example, Embedded Topic Model (ETM) [9] learns interpretable topic vectors and word embedding simultaneously through a Variational Auto Encoder (VAE). It has been shown to outperform LDA-based topic models in finding interpretable topics with large and heavy-tailed vocabularies. Similarly, Contextualized Topic Model (CTM) [5] utilises an encoder-based neural network to map pre-trained contextualized word embeddings (e.g., BERT [8]) to identify topic distributions. Inspired by this success, the proposed GDF-NTM framework employs a neural topic model with contextualised representation that is enhanced by multi-modal representation learning.

Multi-Modal Learning for Topic Modeling. Recently, multi-modal representation learning has gained great popularity in deep learning, especially on downstream tasks such as classification and clustering [3]. A multi-modal representation is generated by a "fusion technique" by learning a joint low-dimensional representation. There exists only a handful of works that introduce multi-modal learning in topic modelling. For instance, Corr-LDA [6] learns the relationship between images and text modalities but without considering the cross-modality interactions. Other models [18] extract 'visual words' from images and use the bag of 'visual words' as the representation of an image for topic modelling. However, the input to the topic model only consists of text data. Different from these approaches, the proposed GDF-NTM framework learns useful features based on cross-modal interactions and projects them into a shared latent space for topic modelling in a pure unsupervised setting.

3 Methodology: The GDF-NTM Framework

Problem Formulation. GDF-NTM aims to model topics from documents in a multi-modal setting by supplementing the text data with another modality. Let each document d consist of a text-image pair, denoted as (d^t, d^v). As an outcome, each document is modelled as a mixture of shared topics where each topic z consists of multinomial distribution over concepts C. Each concept $c \in C$ consists of two parts: a textual part c^t and a visual part c^v. Different from state-of-the-art topic modelling methods [5,7,9,10], each topic depends on both text and image parts of the concept $c \in C$. The probability of a concept $c \in C$ in a topic model with K topics under this setting can be written as follows.

$$p(c_j|d^t, d^v) = \sum_{k=1}^{K} p(c_j|z_k)p(z_k|d^t, d^v) \tag{1}$$

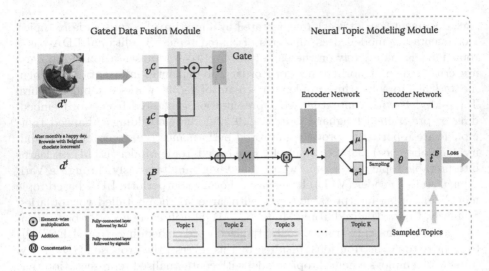

Fig. 1. Architecture of the proposed GDF-NTM framework

We assume that a concept c can be represented with a unigram text (e.g. a single word). That is, each observable concept c_j in a document d can be realised through a representative word w_j in the document. In this multi-modal setting, the objective is to learn text-based topics with the aid of text and visual features. Consequently, to observe a concept c_j, we realise the word discourse $w_j \in d^t$ and use visual cues in d^v to better understand the concept and find more diverse and coherent topics. Therefore, Eq. 1 can be rewritten as:

$$p(w_j \sim c_j^t | d^t, d^v) = \sum_{k=1}^{K} p(c_j | z_k) p(z_k | d^t, d^v) \qquad (2)$$

We use Eq. 2 to implement the framework GDF-NTM. An overview of the framework is given in Fig. 1. GDF-NTM consists of two modules to overcome the two tightly coupled challenges of the feature mismatch problem in multi-model data representation and the context embedding in topic models with multi-model feature learning. Each of these modules and data representation are explained next.

Data Representation. Let the document collection $D = \{d_i = (d_i^t, d_i^v)\}_{i=1}^{N}$ be a set of N number of text-image pairs with the word vocabulary of W. We process the textual part of each document d_i^t to capture the compositional and contextual information. The *BoW* representation presents each document as a vector of length $|W|$, $t^B \in \mathbb{R}^{|W|}$, by counting the Term-Frequency of each unique word in a document. The *Contextualized Text* representation is obtained based on the method described in SentenceBERT [16]. We use the pre-trained RoBERTa model [15], which represents each word in the 768 fixed dimension. We then use the mean pooling strategy to get the embedding vector $t^C \in \mathbb{R}^{768}$ for each $d_i \in D$. The *Contextualized Image* representation is obtained from the well-known pre-trained VGG model [19] which consists of 16 convolution and 3 max-pooling layers. Following [21], we get the representation of each image from the last max-pooling layer. Each image is represented as $(7 \times 7) \times 512$, where 512 represents the number of kernels and (7×7) is the size of each kernel. We then flatten each image into (49×512) and pass it to a fully connected layer. The obtained image representation is a vector with 512 dimensions: $v^C \in \mathbb{R}^{512}$.

Note that the choice of pre-trained embedding models is highly dependent on the dataset. The selection of the best pre-training model is out of the scope of this study. To ensure a fair comparison, we use the same embedding method for each baseline method in the experimental study.

Gated Data Fusion Module. A gating mechanism is commonly used to control the information flow (can be viewed as signals) in deep neural networks [12]. For example, in the long short-term memory (LSTM) model, a forget gate is used to decide which information to drop [11]. In this study, we propose a novel gating mechanism namely Gated Data Fusion Module (GFM) to filter out the irrelevant and inconsistent information from the image modality to solve the **Feature Mismatch** challenge discussed before.

Given the input document d, the objective of GDF is to learn a multimodal representation $\mathcal{M}$ that effectively considers both text (d^t) and image (d^v) modalities to help better understand the concepts before passing them to the next layers of the deep network for topic modelling. We conjecture that the text modality of concepts plays a significant role in topic modelling and the image modality assists in finding concepts for topics. As a result, we use t^C to control the flow of information from v^C. In other words, the text feature t^C is used to decide what information from the image modality v^C to retain, as shown in Fig. 1.

For obtaining controlling vectors, we pass t^C through a single layer NN with a *sigmoid* activation function. This layer transforms t^C into a gate controlling vector t^g, as shown in Eq. 3, by re-weighting each vector component and scaling it between 0 and 1. A value close to 1 (or 0) indicates that an image feature should be retrained (or dropped).

$$t^g = sigmoid(NN(t^C)) \tag{3}$$

To enrich the positive values that encode an image feature, we pass v^C through a single NN layer with a ReLU activation function. This layer enriches v^C to the gate approaching vector v^g (Eq. 4). ReLU activation makes sure that positive values are not altered but the negative values are attenuated.

$$v^g = ReLU(NN(v^C)) \tag{4}$$

We take Hadamard product (i.e. pairwise element multiplication) between v^g and t^g to use text features to determine how much and what part of images to retain for fusing them. As shown in Eq. 5, $\mathcal{G}$ is our retained image information coming out of the gate with the text vector.

$$\mathcal{G} = v^g \odot t^g \tag{5}$$

The superposition principle [13] states that, for all linear systems, the net response obtained by two or more signals is the sum of the responses that would have been caused by each signal individually. Motivated by the superposition principle, we combine the retained (or gate passed) image vector $\mathcal{G}$ and contextual text vector $t^{\mathcal{C}}$ as their superposition and obtain the multi-model vector $\mathcal{M}$ that considers information from both text modality and image modality (Eq. 6). Combining information using the superposition principle also avoids the curse of dimensionality. $\mathcal{M}$ becomes a part of the input to the following NTM module.

$$\mathcal{M} = \mathcal{G} + t^{\mathcal{C}} \tag{6}$$

Neural Topic Modeling (NTM) Module. Recalling the mathematical formulation of the GDF-NTM framework in Eq. 2, the objective of the NTM module is to model and obtain the topic mixture $p(z_k|d_i^t, d_i^v) \propto \theta_{k,i}$ and the corresponding concept distribution $p(c_j|z_k) \propto \phi_{k,j}$, while the only realised observable part is $p(w_j \sim c_j^t|d^t, d^v)$. We propose to use a VAE architecture, a popular approach for neural topic models (NTM) [23], to achieve this objective. In NTM, the hidden variables, θ and ϕ, are modelled by an encoder network and a decoder network, respectively. In this work, we design both the encoder and decoder as a feed-forward neural network.

Encoder Network. We denote $\theta = f_\omega(d)$ by the encoding function where ω is a set of corresponding parameters of the encoder network. We aim to optimize ω to find the best estimation of θ given the input document representation.

For the uni-modal case, it has been empirically demonstrated that using the concatenation of BoW and contextualized text representations produce more coherent topics [5]. Similarly, for the multi-modal case, we encode both the compositional and contextualized information of a document and concatenate $t^{\mathcal{B}}$ and $\mathcal{M}$ (Eq. 7). This becomes input to the encoder network.

$$\hat{\mathcal{M}} = Concat(t^{\mathcal{B}}, \mathcal{M}) \tag{7}$$

The encoder then directly maps the derived input document representation $\hat{\mathcal{M}}$ onto a continuous latent space to approximate $\theta_i \sim (\mu, \sigma^2)$, where μ and σ^2 define the mean and standard deviation of θ_i, respectively. In VAE, this latent space is commonly regularized by a Gaussian Distribution $\mathcal{N}(\mu, \sigma^2)$. Since the sampling process has to be expressed in a way that allows the gradient to be backpropagated through the network, we utilize the reparameterization trick [14] to allow gradient descent possible during optimization, as below.

$$\mu = NN(\hat{\mathcal{M}}; \phi_\mu)$$
$$\sigma = NN(\hat{\mathcal{M}}; \phi_\sigma) \tag{8}$$
$$\theta_i = \mu + \sigma \times \zeta$$

where $\zeta \sim \mathcal{N}(0, I)$.

Decoder Network. Following the structure of the encoder network, let $f_\phi(\theta_i)$ be the decoding function given the encoder network's input θ, where ϕ is the set of learnable parameters. The objective is to find ϕ that gives the best approximation of the original documents. To achieve this, instead of the reverse mapping θ to its input document space, we use the softmax function, as described in Eq. 9, in the decoder network to project the hidden variable θ into individual word over the vocabulary. This output is denoted as the reconstructed BoW representation $\hat{t}^{\mathcal{B}}$, and is used to estimate the concepts distributions $p(w_j \sim c_j | z_k)$ required for topic modeling. We aim to minimize the error between the original BoW ($t^{\mathcal{B}}$) and the reconstructed BoW ($\hat{t}^{\mathcal{B}}$) during optimization.

$$\hat{t}_i^{\mathcal{B}} = softmax(\theta_i^T \times \phi) \tag{9}$$

where $\phi \in \mathbb{R}^{K \times |W|}$ are learnable parameters.

Optimization. The loss function of the GDF-NTM framework consists of two parts, namely KL-divergence loss and reconstruction loss, as shown in Eq. 10. Firstly, VAE regularises the topic distribution $\theta \sim (\mu, \sigma^2)$ for each document by enforcing the distribution to be close to a standard Gaussian distribution by minimising the KL divergence. The KL-divergence loss prevents the model from encoding data far away in the latent space and encourages returned distributions to overlap. Secondly, the reconstruction loss aims to minimize the differences between $t^{\mathcal{B}}$ and $\hat{t}^{\mathcal{B}}$ with the cross entropy loss. The reconstruction loss reassembles the original document's compositional structure to ensure that the produced concepts are meaningful. The overall loss function of GDF-NTM to optimize can be defined as:

$$Loss = \sum_{i=1}^{N} \left\| t^{\mathcal{B}} - \hat{t}^{\mathcal{B}} \right\| + KL[\mathcal{N}(\mu, \sigma^2), \mathcal{N}(0, 1)] \tag{10}$$

Note, in Eq. 9, $\hat{t}^{\mathcal{B}}$ is generated from both the bag-of-word $t^{\mathcal{B}}$ and the contextualized text and image features (i.e. $t^{\mathcal{C}}$ and $v^{\mathcal{C}}$). However, only the bag-of-word part is used in the reconstruction part (Eq. 10). Since, GDF-NTM aims to derive topic words that exist in the $t^{\mathcal{B}}$ given a set of documents, hence recovering only this part by maintaining these token-level features will benefit the downstream task of topic modelling. Reconstructing the contextualized part will not directly benefit the downstream task.

4 Experiments

Experiments are carried out on three datasets with varying lengths and genres. Four state-of-the-art topic modelling methods are chosen for comparison. The choice of hyper-parameter K (Number of topics) is still an open question for topic modelling [23]. We let K vary across different values to compare the model's overall performance for three diverse datasets.

Datasets. Models are evaluated on two open source datasets, *Crisismmd* [1], and *MM-IMDb* [2], and one dataset that is collected by the research team from Twit-

ter and named *Twitter5000*[2] (Table 1). Each data is in the (text-image) pairs format. As there exist limited multi-modal open-source datasets available for topic
modelling, we source *Twitter5000* collected from the Twitter API using the search
tweets endpoint[3] with five keywords including health, movie, music, sports and
cybersecurity. A total of 5110 data instances are collected on 13 Oct. 2022.

Table 1. Statistics of datasets

Dataset	#Instance	#Vocab	Avg length per document
Crisismmd	3802	1972	8
MM-IMDb	25912	5000	36
Twitter5000	5110	1990	15

Evaluation Metrics. (1) *Coherence* estimates the ability of the discovered
topics to be coherent and interpretable. Specifically, C_V is chosen due to its
consistency reported with human judgement [17]. C_V ranges from $[0, 1]$, higher
values indicate more coherent topics. (2) *Diversity* measures the extent of how
diverse the discovered topics are and describes different semantic topical meanings. Specifically, We choose Inversed Rank-Biased Overlap (I-RBO) for topic
diversity measurement [20]. I-RBO ranges from $[0,1]$, higher values indicate more
diverse topics. (3) *Normalized Pair-wise Mutual Information (NPMI)* estimates
the co-occurrence information between two topic words w_i and w_j in a list of
topics [17]. In other words, $NPMI = \frac{p(w_i, w_j)}{p(w_i)p(w_j)}$. NPMI ranges from $[0,1]$, higher
values indicate more coherent and related topics. (4) *Topic Quality* is the product of coherence and diversity to measure the overall quality of a topic [9].

Baseline Models. To the best of our knowledge, there exist no multi-modal
topic modelling methods. Hence, we implemented two multi-modal baselines
to validate the performance of GDF-NTM. (1) We extend BERTopic [10], a
clustering-based topic model, to a multi-modal topic model and name it *Multi-
Modal BerTopic (MM-BERTopic)*. We use concatenation to join text and image
features as the input. We replaced HDBSCAN in BERTopic with K-means to
stay consistent with other parametric baseline methods used in the experiments.
(2) We also compare a variant of GDF-NTM by removing the GDF component
and name it W/o gate. It simply uses concatenation to join text and image
features as the input for topic modelling. (3) Additionally, we use three state-of-
the-art uni-modal baselines, as discussed in the related work section, *LDA* [7],
ETM [9] and *CTM* [5]. The LDA multi-core implementation from Gensim[4] is
used with BoW as the document representation [4]. ETM is implemented from

---

[2] This dataset is available on https://github.com/Duoyi1/GDF-NTM.
[3] https://developer.twitter.com/en/docs/twitter-api.
[4] https://radimrehurek.com/gensim/.

its official GitHub repository[5]. We implement CombinedTM[6] which utilize both BoW and contextualized representations for CTM.

Results and Analysis. Table 2 and Fig. 2 show the experimental results. In most cases, GDF-NTM performed the best or second best in all measures. As shown by the collective measure of Topic Quality, GDF-NTM significantly outperforms all uni-modality and multi-modality baselines. GDF-NTM shows similar performance in all three datasets that contain varied length documents (including short to medium-sized documents). Additionally, as shown in Fig. 2, the performance of GDF-NTM on short, noisy social media dataset is also satisfactory by capturing contextualised multi-modal information rather than the BoW document representation as used in LDA and ETM.

The superior performance of GDF-NTM over MM-BERTopic and W/o gate indicates that the GDF module is essential in learning useful multi-modal features for the topic modelling task. This confirms our prior research [21] in multi-modal learning that using simple concatenation for data fusion results in a high-dimensional feature and does not aid in learning useful cross-modal interactions for downstream tasks. The GDF module employs a gating mechanism that can balance image features based on textual features and helps retain only useful contextualised information for topic modelling.

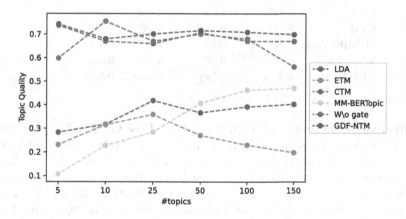

Fig. 2. Topic Quality for the Twitter5000 dataset with K ranging from 5 to 150

CTM also performs well across all datasets, indicating that using contextualised document representation improves topic quality [5]. However, the improved performance of GDF-NTM over CTM (as shown in Fig. 2) suggests that the complementary information from image modality further helps improve the overall topic model performance. In fact, we discovered that learning useful features from text modality can be difficult when the textual information in social media data is noisy. In this case, GDF-NTM is advantageous because it provides additional contextual information from image modality for topic modelling.

---

[5] https://github.com/adjidieng/ETM.
[6] https://github.com/MilaNLProc/contextualized-topic-models.

Table 2. Experimental results averaging K across multiple values (5,10, 25, 50, 100, 150). The top-5 words for each topic are evaluated. The best scores are marked in bold, and the second-best scores are underlined.

	Coherence ↑	Diversity ↑	NPMI ↑	Topic Quality ↑
Crisis-MMD				
LDA	0.464	0.774	-0.006	0.363
ETM	0.313	0.322	-0.113	0.105
CTM	0.464	**0.960**	-0.054	0.445
MM-BERTopic	0.446	0.731	-0.027	0.330
W\o gate	<u>0.510</u>	0.936	<u>-0.002</u>	<u>0.478</u>
GDF-NTM	**0.594**	<u>0.940</u>	**0.103**	**0.560**
MM-IMDb				
LDA	0.550	0.951	0.005	0.521
ETM	<u>0.611</u>	0.932	**0.053**	0.570
CTM	0.594	<u>0.971</u>	0.009	<u>0.576</u>
MM-BERTopic	0.551	0.617	-0.005	0.336
W\o gate	0.561	0.958	-0.005	0.447
GDF-NTM	**0.616**	**0.984**	<u>0.044</u>	**0.606**
Twitter5000				
LDA	0.519	0.698	0.031	0.363
ETM	0.487	0.545	0.022	0.267
CTM	<u>0.693</u>	**0.990**	<u>0.147</u>	<u>0.686</u>
MM-BERTopic	0.412	0.767	-0.098	0.327
W\o gate	0.677	0.976	0.132	0.661
GDF-NTM	**0.723**	<u>0.980</u>	**0.180**	**0.708**

Qualitative Assessment of Topics Obtained by GDF-NMT. To demonstrate the quality of topics obtained by GDF-NTM, we show some of the topics with the highest Topic Quality generated from the Crisismmd dataset. Specifically, we assign the most likely topic to each text-image data instance based on the learnt topic distribution (θ_i). For better visualization, We find the $top - 8$ most probable images based on the topic assignment and the $top - 5$ most probable terms based on the word distribution for each topic accordingly.

Figure 3 presents some of our results. It can be clearly seen that images are highly correlated to their corresponding topic words and they provide a good summary of the given topics from the visual perception. The learned information is coherent for each topic and diverse across different topics. A closer look into Topic 3 and Topic 10 reveals that, despite their similarity, the two topics reflect two distinct events: "Tornado storm" and "Hurricane Maria". This demonstrates that GDF-NTM can effectively discover two related but distinct topics.

In general, GDF-NTM provides an interpretable way to understand the relations between text patterns and image patterns in a large dataset. It has high applicability in practice, e.g., it can assist with multi-modal data exploration and image annotation.

Topic 6
earthquake, iran, iraq, mexico, city

Topic 8
wildfires, california, flooding, fire, wildfire

Topic 3
tornado, warning, storm, tropical, pm

Topic 10
hurricane, maria, track, storm, update

Fig. 3. Illustration of topics by the terms and visual data found by GDF-NTM

5   Conclusion and Future Work

This paper proposes GDF-NTM, a novel neural topic modelling framework that is enhanced by multi-modal representation learning. We extend the topic modelling to allow hidden representation learnt from both text features and image features. We use a gating mechanism to learn the effective cross-modality interactions for data fusion. The experimental results on three datasets demonstrate that GDF-NTM greatly improves the topic quality compared with existing state-of-the-art topic models. In future work, we will explore the use of GDF-NTM for social media data analysis and information retrieval for recommender systems. We will also improve the current approach to make it more robust towards the missing modality from the input data.

References

1. Alam, F., Ofli, F., Imran, M.: Crisismmd: multimodal twitter datasets from natural disasters. In: Proceedings of the 12th International AAAI Conference on Web and Social Media (ICWSM) (June 2018)
2. Arevalo, J., Solorio, T., Montes-y Gómez, M., González, F.A.: Gated multimodal units for information fusion. arXiv preprint arXiv:1702.01992 (2017)
3. Baltrušaitis, T., Ahuja, C., Morency, L.P.: Multimodal machine learning: a survey and taxonomy. IEEE Trans. Pattern Anal. Mach. Intell. **41**(2), 423–443 (2018)

4. Bashar, M.A., Nayak, R., Balasubramaniam, T.: Deep learning based topic and sentiment analysis: Covid19 information seeking on social media. Soc. Netw. Anal. Min. **12**(1), 1–15 (2022)
5. Bianchi, F., Terragni, S., Hovy, D.: Pre-training is a hot topic: Contextualized document embeddings improve topic coherence. arXiv preprint arXiv:2004.03974 (2020)
6. Blei, D.M., Lafferty, J.D.: A correlated topic model of science. Ann. Appl. Stat. **1**(1), 17–35 (2007)
7. Blei, D.M., Ng, A.Y., Jordan, M.I.: Latent Dirichlet allocation. J. Mach. Learn. Res. **3**(Jan), 993–1022 (2003)
8. Devlin, J., Chang, M.W., Lee, K., Toutanova, K.: Bert: pre-training of deep bidirectional transformers for language understanding. arXiv preprint arXiv:1810.04805 (2018)
9. Dieng, A.B., Ruiz, F.J., Blei, D.M.: Topic modeling in embedding spaces. Trans. Assoc. Comput. Linguistics **8**, 439–453 (2020)
10. Grootendorst, M.: Bertopic: neural topic modeling with a class-based tf-idf procedure. arXiv preprint arXiv:2203.05794 (2022)
11. Hochreiter, S., Schmidhuber, J.: Long short-term memory. Neural Comput. **9**(8), 1735–1780 (1997). https://doi.org/10.1162/neco.1997.9.8.1735
12. Huang, Z., Xu, W., Yu, K.: Bidirectional lstm-crf models for sequence tagging. arXiv preprint arXiv:1508.01991 (2015)
13. Illingworth, V.: The Penguin Dictionary of Physics 4e. National Geographic Books (2009)
14. Kingma, D.P., Welling, M.: Auto-encoding variational bayes. arXiv preprint arXiv:1312.6114 (2013)
15. Liu, Y., et al.: Roberta: a robustly optimized bert pretraining approach. arXiv preprint arXiv:1907.11692 (2019)
16. Reimers, N., Gurevych, I.S.B.: Sentence embeddings using siamese bert-networks. arxiv 2019. arXiv preprint arXiv:1908.10084 (1908)
17. Röder, M., Both, A., Hinneburg, A.: Exploring the space of topic coherence measures. In: Proceedings of the Eighth ACM International Conference on Web Search and Data Mining, pp. 399–408 (2015)
18. Roller, S., Im Walde, S.S.: A multimodal lda model integrating textual, cognitive and visual modalities. In: Proceedings of the 2013 Conference on Empirical Methods in Natural Language Processing, pp. 1146–1157 (2013)
19. Simonyan, K., Zisserman, A.: Very deep convolutional networks for large-scale image recognition. arXiv preprint arXiv:1409.1556 (2014)
20. Terragni, S., Fersini, E., Galuzzi, B.G., Tropeano, P., Candelieri, A.: Octis: comparing and optimizing topic models is simple! In: Proceedings of the 16th Conference of the European Chapter of the Association for Computational Linguistics: System Demonstrations, pp. 263–270 (2021)
21. Zhang, D., Nayak, R., Bashar, M.A.: Exploring fusion strategies in deep learning models for multi-modal classification. In: Australasian Conference on Data Mining, pp. 102–117. Springer (2021)
22. Zhang, L., et al.: Pre-training and fine-tuning neural topic model: A simple yet effective approach to incorporating external knowledge. In: Proceedings of the 60th Annual Meeting of the Association for Computational Linguistics (Volume 1: Long Papers), pp. 5980–5989 (2022)
23. Zhao, H., Phung, D., Huynh, V., Jin, Y., Du, L., Buntine, W.: Topic modelling meets deep neural networks: a survey. arXiv preprint arXiv:2103.00498 (2021)

Dynamic Multi-View Fusion Mechanism for Chinese Relation Extraction

Jing Yang, Bin Ji, Shasha Li, Jun Ma, Long Peng, and Jie Yu[✉]

National University of Defense Technology, Changsha, China
yangjing3026@alumni.nudt.edu,
{jibin,shashali,majun,penglong,yj}@nudt.edu.cn

Abstract. Recently, many studies incorporate external knowledge into character-level feature based models to improve the performance of Chinese relation extraction. However, these methods tend to ignore the internal information of the Chinese character and cannot filter out the noisy information of external knowledge. To address these issues, we propose a mixture-of-view-experts framework (MoVE) to dynamically learn multi-view features for Chinese relation extraction. With both the internal and external knowledge of Chinese characters, our framework can better capture the semantic information of Chinese characters. To demonstrate the effectiveness of the proposed framework, we conduct extensive experiments on three real-world datasets in distinct domains. Experimental results show consistent and significant superiority and robustness of our proposed framework. Our code and dataset will be released at: https://gitee.com/tmg-nudt/multi-view-of-expert-for-chinese-relation-extraction

Keywords: Natural Language Processing · Multi-view Learning · Chinese Representation · Chinese Relation Extraction

1 Introduction

Information extraction (IE) is widely considered as one of the most important topics in natural language processing (NLP), which is defined as identifying the required structured information from the unstructured texts. Relation extraction (RE) has a pivotal role in information extraction, which aims to extract semantic relations between entity pairs from unstructured texts. Recently, deep learning-based models have obtained tremendous success in this task. However, research on Chinese RE is quite limited compared to the progress in English corpora. We attribute this to the following main challenge: it is hard to extract semantic information from Chinese texts for the Chinese language makes less use of function words and morphology. Although there has been extensive previous work integrating the external knowledge (i.e. lexicon feature) of Chinese characters is shown to be effective for sequence labeling tasks [1–4], there is room for further investigation to leverage the internal characteristics of Chinese characters.

© The Author(s) 2023
H. Kashima et al. (Eds.): PAKDD 2023, LNAI 13935, pp. 405–417, 2023.
https://doi.org/10.1007/978-3-031-33374-3_32

In Chinese texts, a sentence contains semantic information from different view features including character, word, structure and contextual semantic information. As shown in Fig. 1(a), in order to reduce segmentation errors and increasing the semantic and boundary information of Chinese characters, some methods are proposed to establish a model to learn both character-level and word-level features [3,5]. However, this external knowledge information is limited by the quality of domain lexicons and will inevitably introduce redundant noise. For example, only one of '南京 (Nanjing)','市长 (Mayor)' and '南京市 (Nanjing City)','长江 (YangZi River)' is an appropriate contextual information. In addition, Chinese characters have evolved from pictographs since ancient times, and their structures often reflect more information about the characters. The internal character structures can enrich the semantic representation of Chinese characters. As shown in Fig. 1(b), '氵' is the radical of '江 (River)', and suggests '氵 (water)' that river is water-like liquid. On the contrary, the '南 (south)' can be encoded as a structure consisting of '冇','冂','丷', and '二', but they convey no meaningful semantic information. Previous studies have proven that semantic irrelevant sub-character component information will be noisy for representing a Chinese character [1,4]. Although above methods have achieved reasonable performance, they still suffer from two common issue: (1) The underlying fact that different view feature contains its own specific contribution to the semantic representation is ignored. Existing methods map different view features into a shared space without interaction among views, which is difficult to guarantee that all common semantic information is adequately exploited. (2) Existing methods introduce external knowledge as well as more noisy information, and they suffer from the inability of discriminating the importance of the different features and filtering out the noisy information.

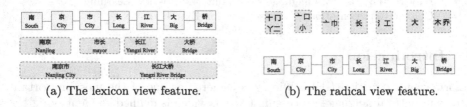

(a) The lexicon view feature.   (b) The radical view feature.

Fig. 1. An example of multi-view features in Chinese language texts.

Is there any other better way to fuse multi-view features of Chinese characters? Inspired by mixture-of-experts [6–8], we propose a novel Mixture-of-View-Experts (MoVE) model to dynamically fuse both internal and external features for Chinese RE. As shown in Fig. 2, MoVE is a method for conditionally computing feature representation, given multiple view expert inputs that can be represented utilizing diverse knowledge sources. In addition, a gating network is designed to dynamically calculate each expert weight per instance based on the multi-view feature. In this way, the knowledge from different view experts can be incorporated to model the inherent ambiguity and enhance the ability to generalize to specific domains. Extensive experiments are conducted on three representative datasets across different domains, Experimental results show that our framework consistently improves the selected baselines

In this paper, we propose a novel multi-view features fusion framework which leverages both external and internal knowledge. The main contributions of this paper can be summarized as follows:

- We design a novel architecture framework capable of acquiring semantic, lexical, and radical feature information from Chinese characters.
- Based on the multi-view features, we propose the MoVE method for dynamically composing the different features for Chinese relation extraction.
- Our method achieves new state-of-the-art performance on three real-world Chinese relation extraction datasets.

2   Related Work

2.1   Chinese Relation Extraction

As a fundamental task in NLP, Relation Extraction (RE) has been studied extensively in the past decade. Here various neural network based models, such as CNNs [9], RNNs [10] or Transformer-based architectures [11] have been investigated. Existing methods for Chinese RE are mostly character-based or word-based implementations of mainstream NRE models. In most cases, these methods only focus on the improvement of the model itself, ignoring the fact that different granularity of input will have a significant impact on the RE models. [12–14]. The character-based model can not utilize the information of words, capturing fewer features than the word-based model. On the other side, the performance of the word-based model is significantly impacted by the quality of segmentation [15]. Then, lexicon enhanced methods are used to combine character-level and word-level information in other NLP tasks like character-bigrams and lexicons information [16–18]. Although, lexicon enhanced models can exploit char and external lexicon information, it still could be severely affected by the ambiguity of polysemy. Therefore, We utilize external linguistic knowledge with the help of HowNet [19], which is a concept knowledge base that annotates Chinese with correlative word synonyms.

2.2   Chinese Character Representation

Existing models of Chinese character representation can be divided into two categories: exploiting the structural information of the characters themselves and injecting external knowledge. JWE [1] is introduced to jointly learn Chinese component, character and word embeddings, which takes character information for improving the quality of word embeddings. LSN [20] is proposed to capture the relations among radicals, characters and words of Chinese and learn their embeddings synchronously. CW2VEC [21] adopts the stroke n-gram of Chinese words and utilizes the fine-grained information associated with word semantics to learn Chinese word embeddings. In order to effectively leverage the external word semantic and enhance character boundary representation, a few models aimed to integrate lexicon information into character-level sequence labeling [3,15,18,22]. Besides, there are models utilize sense-level information with external sememe-based lexical knowledge base ,to handle the polysemy of words with the change of language situation [12,19,23].

2.3 Multi-View Learning

There has been some research to integrate information from different multi-view to achieve better performance. ME-CNER [22] concatenates the character embeddings in radical, character and word levels to form the final character representation, which exploits multiple embeddings together in different granularities for Chinese NER. To fully explore the contribution of each view embedding, FGAT [24] is proposed to discriminate the importance of the different granularities internal semantic features with the help of graph attention network. ReaLiSe [17] leverages the semantic, phonetic and graphic information to tackle Chinese Spell Checking (CSC) task, which introduce the selective fusion mechanism base Transformer [25] to integrate multi-view information. Recent, some efforts incorporate both internal and external multi-view information (such as lattice, glyce, pinyin, n-gram information) with the character token in Chinese Pre-trained language models (PLMs) and design specific pre-train task [26–28]. To the best of our knowledge, this paper is the first work to leverage multi-view information to tackle the Chinese Relation Extraction task.

3   Methodology

An overview of the proposed MoVE framework is depicted in Fig. 2. In this section, we introduce our model architecture from three perspectives: Multi-View Features Representation, Mixture-of-View-Expert, and Relation Classifier.

3.1   Multi-View Features Representation

3.1.1 Internal View Feature

Semantic Embeddings. We adopt BERT [29] as the backbone of the semantic encoder. BERT provides rich contextual word representation with the unsupervised pretraining on large corpora and has been proven superior in building contextualized representations for various NLP tasks [12,17,29,30]. Hence, we utilize BERT as the underlying encoder to yield the basic contextualized character representations. The output of the last layer H_i^c is used as the contextualized semantic embeddings of Chinese characters at the semantic view.

$$H_i^c = \text{BERT}(x_1, x_2, x_3, \ldots, x_n)$$

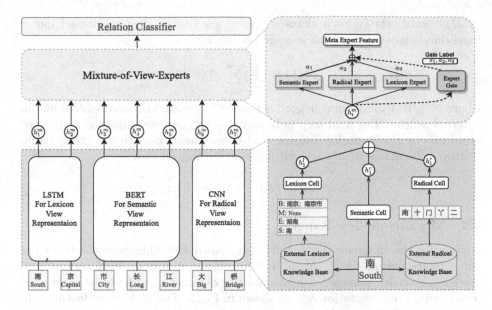

Fig. 2. The architecture of our MoVE framework.

Radical Embeddings. Chinese characters are based on pictographs, and their meanings are expressed in the shape of objects. The radical is often the semantic component and inherently bring with certain levels of semantics regardless of the contexts. In this case, the internal structure of Chinese characters has certain useful information. For example, the radicals such as '月' (body) represents human body parts or organs, and '疒' (disease) represents diseases, which benefits Chinese RE for the medical field. We choose the informative Structural Components (SC) as radical-level features of Chinese characters, which comes from the online XinHua Dictionary[1]. Specifically, we first disassemble the Chinese characters into $SC = (x_1^{c_1}, x_1^{c_2}, x_1^{c_3}, \ldots, x_n^{c_i})$, and then input the radical features into CNN. For example, we can decompose '脚' as '月土厶卩', '疼' as '疒冬丶丶'. Then, we use the max-pooling and fully connection layers to get the feature embedding H_i^r of Chinese characters at the radical view.

$$H_i^r = \text{Max-Pooling}(\text{CNN}(x_1^{c_1}, x_1^{c_2}, x_1^{c_3}, \ldots, x_n^{c_i}))$$

3.1.2 External View Feature

Lexicon Embeddings. Recently, lexical enhancement methods were proposed to enhance character-based models, which have demonstrated the benefits of integrating information from external lexicons for Chinese NER tasks. Following SoftLexicon [3], we retain the segmentation information, all matched words of each character x_i is categorized into four word sets 'BMES'. After obtaining the

---

[1] https://github.com/kfcd/chaizi.

'BMES' word sets for each character, each word set is then condensed into a fixed-dimensional vector with average-pooling method:

$$v^s(S) = \frac{1}{|S|} \sum_{w \in S} e^w(w)$$

where $S \in$ 'BMES' denotes a word set and e^w denotes the word embedding in external lexicon. The final step is to combine the representations of four word sets into one fix-dimension feature, and concatenate them to get the external feature embeddings H_i^l of each character at the lexicon view.

$$H_i^l = \text{concat}(v^s(B); v^s(M); v^s(E); v^s(S))$$

3.2 Mixture-of-View-Experts

After using the aforementioned multi-view feature embeddings methods, we get three representation at h_i^c, h_i^l, h_i^r in semantic-level, lexicon-level and radical-level respectively. Then we concatenate the different view features to get the multi-feature representation h_i^m, as shown in Fig. 2. The multi-view feature representations of Chinese characters h_i^m can capture both internal and external features in different semantic granularity, but they also introduce meaningless and noisy information simultaneously. However, existing methods usually just calculate the unweighted mean of the different view features, or sometimes set the weights as hyper-parameters and calculate the weighted mean of the features [3,24]. Moreover, existing approaches are incapable of distinguishing the significance of the introduced feature. Hence, they are unable to filter out introduced potential noisy information [16,24].

As shown in Fig. 2, we introduce an Mixture-of-View-Experts (MoVE) framework to combine representations generated by experts to produce the final prediction. Specifically, each different feature representation acts as an view expert, which consists of two linear layers. The expert gate consists of a linear layer followed by a softmax layer, which generates the confidence distribution over different view experts. Finally, the meta-expert feature incorporates features from all experts based on the confidential scores from the expert gate. We formulate the MoVE module as follows:

$$[expt_i^1, \cdots, expt_i^E] = [L(h_i^m), \cdots, L^E(h_i^m)]$$
$$[\alpha_1, \cdots, \alpha_E] = \text{softmax}(\text{Linear}(h_i^m))$$
$$h_i^f = \sum_{k=1}^{E} \alpha_k * expt_i^k$$

where h_i^f is the meta-expert feature of h_i, $expt$ is the feature derived from the expert, and L denotes the linear layer. As shown in Fig. 2, the MoVE has three types of experts, namely semantic, radical, and lexicon experts. The expert features are computed based on the multi-view feature representations, and the predictions are conditioned on the meta-expert features and the multi-view features.

3.3 Relation Classifier

After the MoVE model dynamically combines meta-expert feature h_i^f for each token. We first merge h_i^f into a sentence-level feature vector H^f, then the final sentence representation H^f is passed through a softmax classifier to compute the confidence of each relation. We formulate the classifier module as follows:

$$P(y|S) = \text{softmax}(W \cdot H^f + b)$$

where $W \in R^{Y \times d}$ is the transformation matrix and $b \in R^Y$ is a bias vector. Y indicates the total number of relation types, and y is the estimated probability for each type. Finally, given all training examples (S^i, y^i), we define the objective function using the following cross-entropy loss:

$$\mathcal{L}(\theta) = \sum_{i=1}^{T} log P(y^i | S^i, \theta)$$

4 Experiments

4.1 Datasets

We evaluate our approach on two popular Chinese RE datasets: FinRE [12] and SanWen [13]. To increase domain diversity, we manually annotate the SciRE, which is the first Chinese dataset for scientific relation extraction. FinRE is a manual-labeled financial news dataset, which contains 44 distinguished relationships, including a special relation NA. SanWen is a document-level Chinese literature dataset for relation extraction, including 9 relation types specific to Chinese literature articles. The SciRE dataset is collected from 3500 Chinese scientific papers in CNKI[2], which defines scientific terms and relations especially for scientific knowledge graph construction. There are 4 relation types (Used-For, Compare-For, Conjunction-Of, Hyponym-Of) defined in SciRE dataset. Following previous work, we use the same preprocessing procedure and splits for all datasets [12]. Table 1 shows the characteristics of each dataset.

Table 1. Statistics of the three experimental datasets.

Dataset	#Type	Domain	Characteristic	#Train	#Dev	#Test
FinRE	44	Financial	Sentences	4477	500	1219
			Triples	9873	1105	2722
SanWen	9	Literature	Sentences	10754	1108	1376
			Triples	12608	1283	1560
SciRE	4	Scientific	Sentences	7251	1067	1990
			Triples	18548	2778	5148

[2] https://www.cnki.net/.

4.2   Baselines

To investigate the effectiveness of our model, we compare our model with the both character-based and lattice-based variations of the five following models. For the character-based models, we conduct experiments with BLSTM [10], Att-BLSTM [31], PCNN [32] and Att-PCNN [33], which utilize traditional neural network RNN, CNN or Attention mechanism for Chinese relation extraction. We use DeepKE [34], an open-source neural network relation extraction toolkit to conduct the experiments. For the lattice-based models, we compare with Basic-Lattice and MG-Lattice. In Basic-Lattice [15], an extra word cell is employed to encode the potential words, and attention mechanism is used to fuse feature, which can explicitly leverages character and word information. Moreover, MG-Lattice [12] models multiple senses of polysemous words with the help of external linguistic knowledge to alleviate polysemy ambiguity .

We use Chines BERT-wwm [35] as the base semantic encoder for all datasets. We follow the standard evaluation metric and report Precision, Recall, and F1 scores to compare the performance of different models. For each view feature, we implement a linear projection layer with hidden dim 100. We train our model with the AdamW [36] optimizer for 50 epochs. The learning rate is set to 1e-3/5e-5, the batch size is set to 32, and the model is trained with learning rate warmup ratio 10% and linear decay.

4.3   Experimental Results

To conduct a comprehensive comparison and analysis, we conduct experiments on character-based, lattice-based, multiview-based models on three datasets. For a fair comparison, we implement Basic-MultiView semantic encoder by replacing the BERT with a bidirectional LSTM and improving the representation of characters using additional bi-word features. In addition, to verify the capability of our method combined with the pretrained model, we choose our method with the BERT+BLSTM model.

Table 2. F1-scores of Character, Lattice and Multi-View models on all datasets.

Baseline		FinRE	SanWen	SciRE
Character-based	BLSTM	42.87	61.04	87.35
	Att-BLSTM	41.48	59.48	88.47
	PCNN	45.51	81.00	87.86
	Att-PCNN	46.13	60.55	88.78
Lattice-based	Basic-Lattice	47.41	63.88	89.25
	MG-Lattice	49.26	65.61	89.82
MultiView-based	Basic-MultiView	51.01	**68.96**	90.32
	MultiView+biword	**51.56**	68.45	**91.32**
MultiView-(BERT)	BERT+BLSTM	51.43	70.12	91.25
	BERT+MultiView	**53.89**	**72.98**	**92.18**

We report the main results in Table 2, from which we can observe that: (1) The performance of the MultiView-based model that integrates radical feature and lexicon feature into character-based RE models is notably better than the previous RE baseline models. This shows our model can leverage the knowledge from character, radical, and lexicon effectively. (2) We observe that our model consistently outperforms the character-based and lattice-based model on three datasets. (3) We can see that the BERT-MultiView outperforms the BERT-BLSTM, which show that the our method can be effectively combined with pre-trained model. Moreover, the results also verify the effectiveness of our method in utilizing multi-view features information, which means it can complement the information obtained from the pre-trained model. Based on the experimental results above, it makes sense that integrating different granularity and view information and pre-trained model is beneficial for Chinese RE.

4.4 Ablation Studies

We conduct ablation studies to further investigate the effectiveness of the main components in our model on SanWen from two perspectives: Multi-View Features Encoder Layer and Mixture-of-View-Experts (MoVE) fusion mechanism.

Effect Against Multi-View Features Encoder Layer. In this part, we mainly focus on the effect of the different view encoder layer and we conduct following experiment with Basic-MultiView. We consider three model variants setting: w/o semantic layer, w/o lexicon layer, w/o radical layer.

Table 3. Precision, Recall, and F1 score on SanWen.

Model Variant	Precision	Recall	F1
Basic-MultiView	68.55	67.89	68.96
w/o Semantic View	49.47	53.69	51.50
w/o Lexicon View	64.29	67.88	66.04
w/o Radical View	66.78	66.58	66.68

As shown in Table 3, the performance of our model degrades regardless of which view encoder removed. The model without semantic layer has the most significant performance drops, which means that contextual information between entities is the most important feature in our multi-view encoder. The lexicon features are complementary to character-based semantic information in different granularity, which also is consistent with previous studies [3,22]. Finally, we are surprised to find that the less-mentioned radical features, compared to the lexicon feature, appear to bring more benefits. We conclude that radical information that will give more concrete evidence and show more logic patterns to the model.

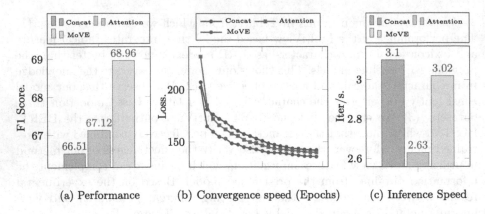

(a) Performance   (b) Convergence speed (Epochs)   (c) Inference Speed

Fig. 3. Compared with different view fusion strategies in Performance, Convergence Speed, Inference Speed.

Effect against MoVE Fusion Layer. In this part, we compare three different multi-view fusion strategies: concat (Concat), attention (Attention) and our mixture-of-view-experts (MoVE). The most intuitive way is Concat method, which just put the different view feature together and feed to classifier layer. In order to interact with the information of different views, the existing work design a fusion layer based on the Attention mechanism [25]. As shown in Fig. 3(a), we find that the performance of our designed fusion module exceeds Concat 1.2% and Attention 0.5% on SanWen, respectively. This is due to the fact that the proposed MoVE can adapt the fusion weight of relevant views based on the properties of datasets, allowing the model to get appropriate composite features.

A potential concern of our model is that the implementation of MoVE brings additional parameters and increases the complexity of the model. To verify the impact of the MoVE module on the efficiency of our model, we use the same hyper-parameters setting to observe the convergence speed of the model during training and the inference speed during prediction under three different view fusion strategies. As show in Fig. 3(b), with the same training settings, the MoVE helps the model to converge faster and be more stable than Concat and Attention strategies. We attribute this to the ability of MoVE to fuse different view information and filter noise information more effectively, which helps the model to learn more efficiently. Figure 3(c) shows the inference speed. Our MoVE does not reduce the inference speed, because in inference stage the gating structure of MoVE only sparsely activated, and will only leverage information from views that have made significant contributions in the training phase.

5   Conclusion

In this paper, we propose a novel multi-view features model for Chinese relation extraction. The proposed model integrates different view representations that fuses semantic, lexicon and radical features through a mixture-of-view-expert mechanism. We conduct extensive experiments on three distinct domain datasets and compare our model with several strong baselines. Experimental results demonstrate that the multi-view Chinese characters can effectively improve the performance for Chinese RE. Furthermore, the ablation studies show mixture-of-view-expert mechanism effectively filters noisy information while improving the efficiency of model training and inference. In further, we will consider how to integrate more knowledge such as part-of-speech, dependency and syntax information and extend it to other language and NLP tasks.

Acknowledgments. This research was supported partly by the science and technology innovation program of Hunan province under grant No. 2021GK2001

References

1. Yu, J., Jian, X., Xin, H., Song, Y.: Joint embeddings of Chinese words, characters, and fine-grained subcharacter components. In: Empirical Methods in Natural Language Processing (2017)
2. Meng, Y., et al.: Glyce: Glyph-vectors for Chinese character representations. In: Neural Information Processing Systems (2019)
3. Ma, R., Peng, M., Zhang, Q., Wei, Z., Huang, X.: Simplify the usage of lexicon in Chinese NER. In: Meeting of the Association For Computational Linguistics (2020)
4. Shi, J., Sun, M., Sun, Z., Li, M., Gu, Y., Zhang, W.: Multi-level semantic fusion network for Chinese medical named entity recognition (2022)
5. Wu, S., Song, X., Feng, Z.H.: MECT: multi-metadata embedding based cross-transformer for Chinese named entity recognition. In: Meeting of the Association for Computational Linguistics (2021)
6. Shazeer, N., et al.: Outrageously large neural networks: the sparsely-gated mixture-of-experts layer. Learning (2017)
7. Ma, J., Zhao, Z., Yi, X., Chen, J., Hong, L., Chi, E.H.: Modeling task relationships in multi-task learning with multi-gate mixture-of-experts. Knowledge Discovery and Data Mining (2018)
8. Liu, Z., Winata, G.I., Fung, P.: Zero-resource cross-domain named entity recognition. In: Meeting of the Association for Computational Linguistics (2020)
9. Zeng, D., Liu, K., Lai, S., Zhou, G., Zhao, J.: Relation classification via convolutional deep neural network. In: International Conference on Computational Linguistics (2014)
10. Zhang, D., Wang, D.: Relation classification via recurrent neural network. arXiv:1508.01006 Computation and Language (2015)
11. Wu, S., He, Y.: Enriching pre-trained language model with entity information for relation classification. In: Conference on Information and Knowledge Management (2019)
12. Li, Z., Ding, N., Liu, Z., Zheng, H.T., Shen, Y.: Chinese relation extraction with multi-grained information and external linguistic knowledge. In: Meeting of the Association for Computational Linguistics (2019)
13. Xu, J., Wen, J., Sun, X., Su, Q.: A discourse-level named entity recognition and relation extraction dataset for Chinese literature text. arXiv:1711.07010 Computation and Language (2017)

14. Zhang, Q.Q., Chen, M.D., Liu, L.Z.: An effective gated recurrent unit network model for Chinese relation extraction. DEStech Transactions on Computer Science and Engineering (2018)
15. Zhang, Y., Yang, J.: Chinese NER using lattice LSTM. In: Meeting of the Association for Computational Linguistics (2018)
16. Zhou, X., Zhang, T., Cheng, C., Song, S.: Dynamic multichannel fusion mechanism based on a graph attention network and BERT for aspect-based sentiment classification (2022)
17. Xu, H.D., et al.: Read, listen, and see: Leveraging multimodal information helps chinese spell checking. In: Meeting of the Association for Computational Linguistics (2021)
18. Wang, B., et al.: Dylex: Incorporating dynamic lexicons into BERT for sequence labeling. In: Empirical Methods in Natural Language Processing (2021)
19. Dong, Z., Dong, Q.: HowNet - a hybrid language and knowledge resource. In: International Conference Natural Language Processing (2003)
20. Song, Y., Shi, S., Li, J.: Joint learning embeddings for Chinese words and their components via ladder structured networks. In: International Joint Conference on Artificial Intelligence (2018)
21. Shaosheng, C., Lu, W., Zhou, J., Li, X.: cw2vec: Learning Chinese word embeddings with stroke n-gram information. National Conference On Artificial Intelligence (2018)
22. Xu, C., Wang, F., Han, J., Li, C.: Exploiting multiple embeddings for chinese named entity recognition. Conference on Information and Knowledge Management (2019)
23. Qi, F., Yang, C., Liu, Z., Dong, Q., Sun, M., Dong, Z.: OpenHowNet: an open sememe-based lexical knowledge base. arXiv:1901.09957 Computation and Language (2019)
24. Wang, X., Xiong, Y., Niu, H., Yue, J., Zhu, Y., Yu, P.S.: Improving Chinese character representation with formation graph attention network. In: Conference on Information and Knowledge Management (2021)
25. Vaswani, A., et al.: Attention is all you need. Neural Information Processing Systems (2017)
26. Sun, Z., et al.: ChineseBERT: Chinese pretraining enhanced by glyph and pinyin information. Meeting of the Association for Computational Linguistics (2021)
27. Chen, Q., Li, F.L., Xu, G., Yan, M., Zhang, J., Zhang, Y.: DictBERT: dictionary description knowledge enhanced language model pre-training via contrastive learning. In: International Joint Conference on Artificial Intelligence (2022)
28. Lai, Y., Liu, Y., Feng, Y., Huang, S., Zhao, D.: Lattice-BERT: leveraging multi-granularity representations in Chinese pre-trained language models. North American Chapter of the Association for Computational Linguistics (2021)
29. Devlin, J., Chang, M.W., Lee, K., Toutanova, K.: BERT: pre-training of deep bidirectional transformers for language understanding (2022)
30. Guan, T., Zan, H., Zhou, X., Xu, H., Zhang, K.: CMeIE: construction and evaluation of Chinese medical information extraction dataset. In: International Conference Natural Language Processing (2020)
31. Zhou, P., et al.: Attention-based bidirectional long short-term memory networks for relation classification. Meeting of the Association for Computational Linguistics (2016)
32. Lin, Y., Shen, S., Liu, Z., Luan, H., Sun, M.: Neural relation extraction with selective attention over instances. Meeting of the Association for Computational Linguistics (2016)

33. Lee, J., Seo, S., Choi, Y.S.: Semantic relation classification via bidirectional LSTM networks with entity-aware attention using latent entity typing. Symmetry (2019)
34. Zhang, N., et al.: DeepKE: a deep learning based knowledge extraction toolkit for knowledge base population (2022)
35. Cui, Y., et al.: Pre-training with whole word masking for Chinese BERT. Speech, and Language Processing, IEEE Transactions on Audio (2021)
36. Loshchilov, I., Hutter, F.: Fixing weight decay regularization in Adam (2018)

Alignment-Aware Word Distance

Guoliang Zhao[1], Jinglei Zhang[1], DongDong Du[2(✉)], Qing Gao[1(✉)],
Shujun Lin[3], Xiongfeng Xiao[4], and Yupeng Wu[1]

[1] National Engineering Research Center for Software Engineering, Peking University,
Beijing, China
{1801120010,gaoqing,wuyupeng}@pku.edu.cn, jingeli.zhang@stu.pku.edu.cn
[2] China Academy of Industrial Internet, Beijing, China
dudongdong@china-aii.com
[3] Institute for Global Health and Development, Peking University, Beijing, China
linshujunpku@126.com
[4] Aegis Information Technology, Nanjing, China
xiaoxiongfeng@pku.edu.cn

Abstract. Recent approaches [6,20] formulated the task of unsupervised Semantic Textual Similarity (STS) as Earth Mover's Distance problem, demonstrating superior interpretability and competitive performance. The main idea behind is using various word distances (or word dissimilarity) as word transportation cost, and then measure text dissimilarity by optimizing the accumulative cost of transporting all words of texts. However, these approaches use static word distance without considering the context of text pairs. Intuitively, the distance of two words tends to contribute more to text dissimilarity if they are well-aligned between texts. Inspired by this observation, we propose **A**lignment-aware **W**ord **D**istance (AWD), which leverages prior word alignment information of sentence pairs to refine word transportation cost. Specifically, we design two simple and effective mechanisms to capture prior alignment knowledge via exploiting word position and syntactic dependency, respectively. By incorporating AWD, our method remarkably outperforms current state-of-the-art models on STS tasks.

Keywords: Earth Mover's Distance · Word Distance · Semantic Textual Similarity

1 Introduction

The semantic textual similarity (STS) task aims to rate the semantic similarity degree of two sentences [2], which is a fundamental language understanding problem related to many natural language processing applications, including machine translation, question answering, semantic search, and dialog systems [3].

Z. Guoliang and Z. Jinglei—Equal Contribution.

H. Kashima et al. (Eds.): PAKDD 2023, LNAI 13935, pp. 418–429, 2023.
https://doi.org/10.1007/978-3-031-33374-3_33

Due to the simplicity and efficiency of unsupervised STS, it has received grow-ing attention in recent years. According to [20], existing approaches for unsuper-vised STS can be divided into two categories. The first one is *alignment-based approaches* [6,20] that measure the degree of semantic overlap between two sentences by considering word alignment. The second one is *sentence vector approaches* [5,13], which exploit word embeddings to generate sentence vec-tors and then calculate their similarity. This paper focuses on alignment-based approaches.

One popular line of alignment-based approach is casting STS to the well-studied Earth Mover's Distance (EMD) problem [14]. As a pioneering work that introduces EMD, Word mover's distance (WMD) [6] measures the dissimilarity between two texts as the minimum cost of transporting the embedded words of one text to the counterparts of another text. In WMD, the mass of each word to transport within a text is equally divided, and Euclidean distance is used as the transportation cost (word dissimilarity) between word vectors. Word rotator's distance (WRD) [20] further improves WMD by utilizing the norm of a word vector to weight the mass and then exploiting the angle between word vectors to calculate transportation cost.

Despite their excellent interpretability and competitive performance, most of the above approaches calculate word dissimilarity statically without consid-ering the context of the sentence pair. In other words, we hypothesize that the transportation cost for the same word pair may diversify given different sen-tence pairs that contain them. Intuitively, for a word pair (e.g., transporting a source word to a target word), if the source word and the target word align well between two texts (e.g., the roles or function of the source and target word in their sentences are similar), the word pair's dissimilarity will contribute more to the text pair's dissimilarity. This observation inspires us that alignment-aware transportation cost is essential for EMD-based STS models. From another per-spective, since solving the EMD problem is actually to yield a word alignment strategy between two texts, leveraging prior alignment information to refine the inputs (transportation cost) of EMD could potentially lead to a better solution.

One straightforward way to measure word alignment degree is by considering word positions in texts. If the source word and target word have similar posi-tions in their texts, the transportation cost can be magnified. Otherwise, we will weaken the transportation cost by a scale factor. Surprisingly, by refining the word transportation cost via this simple position-based alignment strategy, we achieved extensive performance improvement on STS tasks (Sect. 5.1).

By looking into text pair samples, we find a large portion of the sentence pairs in the data set (e.g., about three-quarters of the sample in the STS-Benchmark dataset [3]) share very similar length and structure (Sect. 5.2). Therefore, the improvements brought by the position-based alignment strategy may attribute to the bias of the datasets. We can easily imagine that words with similar positions can play very different roles in their sentences give more complex scenarios. Hence we further propose a syntax-based alignment strategy. The intuition is that we can roughly identify the role or function of one word by checking the syntactical dependency tree, so we can make a better word alignment by mining prior syntax information. If their local structure in the syntactical dependency tree is similar, we can safely assign a better alignment degree. Experimental results demonstrate that the syntax-based mechanism provides a robust and practical approach to handling more complex textual similarity comparison scenarios (Sect. 5.2).

In summary, our main contributions are as follows:

- We propose **A**lignment-aware **W**ord **D**istance (AWD), which leverages prior word alignment information of sentence pairs to measure word dissimilarity (or word transportation cost).
- We design two simple and effective mechanisms to capture prior word alignment knowledge beneficial to unsupervised Semantic Textual Similarity (STS) tasks, demonstrating impressive performance improvements compared with previous SOTA models.

2   Background

2.1   Earth Mover's Distance

EMD [14] aims to minimize the transportation cost from one pile dirty to another. In practice, researchers usually set two distributions u and u' and minimize the transportation cost between them. EMD has the following two inputs:

1) Two probability distributions u (initial distribution) and u' (target distribution).

$$u = \left\{(x_i, p_i)\right\}_1^m, u' = \left\{(x_j', p_j')\right\}_1^n. \tag{1}$$

In which each point x_i has a probability p_i, as shown in Fig., where $p_i \in [0,1]$ and $\sum_{i=1}^m p_i = 1$.

2) The transportation cost function c. c is used to compute the transportation cost between two points x_i and x_j' from different distributions, thus a transportation cost matrix $\mathbf{C} \in \mathbb{R}^{m \times n}$ will be generated.

Let $\mathbf{T} \in \mathbb{R}^{m \times n}$ be a matrix where $T_{ij} \geq 0$ denotes how much of x_i travels to x_j'. To transform x_i entirely into x', EMD ensures that the entire outgoing flow from x_i equals p_i, i.e. $\sum_j T_{ij} = p_i$. Further, the amount of incoming flow to x_j' must match p_j', i.e. $\sum_i T_{ij} = p_j'$.

Transportation Problem. Finally, the minimum cumulative cost of moving u and u' given the constraints is provided by the solution to the following linear program:

$$\min_{\mathbf{T} \geq 0} \sum_{i,j} \mathbf{T}_{ij} c(x_i, x_j')$$

$$s.t. \begin{cases} \sum_{j=1}^n \mathbf{T}_{ij} = u_i, \\ \sum_{i=1}^m \mathbf{T}_{ij} = u_j' \end{cases} \tag{2}$$

2.2 Word Mover's Distance and Word Rotator's Distance

Kusner, M. et al. [6] first introduced EMD into STS tasks and proposed Word Mover's Distance (WMD), which first measures the word moving distance between two words, and then applies EMD to compute the dissimilarity between two sentences. WMD takes the following inputs.

1) Two probability distributions. For a given sentence, WMD sets every word has the same weight (probability).

2) The transportation cost function. WMD uses Euclidean distance between two pre-trained word embeddings as the transportation cost.

Yokoi, S. et al. [20] improved WMD as Word Rotator's Distance (WRD) by using the norm of word as weight, and use the dissimilarity of word vectors' angle to measure the transportation cost.

3 Alignment-Aware Word Degree

By leveraging the alignment information of word position and syntactic dependency, we propose two simple yet effective strategies to capture prior alignment knowledge, named **Alignment-aware Word Degree (AWD)**. Assume that we have a sentence pair $s_1 = (x_1, x_2, ..., x_m)$ and $s_2 = (x_1', x_2', ..., x_n')$, where x_i is the ith word in s_1, and x_j' is the jth word in s_2. The key insight behind AWD is that if x_i and x_j' align well, the word pair's dissimilarity will contribute more to the sentence pair's dissimilarity.

3.1 Position-Based AWD

For the given two words x_i and x_j', the position-based alignment degree leverages the prior alignment knowledge via exploiting word position, and thus refines the transportation cost between x_i and x_j'. If the sentence lengths are different, we first pad the shorter sentence s_1 into the same length with the longer sentence s_2, denoted as $\hat{s}_1$. We simply add the special token $< pad >$ at the beginning and the end of s_1, making the s_1 at the middle of $\hat{s}_1$. Then we compute the position-based alignment degree between x_i and x_j' from different sentences, which considers the exact position-based alignment information between them.

$$d_{ij}^{pd} = \begin{cases} 1 - \frac{|i-j|}{\alpha}, & \text{if } |i-j| < \alpha, \\ 0, & else, \end{cases} \tag{3}$$

where d_{ij}^{pd} is the position-based alignment degree between x_i and x_j', which we named as AWD(P). AWD(P) can be used to refine the transportation cost between words. We simply set the threshold value as $\alpha = 10$ for all datasets.

3.2 Syntax-Based AWD

Though AWD(P) could obtain impressive improvements in various datasets (Sect. 5.1), when the two sentences have different syntactic structures, AWD(P) may misuse the meaningless alignment information. In this subsection, we introduce Syntax-based AWD, denoted as AWD(S), which leverages syntactic dependency to assign a more reasonable alignment degree.

For a given sentence s, every word in s has exactly one dependency relation with its head, and every word may contain multiple children [9]. Thus for word x, the dependency relations between its head and its children describe the contextual information and syntactic dependency.

AWD(S) is computed using the following two types of alignment degree: 1) **Head Alignment Degree**, which represents the alignment degree between x_i and x_j' according to the syntactic relation with their head; 2) **Children Alignment Degree**, which depicts the alignment degree between x_i and x_j' depending on syntactic relations with their children. Finally, the above two alignment degrees will be combined to refine the transportation cost.

The head alignment degree simply indicates if x_i and x_j' have the same dependency relation with their head. Intuitively, if two words x_i and x_j' have the same dependency relation with their heads, the dissimilarity between them tends to contribute more to the sentence pair's dissimilarity. Specifically, given the word x_i, AWD(S) utilizes SDP to obtain the dependency relation with its head, denoted as x_i^h, then the head alignment degree is computed as follows:

$$d_{ij}^{hd} = \begin{cases} 1, & \text{if } x_i^h = x_j^{h'}, \\ 0.5, & else. \end{cases} \tag{4}$$

The children alignment degree represents the similarity between x_i and x_j' in terms of the dependency relations with their children. Given the word x_i, AWD(S) utilizes SDP to obtain the dependency relation set with its children, denoted as $\mathbf{S}^i$. Then we use Jaccard similarity to measure the children alignment degree between x_i and x_j' based on $\mathbf{S}^i$ and $\mathbf{S}^{j'}$.

$$d_{ij}^{cd} = \frac{\mathbf{S}^i \cap \mathbf{S}^{j'}}{\mathbf{S}^i \cup \mathbf{S}^{j'}}, \tag{5}$$

where $\mathbf{S}^i$ and $\mathbf{S}^{j'}$ are two dependency relation sets of their children, $\mathbf{S}^i \cap \mathbf{S}^{j'}$ is the intersection between $\mathbf{S}^i$ and $\mathbf{S}^{j'}$, and $\mathbf{S}^i \cup \mathbf{S}^{j'}$ is the union.

We simply add the d_{ij}^{hd} and d_{ij}^{cd} to generate syntax-based alignment degree.

$$d_{ij}^{dep} = d_{ij}^{hd} + d_{ij}^{cd}, \tag{6}$$

where $d_{ij}^{dep} \in [0.5, 2]$. Compared with AWD(P), AWD(S) could assign better alignment degree for transportation cost under more complex scenarios.

3.3 Imbalanced Alignment Adaptation

When the lengths of two sentences differ extensively, we face an imbalanced word transportation problem. For example, consider the following two sentences:

Sentence 1: *Some good news for a change More lives being saved - cancer death rates drop 20%.*

Sentence 2: *More lives being saved: Cancer death rates drop 20%.*

The ground truth STS score for the sentence pair is 4.8 (ranging from 0 to 5), while existing EMD-based methods usually will predict a lower score. It means that EMD-based techniques may increase dissimilarity improperly in these complex scenarios. As a rough guess, the reason may lie in that every source word will generate transportation cost for every target word, hence accumulating redundancy transportation cost that leads to enlarged semantic dissimilarity in this imbalanced scenario.

To tackle this problem, we design an adaptation mechanism for word transportation costs. In particular, we generate a global imbalance ratio for the sentence pair to scale down word transportation cost in-between. We compute the *imbalance ratio* ρ between two sentences by the following:

$$\rho = \begin{cases} \dfrac{Z}{Z'}, \text{if } Z' \geq Z, \\ \dfrac{Z'}{Z}, \text{if } Z' < Z, \end{cases} \tag{7}$$

where Z and Z' represent the sum of all words' norm in the source sentence and the target sentence, and $\rho \in (0,1]$. Using ρ to adjust the dissimilarity between words, we achieve surprisingly promising results (see Sect. 5.1).

3.4 Transportation Problem

Now we can incorporate our word alignment strategies and imbalanced alignment adaptation into transportation problem solver. We take AWD(S) as an example.

1) Two probability distributions. Following WRD, we use the norm of each word as weighting factor.

$$u = \left\{ (x_i, \tfrac{\lambda_i}{Z}) \right\}_1^m, u' = \left\{ (x'_j, \tfrac{\lambda'_j}{Z'}) \right\}_1^n, \tag{8}$$

where λ_i and λ'_j are the norm of word x_i and x'_j, Z and Z' are normalizing constants $(Z = \sum_i^m \lambda_i, Z' = \sum_j^n \lambda'_j)$

2) The transportation cost function. We first use cosine distance between two pre-trained word embeddings as the transportation cost:

$$c_{cos}(x_i, x'_j) = 1 - cos(w_i, w'_j), \tag{9}$$

where w_i and w'_j are the pre-trained word embeddings of x_i and x'_j, and $c_{cos}(x_i, x'_j) \in [0,2]$. Then we use the c_{ij}^{dep} and ρ to scale the transportation cost for every word pairs, and the final transportation problem is computed as follows:

$$\min_{\mathbf{T}\geq 0} \sum_{i,j} \mathbf{T}_{ij} c_{cos}(x_i, x_j')d_{ij}^{dep} \rho$$

$$s.t. \begin{cases} \sum_{j=1}^{n} \mathbf{T}_{ij} = u_i, \\ \sum_{i=1}^{m} \mathbf{T}_{ij} = u_j' \end{cases} \tag{10}$$

Finally, we can solve the above linear programming problem with constraints and obtains the dissimilarity between two sentences.

4 Experimental Settings

4.1 Task Definition and Evaluation Criterion

The goal of **Semantic Textual Similarity** is to measure the semantic similarity between two sentences, and **Unsupervised STS** only uses the test set without any additional datasets and hyper-parameter tuning.

The golden labels are contiguous real numbers in STS datasets, typically from 0 to 5, e.g., 0 indicates "dissimilar" and 5 means "similar". Thus the evaluation criterion used in previous works is the **Pearson Correlation Coefficient (Pearson's r)** between the predicted and actual similarity scores. It is worth noting that we only need to predict the relative similarity rather than the absolute score.

4.2 Datasets and Pre-trained Word Embeddings

Following the previous works [5,10,12,13,16,20], we use three commonly used datasets to evaluate the performance of our methods.

STS Benchmark (STS-B) [3] is one of the most popular dataset for STS task that contains 3 selected genres and has 1379 sentence pairs in test set. **STS-15** [1] contains sentence pairs on news headlines, image captions, student answers, answers to question and sentences expressing committed belief. **PIT-2015** [19] is used to evaluate the semantic similarity of text written in an informal style in Twitter, and can verify the generalization ability of our method in different scenarios.

Following the previous works [5,10,12,13,16,20], we test our methods with three types of pre-trained word vectors: 1) **GloVe** [11], 2) **fastText**, and 3) **ParaNMT** [16] (domain-specific word vectors). We directly apply our methods on the test set, without any training set, validation set, and external corpus.

4.3 Baselines

We use the following baselines: **WMD** [6] first introduces EMD into unsupervised STS task. **OWMD** [8] adds two global penalty terms to solve the OT problems. **ABT** [10] proposes a simply post-processing method that eliminates the common mean vector and a few top dominating directions from the word vectors. **UP** [5] proposes an angular distance-based random walk model where the probability of a sentence being generated is robust to distortion from word vector length. **TCOT** [15] uses temporally coupled optimal transport for sequences in computer vision datasets, it assumes close sequence pairs should have low transport weight, which is the opposite of our work. We re-implement their work for a comparison **DynaMax** [23] which dynamically extracts and max-pools good features depending on the sentence pair for unsupervised STS task. **WRD** [20] improves the WMD by using the norm of each word as weighting factor, and cosine distance as the dissimilarity. **meta:gcca** [13] is an ensemble method which combines various pre-trained sentence encoders into sentence meta-embeddings.

Table 1. Pearson's $r \times 100$ of our methods and baseline models on three STS datasets. Results with $*$ are retrieved from published paper or re-implemented using the opensource code, others are re-implemented by us. WMD(w. AWD) is the model that integrates syntax-based AWD into WMD. AWD(S) w. IAA is the model that incorporates the Imbalanced Alignment Adaptation mechanism (IAA) into AWD(S). The previous best results are underlined, and results better than state-of-the-art are in **bold**.

	GloVe			fastText		
	PIT2015	STS-15	STS-B	PIT2015	STS-15	STS-B
MaxPooling	45.65	66.59	70.02	57.19	68.10	71.39
UP* [5]	50.22	<u>76.10</u>	71.50	<u>58.50</u>	<u>77.10</u>	74.02
ABT* [10]	49.35	67.23	71.59	58.17	74.50	<u>75.10</u>
DynaMax* [23]	-	70.90	-	-	76.60	-
WMD [6]	48.83	75.34	73.22	49.57	75.15	72.03
OWMD* [8]	46.48	75.83	73.45	53.67	75.62	71.11
TCOT [15]	44.57	72.88	70.02	45.54	73.03	69.16
WRD [20]	<u>50.46</u>	75.12	<u>74.28</u>	58.01	76.94	74.69
AWD(P)	**51.83**	**76.22**	**75.90**	**59.50**	**77.54**	**76.28**
AWD(S)	**51.83**	**76.24**	**76.13**	**60.61**	**78.04**	**77.24**
AWD(S) w. IAA	**52.50**	**76.54**	**76.46**	**62.27**	**78.80**	**77.65**

5   Experimental Results

5.1   Main Results

Table 1 shows the overall performance and we have the following observations.

First, our method consistently outperforms baselines on three datasets with GloVe and fastText word vectors. For example, when using fastText, compared with previous best-performing models (without using task-specific word embeddings), our most robust model (AWD(s) w. IAA) achieves 3.77, 1.70, and 2.55

absolute increments on the PIT2015, STS-15, and STS-B dataset, respectively. We also suppress TCOT and OWMD, two EMD-based methods that are similar to ours. Since TCOT is designed for computer vision, and may not be suitable for STS tasks. As for OWMD, it considers the global penalty, thus our method could be seen as a more fine-grained method, since we directly add our components into the transportation cost matrix for each word pair.

Second, all AWD variants bring notable performance improvements over WRD. These AWD-equipped methods are constructed by incorporating prior alignment information into WRD, clearly verifying our initial intuition that alignment-aware transportation cost is essential for the EMD-based model.

Third, AWD(S) outperforms AWD(P) significantly, e.g., with improvements of 1.11 **Pearson's r $\times$ 100** on PIT2015 with fastText word vectors. The results indicate that (1) prior word alignment information can be better characterized via exploiting text syntax compared with simply using word position, and (2) our syntax-based strategy is capable of capturing alignment knowledge.

Finally, we find the imbalanced alignment adaptation mechanism (AWD(s) w. IAA) achieves further improvement over the powerful AWD(S). Another observation here is that the performance gain on PIT2015 is more significant than STS-15 and STS-B (e.g., 1.66 on PIT 2015 v.s. 0.76 on STS-15 v.s. 0.41 on STS-B with fastText word vectors). According to our statistics, 14.4% of sentence pairs in PIT2015 have a sentence length difference greater than or equal to 3, while in STS-B and STS-15, the proportions are 11.1% and 11.0%, respectively. The statistics suggest that the unbalanced transportation problem will be more prominent in PIT2015. Based on these statistics, the superiority demonstrated on PIT2015 can be attributed to the capability of our alignment adaptation mechanism to address the imbalanced transportation issue.

5.2   AWD(S) v.s. AWD(P)

To further investigate how AWD(S) outperforms ASD(P), we conduct comparisons and analyses on the effect of sentence structure difference. Specifically, we first use word-level Levenshtein Distance as a rough metric to measure structure differences between two sentences. In the most widely-used STS-B dataset, we select the top 25% samples with the largest Levenshtein Distance [17] as hard samples, forming a subset named STS-B(hard). As shown in Table 2 the improvement of AWD(S) over WRD on STS-B(hard) is much more remarkable than its counterpart on STS-B (+3.22 v.s. +1.85). Interestingly, AWD(P)'s improvement demonstrates an opposite pattern (+1.15 v.s. +1.65), reminding us that AWD(P) may heavily depend on similar sentence structures, and the current dataset may contain unintended bias.

5.3   Applying Task-Specific Word Embeddings

Following previous work [5,12,13,16,20,23], we also verify the effectiveness of our methods by using ParaNMT vectors, which can be seen as the task-specific word embeddings to improve unsupervised STS performances [23]. We choose UP and

Table 2. Pearson's $r \times 100$ of WRD and our proposed methods on STS-B(hard) with GloVe vectors. The results with fastText demonstrate similar trends and are omitted due to space limitation. STS-B(hard) is a subset we manual extracted from STS-B according to the edit distance between two sentences, in which the given two sentences tend to have significant semantic structure differences.

	STS-B(hard)	$\triangle$	STS-B	$\triangle$
WRD	66.97		74.28	
AWD(P)	68.12	+1.15	75.90	+1.62
AWD(S)	70.19	+3.22	76.13	+1.85

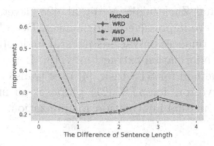

Fig. 1. The Effectiveness of IAA. Due to space limit, we only report the results of *PIT2015* with *fastText*, but the results of different datasets and pre-trained word vectors are similar and exhibit the same finding. The y-axis shows the improvements of different methods compared with WMD.

meta:gcca as baselines since they are the best single and ensemble models with ParaNMT vectors. It shows that AWD achieves about 4.15 absolute increments compared with UP. AWD's superiority can even be observed when compared with the robust ensemble model meta:gcca. The above results verify AWD's generalization capability of working with various kinds of word vectors.

5.4   The Effectiveness of IAA

We also investigate the effect of IAA when the lengths of two sentences are diverse. We choose the hardest dataset *PIT2015* and split it based on the length differences of sentence pairs. Experiments are conducted on five subsets, where the length differences are 0, 1, 2, 3, and 4 words, respectively. We report the improvements of different methods compared with WMD. As shown in Fig. 1, although WRD and AWD achieve much better performances than WMD, these methods do not achieve obvious improvements when the sentence gaps are more than 2. As for AWD w. IAA, we can see significant improvements when the length differences are more evident. This further demonstrates the capability of IAA in dealing with the imbalance scenario.

6   Related Work

A popular line of methods utilizes Earth Mover's Distance [14] as the theoretical foundation. Kusner, M. et al. [6] first introduces EMD into STS task as Word Mover's Distance (WMD), with equal weights for every word in the sentence pair. Yokoi, S. et al. [20] improves WMD as Word Rotator's Distance (WRD) by using the norm of word as weight, and use the dissimilarity of word vectors' angle to measure the transportation cost. [8] adds two global penalty terms to solve the OT problems. Besides, Liu, B. et al. [8] don't consider the case when sentences contain unequal semantic details. Su, B et al. [15] uses temporally coupled optimal transport for sequences in computer vision datasets. In addition, Zhao, R. et al. [22] and Zhelezniak, V. et al. [23] leverage fuzzy set theory for STS tasks and achieve impressive performances. Deep Learning method has been shown great performances on various natural understanding tasks, such as text classification and information extraction [7,18]. Lots of works also leverage deep learning based method for STS task. [4] uses transfer learning to generate high-quality sentence vectors for STS. [21] proposes an unsupervised sentence embedding method by mutual information maximization.

7   Conclusion

We have introduced Alignment-aware Word distance (AWD), a novel technique to measure text dissimilarity for unsupervised Semantic Textual Similarity (STS) tasks. The main characteristic that distinguishes AWD from previous EMD (Earth Mover's Distance)-based approaches is that we adjust word distance by considering the context of text pairs instead of using static word distance.

We exploit prior alignment information to scale word transportation costs fed into STS methods, demonstrating promising superiority in a set of experiments. We wish this work can inspire more sophisticated STS methods that exploit the prior alignment information of sentence pairs.

References

1. Agirre, E., et al.: Semeval-2015 task 2: Semantic textual similarity, english, spanish and pilot on interpretability. In: Proceedings of the 9th International Workshop on Semantic Evaluation (SemEval 2015), pp. 252–263 (2015)
2. Agirre, E., Cer, D., Diab, M., Gonzalez-Agirre, A.: Semeval-2012 task 6: a pilot on semantic textual similarity. In: SemEval, pp. 385–393 (2012)
3. Cer, D., Diab, M., Agirre, E., Lopez-Gazpio, I., Specia, L.: Semeval-2017 task 1: Semantic textual similarity-multilingual and cross-lingual focused evaluation. arXiv preprint arXiv:1708.00055 (2017)
4. Cer, D., et al.: Universal sentence encoder for English. In: Blanco, E., Lu, W. (eds.) EMNLP, pp. 169–174 (2018)
5. Ethayarajh, K.: Unsupervised random walk sentence embeddings: A strong but simple baseline. In: Proceedings of the Third Workshop on Representation Learning for NLP, pp. 91–100 (2018)

6. Kusner, M., Sun, Y., Kolkin, N., Weinberger, K.: From word embeddings to document distances. In: International Conference on Machine Learning, pp. 957–966. PMLR (2015)
7. Li, B., Ye, W., Sheng, Z., Xie, R., Xi, X., Zhang, S.: Graph enhanced dual attention network for document-level relation extraction. In: COLING (2020)
8. Liu, B., Zhang, T., Han, F.X., Niu, D., Lai, K., Xu, Y.: Matching natural language sentences with hierarchical sentence factorization. In: Proceedings of the 2018 World Wide Web Conference, pp. 1237–1246 (2018)
9. Mel'cuk, I.A., et al.: Dependency syntax: theory and practice. SUNY Press (1988)
10. Mu, J., Bhat, S., Viswanath, P.: All-but-the-top: simple and effective postprocessing for word representations. arXiv preprint arXiv:1702.01417 (2017)
11. Pennington, J., Socher, R., Manning, C.D.: Glove: global vectors for word representation. In: Proceedings of the 2014 Conference on Empirical Methods in Natural Language Processing (EMNLP), pp. 1532–1543 (2014)
12. Poerner, N., Schütze, H.: Multi-view domain adapted sentence embeddings for low-resource unsupervised duplicate question detection. In: Proceedings of the 2019 Conference on Empirical Methods in Natural Language Processing and the 9th International Joint Conference on Natural Language Processing (EMNLP-IJCNLP), pp. 1630–1641 (2019)
13. Poerner, N., Waltinger, U., Schütze, H.: Sentence meta-embeddings for unsupervised semantic textual similarity. arXiv preprint arXiv:1911.03700 (2019)
14. Rubner, Y., Tomasi, C., Guibas, L.J.: A metric for distributions with applications to image databases. In: Sixth International Conference on Computer Vision (IEEE Cat. No. 98CH36271). pp. 59–66. IEEE (1998)
15. Su, B., Hua, G.: Order-preserving optimal transport for distances between sequences. IEEE Trans. Pattern Anal. Mach. Intell. **41**(12), 2961–2974 (2018)
16. Wieting, J., Gimpel, K.: Paranmt-50m: pushing the limits of paraphrastic sentence embeddings with millions of machine translations. arXiv preprint arXiv:1711.05732 (2017)
17. Wong, C., Chandra, A.K.: Bounds for the string editing problem. J. ACM (JACM)., 13–16 (1976)
18. Xu, H., Liu, L., Abbasnejad, E.: Progressive class semantic matching for semi-supervised text classification. In: Carpuat, M., de Marneffe, M., Ruíz, I.V.M. (eds.) NAACL, pp. 3003–3013 (2022)
19. Xu, W., Callison-Burch, C., Dolan, W.B.: Semeval-2015 task 1: paraphrase and semantic similarity in twitter (pit). In: Proceedings of the 9th International Workshop on Semantic Evaluation (SemEval 2015), pp. 1–11 (2015)
20. Yokoi, S., Takahashi, R., Akama, R., Suzuki, J., Inui, K.: Word rotator's distance. arXiv preprint arXiv:2004.15003 (2020)
21. Zhang, Y., He, R., Liu, Z., Lim, K.H., Bing, L.: An unsupervised sentence embedding method by mutual information maximization. In: Webber, B., Cohn, T., He, Y., Liu, Y. (eds.) EMNLP (2020)
22. Zhao, R., Mao, K.: Fuzzy bag-of-words model for document representation. IEEE Trans. Fuzzy Syst. **26**(2), 794–804 (2017)
23. Zhelezniak, V., Savkov, A., Shen, A., Moramarco, F., Flann, J., Hammerla, N.Y.: Don't settle for average, go for the max: fuzzy sets and max-pooled word vectors. arXiv preprint arXiv:1904.13264 (2019)

MISNN: Multiple Imputation via Semi-parametric Neural Networks

Zhiqi Bu[1], Zongyu Dai[1], Yiliang Zhang[1], and Qi Long[2(✉)]

[1] Groups of Applied Mathematics and Computational Science,
University of Pennsylvania, Philadelphia, USA
{zbu,daizy,zylthu14}@sas.upenn.edu
[2] Department of Biostatistics, University of Pennsylvania, Philadelphia, USA
qlong@pennmedicine.upenn.edu

Abstract. Multiple imputation (MI) has been widely applied to missing value problems in biomedical, social and econometric research, in order to avoid improper inference in the downstream data analysis. In the presence of high-dimensional data, imputation models that include feature selection, especially ℓ_1 regularized regression (such as Lasso, adaptive Lasso, and Elastic Net), are common choices to prevent the model from underdetermination. However, conducting MI with feature selection is difficult: existing methods are often computationally inefficient and poor in performance. We propose MISNN, a novel and efficient algorithm that incorporates feature selection for MI. Leveraging the approximation power of neural networks, MISNN is a general and flexible framework, compatible with any feature selection method, any neural network architecture, high/low-dimensional data and general missing patterns. Through empirical experiments, MISNN has demonstrated great advantages over state-of-the-art imputation methods (e.g. Bayesian Lasso and matrix completion), in terms of imputation accuracy, statistical consistency and computation speed.

Keywords: Missing value · Imputation · Semi-supervised Learning

1 Introduction

1.1 Missing Value Mechanisms and Imputation

Missing data are commonly encountered in data analyses. It is well-known that inadequate handling of missing data can lead to biased findings, improper statistical inference [11,37] and poor prediction performance. One of the effective remedies is missing data imputation. Existing imputation methods can be mainly

---

Z. Bu, Z. Dai and Y. Zhang—Equal contribution.

Supplementary Information The online version contains supplementary material available at https://doi.org/10.1007/978-3-031-33374-3_34.

H. Kashima et al. (Eds.): PAKDD 2023, LNAI 13935, pp. 430–442, 2023.
https://doi.org/10.1007/978-3-031-33374-3_34

classified as single imputation (SI) and multiple imputation (MI) [26]. The former imputes missing values only once while the latter generates imputation values multiple times from some distribution. In fields such as finance and medical research, linear models are often preferred as it is important to not only predict accurately but also explain the uncertainty of the prediction and the effect of features. In the interest of statistical inference, MI methods, including MISNN proposed in this paper, are more suitable as they adequately account for imputation uncertainty and provide proper inference.

In general, performances of imputation are highly related to the mechanisms that generate missing values, which can be categorized into three types: missing completely at random (MCAR), missing at random (MAR) and missing not at random (MNAR). Missing data are said to be MCAR if the probability of being missing is the same for all entries; MAR means that the missing probability only depends on the observed values; MNAR means that the missing probability depends on the unobserved missing values. Intuitively, imputation is easier under MCAR mechanisms as the missing probability is only a (unknown) constant, and therefore most methods are designed to work under MCAR. However, MAR and MNAR are usually more difficult and fewer methods perform well on these problems.

1.2 Feature Selection in Imputation Models

In many applications including gene expression and financial time series research, we need to analyze high dimensional data with number of features being much larger than number of samples. In such cases, multiple imputation, which estimates the (conditional) distribution of missing data, can be inaccurate due to the overwhelming amount of features. Existing works [11,37] propose to use regularized linear model for feature selection, before building the imputation model. Some representative models include Lasso [31], SLOPE [4], Elastic Net [39], Adaptive Lasso [38], Sparse Group Lasso [13,27], etc.

While the regularized linear models successfully reduces the number of features, they often fail to capture the true distribution of missing data due to the linear dependence on the selected features and information loss in the unselected features when building the imputation model. Hence, the corresponding inference can be significantly biased. MISNN proposed in this paper overcomes the shortcome via semi-parametric neural networks. At a high level, MISNN is a semi-parametric model based on neural networks, which divides predictors into two sets: the first set are used to build a linear model and the other is used to build neural networks, which are often regarded as non-parametric models. We highlight that the outperformance of MISNN is contributed both by its neural network and linear parts. The neural networks effectively capture the non-linear relationship in the imputation model, and the linear model, in addition to capturing the linear relationships, allows efficient MI, through maximum likelihood estimation for the regression parameters.

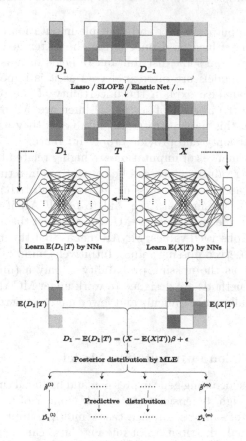

Fig. 1. MISNN framework.

1.3   Our Contribution

This paper makes two contributions. Firstly, we propose MISNN, a novel imputation method that outperforms state-of-the-art imputation methods in terms
of imputation accuracy, statistical consistency, and computation speed. MISNN
is easy to tune, interpretable, and robust to high missing rates and high-
dimensional features. Secondly, MISNN is a flexible imputation framework that
can be used with any appropriate feature selection method, such as Lasso and
forward-selection. Additionally, MISNN is compatible with any neural network,
including under or over-parameterized networks, CNN, ResNet, dropout, and
more (Fig. 1).

2   Related Work

Regarding missing data imputation, SI methods have long history before the
concept of MI [26], of which one representative approach is the mean imputation. Recent work in SI include matrix completion approaches that translate the

imputation into an optimization problem. Existing methods such as SoftImpute [23] and MMMF (Maximum-Margin Matrix Factorization) [28] provably work under MCAR mechanisms. Meanwhile, an increasing number of MI methods are studied: MICE [5,32] imputes missing values through the chained equations; MissForest [30] imputes missing values via bootstrap aggregation of multiple trees. Deep generative models [9,14,19,22,34], including Generative Adversarial Impu-tation Nets (GAIN), are also proposed for imputation. We remark that most of the existing methods only provably work under MCAR (though some methods empirically work well under MAR).

Regularized linear models have been proposed for MI in high-dimensional data. Bayesian Lasso [16,24] estimates the posterior distribution of coefficients, while alternative approaches [11,37] de-bias the estimator from the regularized linear regression. Namely, the direct use of regularized regression (DURR) and the indirect use of regularized regression (IURR). However, linear imputation models fail to capture the potential non-linear relations in the conditional distribution of missing data. MISNN falls into this line of research, is computationally more efficient than Bayesian Lasso, and captures non-linear relations during imputation.

Recent work has highlighted the importance of trustworthiness in missing data imputation, with privacy-preserving [8,10,18] and fairness-aware [6,21,35,36] imputation models drawing attention. MISNN has strong interpretability, allowing for better understanding of the imputation process and greater trust in the results.

3   Data Setup

Denote the data matrix by $\mathbf{D} \in \mathbb{R}^{n \times p}$, where n is the number of samples/cases and p is the number of features/variables. We define the j-th feature by $\mathbf{D}_j$ and its complement features by $\mathbf{D}_{-j} := \mathbf{D}_{2:p}$ for $j \in [p]$. In the presence of missing data, $\mathbf{D}$ can be separated into two submatrices $\mathbf{D}_{cc}$ and $\mathbf{D}_{ic}$, where $\mathbf{D}_{cc}$ denotes all complete cases (i.e. all features are observed) and $\mathbf{D}_{ic}$ denotes all incomplete cases. We let $\mathbf{D}_{cc,j}$ and $\mathbf{D}_{ic,j}$ denote the j-th feature of complete cases and incomplete cases, respectively. We also define $\mathbf{D}_{miss}$, the set of missing features in $\mathbf{D}$, and $\mathbf{D}_{obs}$, the set of observed features for samples in $\mathbf{D}$. Briefly speaking, to impute the missing values, we fit an imputation model g using $\mathbf{D}_{obs}$, and use $\mathbf{D}_{ic,obs}$ as input to give imputation result $\hat{\mathbf{D}}_{miss}$. For the ease of presentation, we start with a single feature missing, in which only the first column in $\mathbf{D}$ (i.e., $\mathbf{D}_1$) contains missing values. We then move on to the general missing pattern with multiple features missing in Sect. 5.

3.1   A Framework for Multiple Imputation

Here we provide a brief discussion about a general framework for multiple imputation, which is also adopted in MISNN. Under the above data setting, MI methods estimate the conditional distribution $\rho(\mathbf{D}_{miss}|\mathbf{D}_{obs})$ and sample imputed values from it multiple times. Assuming the distribution of $\mathbf{D}$ is characterized by unknown parameters $\boldsymbol{\xi}$, then

$$\rho(\mathbf{D}_{\mathrm{miss}}|\mathbf{D}_{\mathrm{obs}}) = \int \rho_{\mathrm{miss}}(\mathbf{D}_{\mathrm{miss}}|\mathbf{D}_{\mathrm{obs}}, \boldsymbol{\xi})\rho_2(\boldsymbol{\xi}|\mathbf{D}_{\mathrm{obs}})d\boldsymbol{\xi}$$

in which ρ, ρ_1, ρ_2 are three conditional distributions. For the m-th imputation, we randomly sample $\boldsymbol{\xi}^{(m)}$ from the posterior distribution of $\boldsymbol{\xi}$, i.e. $\rho_2(\boldsymbol{\xi}|\mathbf{D}_{\mathrm{obs}})$; we then generate the m-th imputed data $\mathbf{D}_{\mathrm{miss}}^{(m)}$ from the predictive distribution $\rho_1(\mathbf{D}_{\mathrm{miss}}|\mathbf{D}_{\mathrm{obs}}, \boldsymbol{\xi})$. With multiple imputed datasets, further analysis and inference can be conducted with the help of Rubin's rule [20,26]. A detailed introduction of Rubin's rule is provided in Appendix A.

4   Multiple Imputation with Semi-parametric Neural Network (MISNN)

At the high level, MISNN imputes the missing data in each column through a partial linear model (PLM), which takes the form

$$\hat{\mathbf{D}}_1 = \mathbf{X}\hat{\beta} + \hat{f}(\mathbf{T})$$

where $(\mathbf{X}, \mathbf{T})$, determined through feature selection, is a partition of the rest $p-1$ columns. While the choice of $\hat{\beta}$ and $\hat{f}(\mathbf{T})$ can be determined in an arbitrary manner, we adopt a partialling out approach [25] (also known as the orthogonalization in [7]) that can provide consistent parameter estimation if the true model takes the form $\mathbf{D}_1 = \mathbf{X}\beta + f(\mathbf{T}) + \epsilon$. To do so, we take the conditional expectation on $\mathbf{T}$, assuming $\mathbb{E}(\epsilon|\mathbf{T}) = 0$:

$$\mathbf{D}_1 = \mathbf{X}\beta + f(\mathbf{T}) + \epsilon$$
$$\mathbb{E}(\mathbf{D}_1|\mathbf{T}) = \mathbb{E}(\mathbf{X}|\mathbf{T})\beta + f(\mathbf{T}) \tag{1}$$
$$\mathbf{D}_1 - \mathbb{E}(\mathbf{D}_1|\mathbf{T}) = (\mathbf{X} - \mathbb{E}(\mathbf{X}|\mathbf{T}))\beta + \epsilon$$

Let $\mathcal{S}$ denote the set of features selected. Notice that $\mathbf{T} := \mathbf{D}_{-1} \setminus \mathbf{D}_{\mathcal{S}}$ is explicitly removed in the last equation. Therefore, if the number of selected features can be controlled (i.e., $|\mathcal{S}|$ is small), we are left with a low-dimensional linear model (as $\mathbf{X} - \mathbb{E}(\mathbf{X}|\mathbf{T}) \in \mathbb{R}^{n \times |\mathcal{S}|}$), as long as we can estimate the mapping $\mathbb{E}(\mathbf{D}_1|\mathbf{T})$ and $\mathbb{E}(\mathbf{X}|\mathbf{T})$ properly. To realize the above approach, MISNN algorithm takes three key steps:

- **Feature Selection:** During imputation of each missing feature, MISNN conducts feature selection to select at most n features. The selected features $\mathbf{X}$ are expected to have significant linear correlation with the missing feature, which later will be fitted in a linear model (e.g., least squares).
- **Fitting Partially Linear Model:** Suppose the remaining features after the selection are denoted by $\mathbf{T}$, MISNN fits two neural networks to learn $\mathbb{E}(\mathbf{D}_{\mathrm{miss}}|\mathbf{T})$ and $\mathbb{E}(\mathbf{X}|\mathbf{T})$, so as to derive a low-dimensional ordinary linear model (1);
- **Multiple Imputation:** MISNN uses maximum likelihood to estimate parameters in (1), then draw M times from the posterior distribution of $\hat{\beta}$ and further draw $\hat{\mathbf{D}}_{\mathrm{miss}}$ from the predictive distribution.

Note that the first two steps in combination is closely related to DebiNet [33], though we do not refine ourselves to over-parameterized neural network, and we utilize two neural networks to learn $(\mathbb{E}(\mathbf{D}_{\mathrm{miss}}|\mathbf{T}), \mathbb{E}(\mathbf{X}|\mathbf{T}))$. In the following, we introduce MISNN in Algorithm 1 and validate the procedure of MISNN rigorously. Here we assume the missing feature is continuous. For non-continuous features, some modifications to the algorithm should be made. See details in Appendix B.

Remark 1 *If one only focuses on the prediction, not the inference, single impu-tation can be conducted in Algorithm 1. In particular, OLS can solve the linear model in step (4) and we impute by*

$$\widehat{\mathbf{D}}_{ic,1} = \left(\mathbf{X}_{cc} - \mathbb{E}(\mathbf{X}_{cc}|\mathbf{T}_{cc})\right)\hat{\beta} + \mathbb{E}(\mathbf{D}_{cc,1}|\mathbf{T}_{cc})$$

We name the imputation algorithm as SISNN (see Algorithm 4 in Appendix D).

4.1 Sampling from Posterior and Predictive Distributions

To conduct multiple imputation in MISNN, we need to sample the parameters from the posterior distribution $\rho_2\left(\beta, \sigma^2 \middle| \mathbf{D}_{\mathrm{obs},1}, \mathbf{X}_{\mathrm{obs}}, \mathbf{T}_{\mathrm{obs}}\right)$ and the predictive

---

noend 1. Multiple Imputation via Semi-parametric Neural Network (MISNN)

Input: Incomplete data $\mathbf{D}$, number of imputation M

1: Fit a regularized regression $\mathbf{D}_{cc,1} \sim \mathbf{D}_{cc,-1}$, with the penalty function P, by

$$(\widehat{a}, \widehat{a}_0) := \underset{(a,a_0)}{\mathrm{argmin}} \frac{1}{2}\|\mathbf{D}_{cc,1} - \mathbf{D}_{cc,-1}a - a_0\|^2 + P(a).$$

2: Obtain the active set $\mathcal{S} := \{i : \widehat{a}_i \neq 0\}$ and split $\mathbf{D}_{-1}$ into sub-matrices $\mathbf{X} = [\mathbf{D}_{-1}]_{\mathcal{S}}$ and $\mathbf{T} = \mathbf{D}_{-1}\backslash\mathbf{X}$.
3: Given the training data $\{\mathbf{T}_{cc}, \mathbf{D}_{cc,1}, \mathbf{X}_{cc}\}$, train neural networks to learn

$$\eta_D(\mathbf{T}) := \mathbb{E}(\mathbf{D}_1|\mathbf{T}), \eta_X(\mathbf{T}) := \mathbb{E}(\mathbf{X}|\mathbf{T})$$

4: Apply standard maximum likelihood technique onto

$$\mathbf{D}_{cc,1} - \mathbb{E}(\mathbf{D}_{cc,1}|\mathbf{T}_{cc}) = (\mathbf{X}_{cc} - \mathbb{E}(\mathbf{X}_{cc}|\mathbf{T}_{cc}))\beta + \epsilon$$

where $\epsilon \sim \mathcal{N}(0, \sigma^2)$ and approximate the distribution $\rho_2\left(\beta, \sigma \middle| \mathbf{D}_{\mathrm{obs},1}, \mathbf{X}_{\mathrm{obs}}, \mathbf{T}_{\mathrm{obs}}\right)$
5: **for** $m \in \{1, \dots, M\}$ **do**
6:    Randomly draw $\hat{\beta}^{(m)}, \hat{\sigma}^{(m)}$ from the conditional distribution $\rho_2\left(\beta, \sigma \middle| \mathbf{D}_{cc,1}, \mathbf{X}_{cc}, \mathbf{T}_{cc}\right)$.

   Subsequently, impute $\mathbf{D}_{ic,1}$ with $\widehat{\mathbf{D}}_{ic,1}^{(m)}$ by drawing randomly from the predictive distribution $\rho_1\left(\mathbf{D}_{ic,1}|\mathbf{X}_{ic}, \mathbf{T}_{ic}, \hat{\beta}^{(m)}, \hat{\sigma}^{(m)^2}\right)$

distribution $\rho_1 \left(\mathbf{D}_{\text{miss},1} | \mathbf{X}_{\text{miss}}, \mathbf{T}_{\text{miss}}, \hat{\beta}^{(m)}, \hat{\sigma}^{(m)^2} \right)$ in MISNN (c.f. Algorithm 1). With the partialling out, we fit a linear regression at step (4),

$$\mathbf{D}_{\text{obs},1} - \mathbb{E}(\mathbf{D}_{\text{obs},1} | \mathbf{T}_{\text{obs}}) = (\mathbf{X}_{\text{obs}} - \mathbb{E}(\mathbf{X}_{\text{obs}} | \mathbf{T}_{\text{obs}})) \, \beta + \epsilon$$

We approximate the posterior distribution of β, σ using

$$\rho_2 \left(\beta, \sigma^2 \Big| \mathbf{D}_{\text{obs},1}, \mathbf{X}_{\text{obs}}, \mathbf{T}_{\text{obs}} \right) = f_1 \left(\beta \Big| \mathbf{D}_{\text{obs},1}, \mathbf{X}_{\text{obs}}, \mathbf{T}_{\text{obs}} \right) \times f_2 \left(\sigma^2 \Big| \mathbf{D}_{\text{obs},1}, \mathbf{X}_{\text{obs}}, \mathbf{T}_{\text{obs}} \right)$$

Suppose the OLS estimate for β and its variance are $\bar{\beta}$ and Σ_β, respectively. We can approximate the distribution of β by a normal distribution:

$$f_1 \left(\beta \Big| \mathbf{D}_{\text{obs},1}, \mathbf{X}_{\text{obs}}, \mathbf{T}_{\text{obs}} \right) \sim \mathcal{N} \left(\bar{\beta}, \Sigma_\beta \right)$$

where the parameters are defined as:

$$\bar{\beta} = \text{argmin}_b \| \mathbf{D}_{\text{obs},1} - \eta_D(\mathbf{T}_{\text{obs}}) - [\mathbf{X}_{\text{obs}} - \eta_X(\mathbf{T}_{\text{obs}})] b \|^2$$

$$\Sigma_\beta = \bar{\sigma}^2 \left((\mathbf{X}_{\text{obs}} - \eta_X(\mathbf{T}_{\text{obs}})^\top (\mathbf{X}_{\text{obs}} - \eta_X(\mathbf{T}_{\text{obs}})) \right)^{-1}$$

Here $\bar{\sigma}^2$ can be estimated as the mean of squared residuals:

$$f_2 \left(\sigma^2 \Big| \mathbf{D}_{\text{obs},1}, \mathbf{X}_{\text{obs}}, \mathbf{T}_{\text{obs}} \right) = \| \mathbf{D}_{\text{obs},1} - \eta_D(\mathbf{T}_{\text{obs}}) - (\mathbf{X}_{\text{obs}} - \eta_X(\mathbf{T}_{\text{obs}})) \bar{\beta} \|^2 / n_{\text{obs}}$$

As for drawing from the predictive distribution, we calculate $\hat{\sigma}^{(m)}$ from f_2 (with $\bar{\beta}$ substituted by $\hat{\beta}^{(m)}$). At last, we can draw $\hat{\mathbf{D}}_{\text{miss},1}^{(m)}$ from

$$\rho_1 \left(\mathbf{D}_{\text{miss},1} | \mathbf{X}_{\text{miss}}, \mathbf{T}_{\text{miss}}, \hat{\beta}^{(m)}, \hat{\sigma}^{(m)^2} \right) = \eta_D(\mathbf{T}_{\text{miss}}) + (\mathbf{X}_{\text{miss}} - \eta_X(\mathbf{T}_{\text{miss}})) \hat{\beta}^{(m)} + \mathcal{N}(0, \hat{\sigma}^{(m)^2})$$

4.2   Flexibility of MISNN Framework

Again, we highlight that the framework of MISNN is flexible in two folds: It can incorporate arbitrary feature selection method and arbitrary neural network models during imputation.

MISNN can incorporate an arbitrary feature selection method. Here, we adopt Lasso to select features $\mathbf{X} = \mathbf{D}_{\mathcal{S}}$ and $\mathbf{T} = \mathbf{D}_{-1} \setminus \mathbf{D}_{\mathcal{S}}$, where $\mathcal{S} = \{i > 0 : \hat{\alpha}_i \neq 0\}$ comes from the non-zero part of lasso estimate

$$(\hat{\alpha}, \hat{\alpha}_0) = \text{argmin}_{a, a_0} \frac{1}{2} \| \mathbf{D}_{\text{cc},1} - \mathbf{D}_{\text{cc},-1} a - a_0 \|_2^2 + \lambda \| a \|_1$$

MISNN works compatibly with all types of networks. Especially, when equipped with over-parameterized neural networks, MISNN can borrow the results from DebiNet [33, Theorem 1&2] to claim $\sqrt{n}$-consistency and exponentially fast convergence.

In practice, MISNN can work with a much richer class of neural networks than those theoretically supported in the neural tangent kernel regime [1,12]. This includes the under-parameterized, moderately wide and deep neural networks. Empirical experiments shows that PLM learned by such neural networks exhibit strong prediction accuracy as well as post-selection inference (see Table 2).

4.3 Other Properties of MISNN

Here we discuss some properties that MISNN enjoys, besides the flexibility of the framework, the consistent estimation of β and the fast training of PLM aforementioned. Numerical evidence can be found in Sect. 5.

Trainability: MISNN can be trained by existing optimizers in an efficient manner, in comparison to Bayesian Lasso (which may require expensive burn-in period, see Table 3), boostrap methods (e.g. DURR, which needs many bootstrapped subsamples to be accurate) or MICE (which fits each feature iteratively and may be slow in high dimension).

Robustness: Empirically, MISNN is robust to hyper-parameter tuning (e.g. the width of hidden layers does not affect the performance much). From the data perspective, in high feature dimension and high missing rate (e.g. when compared to DURR, IURR and GAIN), MISNN still works reasonably well.

4.4 MISNN for General Missing Patterns

The imputation procedure can be naturally extended to the case of general missing patterns, in which the pseudo code is provided in Algorithm 3 in the Appendix D. Suppose the first K columns are missing in $\mathbf{D}$, denoted as $\mathbf{D}_{\text{full},[K]}$ and the k-th column is denoted by $\mathbf{D}_{\text{full},k}$. The set $-[K]$ represents all other columns except those in $[K]$. Similar to the case of single column missingness, to construct a partial linear model, we need to partition the data into $\mathbf{X}$ and $\mathbf{T}$. We fit regularized linear regression for each of the K columns that have missing values and obtained K active sets. Then we propose to use either intersection or union to combine the sets into a single one, which will be treated as $\mathbf{X}$. To estimate the parameters β, during each imputation, for the k-th column, we consider an OLS model that uses $\mathbf{D}_{\text{full},[K]}$ as regressors and the k-th column as response. Maximum likelihood techniques are adopted to generate regression coefficients β_k.

We remark that other proper feature selection methods and set-merging rules can be adopted to replace what we use. It's also possible that we use an iterative approach, following the idea of MICE, to conduct column-wise imputation. Generalization to the case of discrete missing values can be realized with the help of GPLM, which is similar to the discussion in Appendix B.

5 Numerical Results

We compared MISNN with other state-of-the-art methods on various synthetic and real-world datasets. To establish baselines, we included complete data analysis, complete case analysis, and column mean imputation. We also evaluated two MI methods that incorporate regularized linear models for feature selection in high-dimensional settings: MICE-DURR and MICE-IURR. Additionally, we included MissForest, a MICE approach that uses random forest as the regression model, as well as GAIN, a deep-learning-based imputation method, and two matrix completion methods: SoftImpute and MMMF. More details about our experimental setup and results can be found in Appendix C.

Table 1. Multi-feature missing pattern in synthetic data over 500 Monte Carlo datasets. Bias: mean bias $\hat{\beta}_1 - \beta_1$; Imp MSE: $\|\hat{\mathbf{D}}_{\text{miss},1:3} - \mathbf{D}_{\text{miss},1:3}\|^2/n_{\text{miss}}$; Coverage: coverage probability of the 95% confidence interval for β_1; Seconds: wall-clock imputation time; SE: mean standard error of $\hat{\beta}_1$; SD: Monte Carlo standard deviation of $\hat{\beta}_1$. Model settings are in Sect. 5.1 and data generation is left in Appendix C.2.

Method	Style	Bias	Imp MSE	Coverage	Seconds	SE	SD
Complete Data	–	**0.0027**	–	**0.954**	–	0.1126	0.1150
Complete Case	–	0.1333	–	0.854	–	0.1556	0.1605
Mean-Impute	SI	0.1508	12.6215	**0.994**	**0.005**	0.3268	0.1933
MISNN-wide (Lasso)	MI	**−0.0184**	**4.2382**	0.902	0.324	0.1438	0.1713
MISNN-wide (ElasticNet)	MI	**−0.0134**	**4.2191**	0.924	0.286	0.1431	0.1641
MISNN-narrow (Lasso)	MI	**−0.0251**	6.2666	**0.944**	0.370	0.1816	0.1755
MISNN-narrow (ElasticNet)	MI	**−0.0246**	6.2550	**0.956**	0.344	0.1818	0.1647
MICE-DURR (Lasso)	MI	0.1815	12.6704	**0.978**	1.266	0.2275	0.1196
MICE-DURR (ElasticNet)	MI	0.1314	10.8060	**0.990**	0.633	0.2241	0.1219
MICE-IURR (Lasso)	MI	0.2527	15.7803	0.886	1.483	0.2136	0.1150
MICE-IURR (ElasticNet)	MI	0.2445	15.3266	0.892	0.566	0.2153	0.1399
MissForest	MI	0.0579	9.6174	**0.962**	69.948	0.2851	0.2609
GAIN	SI	0.7578	27.3505	0.289	14.812	0.2869	0.4314
SoftImpute	SI	-0.1432	**4.6206**	0.842	**0.019**	0.1804	0.2005
MMMF	SI	-0.1239	**4.0956**	0.782	3.385	0.1491	0.1869

In addition to imputation accuracy, we evaluate the performance of imputation models in statistical inference that are based on imputed datasets. In all the experiments, we specify a set of predictors and a response in the data matrix $\mathbf{D} = (\mathbf{Z}, y)$. A linear regression $\hat{y} = \mathbf{Z}\hat{\boldsymbol{\theta}}$ is fitted using imputed dataset to predict y and we record the regression parameters $\hat{\boldsymbol{\theta}}$. In synthetic datasets, we have access to the ground truth $\boldsymbol{\theta}$, so we focus on inference performance. In real data analysis, we lose access to the true $\boldsymbol{\theta}$ and focus on the prediction error instead.

5.1   Viewpoint of Statistical Inference

In terms of the statistical inference, we consider four statistical quantities: bias of $\hat{\boldsymbol{\theta}}$, coverage rate of the 95% confidence interval (CR) for $\boldsymbol{\theta}$, mean standard error (SE) for $\hat{\boldsymbol{\theta}}$ and Monte Carlo standard deviation (SD) of $\hat{\boldsymbol{\theta}}$. Imputation mean squared error (MSE) is also compared. We study the performance of MISNN under general missing patterns, in which multiple columns (features) in the dataset can contain missing values. We adopt a similar experiment setting to that in [11] and evaluate performance over 500 Monte Carlo datasets. A detailed experiment description can be found in Appendix C.

Potentially, one can combine MICE with MISNN for single-column missingness as well. Nevertheless, we avoid doing so by proposing Algorithm 3 in

Table 2. Multi-feature missing pattern in ADNI dataset over 100 repeats. Estimator: estimated $\hat{\beta}_1$ through OLS using first 5 features as regressors; Imp MSE: imputation mean squared error $\|\widehat{\mathbf{D}}_{\text{miss},1:3} - \mathbf{D}_{\text{miss},1:3}\|^2/n_{\text{miss}}$; Seconds: wall-clock imputation time; SE: mean standard error of $\hat{\beta}_1$; Pred MSE: mean squared error between $\mathbf{A}\hat{\theta}$ and $\mathbf{y}$. Model settings are in Sect. 5.2 and data generation is left in Appendix C.3. MissForest is too slow (more than 5 min per dataset) to be considered.

Method	Style	Estimator	Imp MSE	Seconds	SE	Pred MSE
Complete Data	–	**0.0532**	–	–	0.0676	**0.8695**
Complete Case	–	0.1278	–	–	0.1392	1.3376
Mean-Impute	SI	-0.0374	1.3464	**0.006**	0.0686	0.8938
MISNN (Lasso)	MI	**0.0545**	0.6620	**1.501**	0.0681	**0.8780**
MISNN (ElasticNet)	MI	**0.0521**	0.5140	**0.861**	0.0716	**0.8789**
MICE-DURR (Lasso)	MI	**0.0504**	1.8256	3.946	0.0508	**0.8755**
MICE-DURR (ElasticNet)	MI	0.0426	1.6998	2.709	0.0552	0.8817
MICE-IURR (Lasso)	MI	0.0474	2.0404	4.093	0.0476	**0.8747**
MICE-IURR (ElasticNet)	MI	0.0318	2.0219	2.620	0.0484	0.8803
GAIN	SI	0.0304	0.9902	67.432	0.0504	**0.8749**
SoftImpute	SI	**0.0533**	0.6667	**0.0344**	0.0763	0.8808
MMMF	SI	0.0833	**0.3051**	5.0261	0.0838	**0.8755**

Appendix D, which deals with the general missing patterns differently, in a parallel computing fashion. During the experiments, we use different network structures at step (3) of Algorithm 3: MISNN-wide uses two hidden layers with width 500, each followed by ReLU activation, a Batch Normalization layer [17] and a Dropout layer [29] at rate 0.1. The neural networks in MISNN-narrow are the same as in MISNN-wide, except the hidden layers have width 50 instead.

The results are summarized in Table 1. We highlight that all MISNN give the smallest estimation bias compared with the rest of imputation methods. MISNN also achieves satisfying imputation MSE, statistical coverage and computation speed. In comparison, two matrix completion methods achieve comparable imputation MSE, but their coverage is much worse than MI methods.

It is interesting to note that MISNN-wide tends to have smaller imputation MSE and estimation bias than MISNN-narrow. However, the coverage of the former is not as good as the latter, mainly due to the small SE. We suggest that in practice, if the accuracy of imputation or the parameter estimation is of main interest, MISNN with wide hidden layers should be adopted. If the statistical inference on parameters of interest is emphasized, then MISNN should be equipped with narrow hidden layers.

5.2 Viewpoint of Prediction

We applied MISNN to the Alzheimer's Disease Neuroimaging Initiative (ADNI) gene dataset[1], which includes over 19k genomic features for 649 patients and a response, VBM right hippocampal volume, ranging between [0.4,0.6]. We selected the top 1000 features with the largest correlations with the response, and focused on the linear analysis model between the response and the top 5 features. Since we did not have access to the true coefficients in the linear model, we studied the difference between the estimated coefficients from complete data analysis and the ones from imputed datasets. We artificially generated missing values under MAR in the top 3 features that had the largest correlations with the response, with a missing rate of approximately 65%. We used MISNN, containing a single hidden layer with width 500 and a Batch Normalization layer, and fit a linear regression between the response $\mathbf{y}$ and the top five features $\mathbf{D}_1 \sim \mathbf{D}_5$ for downstream prediction.

Our results, summarized in Table 2, show that MISNN achieved small imputation and prediction MSEs in a computationally efficient manner, particularly when compared to other MI methods. Additionally, the estimators by MISNN (as well as SoftImpute) were closest to the gold criterion from complete data analysis. Further experiment details can be found in Appendix C.

6 Discussion

In this work, we propose MISNN, a novel deep-learning based method for multiple imputation of missing values in tabular/matrix data. We demonstrate that MISNN can flexibly work with any feature selection and any neural network architecture. MISNN can be trained with off-the-shelf optimizers at high computation speed, providing interpretability for the imputation model, as well as being robust against data dimension and missing rate. Various experiments with synthetic and real-world datasets illustrate that MISNN significantly outperforms state-of-the-art imputation models.

While MISNN works for a wide range of analysis models, we have only discussed the case for continuous missing values using the partialling out. We can easily extend MISNN to discrete missing value problems by considering the generalized partially linear models (GPLM, see Sect. 4.1 for details). However, the partialling out technique generally renders invalid for GPLM. Therefore, iterative methods including the backfitting, which can be slow, may be required to learn MISNN.

Acknowledgement. This work was supported in part by National Institutes of Health grant, R01GM124111. The content is solely the responsibility of the authors and does not necessarily represent the official views of the National Institutes of Health.

---

[1] The complete ADNI Acknowledgement is available at http://adni.loni.usc.edu/wp-content/uploads/how_to_apply/ADNI_Acknowledgement_List.pdf.

References

1. Allen-Zhu, Z., Li, Y., Song, Z.: A convergence theory for deep learning via over-parameterization. In: International Conference on Machine Learning, pp. 242–252. PMLR (2019)
2. Barnard, J., Rubin, D.B.: Miscellanea small-sample degrees of freedom with multiple imputation. Biometrika **86**(4), 948–955 (1999)
3. Belloni, A., Chernozhukov, V., et al.: Least squares after model selection in high-dimensional sparse models. Bernoulli **19**(2), 521–547 (2013)
4. Bogdan, M., Van Den Berg, E., Sabatti, C., Su, W., Candès, E.J.: Slope-adaptive variable selection via convex optimization. Ann. Appl. Stat. **9**(3), 1103 (2015)
5. Buuren, S.V., Groothuis-Oudshoorn, K.: mice: Multivariate imputation by chained equations in R. J. Stat. Software 1–68 (2010)
6. Caton, S., Malisetty, S., Haas, C.: Impact of imputation strategies on fairness in machine learning. J. Artif. Intell. Res. **74**, 1011–1035 (2022)
7. Chernozhukov, V., et al.: Double/debiased machine learning for treatment and structural parameters. Economet. J. **21**(1), C1–C68 (2018)
8. Clifton, C., Hanson, E.J., Merrill, K., Merrill, S.: Differentially private k-nearest neighbor missing data imputation. ACM Trans. Privacy Secur. **25**(3), 1–23 (2022)
9. Dai, Z., Bu, Z., Long, Q.: Multiple imputation via generative adversarial network for high-dimensional blockwise missing value problems. In: 2021 20th IEEE International Conference on Machine Learning and Applications (ICMLA), pp. 791–798. IEEE (2021)
10. Das, S., Drechsler, J., Merrill, K., Merrill, S.: Imputation under differential privacy. arXiv preprint arXiv:2206.15063 (2022)
11. Deng, Y., Chang, C., Ido, M.S., Long, Q.: Multiple imputation for general missing data patterns in the presence of high-dimensional data. Sci. Rep. **6**(1), 1–10 (2016)
12. Du, S., Lee, J., Li, H., Wang, L., Zhai, X.: Gradient descent finds global minima of deep neural networks. In: International Conference on Machine Learning, pp. 1675–1685. PMLR (2019)
13. Friedman, J., Hastie, T., Tibshirani, R.: A note on the group lasso and a sparse group lasso. arXiv preprint arXiv:1001.0736 (2010)
14. Gondara, L., Wang, K.: MIDA: multiple imputation using denoising autoencoders. In: Phung, D., Tseng, V.S., Webb, G.I., Ho, B., Ganji, M., Rashidi, L. (eds.) PAKDD 2018. LNCS (LNAI), vol. 10939, pp. 260–272. Springer, Cham (2018). https://doi.org/10.1007/978-3-319-93040-4_21
15. Gramacy, R.B.: MONOMVN: estimation for multivariate normal and student-t data with monotone missingness. R package version pp. 1–8 (2010)
16. HANS, C.: Bayesian lasso regression. Biometrika 1–11 (2009)
17. Ioffe, S., Szegedy, C.: Batch normalization: accelerating deep network training by reducing internal covariate shift. In: International Conference on Machine Learning, pp. 448–456. PMLR (2015)
18. Jagannathan, G., Wright, R.N.: Privacy-preserving imputation of missing data. Data Knowl. Eng. **65**(1), 40–56 (2008)
19. Li, S.C.X., Jiang, B., Marlin, B.: MISGAN: learning from incomplete data with generative adversarial networks. arXiv preprint arXiv:1902.09599 (2019)
20. Little, R.J., Rubin, D.B.: Statistical Analysis with Missing Data. Wiley (2002)
21. Martínez-Plumed, F., Ferri, C., Nieves, D., Hernández-Orallo, J.: Fairness and missing values. arXiv preprint arXiv:1905.12728 (2019)

22. Mattei, P.A., Frellsen, J.: MIWAE: deep generative modelling and imputation of incomplete data. arXiv preprint arXiv:1812.02633 (2018)
23. Mazumder, R., Hastie, T., Tibshirani, R.: Spectral regularization algorithms for learning large incomplete matrices. J. Mach. Learn. Res. **11**, 2287–2322 (2010)
24. Park, T., Casella, G.: The Bayesian lasso. J. Am. Stat. Assoc. **103**(482), 681–686 (2008)
25. Robinson, P.M.: Root-n-consistent semiparametric regression. Econometrica: J. Econ. Soc. 931–954 (1988)
26. Rubin, D.B.: Multiple Imputation for Nonresponse in Surveys, vol. 81. Wiley, New York (2004)
27. Simon, N., Friedman, J., Hastie, T., Tibshirani, R.: A sparse-group lasso. J. Comput. Graph. Stat. **22**(2), 231–245 (2013)
28. Srebro, N., Rennie, J., Jaakkola, T.: Maximum-margin matrix factorization. Adv. Neural. Inf. Process. Syst. **17**, 1329–1336 (2004)
29. Srivastava, N., Hinton, G., Krizhevsky, A., Sutskever, I., Salakhutdinov, R.: Dropout: a simple way to prevent neural networks from overfitting. J. Mach. Learn. Res. **15**(1), 1929–1958 (2014)
30. Stekhoven, D.J., Bühlmann, P.: Missforest-non-parametric missing value imputation for mixed-type data. Bioinformatics **28**(1), 112–118 (2012)
31. Tibshirani, R.: Regression shrinkage and selection via the lasso. J. Roy. Stat. Soc.: Ser. B (Methodol.) **58**(1), 267–288 (1996)
32. Van Buuren, S.: Multiple imputation of discrete and continuous data by fully conditional specification. Stat. Methods Med. Res. **16**(3), 219–242 (2007)
33. Xu, S.: DEBINET: debiasing linear models with nonlinear overparameterized neural networks. arXiv preprint arXiv:2011.00417 (2020)
34. Yoon, J., Jordon, J., Van Der Schaar, M.: Gain: missing data imputation using generative adversarial nets. arXiv preprint arXiv:1806.02920 (2018)
35. Zhang, Y., Long, Q.: Fairness-aware missing data imputation. In: Workshop on Trustworthy and Socially Responsible Machine Learning, NeurIPS 2022 (2022)
36. Zhang, Y., Long, Q.: Fairness in missing data imputation. arXiv preprint arXiv:2110.12002 (2021)
37. Zhao, Y., Long, Q.: Multiple imputation in the presence of high-dimensional data. Stat. Methods Med. Res. **25**(5), 2021–2035 (2016)
38. Zou, H.: The adaptive lasso and its oracle properties. J. Am. Stat. Assoc. **101**(476), 1418–1429 (2006)
39. Zou, H., Hastie, T.: Regularization and variable selection via the elastic net. J. Roy. Stat. Soc. Ser. B (Stat. Methodol.) **67**(2), 301–320 (2005)

LSG Attention: Extrapolation of Pretrained Transformers to Long Sequences

Charles Condevaux[1][(✉)] and Sébastien Harispe[2]

[1] CHROME, University of Nîmes, Nîmes, France
charles.condevaux@unimes.fr
[2] EuroMov Digital Health in Motion, Univ Montpellier, IMT Mines, Ales, France
sebastien.harispe@mines-ales.fr

Abstract. Transformer models achieve state-of-the-art performance on a wide range of NLP tasks. They however suffer from a prohibitive limitation due to the self-attention mechanism, inducing $O(n^2)$ complexity with regard to sequence length. To answer this limitation we introduce the LSG architecture which relies on Local, Sparse and Global attention. We show that LSG attention is fast, efficient and competitive in classification and summarization tasks on long documents. Interestingly, it can also be used to adapt existing pretrained models to efficiently extrapolate to longer sequences with no additional training. Along with the introduction of the LSG attention mechanism, we propose a PyPI package to train new models and adapt existing ones based on this mechanism.

Keywords: Attention mechanism · Long sequences · Extrapolation

1 Introduction

Transformer models [33] are nowadays state-of-the-art in numerous domains, and in particular in NLP where they are used in general language models, and to successfully tackle several specific tasks such as document summarization, machine translation and speech processing to cite a few [13, 26]. The cornerstone of Transformer models is the Attention mechanism used to iteratively build complex context-dependent representations of sequence elements, e.g. tokens, by dynamically aggregating prior representations of these elements. Using self-attention, a popular Attention flavour, this is made by computing full attention scores defining how each prior element representation will contribute to building the new representation of an element. Considering a sequence of n elements, the computation of the attention scores is therefore of complexity $O(n^2)$ which is prohibitive when large sequences have to be processed. In the current context where a large number of models based on full attention have been trained on various datasets and tasks, we are therefore interested in extrapolating those models to long sequences by simply substituting full attention by new attention mechanisms post training on shorter input sequences. Common pretrained models (e.g RoBERTa) are indeed known to underperform when extrapolated

H. Kashima et al. (Eds.): PAKDD 2023, LNAI 13935, pp. 443–454, 2023.
https://doi.org/10.1007/978-3-031-33374-3_35

to sequences of length exceeding the 512 tokens considered during training. This is due to the nature of the attention mechanism which largely impacts extrapolation capabilities: full attention usually fails to extrapolate, even considering post hoc adaptations, e.g. using a relative positional embedding [30] or duplicating the positional embedding [3]. Defining new attention mechanisms that can efficiently substitute full attention in pretrained models that are not originally capable of handling long sequences would avoid the costs induced by training large language models from scratch. In this context, the main contributions of this paper are:

1. LSG (Local Sparse Global) attention, an efficient $O(n)$ approach to approximate self-attention for processing long sequences.[1]
2. Results demonstrating that LSG is fast, efficient and competitive on classification and summarization tasks applied to long documents. It is also shown that LSG can adapt and extrapolate existing pretrained models not based on LSG, with minimal to no additional training.
3. A procedure and a PyPI package are proposed to convert existing models and checkpoints (e.g. RoBERTa, DistilBERT, BART) to their LSG variant.[2]

Compared to several contributions aiming at reducing the complexity of self-attention introduced hereafter, a specific focus is given in our work on the extrapolation of existing Transformer models, i.e. reuse, to longer sequences.

2   Related Works

Several contributions have been devoted to the optimization of the Attention mechanism. Four categories of approaches can be distinguished in the literature: (i) recurrent models such as Transformers-XL [12] and Compressive Transformers [25] which maintain a memory of past activation at each layer to preserve long-range contextual information; (ii) factorization or kernels aiming at compressing attention score matrices, such as Linformer [34] or Performer [9]; (iii) models based on clustering such as Reformer [21] that dynamically define eligible attention patterns (i.e. where attention may be made); and (iv) models based on fixed or adaptative attention patterns, e.g. Longformer [3] or Big Bird [37].

Recurrent approaches iteratively process the sequence by maintaining a memory to enable long-range dependencies. They generally suffer limitations induced by specific, slow, and difficult to implement forward and back propagation procedures. Alternatively, one of the main line of study for reducing the complexity of Attention is thus to perform sparsity by limiting the number of elements on which new representations will be based, i.e. reducing the number of elements with non-null attention scores. This approach is motivated by the observation of global or data-dependent positional patterns of non-null attention scores depending on the task [7]. The sparsity of attention scores in the traditional Attention mechanism

---

[1] Checkpoints and datasets are available at https://huggingface.co/ccdv.

[2] https://github.com/ccdv-ai/convert_checkpoint_to_lsg.

is indeed documented in the literature. It has for instance been shown that in practice, full attention tends to overweight close elements in average, in particular for MLM, machine translation, and seq-to-seq tasks in general [10]. Moreover, according to analyses on the use of multi-head full attention on specific tasks, e.g. machine translation, numerous heads learn similar simple patterns [27]. Such redundant patterns may be hardcoded implementing fixed-positional patterns, eventually in a task-dependent manner.

Two main approaches are discussed in the literature for implementing sparsity: fixed or adaptative patterns based on whether attention scores are computed considering (1) predefined fixed elements based on their location in the sequence, or (2) elements selected from a given procedure. As an example, [35] have shown that fixed $O(n)$ convolutions can perform competitively on machine translation. Longformer proposes an alternative $O(n)$ approach based on sliding and global patterns [3]. In the context of image, audio, and text processing, [7] propose sparse Transformer, an $O(n\sqrt{n})$ model based on sparse factorization of the attention matrix relying on specific 2D factorized attention schemes. Those approaches however prevent the use of task-dependent dynamic patterns. Considering adaptative patterns, [35] also introduced dynamic convolutions as an $O(n)$ complexity substitute to self-attention. Kernels defining the importance of context elements are specified at inference time rather than fixed after training. Another example is Reformer [21], an $O(n \log n)$ approach based on locality-sensitive hashing (LSH) based on random projections.

In a transverse manner, several authors, explicitly or implicitly motivated by the compositional nature of language have studied structured approaches in which subsequences (i.e. blocks) are processed independently and then aggregated. This aims at implementing a local or global dynamic memory for considering close to long-range dependencies. Some approaches use a blockwise approach to reduce the quadratic complexity induced by large sequences in encoder-decoder architectures [4]. Other propose a chunkwise attention in which attention is performed in a blockwise manner adaptively splitting the sequence into small chunks over which soft attention is computed [8]. This idea is also used in Transformer-XL [12]. Masked block self-attention mechanism in which the entire sequence is divided into blocks, to further 1) apply self-attention intra-block for modeling local contexts, to further 2) apply self-attention inter-block for capturing long-range dependencies, as also been proposed [31]. Such an approach enables implementing some forms of connectivity between all positions over several steps without being restricted by full attention limitations. This can also be achieved by factorization techniques, e.g. [7]. More recently authors have proposed global attention mechanisms encoding information related to blocks on which attention is based [1,16,39].

This paper presents the LSG (Local, Sparse and Global) attention based on block local attention to capture local context, sparse attention to capture extended context, and global attention to improve information flow. Contrary to prior work mostly focusing on defining new models, the proposed LSG Attention mechanism is model agnostic and aims at facilitating adapting existing (pretrained) models for them to be used on long sequences.

3   LSG: Mixing Local, Sparse and Global Attentions

LSG assumes (1) that locally, a token needs to capture precise low level information using dense attention, (2) as the context grows, higher level information is sufficient, i.e. a limited number of tokens specifically selected are sufficient. LSG therefore relies on block local attention to capture local context, sparse attention to capture extended context, and global attention to improve information flow.
Local Attention. LSG take advantage of a block-based processing of the input. The sequence is split into n_b non-overlapping chunks of size b_t. For a given block, each token attends to the tokens inside the block, as well as to those in the previous and next blocks. The local attention window is asymmetrical since a token can connect up to $2 \times b_t - 1$ tokens on the left or on the right.
Sparse Attention. Sparse connections are used to expand the local context by selecting additional tokens. These tokens can be directly selected based on a specific metric or using some computation such as a pooling method. In the proposed approach, each attention head can process different sparse tokens independently. Sparse attention also relies on a block structure where the sparse selection is done inside each block. Five alternative criteria can be used in LSG.
1. *Head-wise strided:* Each attention head attend to a set of tokens following a specific stride defined as the sparsify factor. Figure 1 shows the selection pattern.
2. *Head-wise block strided* selects consecutive tokens, see Fig. 2.

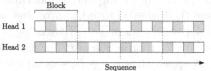

Fig. 1. Head-wise selection (stride 2).   **Fig. 2.** Block selection (stride 2).

3. *Average pooling:* sparse tokens are computed using average pooling on blocks. For a block of size b_t and a sparsify factor f, pooling is applied to each block with a window of f and a stride of f to produce b_t/f tokens.
4. *Max norm:* selects tokens that are most likely to have high scores. Finding those keys efficiently is difficult in practice so we use a simple and deterministic heuristic selecting inside each block and each head b_t/f tokens with the highest key norm. Indeed, note that for a query and a key $q, k \in \mathbb{R}^d$, $qk^\top = \cos(\theta)\|q\|\|k\|$. If $\cos(\theta)$ is positive and $\|k\|$ is high, the key will likely dominate the softmax regardless of the query.
5. *LSH Clustering:* non deterministic approach relying on the LSH algorithm [2]. For each block, b_t/f clusters are built using a single round LSH. To get $c = b_t/f$ hashes and for an input $x \in \mathbb{R}^d$, a random matrix $R \in \mathbb{R}^{d \times c/2}$ is generated, such that $h(x) = \arg \max([xR; -xR])$ with $[a; b]$ the concatenation of two vectors. Using the key matrix as input, each token inside the block gets a cluster index from $h(x)$. Tokens inside a cluster are averaged.

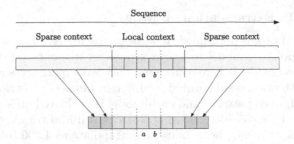

Fig. 3. Local and sparse contexts with a block size of 2 and a sparsity factor of 4. Queries a and b will attend to 6 local keys and 4 sparse keys.

Global Attention . Global tokens improve the flow of information inside the model. They attend to every tokens across the sequence and all tokens attend to them. Rather than picking a subset of tokens, additional tokens are prepended to the sequence and trained using their own embedding matrix (their number is an hyperparameter). When a model is converted to its LSG version, the first global token is initialized as the sum of the [CLS] token and the first position from the positional embedding. The other global tokens are initialized as the sum of [MASK] token and the other positions from the positional embedding. Thus, they can be trained and fine-tuned independently.

Positional Embedding . It is necessary to modify the positional embedding matrix to reuse existing models to process long sequences. In LSG, instead of randomly initializing the new positions, the original matrix is duplicated and concatenated until the desired max sequence length is reached.

4   Experiments

We evaluate LSG in the context of model extrapolation by replacing full attention by the LSG attention in various architectures. The official RoBERTa-base checkpoint for classification tasks and BART-base checkpoint for summarization tasks are extrapolated using LSG attention. All metrics are reported for the test set except in the case where only the validation set is available and datasets are all available on the HuggingFace hub. We use a batch size of 32, a linear decaying learning rate, a dropout rate of 0.10 and Adam (0.9, 0.999) optimizer [20] for classification and summarization experiments. An experiment comparing several attention approximations to extrapolate RoBERTa in an MLM task is first discussed as it is used to limit the number of tested alternatives, and therefore reduce the cost of the proposed evaluations. All experiments are conducted on NVIDIA Quadro RTX 8000 48 Gb GPUs.

4.1   RoBERTa Extrapolation on MLM

A test on a MLM task is performed to question the ability of an attention mechanism to extrapolate a model to longer sequences without additional training. A RoBERTa-base model is here considered and two experiments are conducted. First, the full attention is substituted by different kinds of attention (kernel, factorization, local, fixed pattern) and each model is evaluated on sequences of the same length as those considered during RoBERTa initial training (512 tokens). In the second experiment, their ability to extrapolate to 4,096 tokens sequences without additional training is tested (positional embedding duplicated 8 times).

A random sample from Wikipedia + BookCorpus + CC_News is used; BPC and MLM accuracy are in Table 1. RoBERTa's author report a 1.880 BPC loss; we obtain a comparable loss of 1.881 on this random sample.

Only Longformer, Big Bird and LSG obtain competing BPC while processing sequences of the same length as those considered during the original RoBERTa training. Other approaches such as Linformer, Performer or Reformer requires additional MLM fine-tuning to leverage an existing checkpoint. It can be seen that RoBERTa fails to extrapolate to longer sequences (+2.454 BPC), which highlights that full attention is not suitable for extrapolation. Longformer and Big Bird are able to perform some form of extrapolation. Therefore, we restrict our comparison to these two approaches in order to limit experimentation costs.

Table 1. BPC and MLM accuracy of RoBERTa-base with various Attention.

Attention	512 length		4,096 length	
	BPC	Accuracy	BPC	Accuracy
RoBERTa (full) [23]	1.881	**0.732**	4.335	0.359
Linear Attn. [19]	11.324	0.061	11.474	0.058
Efficient Attn. [32]	21.022	0.102	20.574	0.097
Performer [9]	10.382	0.107	10.556	0.102
Linformer (128 proj.) [34]	22.176	0.098	20.386	0.032
Reformer [21]	17.602	0.003	18.608	0.002
Longformer (512) [3]	1.929	0.726	2.051	0.708
Big Bird (64) [37]	1.881	**0.732**	2.439	0.659
LSG-Norm (128/2) (block size/sparsity)	1.919	0.727	2.032	**0.712**
LSG-Stride (128/2)	1.938	0.724	2.046	0.710
LSG-BlockStride (128/2)	1.940	0.724	2.048	0.709
LSG-Pooling (128/2)	1.968	0.720	2.064	0.706
LSG-LSH (128/2)	1.969	0.719	2.065	0.705

4.2   Classification Tasks

We compare LSG to Longformer [3] and Big Bird [37], two approaches able to process long sequences with a similar number of parameters. Tests are performed considering sparse attentions with a block size of 128 and a sparsify factor of 4.

Datasets. Standard NLP datasets are used. *IMDb* [24]: binary sentiment analysis classification task from movie reviews. *ArXiv* [17]: set of documents from ArXiv where the objective is to predict a topic from 11 available classes. Because there is no official split, a random one is made of 28K, 2.5K and 2.5K documents for train, validation and test. *Patent* [29]: subset of the Big Patent summarization dataset. The task is redefined as a classification task where the objective is to predict the patent category using the full document (9 classes, random split of 25K, 5K and 5K documents for train, validation and test). Some specific domains are highly dependent on processing long sequences, e.g. legal domain in which sentences tend to be long and complex. To demonstrate the ability of LSG to leverage pretrained models in such cases, the following three datasets are chosen from LexGlue [6], a benchmark focused on legal documents. Tasks where the input is on average significantly longer than 512 tokens have been selected. *Scotus*: given a court opinion, the task is to predict the relevant issue area among 14 choices. *ECtHRa* and *ECtHRb*: the objective is to predict which articles of the European Court of Human Rights (ECHR) have been violated (if any) from case description: multi-label task (10 + 1 labels).

Training Setup and Architecture. To make a fair comparison between models and architectures, fine-tuning is done with the same learning rate, number of steps and batch size. To show that LSG is compatible with different architectures, the LexGlue tasks are also run with an LSG version of LEGAL-BERT [5].

Results. Reported metrics (Table 2) show that LSG outperforms most of the time Longformer and Big Bird models with input sequences up to 4096 tokens long. A major difference lies in the implementation itself since the LSG variant is twice as fast to train on these lengths with no additional memory cost.[3]

Table 2. Micro/Macro F-1 on classification datasets.

	IMDb	Arxiv	Patent	Scotus	ECtHRa	ECtHRb
Epochs	3	3	3	7	5	5
Learning rate	2e-5	5e-5	2e-5	1e-4	1e-4	1e-4
RoBERTa (512-length)	95.5	87.2/86.8	66.6/61.8	69.4/60.8	62.9/58.2	72.0/65.9
Longformer	95.9	**88.2/87.9**	69.8/63.8	72.9/62.6	68.3/59.7	78.9/72.2
Big Bird ETC	95.4	85.9/85.5	69.4/63.9	69.4/58.2	68.3/60.3	**80.0/70.6**
LSG-Local (256/0)	**96.0**	87.5/87.1	69.9/64.8	**73.3/63.7**	68.8/63.7	79.9/73.4
LSG-Stride (128/4)	95.6	**88.2/87.9**	69.2/64.0	70.5/60.0	69.5/62.3	79.3/71.6
LSG-BlockStride (128/4)	95.7	87.7/87.4	69.6/64.1	72.5/63.1	69.1/58.6	79.5/71.8
LSG-Norm (128/4)	95.7	87.0/86.6	**70.0/64.4**	71.3/60.8	70.1/61.9	79.4/72.1
LSG-Pooling (128/4)	95.9	87.5/87.3	69.4/64.1	72.6/60.9	70.2/61.4	79.0/73.1
LSG-LSH (128/4)	95.8	**88.2/87.9**	69.5/64.2	70.3/54.6	**71.0/60.3**	78.9/71.0
Legal-BERT (512-length)	-	-	-	73.5/60.5	64.2/58.2	73.2/65.9
LSG-Legal-BERT (256/0)	-	-	-	74.5/62.6	71.7/63.9	81.0/75.1

On Patent, ECtHRa and ECtHRb tasks, the ability to process longer sequences improves significantly the F-measures compared to a vanilla (full

---

[3] See https://github.com/ccdv-ai/convert_checkpoint_to_lsg for a benchmark.

attention) RoBERTa model. We also observe that Big Bird model is in general slightly under its counterpart except for the ECtHRb dataset. This probably comes from the random attention mechanism which may require additional training steps. LSG-LSH and Big Bird models are affected by randomness during inference, thus their performance can differ between runs.

Extrapolating LEGAL-BERT with LSG to handle longer sequences improves predictions. The choice of the sparse attention is likely task specific. Using local attention only with a large block size is also a viable option. The role of global tokens is not discussed here since we only use one for all experiments. We show in the next section with summarization tasks the utility of such tokens.

4.3 Summarization Tasks

We evaluate our models on summarization tasks where the input is significantly longer than 1k tokens only. The models have been fine-tuned on each dataset.[4]

Datasets. In both *ArXiv and Pubmed* [11], the goal is to generate an abstract using a document as input. *MultiNews* [14] involves generating human-written summaries from sets of news documents. *MediaSum* [40] consists of using interview transcripts from CNN and NPR media to generate a summary.

Table 3. Parameters count of summarization models.

Models	Params
PRIMERA [36]	447M
LED [3]	460M
HAT-BART [28]	471M
Pegasus [38]	577M
Big Bird-Peg. [37]	577M
Hepos [18]	406M
LongT5-Base [15]	220M
LongT5-L	770M
LongT5-XL	3B
Ours, LSG-BART-base (256/0)	145M

Training Setup and Architecture. The BART-base model [22] is converted to its LSG version by replacing the full attention in the encoder part and adding global tokens. The model is then fine-tuned on 4,096-length inputs and evaluated. To reduce computational costs, experiments on 16,384-length inputs are warm started from the 4,096-length experiments. The model is then fine-tuned during a single epoch if necessary using the same training parameters. We propose 3

---

[4] All summarization experiments are run using a 8e-5 learning rate, a 10% warmup, a length penalty of 2.0 and a beam size of 5 for beam search.

setups for the 16,384-length. First we evaluate the model with pure extrapolation from 4,096-length (no additional training). In the second setup, we extrapolate and add 64 global tokens we choose to fine-tune. In the last setup, we extrapolate, we add 64 global tokens and we fine-tune the full model. Extrapolation is done by concatenating 4 copies of the positional embedding matrix (4 × 4096).

The tested model - LSG-BART-base - is significantly smaller than common models from the existing literature (Table 3. An input sequence of 16384 tokens can fit on a 32Gb GPU (without attention dropout) during training without a specific memory reduction tool (i.e. gradient checkpointing).

Results. LSG-BART is compared to state-of-the-art models by reporting the results from their respective papers. We use ROUGE-1, ROUGE-2 and ROUGE-L evaluation metrics as comparison points.

Table 4. ROUGE on PubMed dataset.

Models	R1	R2	RL
Pegasus (1K)	45.49	19.90	27.69
Big Bird-Peg. (4K)	46.32	20.65	42.33
HAT-BART (4K)	48.36	21.43	37.00
Hepos-LSH (7.2K)	48.12	21.06	42.72
Hepos-SKN (10.2K)	47.93	20.74	42.58
LongT5-Base (4K)	47.77	22.58	44.38
LongT5-L (16K)	49.98	24.69	46.46
LongT5-XL (16K)	50.23	24.76	46.67
Ours (4K)	47.37	21.74	43.67
Ours (16K)	48.03	22.42	44.32
+ global tuning	48.12	20.46	44.40
+ full tuning	48.32	22.52	44.57

Table 5. ROUGE on MultiNews.

Models	R1	R2	RL
TG-MultiSum	47.10	17.55	20.73
PRIMERA (4K)	49.90	21.10	25.9
LongT5-Base (4K)	46.01	17.37	23.50
LongT5-L (4K)	46.99	18.21	24.08
LongT5-L (8K)	47.18	18.44	24.18
LongT5-XL (8K)	48.17	19.43	24.90
Ours (4K)	47.10	18.94	25.22
Ours (16K)	47.30	19.19	25.38
+ global tuning	47.23	19.18	25.29
+ full tuning	47.07	19.04	25.35

Table 6. ROUGE on ArXiv dataset.

Models	R1	R2	RL
Pegasus (1K)	44.70	17.27	25.80
Big Bird-Peg. (4K)	46.63	19.02	41.77
LED (16K)	46.63	19.62	41.83
PRIMERA (4K)	47.58	20.75	42.57
HAT-BART (4K)	46.68	19.07	42.17
Hepos-LSH (7.2K)	48.24	20.26	41.78
Hepos-SKN (10.2K)	47.87	20.00	41.50
LongT5-Base (4K)	44.87	18.54	40.97
LongT5-L (16K)	48.28	21.63	44.11
LongT5-XL (16K)	48.35	21.92	44.27
Ours (4K)	46.65	18.91	42.18
Ours (16K)	47.03	20.19	42.69
+ global tuning	48.08	20.42	43.65
+ full tuning	48.74	20.88	44.23

Table 7. ROUGE on MediaSum.

Models	R1	R2	RL
BART-Large (1K)	35.09	18.05	31.44
T5-large (1K)	30.68	14.88	27.88
LongT5-Base (4K)	35.09	18.35	31.87
LongT5-L (4K)	35.54	19.04	32.20
LongT5-XL (4K)	36.15	19.66	32.80
Ours (4K)	35.16	18.13	32.20
Ours (16K)	35.17	18.13	32.21
+ global tuning	35.22	18.08	32.22
+ full tuning	35.31	18.35	32.47

As shown in Tables 4, 5, 6 and 7, LSG achieve very competitive performances by enabling adapting existing pretrained models to longer sequences. On the

ArXiv dataset (Table 6), LSG is competitive with every size of the LongT5 model, despite the limited number of model parameters. On the PubMed dataset (Table 4), LSG also outperforms Pegasus and Big Bird Pegasus, and is close to Hepos models. On the MultiNews dataset (Table 5), LSG is close to the large L and XL LongT5 models. We note that while extrapolation improves metrics, additional fine-tuning has a negative impact in this case. Since this dataset is rather small (45K examples, 1,400 steps), fine-tuning a single epoch is not enough for the model to converge properly, longer training is required. On the MediaSum dataset (Table 7), LSG is close to the LongT5-base model again. This dataset has the shortest inputs, thus processing a maximum of 16,384 tokens has a marginal impact on performances. These results underline the ability of LSG to efficiently substitute full-attention mechanisms to process long sequences.

The second surprising and important finding is the ability of LSG to improve metrics from 4.096 to 16.384-length inputs without additional fine-tuning. This is especially true on ArXiv and PubMed datasets which have the longest input sequences. Fine tuning additional global tokens further improves metrics while limiting cost and training time compared to a fully tuned model.

5 Conclusion

We have presented LSG attention, a novel efficient $O(n)$ alternative to the full attention mechanism relying on local, sparse and global attentions. Our results on MLM, classification and summarization tasks show that LSG is a fast and very competitive full attention substitute for pretrained Transformers to efficiently extrapolate to long input sequences. We also proposed an optimized implementation of the LSG attention mechanism on HuggingFace, improving training speed by a factor of 2 without additional memory cost compared to Longformer and Big Bird models. By providing a PyPI package conversion tool to leverage existing models and checkpoints (BERT, RoBERTa, DistilBERT, BART), the proposed approach removes the need of a costly re-training of existing models to handle long sequences.[5]

References

1. Ainslie, J., et al.: ETC: encoding long and structured inputs in transformers. arXiv preprint arXiv:2004.08483 (2020)
2. Andoni, A., Indyk, P., Laarhoven, T., Razenshteyn, I.P., Schmidt, L.: Practical and optimal LSH for angular distance. CoRR arXiv:abs/1509.02897 (2015)
3. Beltagy, I., Peters, M.E., Cohan, A.: Longformer: the long-document transformer. arXiv:2004.05150 (2020)
4. Britz, D., Guan, M.Y., Luong, M.T.: Efficient attention using a fixed-size memory representation. arXiv preprint arXiv:1707.00110 (2017)

---

[5] This work has benefited from LAWBOT (ANR-20-CE38-0013) grant and HPC resources from GENCI-IDRIS (Grant 2023-AD011011309R3).

5. Chalkidis, I., Fergadiotis, M., Malakasiotis, P., Aletras, N., Androutsopoulos, I.: LEGAL-BERT: the muppets straight out of law school. In: Findings of the Association for Computational Linguistics: EMNLP 2020, pp. 2898–2904 (2020)

6. Chalkidis, I., et al.: LexGLUE: a benchmark dataset for legal language understanding in english. In: Proceedings of the 60th Annual Meeting of the Association for Computational Linguistics. Dubln, Ireland (2022)

7. Child, R., Gray, S., Radford, A., Sutskever, I.: Generating long sequences with sparse transformers. arXiv preprint arXiv:1904.10509 (2019)

8. Chiu, C.C., Raffel, C.: Monotonic chunkwise attention. arXiv preprint arXiv:1712.05382 (2017)

9. Choromanski, K., et al.: Rethinking attention with performers. arXiv:2009.14794 (2021)

10. Clark, K., Khandelwal, U., Levy, O., Manning, C.D.: What does BERT look at? an analysis of bert's attention. arXiv preprint arXiv:1906.04341 (2019)

11. Cohan, A., et al.: A discourse-aware attention model for abstractive summarization of long documents. Proceedings of the 2018 Conference of the North American Chapter of the Association for Computational Linguistics: Human Language Technologies, Volume 2 (Short Papers) (2018)

12. Dai, Z., Yang, Z., Yang, Y., Carbonell, J., Le, Q.V., Salakhutdinov, R.: Transformer-XL: attentive language models beyond a fixed-length context. arXiv preprint arXiv:1901.02860 (2019)

13. Devlin, J., Chang, M.W., Lee, K., Toutanova, K.: BERT: pre-training of deep bidirectional transformers for language understanding. arXiv preprint arXiv:1810.04805 (2018)

14. Fabbri, A.R., Li, I., She, T., Li, S., Radev, D.R.: Multi-news: a large-scale multi-document summarization dataset and abstractive hierarchical model (2019)

15. Guo, M., et al.: Longt5: efficient text-to-text transformer for long sequences. CoRR arXiv:abs/2112.07916 (2021)

16. Guo, Q., Qiu, X., Liu, P., Shao, Y., Xue, X., Zhang, Z.: Star-transformer. arXiv preprint arXiv:1902.09113 (2019)

17. He, J., Wang, L., Liu, L., Feng, J., Wu, H.: Long document classification from local word glimpses via recurrent attention learning. IEEE Access **7**, 40707–40718 (2019)

18. Huang, L., Cao, S., Parulian, N., Ji, H., Wang, L.: Efficient attentions for long document summarization. In: Proceedings of the 2021 Conference of the North American Chapter of the Association for Computational Linguistics: Human Language Technologies. Association for Computational Linguistics, Online (Jun 2021)

19. Katharopoulos, A., Vyas, A., Pappas, N., Fleuret, F.: Transformers are RNNs: fast autoregressive transformers with linear attention. CoRR arXiv:abs/2006.16236 (2020)

20. Kingma, D.P., Ba, J.: Adam: a method for stochastic optimization (2014)

21. Kitaev, N., Kaiser, L., Levskaya, A.: Reformer: the efficient transformer. CoRR arXiv:abs/2001.04451 (2020)

22. Lewis, M., et al.: BART: denoising sequence-to-sequence pre-training for natural language generation, translation, and comprehension. In: Proceedings of the 58th Annual Meeting of the Association for Computational Linguistics, pp. 7871–7880. Association for Computational Linguistics, Online (Jul 2020)

23. Liu, Y., et al.: Roberta: a robustly optimized BERT pretraining approach. CoRR arXiv:abs/1907.11692 (2019)

24. Maas, A.L., Daly, R.E., Pham, P.T., Huang, D., Ng, A.Y., Potts, C.: Learning word vectors for sentiment analysis. In: Proceedings of the 49th Annual Meeting of the Association for Computational Linguistics: Human Language Technologies, pp. 142–150. Association for Computational Linguistics (Jun 2011)
25. Rae, J.W., Potapenko, A., Jayakumar, S.M., Lillicrap, T.P.: Compressive transformers for long-range sequence modelling. arXiv preprint arXiv:1911.05507 (2019)
26. Raffel, C., et al.: Exploring the limits of transfer learning with a unified text-to-text transformer. J. Mach. Learn. Res. **21**(140), 1–67 (2020)
27. Raganato, A., Scherrer, Y., Tiedemann, J.: Fixed encoder self-attention patterns in transformer-based machine translation. arXiv preprint arXiv:2002.10260 (2020)
28. Rohde, T., Wu, X., Liu, Y.: Hierarchical learning for generation with long source sequences. CoRR arXiv:abs/2104.07545 (2021)
29. Sharma, E., Li, C., Wang, L.: BIGPATENT: a large-scale dataset for abstractive and coherent summarization. In: Proceedings of the 57th Annual Meeting of the Association for Computational Linguistics, pp. 2204–2213. Association for Computational Linguistics, Florence, Italy (Jul 2019)
30. Shaw, P., Uszkoreit, J., Vaswani, A.: Self-attention with relative position representations. In: Proceedings of the 2018 Conference of the North American Chapter of the Association for Computational Linguistics: Human Language Technologies, Volume 2 (Short Papers), pp. 464–468. New Orleans, Louisiana (Jun 2018)
31. Shen, T., Zhou, T., Long, G., Jiang, J., Zhang, C.: Bi-directional block self-attention for fast and memory-efficient sequence modeling. arXiv preprint arXiv:1804.00857 (2018)
32. Shen, Z., Zhang, M., Yi, S., Yan, J., Zhao, H.: Factorized attention: self-attention with linear complexities. CoRR arXiv:abs/1812.01243 (2018)
33. Vaswani, A., et al.: Attention is all you need. In: Advances in Neural Information Processing Systems, vol. 30 (2017)
34. Wang, S., Li, B.Z., Khabsa, M., Fang, H., Ma, H.: Linformer: self-attention with linear complexity. CoRR arXiv:abs/2006.04768 (2020)
35. Wu, F., Fan, A., Baevski, A., Dauphin, Y.N., Auli, M.: Pay less attention with lightweight and dynamic convolutions. arXiv preprint arXiv:1901.10430 (2019)
36. Xiao, W., Beltagy, I., Carenini, G., Cohan, A.: PRIMERA: pyramid-based masked sentence pre-training for multi-document summarization. In: Proceedings of the 60th Annual Meeting of the Association for Computational Linguistics (Volume 1: Long Papers), pp. 5245–5263. Dublin, Ireland (May 2022)
37. Zaheer, M., et al.: Big bird: transformers for longer sequences. In: Advances in Neural Information Processing Systems, vol. 33 (2020)
38. Zhang, J., Zhao, Y., Saleh, M., Liu, P.J.: Pegasus: pre-training with extracted gap-sentences for abstractive summarization (2019)
39. Zhang, X., Wei, F., Zhou, M.: Hibert: document level pre-training of hierarchical bidirectional transformers for document summarization. arXiv preprint arXiv:1905.06566 (2019)
40. Zhu, C., Liu, Y., Mei, J., Zeng, M.: Mediasum: a large-scale media interview dataset for dialogue summarization. arXiv preprint arXiv:2103.06410 (2021)

Dimensionality Detection and Feature Selection

Compressing the Embedding Matrix by a Dictionary Screening Approach in Text Classification

Jing Zhou[1]([✉]), Xinru Jing[1], Muyu Liu[1], and Hansheng Wang[2]

[1] Center for Applied Statistics and School of Statistics, Renmin University of China, Beijing, China
`jing.zhou@ruc.edu.cn`, `jingxinru0207@ruc.edu.cn`, `Muyu.Liu@ruc.edu.cn`
[2] Guanghua School of Management, Peking University, Beijing, China
`hansheng@pku.edu.cn`

Abstract. In this paper, we propose a dictionary screening method for embedding compression in text classification. The key point is to evaluate the importance of each keyword in the dictionary. To this end, we first train a pre-specified recurrent neural network-based model using a full dictionary. This leads to a benchmark model, which we use to obtain the predicted class probabilities for each sample in a dataset. Next, to evaluate the impact of each keyword in affecting the predicted class probabilities, we develop a novel method for assessing the importance of each keyword in a dictionary. Consequently, each keyword can be screened, and only the most important keywords are reserved. With these screened keywords, a new dictionary with a considerably reduced size can be constructed. Accordingly, the original text sequence can be substantially compressed. The proposed method leads to significant reductions in terms of parameters, average text sequence, and dictionary size. Meanwhile, the prediction power remains very competitive compared to the benchmark model. Extensive numerical studies are presented to demonstrate the empirical performance of the proposed method.

Keywords: Embedding Compression · Dictionary Screening · Text Classification

1 Introduction

Over the past few decades, natural language processing (NLP) has become a popular research field. Among the applications of this filed, text classification is considered to be a problem of great importance. Many successful applications exist, such as news classification [14], topic labeling [2], sentiment analysis [8], and many others. Note that, for most text classification tasks, the inputs are documents constructed from word sequences. Therefore, a standard RNN-based model can be readily applied. Remarkably, the model complexity of an RNN-based model is mainly determined by two factors. They are, respectively, the

H. Kashima et al. (Eds.): PAKDD 2023, LNAI 13935, pp. 457–468, 2023.
https://doi.org/10.1007/978-3-031-33374-3_36

model structure and dictionary size. Obviously, large dictionaries lead to more complicated RNN models with a large number of parameters. To illustrate this idea, consider, for example, a standard RNN model with one embedding layer and one recurrent hidden layer. The total number of parameters needed is $dk + 2k^2$, where d is the dictionary size and k is the dimension for both the hidden layer and embedding space. In the AG's News dataset [18], for instance, the total number of keywords contained in the dictionary could be as large as $d = 93,994$. If, following [3], we set the dimensions of both the hidden layer and embedding space to $k = 128$, then the total number of parameters is more than 12 million. As to be demonstrated later, we find that the dictionary size can be effectively reduced to be $d = 3,000$, if the method developed in this work is used. As a result, the total number of parameters can be reduced to about 0.58 million. This accounts for only about 4.76% of the original model complexity with limited sacrifice of prediction accuracy.

For text classification, most of the prior work focuses on compressing the embedding matrix. For example, a number of researchers have adopted hashing or quantization-based approaches to compress the embedding matrix [9,13,15]. [1] proposed a low rank matrix factorization for the embedding layer. In addition to compressing the embedding matrix, there is another branch of research that shows training with character-level inputs can achieve several benefits over word-level approaches, and it does so with fewer parameters [17]. For a more detailed review paper, we refer to [5]. Despite the excellent research that has been done on model compressing, it seems that most studies focus on simplifying the model structure in one way or another. Little research has been done on dictionary screening. As discussed in previous paragraph, we can see that the size of the dictionary can have an important impact on model size. As a consequence, we are motivated to fill this gap by proposing a dictionary screening method for text classification applications.

The proposed dictionary screening method aims to exclude less useful keywords from a dictionary. It can be used to effectively reduce dictionary size, leading to a significant reduction in model complexity. Specifically, we develop here a novel dictionary screening method as follows. First, we train a pre-specified RNN-based model using the full dictionary on a training dataset. This leads to a benchmark model. With the help of the benchmark model, we obtain the predicted class probabilities for every sample in the validation set. Next, for a given keyword in the dictionary, we consider whether it should be excluded from the dictionary. Obviously, keywords that significantly impact the predicted class probabilities should be kept, while those that do not should be excluded. Thus, the key here is determining how to evaluate the impact of the target keyword in affecting the estimated class probabilities.

To this end, for each document, we replace the target keyword with a meaningless substitute, which is often an empty space. By doing so, the input of the target keyword is excluded. We then apply the pretrained benchmark model to this new document. This leads to a new set of estimated class probabilities. Thereafter, for each document, we obtain two sets of estimated class probabilities. One is computed based on the full dictionary, and the other is computed based on the dictionary with the target keyword excluded. Next, the difference between the two sets of probability estimators is evaluated and summarized. Keywords with large differences in class probability estimators should be kept. By selecting an appropriate threshold value, a new dictionary with a substantially compressed size can be obtained. Finally, with the compressed dictionary, each document can be re-constructed. The associated RNN-based model can be re-trained on the re-constructed documents, and its prediction accuracy can be evaluated. Our extensive numerical experiments suggest that the proposed method can compress the parameter quantity by more than 90%, on average, with little accuracy sacrificed.

The main contribution of our work is the development of a compression method for text classification using dictionary screening. There has been relatively little work on compressing dictionary size in the previous literature. A second contribution is that we provide a novel method for evaluating the importance of the keywords in a dictionary. We empirically show that our method outperforms popular baselines like term frequency-inverse document frequency (TF-IDF) and the t-test for keyword importance analysis.

2   Methodology

2.1   Problem Set up

Let $\mathcal{D} = \{w_d : 0 \leq d \leq D\}$ be a dictionary containing a total of D keywords with w_d representing the dth keyword. We define w_0 to be an empty space. Then, assume a total of N documents indexed by $1 \leq i \leq N$. Let $X_i = \{X_{it} \in \mathcal{D} : 1 \leq t \leq T\}$ be the ith document. The document is constructed by a sequence of keywords, X_{it}, which is indexed by t and is generated from $\mathcal{D}$. If the actual document length, T^*, is less than T, we then define $X_{it} = w_0$ for $T^* < t \leq T$. Next, let $Y_i \in \{1, 2, \cdots, K\}$ be the class label associated with the ith document. The goal is then to train a classifier so that we can accurately predict Y_i. To this end, various deep learning models can be used. For illustration purposes, we consider here a simple RNN model with four layers: one input layer of a word sequence with dimension T, one embedding layer with $d_1 = 128$ hidden nodes, one simple RNN layer with $d_2 = 64$ hidden nodes, and one fully connected layer with K nodes (i.e., the number of class labels). Suppose the dictionary size is $D = 10,000$ and the number of class labels is $K = 10$, then for the above simple RNN model, the total number of parameters is given by $df_a = 10,000 \times 128 + 64 \times 128 + 64 \times 64 + 64 \times 10 = 1,292,928$ with the bias term ignored. However, this number will be much reduced if the dictionary size can be significantly decreased. For example, the total number of parameters will be reduced to $df_b = 1,000 \times 128 + 64 \times 128 + 64 \times 64 + 64 \times 10 = 140,928$ if $D = 1,000$. This represents a model complexity reduction as large as $(1 - df_b/df_a) \times 100\% = 89.1\%$. We are then motivated to develop a method for dictionary screening.

2.2   Dictionary Screening

As discussed in the introduction, one of the key tasks for dictionary screening is to evaluate the importance of each keyword in the dictionary. We can formulate the problem as follows. First, we train a pre-specified RNN-based model using the full dictionary, $\mathcal{D}$, on the training dataset. Mathematically, we can write this model as $f(X_i, \theta) = \{f_k(X_i, \theta)\} \in \mathbb{R}^K$, where X_i is the input document, with θ as the unknown parameters that need to be estimated. Note that $f(X_i, \theta)$ is a K-dimensional vector, its kth element, $f_k(X_i, \theta) \in [0, 1]$, is a theoretical assumed function to approximate the class probability. That is, $P(Y_i = k | X_i) \approx f_k(X_i, \theta)$. By the universal approximation theorem [4,7], we know that this approximation can be arbitrarily accurate as long as the approximation function, $f(\cdot, \theta)$, can be sufficiently flexible. To estimate θ, an approximately defined loss function (e.g., the categorical cross entropy) is usually used. Denote the loss function as $\mathcal{L}_N(\theta) = N^{-1} \sum_{i=1}^{N} \ell(X_i, \theta)$, where $\ell(X_i, \theta)$ is the loss function evaluated on the ith document. The parameter estimators can then be obtained as $\hat{\theta} = \text{argmax} \mathcal{L}_N(\theta)$. This leads to the pretrained model as $f(X_i, \hat{\theta})$, which serves as the benchmark model.

Next, with the help of the pretrained model, we consider how to evaluate the importance of each keyword in $\mathcal{D}$. Specifically, consider the dth keyword, w_d, in $\mathcal{D}$ with $1 \leq d \leq D$. Define $\mathcal{S}_{\mathcal{F}} = \{1, 2, \cdots, N\}$ as the indices for the full training document. Then, for every $i \in \mathcal{S}_{\mathcal{F}}$, we compute its estimated class probability vector as $\hat{p}_i = f(X_i, \hat{\theta})$. Next, for the document $X_i = \{X_{it} \in \mathcal{D} : 1 \leq t \leq T\}$, we generate a document copy as $X_i^{(d)} = \{X_{it}^{(d)} \in \mathcal{D} : 1 \leq t \leq T\}$, where $X_{it}^{(d)} = X_{it}$ if $X_{it} \neq w_d$, and $X_{it}^{(d)} = w_0$ if $X_{it} = w_d$. In other words, $X_i^{(d)}$ is a document that is almost the same as X_i. The only difference is that keyword w_d is replaced by an empty space, w_0. We next apply the benchmark model to $X_i^{(d)}$, so an update class probability vector, $\hat{p}_i^{(d)} = f(X_i^{(d)}, \hat{\theta})$, can be obtained. The difference between $\hat{p}_i$ and $\hat{p}_i^{(d)}$ is evaluated by their ℓ_2-distance as $||\hat{p}_i - \hat{p}_i^{(d)}||^2$. We then summarize the difference for every $w_d \in \mathcal{D}$ as $\hat{\lambda}(d) = |\mathcal{S}_{\mathcal{F}}|^{-1} \sum_{i \in \mathcal{S}_{\mathcal{F}}} ||\hat{p}_i - \hat{p}_i^{(d)}||^2$, where $|\mathcal{S}_{\mathcal{F}}|$ is the size of $\mathcal{S}_{\mathcal{F}}$. This is further treated as the important score for each keyword, $w_d \in \mathcal{D}$. Because this important score is obtained by evaluating the differences in class probability estimators, we name it as the CPE method for simplicity.

Finally, with a carefully selected threshold value, λ, a new dictionary can be constructed as $\mathcal{D}_\lambda = \{w_d \in \mathcal{D} : \hat{\lambda}(d) \geq \lambda\} \cup \{w_0\}$. To this end, each document X_i can be reconstructed as $X_{\lambda_i} = \{X_{\lambda_i t} \in \mathcal{D}_\lambda : 1 \leq t \leq T\}$, where $X_{\lambda_i t} = X_{it}$ if $X_{it} \in \mathcal{D}_\lambda$, and $X_{\lambda_i t} = w_0$ otherwise. Then, by replacing X_is in the loss function with X_{λ_i}, a new set of parameter estimators can be obtained as $\hat{\theta}_\lambda = \text{argmax} \mathcal{L}_\lambda(\theta)$, where $\mathcal{L}_\lambda(\theta) = N^{-1} \sum_{i=1}^{N} \ell(X_{\lambda_i}, \theta)$. Once $\hat{\theta}_\lambda$ is obtained, the prediction accuracy of the resulting model, $f(X_{\lambda_i}, \hat{\theta}_\lambda)$, can be evaluated on the testing dataset. Thereafter, the resulting model, $f(X_{\lambda_i}, \hat{\theta}_\lambda)$, serves as the reduced model after applying dictionary screening. The algorithm details are presented as follows.

Algorithm 1: Dictionary screening method

---

1 **Procedure1 Train a benchmark model:**

    Input : Document $X_i = \{X_{it} \in \mathcal{D} : 1 \le t \le T\}$; $Y_i \in \{1, 2, \cdots, K\}$
            for $1 \le i \le N$

    Output: Benchmark model $f(X_i, \hat{\theta})$ with estimated parameters $\hat{\theta}$ for
            $1 \le i \le N$

2    $\hat{\theta} = argmax \mathcal{L}_N(\theta)$ with $\mathcal{L}_N(\theta) = N^{-1} \sum_{i \in 1}^{N} \ell(X_i, \theta)$

3   where$\ell(X_i, \theta)$ is the loss evaluated on the ith document

4 **Procedure2 Evaluate the importance of each keyword in $\mathcal{D}$:**

    Input : $\mathcal{D} = \{w_d : 0 \le d \le D\}$, $f(X_i, \hat{\theta})$ for $1 \le i \le N$

    Output: $\mathcal{D}_\lambda = \{w_d \in \mathcal{D} : \hat{\lambda}(d) \ge \lambda\} \cup \{w_0\}$

5   **for** $d = 1$ to D **do**

6   **for** $i \in \mathcal{S}_\mathcal{F}, \mathcal{S}_\mathcal{F} = \{1, 2, \cdots, N\}$ *is the indices for the full training*
         document. **do**

7          $X_i = \{X_{it} \in \mathcal{D} : 1 \le t \le T\}$

8   Generate a copy as $X_i^{(d)} = \{X_{it}^{(d)} \in \mathcal{D} : 1 \le t \le T\}$ following:

9   If $X_{it} \ne w_d$ then $X_{it}^{(d)} = X_{it}$ else $X_{it}^{(d)} = w_0$

10   Uncompressed class probability : $\hat{p}_i = f(X_i, \hat{\theta})$

11   Compressed class probability : $\hat{p}_i^{(d)} = f(X_i^{(d)}, \hat{\theta})$

12   Difference computed by ℓ_2-distance as $||\hat{p}_i - \hat{p}_i^{(d)}||^2$

13   **end**

14   Summarize the difference:

15       $\hat{\lambda}(d) = |\mathcal{S}_\mathcal{F}|^{-1} \sum_{i \in \mathcal{S}_\mathcal{F}} ||\hat{p}_i - \hat{p}_i^{(d)}||^2$ where $|\mathcal{S}_\mathcal{F}|$ is the size of $\mathcal{S}_\mathcal{F}$

16   **end**

17   Select a threshold value λ to finally obtain

     $\mathcal{D}_\lambda = \{w_d \in \mathcal{D} : \hat{\lambda}(d) \ge \lambda\} \cup \{w_0\}$

---

3 Experiments

3.1 Task Description and Datasets

To demonstrate its empirical performance, the proposed dictionary screening method is evaluated on four large-scale datasets covering various text classification tasks. These are, respectively, news classification (AG's News and Sougou News), sentiment analysis (Amazon Review Polarity, ARP), and entity classification (DBPedia). These datasets are popularly studied in previous literature [17,18]. Summary statistics of the four large-scale datasets are presented in Table 1. For all the datasets (except for Sougou News), the sample size of each category is equal in both the training and testing sets. Take AG's News for example, it has 30,000 samples and 1,900 samples per class in the training set and testing set, respectively. For more detailed information about the four datasets, see [18]. It should be noted that, to make the experiments more diverse, the Sougou News data used in this paper are different from those in [18]. Particularly, we used the original Chinese characters of Sougou News to test the proposed method, not the Pinyin style used in [18]'s work. Moreover, unlike the other three datasets, the sample size of each category (e.g., sports, entertainment, business, and the Internet) in Sougou News is not equal. The proportions of the four categories in the training and testing sets are 48%, 15%, 25%, and 12%, respectively.

Table 1. Summary of four large-scale datasets.

Dataset	Classes	Task	TrainingSize	TestingSize
AG's News	4	news classification	120,000	7,600
Sougou News	4	news classification	63,146	15,787
DBPedia	14	entity classification	560,000	70,000
Amazon Review Polarity	2	sentiment analysis	3,600,000	400,000

3.2 Model Settings

We consider here two different types of deep learning models for text classification. They are, respectively, TextCNN [10] and TextBiLSTM [11]. We follow their network structures but with some modifications to adapt to our experiments. In the task of text classification, the input is text sequence X_i with length T. It should be noted that T is different for different datasets. In the current experiment, to train the benchmark models, T is set to 60, 300, 50, and 100 for AG's News, Sougou News, DBPedia, and Amazon Review Polarity, respectively. Practically, each keyword $w_d \in \mathcal{D}$ in the text sequence will be converted to a high dimensional vector of d_1 via an embedding layer [12]. For all three models, the embedding size, d_1, is set to 128. Next, we briefly describe the construction details for the two models.

TextCNN. After the embedding layer, we use three convolutional layers to extract text information. Each convolutional layer has $d_1 = 128$ filters with kernel size $k \in \{3, 4, 5\}$, followed by a max pooling with receptive field size $r = 1$. Rectified linear units (ReLUs) [6] are used as activation functions in the convolutional layers. Then, we concatenate the max pooling results of the three layers and pass it to the final dense layer through a softmax function for classification.

TextBiLSTM. The bidirectional LSTM (Bi-LSTM) can be seen as an improved version of the LSTM. This model structure can consider not only forward encoded information, but also reverse encoded information [11]. We apply a Bi-LSTM layer with the hidden states dimension, d_2, set to 128. We then use the representation obtained from the final timestep (e.g., X_{iT}) of the Bi-LSTM layer and pass it through a softmax function for text classification.

3.3 Tuning Parameter Specification

The implementation of the proposed dictionary screening method involves a tuning parameter, which is the threshold value λ. For a given classification model, this tuning parameter should be carefully selected to achieve the best empirical performance. Here, the best empirical performance means that the proposed dictionary screening method can reduce the number of model parameters as much as possible under the condition of ensuring little or no loss of accuracy. Generally, the larger the λ value is, the smaller the size of the screened dictionary, and thus the higher the reduction rate that can be achieved. However, considerable prediction accuracy might be lost. In our experiments, different λ values indicate that different numbers of keywords can be reserved for subsequent text compression. To this end, for analysis simplicity, we investigate the number of important keywords reserved, denoted as K. Specifically, we rearrange the keywords in descending order according to the importance score (e.g., $\widehat{\lambda}(d)$), and we select the top 1000, 3000, and 5000 keywords, respectively. That is, $K = \{1000, 3000, 5000\}$. Therefore, we can evaluate the impact of different tuning parameters on the performance of the proposed dictionary screening method.

3.4 Competing Methods

For comparison purposes, two other methods for evaluating the importance of the keywords in a dictionary are studied. The first one is to calculate the TF-IDF [16] value of each keyword $w_d \in \mathcal{D}$. For the kth keyword, we use the word counts as the term-frequency (TF). The inverse document frequency (IDF) is the logarithm of the division between the total number of documents and the number of documents with the kth word in the whole dataset. To this end, the TF-IDF value for each $w_d \in \mathcal{D}$ can be obtained by multiplying the values of TF and IDF. It is remarkable that the larger the TF-IDF value is, the more important the keyword is. The second method is to compute a t-test type statistic. Recall that, for the dth keyword, $w_d \in \mathcal{D}$, we have two sets of class probabilities, $\widehat{p}_i$ and $\widehat{p}_i^{(d)}$. Both are K-dimensional vectors. Then, for each dimension $k \in K$, a standard paired two sample t-test can be constructed to test for the statistical significance. The resulting p values obtained from different ks (e.g., different categories) are then summarized, and the smallest one is selected as the final t-test type measure for the target keyword, denoted as $P_{i,d}$. In this case, the smaller the $P_{i,d}$ value is, the more important the keyword is.

In summary, we have three methods to evaluate the importance of each keyword in $\mathcal{D}$. These are the proposed method for evaluating the differences in class probability estimators (CPE), the method for evaluating the TF-IDF values (TF-IDF), and the method for evaluating the t-test type statistics (t-statistic). To make a fair comparison, the new dictionaries constructed by the three methods are of equal size (e.g., with same tuning parameter K). Then, following the procedure described in Sect. 3.2, we can obtain three different prediction accuracies based on the screened dictionary.

3.5   Performance Measures and Implementation

Following the existing literature [1,17,18], and our own concerns, we adopt six measures to gauge the empirical performances of the different compression methods: the parameter reduction ratio (Prr), dictionary reduction ratio (Drr), storage reduction ratio (Srr), FLOP reduction ratio (Frr), and reduction ratio for averaged text sequence (Trr). Meanwhile, the out-of-sample prediction accuracy (Acc) is also monitored.

Both text classification models (e.g., TextCNN and TextBiLSTM) are trained on the four large-scale datasets. This leads to a total of eight working models. All the working models are trained using the AdaDelta (Zeiler, 2012) with $\rho = 0.95, \epsilon = 10^{-5}$, and a batch size of 128. The weight decay is set to 5×10^{-4} with an ℓ_2-norm regularizer. To prevent overfitting, the dropout and early stopping strategies are used for different working models. Finally, a total of 200 epochs are conducted for each working model. For each working model, we choose the epoch with the maximum prediction accuracy on the validation set as the baseline model. All the experiments were run on a Tesla P100 GPU with 64 GB memory.

4   Results Analysis

4.1   Tuning Parameter Effects

In this subsection, we study the impact of the tuning parameter, K, which determines the number of keywords reserved. Three measures are used to gauge the finite sample performance: Acc, Prr, and Trr. For illustration purposes, we use the AG's News dataset as an example. For this experiment, three different K values are studied: $K = \{1000, 3000, 5000\}$. The detailed results are given in Fig. 1. The top panel of Fig. 1 displays the performance of the TextCNN model. The red line in the first barplot is the prediction accuracy for the benchmark model. We find the resulting prediction accuracy (Acc) of the reduced model increases as K becomes larger, while the parameter reduction ratio (Prr) and the reduction ratio for averaged text sequence (Trr) decrease. In the case of $K = 3000$, we can see the parameter reduction ratio (Prr) is more than 95%, but there is almost no accuracy loss. This suggests that the benchmark model can be substantially compressed with little sacrifice in predictive power. Additionally, the averaged text sequence is substantially reduced based on the dictionary screening. This indicates that there might be some redundant information in the original text that contributes less to the text classification. The bottom panel of Fig. 1 presents the results for the TextBiLSTM model, which are very similar to the findings of TextCNN.

4.2   Performance of Compression Results

On the one hand, the proposed dictionary screening method aims to compress a text classification model as much as possible. On the other hand, an over compressed model might suffer from a significant loss of prediction accuracy. Thus, it

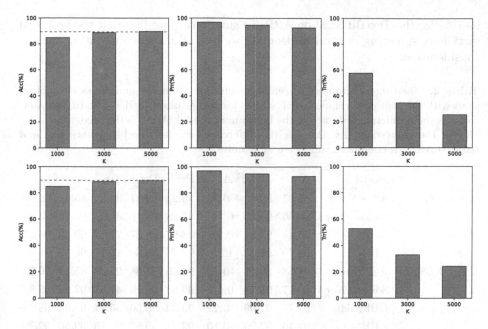

Fig. 1. Detailed experimental results for TextCNN and TextBiLSTM on the AG's News dataset. Three different K values are considered ($K = 1000,3000,5000$). Three performance measures are summarized: prediction accuracy (Acc), parameter reduction ratio (Prr), and reduction ratio for averaged text sequence (Trr). The red dashed line in the left panel represents the accuracy of the benchmark model. (Color figure online)

is of great importance to understand the trade-off between prediction accuracy and model compression. Obviously, they should be appropriately balanced. In this subsection, we report the fine-tuned compression results so that their best performance can be demonstrated. For the best empirical performance, we try to extend the search scope for more tuning parameters. Accordingly, every value in $\{1000, 2000, \cdots, 10000\}$(e.g., with an interval of 1000) is tested for K in this subsection. In our case, we expect that the parameter reduction ratio (Prr) will be no less than 50% and the accuracy loss will be no more than 2%. The best results in terms of the above criteria are summarized in Table 2. From Table 2, we can draw the following conclusions. First, for all cases, the benchmark models can be compressed substantially using the dictionary screening method with little sacrifice of accuracy. For example, the value of Prr is more than 95% in the case of TextCNN on Sougou News with only 0.31% sacrifice of accuracy. Second, we report that for half of the cases, the prediction accuracy of the reduced model is higher than that of the benchmark model (e.g., ΔAcc is smaller than zero). For instance, for the TextBiLSTM model on DBPedia, the prediction accuracy of the baseline model is as high as 97.77%, while that of the reduced model is further improved to 97.99%. Finally, we find the reduction in storage and FLOPs are also quite substantial. For instance, the Srr is 99.56% and the Prr is

99.70% for the TextBiLSTM on ARP. To summarize, we find that the proposed dictionary screening method works quite well on all models and datasets under consideration.

Table 2. Fine-tuned dictionary screening results for all model and dataset combinations with the best performance. ARP stands for the Amazon Review Polarity dataset. Acc-1 is the prediction accuracy of the benchmark model. Acc-2 is the prediction accuracy of the reduced model. ΔAcc is the difference between the benchmark Acc and reduced Acc. All computed values are in % units.

	Model	Acc-1	Acc-2	ΔAcc	Prr	Drr	Trr	Srr	Frr
TextCNN	AG's News	89.37	89.43	-0.06	95.24	96.81	34.54	85.61	89.49
	Sougou News	95.56	95.25	0.31	97.28	98.19	48.35	97.26	93.83
	DBPedia	98.41	97.88	0.53	98.97	99.27	27.17	99.40	98.97
	ARP	92.70	91.45	1.25	99.74	99.66	13.75	98.77	99.33
TextBiLSTM	AG's News	89.58	89.72	-0.14	92.65	94.68	24.25	85.19	90.72
	Sougou News	95.77	95.34	0.43	96.98	98.19	47.53	97.96	95.81
	DBPedia	97.77	97.99	-0.22	99.33	99.56	35.14	98.95	98.44
	ARP	92.16	92.26	-0.10	99.78	99.87	24.75	99.56	99.70

4.3   Competing Methods

Table 3. Results of the three competing methods (e.g., CPE, TF-IDF, and t-statistics). ARP stands for the Amazon Review Polarity dataset. Reduced Acc is the prediction accuracy of the reduced model. For each model and dataset combination, the reduced models were trained with dictionaries of equal size. Trr is the reduction ratio for averaged text sequence. The values outside of brackets are the results of the propose CPE method. The first value in brackets is the result obtained by TF-IDF, and the second value is the result by t-statistic. All computed values are in % units.

	Model	Reduced Acc	Trr
TextCNN	AG's News	89.66 (89.33, 87.91)	34.54 (29.19, 46.07)
	Sogou News	94.71 (94.66, 94.68)	57.54 (44.39, 55.56)
	DBPedia	97.89 (97.89, 17.23)	27.17 (25.36, 93.12)
	ARP	90.73 (90.49, 89.77)	25.50 (24.25, 32.50)
TextBiLSTM	AG's News	90.49 (90.09, 89.89)	18.07 (13.54, 23.84)
	Sogou News	95.47 (95.14, 95.31)	47.53 (37.53, 52.54)
	DBPedia	97.99 (98.00, 17.39)	35.14 (30.80, 99.82)
	ARP	92.32 (92.24, 91.55)	24.75 (24.25, 30.00)

Because the key step of the proposed dictionary screening method is to evaluate the importance of each keyword in the dictionary. We compare the performance of the proposed evaluating method, CPE, with two other competing methods. These are the TF-IDF and t-statistics methods, which are described in Subsect. 4.4. For a fair comparison, the new dictionaries constructed with the three methods are of equal size (e.g., the dictionary reduction ratio, Drr, is the same) for each model and dataset combination. As a result, only the reduced prediction accuracy (Acc) and reduction ratio for averaged text sequence (Trr) are presented in Table 3. From Table 3, we can obtain the following conclusions. First, we can see that the t-statistics method is not stable in evaluating the importance of keywords because its results for the DBPedia dataset were not comparable with the other methods. In the case of DBPedia, the t-statistic method ceases to be effective because its Trr value is nearly 100%. This indicates that it cannot filter the important keywords from the dictionary, leading to a very low reduced Acc value. Second, in all cases, the proposed CPE method achieved a slightly higher reduced accuracy value compared with the TF-IDF method. Moreover, for most cases, by using the proposed CPE method, we can achieve a substantially reduced ratio for averaged text sequence (Trr). This indicates that the proposed CPE method can achieve a better performance in terms of predicted accuracy when keeping a relatively short text sequence.

5   Conclusions

In this paper, we propose a dictionary screening method for embedding compression in text classification tasks. The goal of this method is to evaluate the importance of each keyword in the dictionary. To this end, we develop a method called CPE to evaluate the differences in class probability estimators. With the CPE method, each keyword in the original dictionary can be screened, and only the most important keywords can be reserved. The proposed method leads to a significant reduction in terms of parameters, average text sequence, and dictionary size. Meanwhile, the prediction power remains competitive. Extensive numerical studies are presented to demonstrate the empirical performance of the proposed method.

To conclude this article, we present here a number of interesting topics for future study. First, the proposed dictionary screening method involves a tuning parameter (e.g., K), and its optimal value for balancing prediction accuracy and parameter reduction needs to be learned. This is an important topic for future study. Second, the proposed method is only used for a text classification task. However, there are other natural language tasks, such as machine translation, question answering, and so on. The scalability of the proposed method in these tasks is also a very worthy study. Lastly, the proposed method is only conducted on English and Chinese, other language types should be investigated to test its validity.

Acknowledgements. Zhou's research is supported in part by the National Natural Science Foundation of China (Nos. 72171226, 11971504), the Beijing Municipal Social Science Foundation (No. 19GLC052). Wang's research is partially supported by the National Natural Science Foundation of China (No. 12271012, 11831008) and the Open

Research Fund of the Key Laboratory of Advanced Theory and Application in Statistics and Data Science (KLATASDS-MOE-ECNU-KLATASDS2101).

References

1. Acharya, A., Goel, R., Metallinou, A., Dhillon, I.: Online embedding compression for text classification using low rank matrix factorization. In: Proceedings of the AAAI Conference on Artificial Intelligence. vol. 33, pp. 6196–6203 (2019)
2. Auer, S., Bizer, C., Kobilarov, G., Lehmann, J., Cyganiak, R., Ives, Z.: DBpedia: a nucleus for a web of open data. In: Aberer, K., et al. (eds.) ASWC/ISWC -2007. LNCS, vol. 4825, pp. 722–735. Springer, Heidelberg (2007). https://doi.org/10.1007/978-3-540-76298-0_52
3. Cho, K., Merrienboer, B.V., Gulcehre, C., Schwenk, D., Bougares, F., Schwenk, H., Bengio, Y.: Learning phrase representations using RNN encoder-decoder for statistical machine translation. Computer Science (2014)
4. Cybendo, G.: Approximations by superpositions of a sigmoidal function. Math. Control Signals Systems **2**, 183–192 (1989)
5. Deng, L., Li, G., Han, S., Shi, L., Xie, Y.: Model compression and hardware acceleration for neural networks: a comprehensive survey. Proc. IEEE **108**(4), 485–532 (2020)
6. Glorot, X., Bordes, A., Bengio, Y.: Deep sparse rectifier neural networks. J. Mach. Learn. Res. **15**, 315–323 (2011)
7. Hornik, K., Stinchcombe, M.B., White, H.: Multilayer feedforward networks are universal approximators. Neural Netw. **2**, 359–366 (1989)
8. Hossain, E., Sharif, O., Hoque, M.M., Sarker, I.H.: SentiLSTM: a deep learning approach for sentiment analysis of restaurant reviews. In: Proceedings of 20th International Conference on Hybrid Intelligent Systems (2020)
9. Joulin, A., Grave, E., Bojanowski, P., Douze, M., Jegou, H., Mikolov, T.: Fasttext.zip: compressing text classification models. arXiv preprint arXiv:1612.03651 (2016)
10. Kim, Y.: Convolutional neural networks for sentence classification. Eprint Arxiv (2014)
11. Li, F., Zhang, M., Fu, G., Qian, T., Ji, D.: A Bi-LSTM-RNN model for relation classification using low-cost sequence features (2016)
12. Mikolov, T., Chen, K., Corrado, G., Dean, J.: Efficient estimation of word representations in vector space. Computer Science (2013)
13. Raunak, V.: Effective dimensionality reduction for word embeddings. arXiv preprint arXiv:1708.03629 (2017)
14. Sachan, D.S., Zaheer, M., Salakhutdinov, R.: Revisiting LSTM networks for semi-supervised text classification via mixed objective function. Proceedings of the AAAI Conference on Artificial Intelligence (2019)
15. Shu, R., Nakayama, H.: Compressing word embeddings via deep compositional code learning. arXiv preprint arXiv:1711.01068 (2017)
16. Sparck-Jones, K.: A statistical interpretation of term specificity and its application in retrieval. J. Document. **28**(1), 11–21 (1972)
17. Xiao, Y., Cho, K.: Efficient character-level document classification by combining convolution and recurrent layers (2016)
18. Zhang, X., Zhao, J., Lecun, Y.: Character-level convolutional networks for text classification. MIT Press (2015)

Ethics and Fairness

Disentangled Representation with Causal Constraints for Counterfactual Fairness

Ziqi Xu[1(✉)], Jixue Liu[1], Debo Cheng[1], Jiuyong Li[1(✉)], Lin Liu[1], and Ke Wang[2]

[1] University of South Australia, Adelaide, Australia
{Ziqi.Xu,Debo.Cheng}@mymail.unisa.edu.au,
{Jixue.Liu,Jiuyong.Li,Lin.Liu}@unisa.edu.au
[2] Simon Fraser University, Burnaby, Canada
wangk@sfu.ca

Abstract. Much research has been devoted to the problem of learning fair representations; however, they do not explicitly state the relationship between latent representations. In many real-world applications, there may be causal relationships between latent representations. Furthermore, most fair representation learning methods focus on group-level fairness and are based on correlation, ignoring the causal relationships underlying the data. In this work, we theoretically demonstrate that using the structured representations enables downstream predictive models to achieve counterfactual fairness, and then we propose the Counterfactual Fairness Variational AutoEncoder (CF-VAE) to obtain structured representations with respect to domain knowledge. The experimental results show that the proposed method achieves better fairness and accuracy performance than the benchmark fairness methods.

Keywords: Counterfactual Fairness · Representation Learning · Variational AutoEncoder

1 Introduction

Machine learning algorithms have gradually penetrated into our life [23] and have been applied to decision-making for credit scoring [16], crime prediction [14] and loan assessment [5]. The fairness of these decisions and their impact on individuals or society have become an increasing concern. Some extreme unfair incidents have appeared in recent years. For example, COMPAS, a decision support model that estimates the risk of a defendant becoming a recidivist was found to predict higher risk for black people and lower risk for white people [1]; Facebook users receive a recommendation prompt when watching a video featuring blacks, asking them if they'd like to continue to watch videos about primates [21]. These incidents indicate that the machine learning models become a source of unfairness, which may lead to serious social problems. Since most models are trained with data, which will lead to unfair decisions due to discrimination in

H. Kashima et al. (Eds.): PAKDD 2023, LNAI 13935, pp. 471–482, 2023.
https://doi.org/10.1007/978-3-031-33374-3_37

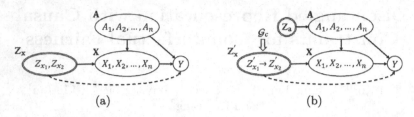

Fig. 1. (a) The process of existing works on learning fair representations to make predictions. (b) The process of our work. $\mathbf{A}$ is the set of sensitive attributes; $\mathbf{X}$ is the set of other observed attributes; $\mathbf{Z_a}$ is the representation of $\mathbf{A}$; Y is the target attribute; $\mathbf{Z_x}$ is the representation of $\mathbf{X}$; $\mathbf{Z_x'}$ is the structured representation of $\mathbf{X}$ with respect to the conceptual level causal graph $\mathcal{G}_c$. The dotted line denotes the prediction process.

the training data. Therefore, the key issue for solving unfair decisions becomes whether we can eliminate these discrimination embedded in the data through algorithms [23].

To obtain fair decisions, many methods [6,10,20,22,25,31] are proposed to learn fair representations through two competing goals: encoding data as much as possible, while eliminating any information that transfers through the sensitive attributes. To separate the information from sensitive attributes, various extensions of Variational Autoencoder (VAE) consider minimising the mutual information among latent representations [6,20,25]. For example, Creager et al. [6] introduced disentanglement loss into the VAE objective function to decompose observed attributes into sensitive latents and non-sensitive latents to achieve subgroup level fairness; Park et al. [25] improved the above methods and proposed the mutual attribute latent (MAL) to retain only beneficial information for fair predictions.

The existing methods [6,20] follow Fig. 1a to achieve fair predictions. Specifically, these methods learn fair representations $\mathbf{Z_x}$ without stating any relationships between Z_{x1} and Z_{x2}, which may not satisfy the domain knowledge. Let us consider an example where we aim to predict a person's salary using some observed attributes. Following the domain knowledge, we know that people's salary is determined by two semantic concepts, intelligence and career respectively. We also note that people's intelligence determines their career with high probability, which can be expressed as a conceptual level causal graph $\mathcal{G}_c$, i.e., *Intelligence* → *Career*. Therefore, we need a method as shown in Fig. 1b that not only ensures the representation of observed attributes with no sensitive information but also retains causal relationships with respect to domain knowledge.

On the measurement of fairness, all fair representation learning methods use fairness metrics based on correlation, including the VAE-based methods [6,20,25]. It is well known that correlation does not imply causation. Recent studies [26,32] have shown that quantifying fairness based on correlation may produce higher deviations. Counterfactual fairness is a fundamental framework based on causation. With counterfactual fairness, a decision is fair towards an individual if it is the same in the actual world and in the counterfactual world when the individual belongs to a different demographic group [17].

In this paper, we follow the counterfactual fairness and propose a VAE-based unsupervised fair representation learning method, namely Counterfactual Fairness Variational AutoEncoder (CF-VAE). We make the following contributions in this paper:

- We propose CF-VAE, a novel VAE-based unsupervised counterfactual fairness method. CF-VAE can learn structured representations with no sensitive information and retain causal relationships with respect to the conceptual level causal graph determined by domain knowledge.
- We theoretically demonstrate that the structured representations obtained by CF-VAE are suitable for training counterfactually fair predictive models.
- We evaluate the effectiveness of the CF-VAE method on real-world datasets. The experiments show that CF-VAE outperforms existing benchmark fairness methods in both accuracy and fairness.

2   Background

We use upper case letters to represent attributes and boldfaced upper case letters to denote the set of attributes. We use boldfaced lower case letters to represent the values of the set of attributes. The values of attributes are represented using lower case letters.

Let $\mathbf{A}$ be the set of sensitive attributes; $\mathbf{X}$ be the set of other observed attributes; $\mathbf{V}$ be the set of all observed attributes, i.e., $\mathbf{V} = \{\mathbf{A}, \mathbf{X}\}$; Y be the target attribute. We use $\widehat{Y}(\cdot)$ to represent the predictor. $\mathcal{G}_c$ is the conceptual level causal graph and represents domain knowledge. The nodes shown in $\mathcal{G}_c$ are "concepts", each of which represents a set of observed attributes that have similar meanings. Each "concept" has causal relationships with the other "concepts".

In this paper, a causal graph is used to represent a causal mechanism. In a causal graph, a directed edge, such as $V_j \rightarrow V_i$ denotes that V_j is a parent (i.e., direct cause) and we use pa_i to denote the set of parents of V_i. We follow Pearl's [26] notation and define a causal model as a triple $(\mathbf{U}, \mathbf{V}, \mathbf{F})$: $\mathbf{U}$ is a set of the latent background attributes, which are the factors not caused by any attributes in the set $\mathbf{V} = \{\mathbf{A}, \mathbf{X}\}$; $\mathbf{F}$ is a set of deterministic functions, $V_i = f_i(pa_i, U_{pa_i})$, such that $pa_i \subseteq \mathbf{V} \backslash \{V_i\}$ and $U_{pa_i} \subseteq \mathbf{U}$. Besides, some commonly used definitions in graphical causal modelling, such as faithfulness, d-separation and causal path can be found in [26, 27].

With the causal model $(\mathbf{U}, \mathbf{V}, \mathbf{F})$, we have the following definition of counterfactual fairness:

Definition 1. (Counterfactual Fairness [17]). Predictor $\widehat{Y}(\cdot)$ is counterfactually fair if under any context $\mathbf{X} = \mathbf{x}$ and $\mathbf{A} = \mathbf{a}$, $P(\widehat{Y}_{\mathbf{A} \leftarrow \mathbf{a}}(\mathbf{U}) = y \mid \mathbf{X} = \mathbf{x}, \mathbf{A} = \mathbf{a}) = P(\widehat{Y}_{\mathbf{A} \leftarrow \bar{\mathbf{a}}}(\mathbf{U}) = y \mid \mathbf{X} = \mathbf{x}, \mathbf{A} = \mathbf{a})$, for all y and for any value $\bar{\mathbf{a}}$ attainable by $\mathbf{A}$.

Counterfactual fairness is considered to be related to individual fairness [17]. Individual fairness means that similar individuals should receive similar predicted outcomes. The concept of individual fairness when measuring the similarity of the individual is unknowable, which is similar to the unknowable distance between the real-world and the counterfactual world in counterfactual fairness [18].

3   Proposed Method

In this section, we first theoretically demonstrate that learning counterfactually fair representations are feasible. Then, we propose the Counterfactual Fairness Variational AutoEncoder (CF-VAE) to obtain the structured representations for predictors to achieve counterfactual fairness.

3.1   The Theory of Learning Counterfactually Fair Representations

We discuss what types of representations enable downstream predictive models to achieve counterfactual fairness. Following the work in [17], the implication of counterfactual fairness is described as follows:

Definition 2. (Implication of Counterfactual Fairness [17]). Let $\mathcal{G}$ be the causal graph of the given model $(\mathbf{U}, \mathbf{V}, \mathbf{F})$. If there exists $\mathbf{W}$ be any non-descendant of $\mathbf{A}$, then downstream predictor $\widehat{Y}(\mathbf{W})$ will be counterfactually fair.

We extend Definition 2 to the fair representation learning and present the following theorem.

Theorem 1. *Given the causal graph $\mathcal{G}$, $\mathbf{Z}_\mathbf{a}$ is the representation of sensitive attributes $\mathbf{A}$, $\mathbf{Z}_\mathbf{x}'$ is the structured representation of the other observed attributes $\mathbf{X}$ with respect to the conceptual level causal graph $\mathcal{G}_c$. We have $\widehat{Y}(\mathbf{Z}_\mathbf{x}')$ satisfy counterfactual fairness.*

Proof. Given the causal graph $\mathcal{G}$ as shown in Fig. 2, there is not a parent node of $\mathbf{A}$ in $\mathbf{X}$, and there is not a child node of Y in $\mathbf{X}$. $\mathbf{X}$ contains four subsets: $\mathbf{X}_Y^A$ is the subset of other observed attributes that are descendants of $\mathbf{A}$ and parents of Y; $\mathbf{X}_Y^N$ is the subset of other observed attributes that are only parents of Y; $\mathbf{X}_N^N$ is the subset of other observed attributes that are no relationships with $\mathbf{A}$ and Y; $\mathbf{X}_N^A$ is the subset of other observed attributes that are only descendants of $\mathbf{A}$. After perfect representation learning, we obtain $\mathbf{Z}_\mathbf{a}$ and $\mathbf{Z}_\mathbf{x}'$.

We proof that $\mathbf{Z}_\mathbf{x}'$ is not the descendant of $\mathbf{A}$ with the following two subsets. For the first subsets $\{\mathbf{X}_Y^A, \mathbf{X}_Y^N, \mathbf{X}_N^A\}$, there are seven paths between $\mathbf{A}$ and $\mathbf{Z}_\mathbf{x}'$, including $\mathbf{A} \rightarrow \mathbf{X}_Y^A \leftarrow \mathbf{Z}_\mathbf{x}'$, $\mathbf{A} \rightarrow \mathbf{X}_Y^A \rightarrow Y \leftarrow \mathbf{Z}_\mathbf{x}'$, $\mathbf{A} \rightarrow \mathbf{X}_Y^A \rightarrow Y \leftarrow \mathbf{X}_Y^N \leftarrow \mathbf{Z}_\mathbf{x}'$, $\mathbf{A} \rightarrow Y \leftarrow \mathbf{X}_Y^A \leftarrow \mathbf{Z}_\mathbf{x}'$, $\mathbf{A} \rightarrow Y \leftarrow \mathbf{Z}_\mathbf{x}'$, $\mathbf{A} \rightarrow Y \leftarrow \mathbf{X}_Y^N \leftarrow \mathbf{Z}_\mathbf{x}'$ and $\mathbf{A} \rightarrow \mathbf{X}_N^A \leftarrow Y$. These seven paths are blocked by $\emptyset$ (i.e., $\mathbf{A}$ and $\mathbf{Z}_\mathbf{x}'$ are d-separated by $\emptyset$), since each path contains a collider either $\mathbf{X}_Y^A$ or Y or $\mathbf{X}_N^A$. For second subset $\mathbf{X}_N^N$, there is no path connecting $\mathbf{X}_N^N$ and Y. Hence, $\mathbf{Z}_\mathbf{x}'$ is not the descendant of $\mathbf{A}$. Therefore, $\widehat{Y}(\mathbf{Z}_\mathbf{x}')$ is counterfactually fair based on Definition 2.   □

We use Fig. 2 to show whether the following predictors satisfy counterfactual fairness.

– $\widehat{Y}(\mathbf{A}, \mathbf{X})$: This model is unfair since it uses sensitive attributes to make prediction.

– $\widehat{Y}(\mathbf{X})$: This model satisfies fairness through awareness [8] but fails to achieve counterfactual fairness. Since it uses $\mathbf{X}_Y^A$ and $\mathbf{X}_N^A$ which are the descendants of $\mathbf{A}$.

– $\widehat{Y}(\mathbf{Z_a}, \mathbf{Z_x'})$: This model is unfair

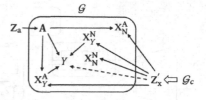

Fig. 2. $\mathcal{G}$ is the causal graph that represents the causal relationship between $\mathbf{A}$, $\mathbf{X} = \{\mathbf{X}_Y^A, \mathbf{X}_Y^A, \mathbf{X}_N^A, \mathbf{X}_N^N\}$ and Y. The dotted line represents the prediction process that uses $\mathbf{Z}_x'$.

because it uses sensitive attributes for prediction. The reason is that $\mathbf{Z_a}$ is the representation of $\mathbf{A}$, which should be consider as sensitive attributes either.

– $\widehat{Y}(\mathbf{X}_Y^N, \mathbf{X}_N^N)$: This model satisfies counterfactual fairness since both $\mathbf{X}_Y^N$ and $\mathbf{X}_N^N$ are non-descendants of $\mathbf{A}$. However, this predictor losses a lot of useful information that embeds in other observed attributes.

– $\widehat{Y}(\mathbf{Z}_x')$: This model is counterfactually fair based on Theorem 1 and achieves higher accuracy than $\widehat{Y}(\mathbf{X}_Y^N, \mathbf{X}_N^N)$ as shown in our experiments.

3.2 CF-VAE

We first discuss the causal constraints and then explain the loss function of CF-VAE in detail. The architecture of CF-VAE is shown in Fig. 3.

Learning Representations with Causal Constraints. We aim to retain causal relationships between "concepts" through a more easily accessible conceptual level causal graph $\mathcal{G}_c$ and embed these relationships in representations.

To formalise causal relationships, we consider n "concepts" in the dataset, which means $\mathbf{Z}_x'$ should have the same dimension as "concepts". The "concepts" in observations are causally structured by $\mathcal{G}_c$ with an adjacency matrix $\mathbf{C}$. For simplicity, in this paper, the causal constraints are exactly implemented by a linear structural equation model: $\mathbf{Z}_x' = (\mathbf{I} - \mathbf{C}^T)^{-1}\mathbf{Z_x}$, where $\mathbf{I}$ is the identity matrix, $\mathbf{Z_x}$ is obtained from the encoder, $\mathbf{Z}_x'$ is constructed from $\mathbf{Z_x}$ and $\mathbf{C}$. $\mathbf{C}$ is obtained from $\mathcal{G}_c$ with respect to domain knowledge. The parameters in $\mathbf{C}$ indicate that there are corresponding edges, and the values of the parameters indicate the weight of the causal relationships. It is worth noting that if the parameter value is zero, it means that such an edge does not exist, i.e., no causal relationship between these two "concepts".

As mentioned above, $\mathbf{Z_x}$ is obtained from the encoder, we cannot guarantee that each attribute inside is independent. To ensure the independence of each attribute in $\mathbf{Z_x}$, we employ the total correction regularisation (TCR) in our loss function. TCR also encourages the correctness of structured $\mathbf{Z}_x'$ with respect to domain knowledge since it guarantees that there are no relationships between

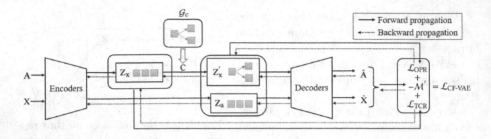

Fig. 3. The architecture of CF-VAE.

each attribute in $\mathbf{Z_x}$ before adding causal constraints. The TCR for our proposed CF-VAE is defined as, $\mathcal{L}_{\text{TCR}} = \gamma D_{KL}[q(\mathbf{Z_x})||\prod_{i=1}^{D_{\mathbf{Z_x}}} q(Z_{\mathbf{x}_i})]$, where γ is the weight value, $D_{\mathbf{Z_x}}$ is dimension of $\mathbf{Z_x}$.

Learning Strategy. We first explain the Evidence lower bound (ELBO) with causal constraints. Then, we add orthogonality promoting regularisation (OPR) to obtain the loss function of CF-VAE. Given the training samples, the parameters can be optimised by maximising the following ELBO:

$$\mathcal{M} = \mathbb{E}_{q(\mathbf{Z_a}|\mathbf{A})}[\log p(\mathbf{A}|\mathbf{Z_a})] + \mathbb{E}_{q(\mathbf{Z'_x}|\mathbf{X})}[\log p(\mathbf{X}|\mathbf{Z'_x})]$$
$$- D_{KL}[q(\mathbf{Z_a}|\mathbf{A})||p(\mathbf{Z_a})] - D_{KL}[q(\mathbf{Z'_x}|\mathbf{X})||p(\mathbf{Z'_x})],$$

where $p(\mathbf{Z'_x}) = (\mathbf{I} - \mathbf{C}^T)^{-1} p(\mathbf{Z_x})$; $p(\mathbf{X}|\mathbf{Z'_x}) = \prod_{i=1}^{D_{\mathbf{x}}} p(X_i|\mathbf{Z'_x})$; $q(\mathbf{Z'_x}|\mathbf{X}) = \prod_{i=1}^{D_{\mathbf{Z'_x}}} \mathcal{N}(\mu = \hat{\mu}_{Z'_{\mathbf{x}_i}}, \sigma^2 = \hat{\sigma}^2_{Z'_{\mathbf{x}_i}})$.

Then, we introduce orthogonality to encourage disentanglement between $\mathbf{Z_a}$ and $\mathbf{Z'_x}$. We employ orthogonality promoting regularisation based on the pairwise cosine similarity among latent representations: if the cosine similarity is close to zero, then the latent representations are closer to being orthogonal and independent [29]. The orthogonality promoting regularisation (OPR) for our proposed CF-VAE is defined as, $\mathcal{L}_{\text{OPR}} = \frac{1}{B}\sum_{i=1}^{B} \frac{\mathbf{Z_{a}}_i{}^T \mathbf{Z'_{x}}_i}{\|\mathbf{Z_{a}}_i\|_2 \, \|\mathbf{Z'_{x}}_i\|_2}$, where B denotes the batch size for neural network, $\|\cdot\|_2$ is the l_2 norm.

In conclusion, the loss function of our proposed CF-VAE is defined as:

$$\mathcal{L}_{\text{CF-VAE}} = -\mathcal{M} + \mathcal{L}_{\text{TCR}} + \mathcal{L}_{\text{OPR}}.$$

4   Experiments

In this section, we conduct extensive experiments to evaluate CF-VAE on real-world datasets. Before showing the detailed results, we first present the details of selected methods and the evaluation metrics. The code is available at https://github.com/IRON13/CF-VAE.

4.1   Framework Comparison

The proposed CF-VAE is considered as a pre-processing technique to address fairness issues. Hence, we compare CF-VAE with traditional and VAE-based pre-processing methods. For traditional methods, we select baselines including ReWeighting (RW) [13], Disparate Impart Remover (DIR) [9] and Optimized Preprocessing (OP) [2]. For VAE-based methods, we compare with VFAE [20] and FFVAE [6]. We also obtain the Full model for comparison, which uses all attributes in the dataset to make predictions.

We select several well-known predictive models to simulate the downstream prediction process. Linear Regression (LR_R), Stochastic Gradient Descent Regression (SGD_R) and Multi-layer Perceptron Regression (MLP_R) are used for regression tasks; Logistic Regression (LR_C), Stochastic Gradient Descent Classification (SGD_C) and Multi-layer Perceptron Classification (MLP_C) are used for classification tasks. For each predictive model, we run 10 times and record the mean and variance of the results for evaluation metrics.

4.2   Evaluation Metrics

Fairness. There are no metrics to quantify counterfactual fairness since we can only obtain real-world data. Thus, we propose the situation test to measure fairness for different predictive models. In our experiment, we define unfairness score (UFS) to measure the result of the situation test. Specifically, the form of score differs for different predictive models. For regression tasks, we define

$$\text{UFS}_R = \sqrt{\frac{1}{N} \sum_{i=1}^{N} \left(\widehat{Y}_{A \leftarrow a}(\mathbf{Z}'_{\mathbf{x}_i}) - \widehat{Y}_{A \leftarrow \bar{a}}(\mathbf{Z}'_{\mathbf{x}_i}) \right)^2};$$ For classification tasks, we define $\text{UFS}_C = \frac{1}{N} \sum_{i=1}^{N} \text{xor} \left(\widehat{Y}_{A \leftarrow a}(\mathbf{Z}'_{\mathbf{x}_i}), \widehat{Y}_{A \leftarrow \bar{a}}(\mathbf{Z}'_{\mathbf{x}_i}) \right)$ (N is the number of samples for evaluation). The lower UFS value means that the predictive models achieve higher fairness performance.

Accuracy. We evaluate the performance on prediction with the following metrics. For regression tasks, we use Root Mean Square Error (RMSE) to compare the error between prediction results and target attributes' values. For classification tasks, we use accuracy to evaluate various predictive models.

4.3   Law School

The law school dataset comes from a survey [28] of admissions information from 163 law schools in the United States. It contains information of 21,790 law students, including their entrance exam scores (LSAT), their grade point average (GPA) collected prior to law school, and their first-year average grade (FYA). The school expects to predict if the applicants will have a high FYA. Gender and race are sensitive attributes in this dataset, and the school also wants to ensure that predictions are not affected by sensitive attributes. However, LSAT, GPA and FYA scores may be biased due to socio-environmental factors. We use the same $\mathcal{G}_c$ as shown in work [17] to model latent "concepts" of GPA and $LSAT$. The process of CF-VAE for the Law school dataset is shown in Fig. 4a.

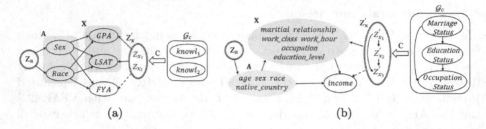

Fig. 4. (a) The process of CF-VAE for Law school dataset. (b) The process of CF-VAE for Adult dataset.

Table 1. The results for Law School dataset. The best fairness aware RMSE and the best UFS_R are shown in bold.

Model	Accuracy (RMSE) ↓			Fairness (UFS_R) ↓		
	LR_R	SGD_R	MLP_R	LR_R	SGD_R	MLP_R
Full	0.865(0.007)	0.867(0.007)	0.865(0.007)	0.660(0.019)	0.762(0.019)	0.760(0.045)
RW	0.955(0.013)	0.956(0.012)	0.953(0.012)	0.067(0.002)	0.067(0.001)	0.079(0.003)
DIR	0.943(0.009)	0.944(0.009)	0.941(0.010)	0.060(0.001)	0.060(0.001)	0.070(0.002)
OP	0.959(0.011)	0.960(0.011)	0.956(0.010)	0.047(0.001)	0.046(0.001)	0.055(0.003)
VFAE	0.932(0.007)	0.933(0.007)	0.934(0.007)	0.035(0.010)	0.074(0.017)	0.096(0.010)
FFVAE	0.933(0.005)	0.934(0.004)	0.935(0.005)	0.032(0.007)	0.060(0.022)	0.097(0.008)
CF-VAE	**0.931(0.006)**	**0.932(0.006)**	**0.932(0.006)**	**0.013(0.006)**	**0.025(0.011)**	**0.044(0.006)**

Results. As shown in Table 1, since the Full model uses sensitive attributes to make predictions, inverting sensitive attributes has the highest impact on the individual's prediction results, which means that the model is unfair. RW, DIR and OP achieves fair predictions by modifying the dataset compared to the Full model. Both VFAE and FFVAE disentangle the sensitive attributes with latent representations, so the influence of inverting the sensitive attributes on the prediction results is small. Our method achieves the lowest UFS_R, 0.013, 0.025, and 0.044 for LR_R, SGD_R, and MLP_R respectively, which means CF-VAE disentangle $\mathbf{Z}'_x$ and $\mathbf{Z_a}$ more precisely.

For accuracy results, the Full model uses sensitive information to more accurately predict FYA and thus achieves the highest accuracy. The proposed CF-VAE achieves the best fairness aware accuracy in all predictive models than other methods.

4.4 Adult

The Adult dataset comes from the UCI repository [7] contains 14 attributes including race, age, education information, marital information as well as capital gain and loss for 48,842 individuals. We use the same $\mathcal{G}_c$ as shown in previous research [4,24] to model the latent "concepts". The adjacency matrix **C** is

defined as: $\mathbf{C} = \begin{vmatrix} 0 & \lambda_{12} & \lambda_{13} \\ 0 & 0 & \lambda_{23} \\ 0 & 0 & 0 \end{vmatrix}$. Then, we construct $\mathbf{Z}'_x$ from $\mathbf{Z_x}$ and **C** as follows:

Table 2. The results for Adult dataset. The best fairness aware accuracy and the best UFS_C are shown in bold.

Model	Accuracy ↑			Fairness (UFS_C) ↓		
	LR_C	SGD_C	MLP_C	LR_C	SGD_C	MLP_C
Full	0.802(0.002)	0.803(0.004)	0.831(0.004)	0.068(0.003)	0.060(0.018)	0.034(0.009)
RW	0.797(0.001)	0.792(0.002)	0.819(0.001)	0.038(0.001)	0.029(0.002)	0.052(0.001)
DIR	0.800(0.001)	0.793(0.003)	0.817(0.001)	0.035(0.001)	0.027(0.002)	0.046(0.001)
OP	0.780(0.002)	0.779(0.003)	0.783(0.002)	0.032(0.003)	0.030(0.004)	0.033(0.005)
VFAE	0.785(0.001)	0.781(0.003)	0.819(0.004)	0.062(0.002)	0.041(0.010)	0.025(0.003)
FFVAE	0.785(0.003)	0.782(0.001)	0.814(0.005)	0.062(0.001)	0.044(0.010)	0.032(0.010)
CF-VAE	**0.801(0.002)**	**0.794(0.004)**	**0.820(0.002)**	**0.031(0.002)**	**0.020(0.006)**	**0.024(0.004)**

$Z'_{x_1} = Z_{x_1}$; $Z'_{x_2} = \lambda_{12}Z_{x_1} + Z_{x_2}$; $Z'_{x_3} = \lambda_{13}Z_{x_1} + \lambda_{23}Z_{x_2} + Z_{x_3}$. We set parameter $\{\lambda_{12} = 1, \lambda_{13} = 1, \lambda_{23} = 1\}$ to denote that edges within latent representations, i.e., $Z'_{x_1} \rightarrow Z'_{x_2}, Z'_{x_1} \rightarrow Z'_{x_3}, Z'_{x_2} \rightarrow Z'_{x_3}$. The process of CF-VAE is shown in Fig. 4b.

Results. The fairness results are shown in Table 2, the Full model achieves the worst UFS_C, since it use **A** to predict *income*. Both baseline fairness models and other VAE-based methods improve fairness to a certain extent. The proposed CF-VAE achieves the best UFS_C, only 3.1%, 2.0% and 2.4% of individuals' results are affected by sensitive attributes' values inversions in LR_C, SGD_C and MLP_C, respectively. Our method achieves better fairness performance than other methods, since it remains causal relationships in latent representations with respect to $\mathcal{G}_c$ and disentangles structured representations with sensitive attributes.

In order to achieve fairness, VFAE and FFVAE lose about 2% of their accuracy performance. RW, DIR and OP modify the dataset resulting in a loss of predictive performance. The proposed CF-VAE not only guarantees the fairness performance but also retains the causal relationships to improve accuracy. CF-VAE loses less information than other VAE-base methods and achieves the best fairness aware accuracy performance in all predictive models, i.e., 80.1%, 79.4% and 82.0% in LR_C, SGD_C and MLP_C, respectively.

4.5 Ablation Study

We follow the same procedure in [3] to generate synthetic datasets and conduct an ablation study to validate the contribution of each component in our method as shown in Table 3.

The Full model (Model i) uses all the observed attributes to train the predictors. The predictors achieve the best accuracy but the worst fairness performance. VFAE (Model ii) is the basic VAE-based unsupervised fair representation learning method. We set it to be the baseline. Model iii is CF-VAE without adding causal constraints, which achieves similar results as VFAE since both methods remove sensitive information from the learnt representations.

Table 3. The results of ablation study. The best fairness aware RMSE and the best UFS$_R$ are shown in bold, and the runner-up results are underlined.

Model	Accuracy (RMSE) ↓			Fairness (UFS$_R$) ↓		
	LR$_R$	SGD$_R$	MLP$_R$	LR$_R$	SGD$_R$	MLP$_R$
i	0.078(0.001)	0.081(0.001)	0.081(0.001)	0.102(0.001)	0.098(0.001)	0.106(0.002)
ii	0.126(0.002)	0.126(0.002)	0.145(0.002)	0.006(0.001)	0.010(0.002)	0.104(0.005)
iii	0.125(0.001)	0.125(0.001)	0.145(0.001)	0.007(0.001)	0.011(0.003)	0.105(0.003)
iv	**0.109(0.001)**	0.111(0.001)	0.122(0.002)	0.003(0.001)	**0.004(0.002)**	0.071(0.002)
v	**0.109(0.001)**	**0.110(0.001)**	**0.121(0.001)**	**0.002(0.001)**	0.005(0.002)	**0.070(0.002)**

Then, we employ causal constraints and add TCR in the loss function as Model iv, which retains causal relationships in latent representations and improves both accuracy and fairness performance than previous models. Model v (a.k.a. CF-VAE) is to encourage $\mathbf{Z}'_x$ and $\mathbf{Z}_a$ are disentangled by adding OPR. As shown in Table 3, CF-VAE achieves the best accuracy performance and UFS$_R$ among most predictive.

5   Related Works

The machine learning literature has increasingly focused on exploring how algorithms can protect marginalised populations from unfair treatment. An important research area is how to quantify fairness, which can be divided into two categories, the statistical framework and the causal framework.

In the statistical framework, Demographic parity was defined by [31], which is used to measure group-level fairness. Other similar metrics include equalised odds [11], predictive rate parity [30]. Dwork et al. [8] proposed a measurement to quantify individual-level fairness, that is, similar individuals should have similar treatments, and they use distance functions to measure how similar between individuals. In the causal framework, the (conditional) average causal effect is used to quantify fairness between groups [19]; Natural direct and natural indirect effects are used to quantify specific fairness [24,33]; When unfair causal paths are identified by domain knowledge, Chiappa [4] used the path-specific causal effects to quantify fairness on approved paths. For more related works, please refer to the literature review [23,32].

Our work is related to learning fair representations, which aims to encode data information into a lower space while removing sensitive information. VAE [15] and β-VAE [12] have inspired several studies in fair representation learning. Louizos et al. [20] first introduced VAE for learning fair representation to disentangle the sensitive information and non-sensitive information, they proposed a semi-supervised method to encourage disentanglement by using "Maximum Mean Discrepancy" (MMD). However, the organisations that collect the data cannot predict the downstream uses of the data and the models that might be used [10,31]. Due to this, many following up works [6,20] focus on unsupervised learning fair representation. But these works only focus on correlation-based constraints to ensure fairness. Our approach combines counterfactual fairness and unsupervised fair representation learning to provide the proper representations. Furthermore, we innovatively embed domain knowledge into representations by adding causal constraints with respect to domain knowledge.

6 Conclusion

In this paper, we investigate unsupervised counterfactually fair representation learning and propose a novel method named CF-VAE which considers causal relationships with respect to domain knowledge. We theoretically demonstrate that the structured representations obtained by CF-VAE enable predictive models to achieve counterfactual fairness. Experimental results on real-world datasets show that CF-VAE achieves better accuracy and fairness performance on downstream predictive models than the benchmark fairness methods. Ablation study on synthetic datasets shows that causal constraints with total correction regularisation achieve better accuracy performance and orthogonality promoting regularisation encourages disentanglement with sensitive attributes.

Acknowledgements. This work has received partial support from the Australian Research Council Discovery Project (DP200101210) to J. Li, J. Liu and K. Wang, the discovery grant from the Natural Sciences and Engineering Research Council of Canada to K. Wang, and the University Presidents Scholarship (UPS) of the University of South Australia to Z. Xu.

References

1. Brennan, T., Dieterich, W., Ehret, B.: Evaluating the predictive validity of the compass risk and needs assessment system. Crim. Just. Behav. **36**(1), 21–40 (2009)
2. Calmon, F.P., Wei, D., Vinzamuri, B., Ramamurthy, K.N., Varshney, K.R.: Optimized pre-processing for discrimination prevention. In: NeurIPS, pp. 3992–4001 (2017)
3. Cheng, D., Li, J., Liu, L., Yu, K., Le, T.D., Liu, J.: Toward unique and unbiased causal effect estimation from data with hidden variables. IEEE Trans. Neural Netw. Learn. Syst., 1–13 (2022)
4. Chiappa, S.: Path-specific counterfactual fairness. In: AAAI, pp. 7801–7808 (2019)
5. Coşer, A., Maer-matei, M.M., Albu, C.: Predictive models for loan default risk assessment. Econ. Comput. Econ. Cybern. Stud. Res. **53**(2), 149–165 (2019)
6. Creager, E., et al.: Flexibly fair representation learning by disentanglement. In: ICML, pp. 1436–1445 (2019)
7. Dua, D., Graff, C.: UCI machine learning repository (2017)
8. Dwork, C., Hardt, M., Pitassi, T., Reingold, O., Zemel, R.S.: Fairness through awareness. In: ITCS, pp. 214–226 (2012)
9. Feldman, M., Friedler, S.A., Moeller, J., Scheidegger, C., Venkatasubramanian, S.: Certifying and removing disparate impact. In: SIGKDD, pp. 259–268 (2015)
10. Gitiaux, X., Rangwala, H.: Learning smooth and fair representations. In: AISTATS, pp. 253–261 (2021)

11. Hardt, M., Price, E., Srebro, N.: Equality of opportunity in supervised learning. In: NeurIPS, pp. 3315–3323 (2016)
12. Higgins, I., et al.: beta-vae: learning basic visual concepts with a constrained variational framework. In: ICLR, pp. 1–22 (2017)
13. Kamiran, F., Calders, T.: Data preprocessing techniques for classification without discrimination. Knowl. Inf. Syst. **33**(1), 1–33 (2012)
14. Kim, S., Joshi, P., Kalsi, P.S., Taheri, P.: Crime analysis through machine learning. In: IEMCON, pp. 415–420 (2018)
15. Kingma, D.P., Welling, M.: Auto-encoding variational bayes. In: ICLR, pp. 1–14 (2014)
16. Kruppa, J., Schwarz, A., Arminger, G., Ziegler, A.: Consumer credit risk: individual probability estimates using machine learning. Expert Syst. Appl. **40**(13), 5125–5131 (2013)
17. Kusner, M.J., Loftus, J.R., Russell, C., Silva, R.: Counterfactual fairness. In: NeurIPS, pp. 4066–4076 (2017)
18. Lewis, D.: Counterfactuals. John Wiley & Sons, Hoboken (2013)
19. Li, J., Liu, J., Liu, L., Le, T.D., Ma, S., Han, Y.: Discrimination detection by causal effect estimation. In: IEEE BigData, pp. 1087–1094 (2017)
20. Louizos, C., Swersky, K., Li, Y., Welling, M., Zemel, R.S.: The variational fair autoencoder. In: ICLR, pp. 1–11 (2016)
21. Mac, R.: Facebook apologizes after ai puts 'primates' label on video of black men (2021). https://www.nytimes.com/2021/09/03/technology/facebook-ai-race-primates.html
22. Madras, D., Creager, E., Pitassi, T., Zemel, R.S.: Fairness through causal awareness: learning causal latent-variable models for biased data. In: FAT*, pp. 349–358 (2019)
23. Mehrabi, N., Morstatter, F., Saxena, N., Lerman, K., Galstyan, A.: A survey on bias and fairness in machine learning. ACM Comput. Surv. **54**(6), 115:1–115:35 (2021)
24. Nabi, R., Shpitser, I.: Fair inference on outcomes. In: AAAI, pp. 1931–1940 (2018)
25. Park, S., Hwang, S., Kim, D., Byun, H.: Learning disentangled representation for fair facial attribute classification via fairness-aware information alignment. In: AAAI, pp. 2403–2411 (2021)
26. Pearl, J.: Causality. Cambridge University Press, Cambridge (2009)
27. Spirtes, P., Glymour, C.N., Scheines, R., Heckerman, D.: Causation, Prediction, and Search. MIT press, Cambridge (2000)
28. Wightman, L.F.: Lsac national longitudinal bar passage study. lsac research report series (1998)
29. Xie, P., Wu, W., Zhu, Y., Xing, E.P.: Orthogonality-promoting distance metric learning: convex relaxation and theoretical analysis. In: ICML, pp. 5399–5408 (2018)
30. Zafar, M.B., Valera, I., Gomez-Rodriguez, M., Gummadi, K.P.: Fairness beyond disparate treatment & disparate impact: Learning classification without disparate mistreatment. In: WWW, pp. 1171–1180 (2017)
31. Zemel, R.S., Wu, Y., Swersky, K., Pitassi, T., Dwork, C.: Learning fair representations. In: ICML, pp. 325–333 (2013)
32. Zhang, L., Wu, X.: Anti-discrimination learning: a causal modeling-based framework. Int. J. Data Sci. Anal. **4**(1), 1–16 (2017)
33. Zhang, L., Wu, Y., Wu, X.: A causal framework for discovering and removing direct and indirect discrimination. In: IJCAI, pp. 3929–3935 (2017)

F3: Fair and Federated Face Attribute Classification with Heterogeneous Data

Samhita Kanaparthy[ID], Manisha Padala[✉][ID], Sankarshan Damle[ID],
Ravi Kiran Sarvadevabhatla[ID], and Sujit Gujar[ID]

International Institute of Information Technology (IIIT), Hyderabad, India
{s.v.samhita,manisha.padala,sankarshan.damle}@research.iiit.ac.in
{ravi.kiran,sujit.gujar}@iiit.ac.in

Abstract. Fairness across different demographic groups is an essential criterion for face-related tasks, Face Attribute Classification (FAC) being a prominent example. Simultaneously, federated Learning (FL) is gaining traction as a scalable paradigm for distributed training. In FL, client models trained on private datasets get aggregated by a central aggregator. Existing FL approaches require data homogeneity to ensure fairness. However, this assumption is restrictive in real-world settings. E.g., geographically distant or closely associated clients may have heterogeneous data. In this paper, we observe that existing techniques for ensuring fairness are not viable for FL with data heterogeneity. We introduce F3, an FL framework for fair FAC under data heterogeneity. We propose two methodologies in F3, (i) Heuristic-based and (ii) Gradient-based, to improve fairness across demographic groups without requiring data homogeneity assumption. We demonstrate the efficacy of our approaches through empirically observed fairness measures and accuracy guarantees on popular face datasets. Using Mahalanobis distance, we show that F3 obtains a practical balance between accuracy and fairness for FAC. The code is available at: github.com/magnetar-iiith/F3.

Keywords: Fairness · Federated Learning · Data Heterogeneity

1 Introduction

Face Attribute Classification (FAC) finds prominence for tasks such as gender classification, face verification, and face identification [25]. Recently, researchers have highlighted a critical issue in FAC: attribute prediction may be biased towards specific demographic groups [17]. E.g., for gender classification, the error rate for 'darker' faces is greater than that on 'lighter' faces. Further, face recognition-based criminal detection systems are prone to classify innocent people with 'darker' faces as suspects. This bias in predictions is *unfairness*. It is often associated with the unavailability of balanced datasets [18]. To overcome this issue, researchers have introduced balanced, large-scale datasets [9].

Federated Learning (FL) has emerged as a popular paradigm for scalable distributed training for large-scale data [11]. FL comprises (i) independent clients

H. Kashima et al. (Eds.): PAKDD 2023, LNAI 13935, pp. 483–494, 2023.
https://doi.org/10.1007/978-3-031-33374-3_38

that train local models on their private data and (ii) a central aggregator which combines these local models (e.g., through a random weighted average aka FedAvg [13]), to derive a generalised global model. Unfortunately, traditional FL models typically focus on standard performance measures (e.g., accuracy) and inherit the fairness related drawbacks of non-FL approaches [6].

To address the unfairness in FL several methods exist [4,6,14,21]. However, these methods inherently assume FL clients with homogeneous data, i.e., they assume that FL clients' data contains samples from all the demographic groups of a particular *sensitive attribute*. For example, with 'age' as the sensitive attribute, the client's local training data would have samples from both 'young' and 'adult' demographic groups. However, clients' data is likely to be heterogeneous in many FL settings. A smartphone belonging to a 'young' user may have content belonging majorly to its peers [16], i.e., inter-client heterogeneity in terms of age. Similarly, geographically separated clients may exhibit inter-client heterogeneity in demographics. Such data heterogeneity may, in turn, reduce fairness for tasks such as FAC.

Our Approach: In this paper, we introduce and study the fair Face Attribute Classification (FAC) problem in FL under data heterogeneity (FL with DH). We first prove that existing approaches to ensure fairness [7,15] are not applicable in this setting (Proposition 1). Consequently, we introduce *F3*, a FL framework for Fair Face Attribute Classification. Under the F3 framework, we propose two different methodologies (i) *Heuristic-based F3* and (ii) *Gradient-based F3*.

- Heuristic-based F3 includes novel aggregation heuristics: (i) FairBest, (ii) α-FairAvg, and (iii) α-FairAccAvg which prioritize specific local client model(s) to improve the accuracy and fairness trade-off (Sect. 3.1).
- Gradient-based F3 introduces FairGrad, where the client training is modified to include fairness through gradients communicated by the aggregator, to train a fair and accurate global model (Sect. 3.2).
- To validate the efficacy of both our methodologies, we conduct extensive experiments on three popular face datasets, namely FairFace [9], FFHQ [10], and UTK [23] (Sect. 4). Our results highlight that F3, through its aggregation and gradient-based methodologies, outperforms the standard approach, FedAvg-DH [13]. More concretely, F3 ensures 25%–82% improvement in terms of fairness with an accuracy drop of 0.4%–17% compared to FedAvg-DH.

Due to space constraints, we place our work w.r.t the existing literature in terms of (i) fairness and (ii) data heterogeneity in FL in the full version [8]. Also, note that the methods proposed in this work address unfairness in FL with DH setting, irrespective of the classification problem. However, we choose Face Attribute Classification to demonstrate our methods' performances.

2 Preliminaries

We consider the Face Attribute Classification (FAC) task, where $\mathcal{X}$ is the universal set of face images, with binary labels from $\mathcal{Y} = \{0, 1\}$ (e.g., male or female), and sensitive attribute $A \in \mathcal{A}$. Here, A can be age, race, or gender. The sensitive attribute takes a finite set of values, $A = \{a_1, \ldots, a_s\}$. E.g., age can take values such as 'young' or 'adult'. We next describe our FL setting for federated FAC.

2.1 Federated Learning (FL) Setting

In FL, the data is distributed across multiple parties referred to as clients. Let $\mathcal{C} = \{C_1, \ldots, C_m\}$ represent the set of clients; each C_i owns the set $D_i \subset \mathcal{X} \times \mathcal{Y} \times A$ containing n_i samples. Each C_i trains its local model $h_{\theta_{i,t}} : \mathcal{X} \to \mathcal{Y}$ parameterized by $\theta_{i,t}$ at round t. The aggregator holds small amount of data not enough to train a good model. Generally, the data is split into a test set and a validation set (D_a) [22]. We assume that D_a comprises samples from each demographic group.

At each round t, a random subset of clients $S_t \subseteq \mathcal{C}$ communicate their locally updated model parameters $\Theta_t = \{\theta_{i,t} \mid C_i \in S_t\}$ to the aggregator. The aggregator combines all communicated model parameters to obtain the global parameters at round t, ϕ_t, using a heuristic choice function $\mu : \Theta_t \to \phi_t$. The Weighted average (or FedAvg) [13] is the most used heuristic, defined as: $\mu_{\texttt{FedAvg}}(\Theta_t) \triangleq \phi_t = \sum_{C_i \in S_t} \frac{n_i}{\sum_j n_j} \theta_{i,t}$. The aggregator then communicates the model parameters back to the clients. Then clients initialise their local model with these parameters and train further. This back and forth process is repeated multiple times till convergence.

2.2 Fairness Notions

The standard notions for fair classification depend on the error rates: False negative rate (FNR) and False positive rate (FPR). For a face attribute classifier h, given a face image x with true label y and sensitive attribute $a \in A$, we have $FNR = \Pr(h(x) = 0|y = 1)$ and $FNR_a = \Pr(h(x) = 0|A = a, y = 1), \forall a \in A$. Likewise, $FPR = \Pr(h(x) = 1|y = 0)$, and $FPR_a = \Pr(h(x) = 1|A = a, y = 0), \forall a \in A$. FNR_a and FPR_a are the error rates observed on the data samples belonging to a particular demographic group with sensitive attribute $a \in A$. E.g., consider an FAC task for 'gender' classification with 'age' as the sensitive attribute. The attribute comprises {'young', 'adult'} as the demographic groups. Now, consider the following group-fairness notions.

Equality of Opportunity (EOpp) [3]: A classifier h satisfies EOpp for a distribution over $(\mathcal{X}, \mathcal{Y}, A)$ if: $FNR_a = FNR, \forall a$. We denote the violation in EOpp as $\Delta_{EOpp} = \max(\{|FNR_a - FNR| | \forall a \in A\})$. That is, Δ_{EOpp} is the maximum disparity in FNR across the groups. Intuitively, EOpp ensures that the probability of predicting a 'male' face as 'female' is the same across age groups.

Equalized Odds (EO) [5]: A classifier h satisfies EO over $(\mathcal{X}, \mathcal{Y}, A)$ if: $FNR_a = FNR$ and $FPR_a = FPR \; \forall a$. Let $\Delta_{EO} = \max(\max(\{|FPR_a - FPR| | \forall a \in A\}), \max(\{|FNR_a - FNR| | \forall a \in A\}))$ denote the violation in EO. EO states that the probability of miss-predicting the gender must be independent of age.

Accuracy Parity (AP) [24]: A classifier h satisfies AP for a distribution over $(\mathcal{X}, \mathcal{Y}, A)$ if: $FPR + FNR = FPR_a + FNR_a, \forall a$. Let $\Delta_{AP} = \max(\{|FPR_a + FNR_a - (FPR + FNR)| \| \forall a \in A\})$ denote violation in AP. AP ensures that the overall classification error is equal across the age groups.

Lagrangian Multiplier Method (LMM) [12]: To incorporate these fairness notions in FAC, the standard technique is to train a model that maximises accuracy while minimising the violation in these fairness notions. LMM adopts a loss function that simultaneously incorporates cross-entropy loss l_{CE} and the violation in fairness constraint $(\Delta_{EOpp}, \Delta_{EO}, \Delta_{AP})$, weighted by the lagrangian multiplier $\lambda \in \mathbb{R}^+$. Formally, in LMM, the loss function $L_{LMM}(h(X), Y, A)$ for a classifier h, for $k \in \{EOpp, EO, AP\}$ and $(X, Y) \subseteq \mathcal{X} \times \mathcal{Y}$, is as follows.

$$L_{LMM}(\cdot) = \mathbb{E}_{(x,y)\sim(X,Y)}[l_{CE}(h(x), y)] + \lambda\Delta_k. \tag{1}$$

3   Methodology

We first motivate the problem by showing that existing fair FL approaches are not applicable in the data heterogenous setting. We then introduce our novel methodologies, (i) Heuristic-based F3 and (ii) Gradient-based F3.

Motivation: In a practical FL setting, each client might only possess samples from an individual demographic group. E.g., samples belonging only to the 'young' age group when age is the sensitive attribute. We refer to this scenario as Federated Learning with Data Heterogeneity (FL with DH). The existing approaches for fair FL typically compute fairness violation locally [12,15]. Proposition 1 shows that with DH, this fairness violation component for the demographic groups not present in a particular client's data is not defined.

Proposition 1. *In the Lagrangian Multiplier Method, the loss L_{LMM} (Eq. 1) is not defined for FL with Data Heterogeneity (FL with DH).*

The formal proof is available in [8]. Proposition 1 holds for any fairness violation function (such as Δ_{EO}, Δ_{AP}) that requires samples belonging to all the demographic groups. E.g., the loss functions defined in [1,19,20]. Thus, we cannot use these functions to train for fairness in FL with DH. Additionally, training only for accuracy compromises fairness [3], implying that standard approaches such as FedAvg may not suffice. As a result, we next propose novel methodologies curated for FL with DH.

3.1   Heuristic-Based F3

Observe that as FedAvg aggregates a random (sub)set of models at each round, it fails to ensure fairness as the local models for aggregation may potentially be biased. In turn, they may amplify the unfairness of the global model. In Heuristic-based F3, we propose novel aggregation heuristics that prioritize the local client models, which perform desirably in terms of fairness. The method comprises the following steps.

Algorithm 1. Heuristic-based F3

Input: (1) Each client $C_k \in \mathcal{C}$ with D_k (2) Hyperparameters: maximum number of communication rounds T, number of local epochs E, learning rate η, accuracy threshold a, epoch threshold τ (3) A heuristic choice function $\mu_\varsigma(\Theta_t)$ s.t. $\varsigma = \{$FedAvg, FAIRBEST, α-FAIRAVG, α-FAIRACCAVG$\}$

Output: Model ϕ

1: $\phi_0 \leftarrow$ randomly initialized weights
2: **for** each round $t = 0, 1 \ldots, T - 1$ **do**
3:    **for** each client $C_k \in S_t$ (in parallel) **do**
4:       (Client C_k) $\theta_{k,t} \leftarrow$ LOCALTRAINING(k, ϕ_t)
5:    **end for**
6:    (Aggregator) $\phi_{t+1} \leftarrow \mu_\varsigma(\Theta_t)$; $\Theta_t = \{\theta_{k,t} \mid C_k \in S_t\}$
7:    $\phi_{best} \leftarrow$ StoppingCondition$(t + 1, \phi_{t+1}, \phi_{best}, a, \tau)$
8: **end for**
9: **return** ϕ_{best}
10: **procedure** LOCALTRAINING(k, ϕ_t)
11:    $\theta_{k,t} \leftarrow \phi_t$
12:    **for** each local epoch $i = 1, 2, \ldots, E$ **do**
13:       (Per Batch) $\theta_{k,t} \leftarrow \theta_{k,t} - \eta \cdot \nabla_{\theta_{k,t}} L_k \left(h_{\theta_{k,t}}(\cdot), D_k \right)$
14:    **end for**
15:    **return** $\theta_{k,t}$
16: **end procedure**

---

1. <u>Local Training.</u> Each client C_i trains its model h_{θ_i} only for maximising accuracy, i.e., minimising $L_i(h_{\theta_i}, D_i) = \mathbb{E}_{(x,y) \sim D_i}[l_{CE}(h_{\theta_i}(x), y)]$. At each round t, a random subset of clients $S_t \subseteq \mathcal{C}$ communicate their model parameters to the aggregator.
2. <u>Model Aggregation.</u> To better control the accuracy and fairness trade-off, we propose novel heuristics for aggregation. These heuristics derive the global model based on accuracy and fairness values for the models in S_t computed on the aggregator's set D_a.
3. <u>Model Communication.</u> The aggregator then communicates the global model parameters to each client. The clients adopt these parameters and further train on them to maximise accuracy.

Figure 1a depicts Heuristic-based F3 and Algorithm 1 provides a procedural outline.

Stopping Criteria. In Algorithm 1, StoppingCondition controls the training's stoppage and model updates (for details see [8]). The procedure records the improvement in accuracy and fairness values across epochs. It updates the "best" model observed so far if: (i) the current epoch is less than a threshold τ, or (ii) ϕ produces a lesser fairness violation than the previous best model. The training stops if the change in accuracy does not exceed a threshold "a" across τ rounds.

Heuristics for Fair FL: With our novel heuristics, we aim to ensure fairness in FL with DH by deliberately aggregating only the subset of local models that perform desirably w.r.t. to fairness and accuracy. The aggregator quantifies the

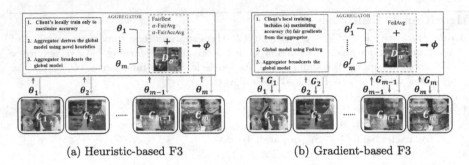

(a) Heuristic-based F3   (b) Gradient-based F3

Fig. 1. Overview of our Methodologies.

performance of local client models based on their empirical fairness violation and accuracy computed on the aggregator set D_a. Specifically, let $\Delta_{loss}(h_i(\theta_{i,t}))$ denote any fairness violation for client's $C_i \in S_t$ model on D_a at any round t. Denote $Acc(h_i(\theta_{i,t}))$ as the accuracy of C_i's model on D_a at t. With this, consider the following novel heuristics that aim to strike a practical balance between fairness and accuracy.

1. FAIRBEST: Aggregator selects a specific model among local models that provides the least fairness violation on D_a. Formally, the global model parameters ϕ_t at round t are: $\mu_{\text{FAIRBEST}}(\Theta_t) \triangleq \phi_t = \theta_{i^*,t}$ s.t. $i^* = \arg\min_i \{\Delta_{loss}(h_i(\theta_{i,t}))\}$.

2. α-FAIRAVG: This heuristic generalizes FAIRBEST by selecting the top $\alpha\%$ of local models followed by their weighted average. Formally, consider the set F_t, at a round t, which comprises the top-$\alpha\%$ of clients in *increasing* order of the ratio $\Delta_{loss}(h_i(\theta_{i,t}))$. Now,

$$\mu_{\alpha\text{-FAIRAVG}}(\Theta_t) \triangleq \phi_t = \sum_{i \in F_t} \frac{n_i}{\sum_{j \in F_t} n_j} \theta_{i,t} \qquad (2)$$

3. α-FAIRACCAVG: Aggregator selects the top-$\alpha\%$ of local model parameters that give the best ratio of accuracy with fairness violation on D_a and take their weighted average. Consider the set F_t, at a round t, comprising the top-$\alpha\%$ of clients in *decreasing* order of the ratio $\frac{Acc(h_i(\theta_{i,t}))}{\Delta_{loss}(h_i(\theta_{i,t}))}$.

$$\mu_{\alpha\text{-FAIRACCAVG}}(\Theta_t) \triangleq \phi_t = \sum_{i \in F_t} \frac{n_i}{\sum_{j \in F_t} n_j} \theta_{i,t} \qquad (3)$$

As α increases, more local models get aggregated akin to `FedAvg` with heterogeneous data. That is, with an increase in α, Eq. 2 and Eq. 3 tend to $\sum_{i \in C} \frac{n_i}{\sum_j n_j} \theta_{i,t}$ (the standard `FedAvg` aggregation), such that α-FAIRAVG and α-FAIRACCAVG tend to mimic `FedAvg`. We also show this behavior empirically in the full version [8].

Algorithm 2. FAIRGRAD: Gradient-based F3

Input: (1) Each client $C_k \in \mathcal{C}$ with D_k (2) Hyperparameters: maximum number of communication rounds T, number of local epochs E, learning rate η, accuracy threshold a, epoch threshold τ and $\beta \in [0,1]$.

Output: Model ϕ

1: $\phi_0 \leftarrow$ randomly initialized weights
2: **for** each round $t = 0, 1 \ldots, T-1$ **do**
3:     **for** each client $C_k \in S_t$ (in parallel) **do**
4:         $\theta_{k,t} \leftarrow \phi_t$
5:         **for** each local epoch $i = 1, \ldots, E$ **do**
6:             (Aggregator) $G_{k,t} \leftarrow \nabla_{\theta_{k,t}} \Delta_{EO}\left(h_{\theta_{k,t}}(\cdot), D_a\right)$
7:             For a fixed $G_{k,t}$, the client updates for all batches,
8:             (Client C_k) $\begin{cases} g_{k,t} & \leftarrow \nabla_{\theta_{k,t}} L_k\left(h_{\theta_{k,t}}(\cdot), B\right); \; g^*_{k,t} \leftarrow \beta \cdot g_{k,t} + (1-\beta) \cdot G_{k,t} \\ \theta_{k,t} & \leftarrow \theta_{k,t} - \eta g^*_{k,t} \end{cases}$
9:         **end for**
10:     **end for**
11:     (Aggregator) $\phi_{t+1} \leftarrow \mu_{\texttt{FedAvg}}(\Theta_t), \Theta_t = \{\theta_{k,t} \mid C_k \in S_t\}$
12:     $\phi_{best} \leftarrow \texttt{StoppingCondition}(t+1, \phi_{t+1}, \phi_{best}, a, \tau)$
13: **end for**
14: **return** ϕ_{best}

3.2   FairGrad: A Gradient-Based F3

In Heuristic-based F3, it is only possible to explore a limited set of models that provide different trade-offs between accuracy and fairness. While each client trains to maximize accuracy, the client models may diverge from the "fair" aggregated model. Hence aggregation of these individual client models may not always provide a good trade-off.

Based on these observations, and motivated from [2], we now propose a Gradient-based approach curated for the FL with DH setting, namely FAIR-GRAD. Informally, in FAIRGRAD we train the individual client models for accuracy and w.r.t. fair gradients obtained from the aggregator. As a result, the local client models comprise fairness information and subsequent aggregation through FedAvg provides a balance between accuracy and fairness. Note that clients communicate with the aggregator even during their local training. Formally, FAIRGRAD comprises the following steps.

1. Local Training. At every epoch, we compute the following gradients.
   (i) Client Level: For each client C_i, we compute the gradients $g_i(\theta_i, D_i)$ w.r.t. client weights θ_i and local dataset D_i for maximizing accuracy. That is, minimizing the cross-entropy loss l_{CE}, $L_i(h_{\theta_i}, D_i) = \mathbb{E}_{(x,y) \sim D_i}[l_{CE}(h_{\theta_i}(x), y)]$.
   (ii) Aggregator Level: For each client C_i, the aggregator computes the gradients $G_i(\theta_i, D_a)$ w.r.t. client weights θ_i and aggregator dataset D_a for minimizing fairness, i.e. Δ_k, for $k \in \{EOpp, EO, AP\}$. Aggregator communicates G_i to each C_i.
   Upon receiving G_i, each client C_i must now judiciously aggregate the two gradients. To this end, we fix G_i for an entire epoch while updating g_i for

each batch-wise training. Further, C_i performs weighted aggregation with weight $\beta \in (0,1)$ to determine the aggregated gradients g_i^* as follows: $g_i^* = \beta g_i + (1 - \beta)G_i$. Next, C_i performs the SGD update: $\theta_i \leftarrow \theta_i - \eta g_i^*$.

2. Model Aggregation. After a few epochs of local training, a random subset of clients send their models to the aggregator, who then aggregates them using FedAvg to derive the global model ϕ.

3. Model Communication. The aggregator communicates the aggregated global model parameters to each client. The clients adopt these parameters to perform further local training.

Figure 1b depicts Heuristic-based F3 and Algorithm 2 provides a procedural outline. Given our novel heuristics and FairGrad, we next conduct experiments to compare their empirical performance and highlight the efficacy of F3 for FL under data heterogeneity.

4   Experiments

We conduct our experiments on the following face datasets: FairFace [9], FFHQ [10], and UTK [23]. In this section, we first define our baseline FedAvg-DH for an appropriate comparison. Then, we provide our FL setup, training details, and network model. Finally, we present our results and the key takeaways.

Baseline: To compare our methods' performance, we create the baseline FedAvg-DH. FedAvg-DH is simply FedAvg [13] for our FL setting with Data Heterogeneity.

FL Setup: We consider 50 clients, i.e., $\mathcal{C} = \{C_1, \ldots, C_{50}\}$. We randomly distribute the training data so that each client has samples of only a particular demographic group to ensure data heterogeneity. Each client locally trains its model on its private data. The global model aggregation is performed periodically till convergence. At each aggregation round t, we let $S_t = \mathcal{C}$. The training details specific to each dataset follow next.

Training Details: We focus on popular face datasets: FairFace [9], FFHQ [10], and UTK [23]. For each of these, we consider 'age' as the sensitive attribute and 'gender' as the predicting label. Further, we divide the samples into two age groups, ≤ 30 and > 30 years. We distribute the data among the clients such that 50% of the clients have access to data samples belonging to the age group ≤ 30 and others have access to samples belonging to age group > 30. Each client's local data comprises $\approx 1K$ training samples for all three datasets.

For FairFace [9] and FFHQ [10], we use a batch size of 256 and train the models for $T = 50$ communication rounds with clients training their local models for $E = 4$ epochs (per round). We use learning rates of $\eta = 0.05$ and $\eta = 0.01$ for FairFace and FFHQ, respectively. We set the accuracy tolerance at $a = 1\%$ and threshold round at $\tau = 20$. For UTK [23], we train using the learning rate $\eta = 0.01$ and batch size 64. We also set $T = 80$, $E = 2$, $a = 1\%$ and $\tau = 20$.

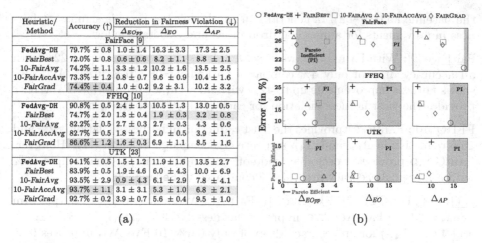

Heuristic/ Method	Accuracy (↑)	Reduction in Fairness Violation (↓)		
		Δ_{EOpp}	Δ_{EO}	Δ_{AP}
FairFace [9]				
FedAvg-DH	79.7% ± 0.8	1.0 ± 1.4	16.3 ± 3.3	17.3 ± 2.5
FairBest	72.0% ± 0.8	0.6 ± 0.6	8.2 ± 1.1	8.8 ± 1.1
10-FairAvg	74.2% ± 1.1	3.3 ± 1.2	10.2 ± 1.6	13.5 ± 2.5
10-FairAccAvg	73.3% ± 1.2	0.8 ± 0.7	9.6 ± 0.9	10.4 ± 1.6
FairGrad	74.4% ± 0.4	1.0 ± 0.2	9.2 ± 3.1	10.2 ± 3.2
FFHQ [10]				
FedAvg-DH	90.8% ± 0.5	2.4 ± 1.3	10.5 ± 1.3	13.0 ± 0.5
FairBest	74.7% ± 2.0	1.8 ± 0.4	1.9 ± 0.3	3.2 ± 0.8
10-FairAvg	82.2% ± 0.5	2.7 ± 0.3	2.7 ± 0.3	4.3 ± 0.6
10-FairAccAvg	82.7% ± 0.5	1.8 ± 1.0	2.0 ± 0.5	3.9 ± 1.1
FairGrad	86.6% ± 1.2	1.6 ± 0.3	6.9 ± 1.1	8.5 ± 1.6
UTK [23]				
FedAvg-DH	94.1% ± 0.5	1.5 ± 1.2	11.9 ± 1.6	13.5 ± 2.7
FairBest	83.9% ± 0.5	1.9 ± 4.6	6.0 ± 4.3	10.0 ± 6.9
10-FairAvg	93.5% ± 2.9	0.9 ± 4.3	6.1 ± 2.9	7.8 ± 4.1
10-FairAccAvg	93.7% ± 1.1	3.1 ± 3.1	5.3 ± 1.0	6.8 ± 2.1
FairGrad	92.7% ± 0.2	3.9 ± 0.7	5.6 ± 0.4	9.5 ± 1.0

(a)   (b)

Fig. 2. Accuracy and Fairness Violation (Δ_k, $\forall k \in \{EOpp, EO, AP\}$) values for the baseline `FedAvg-DH` [13] and our approaches. In Fig. 2a, the numbers highlighted in green represent the least value of Δ_k. The numbers highlighted in magenta provide the highest accuracy out of our proposed approaches. In Fig. 2b, the optimum point is bottom left, i.e., low %-Error and low Δ_k.

Model: We adopt PyTorch's implementation of the standard ResNet-18 architecture for the base model [12]. We run our experiments on 8 NVIDIA GeForce GTX 1080 with 10 GB RAM.

Method: We run every experiment 5 times and report the approaches' average and standard deviation. For each instance, we randomly generate an aggregator set D_a with samples between 10%–20% of the overall dataset size. For α-FAIRAVG and α-FAIRACCAVG, we choose $\alpha = 10$, i.e., 10% of the total local models.

We restate that fairness guarantees often come at the cost of accuracy [3]. However, our methodologies aim to strike an effective balance between fairness improvements and accuracy.

4.1 Results

First, we observe a maximum Coefficient of Variation (CoV) of 0.96 across all experiments. For 70% of our experiments, we observe a CoV < 0.2 indicating the stability of our approaches and the results presented. We next discuss accuracy and fairness violations of our approaches for the three datasets.

Fairness Improvements: Figure 2a provides the accuracy and fairness violation (Δ_k for $k \in \{EOpp, EO, AP\}$) of our approaches compared to the baseline `FedAvg-DH`. Recall that lower Δ_k implies lesser fairness violation. In Fig. 2a, the numbers highlighted in green represent the least value of Δ_k (obtained often by one of our approaches) for each dataset. The numbers highlighted in magenta represent the highest accuracy out of our proposed approaches (i.e., excluding the baseline `FedAvg-DH`). In general, our methodologies provide significant fairness improvements for a marginal drop in accuracy. Details follow.

FairFace [9]: With FAIRGRAD, we observe fairness improvement up to 22% with an accuracy drop of only 4% compared to `FedAvg-DH`. 10-FAIRACCAVG also shows fairness improvements up to 42% with an accuracy drop of 7%. FAIRBEST provides the least fairness violation for each of the three fairness notions.

FFHQ [10]: FAIRBEST provides the least reduction in fairness violation: 82% and 75% reduction in Δ_{EO} and Δ_{AP} respectively, compared to `FedAvg-DH`. However, FAIRGRAD provides a desirable trade-off with fairness improvement upto 20% with a marginal accuracy drop of 4% compared to `FedAvg-DH`.

UTK [23]: For UTK, 10-FAIRAVG, 10-FAIRACCAVG and FAIRGRAD outperform `FedAvg-DH`. 10-FAIRACCAVG improves fairness by 40% (Δ_{EOpp}), 48% (Δ_{EO}) and 42% (Δ_{AP}) for an accuracy drop of only 0.6%. 10-FAIRAVG improves fairness by 55% in Δ_{EO} and 49% in Δ_{AP} for an accuracy drop of 0.4%. Similarly, FAIRGRAD provides improvement in fairness by 53% in Δ_{EO} and 30% in Δ_{AP} with 1.7% accuracy drop.

Visualizing Accuracy and Fairness Trade-Off: Figure 2b depicts the accuracy and fairness trade-offs of our heuristics with the baseline `FedAvg-DH`. Note that the optimum point is bottom left, i.e., low %-Error and low fairness violation (Δ_k). The red circle marker for `FedAvg-DH` appears at the bottom right on most of the plots, showing low error (higher accuracy) at the cost of fairness. Markers in the bottom-left corner of the plot assure the least fairness violation while maintaining high accuracy. FAIRBEST, 10-FAIRAVG and 10-FAIRACCAVG provide lower fairness violations, for a marginal decrease in accuracy. FAIRGRAD also shows a better accuracy and fairness trade-off by obtaining accuracy values approximately equivalent to `FedAvg-DH` and fairness violations closer to our heuristics. The highlighted "Pareto Inefficient" region is the area which is *pareto dominated* by the baseline `FedAvg-DH`. Observe that, in general, our approaches lie outside the Pareto inefficient region.

Additional Experiments: In the full version [8], we provide additional results, including (i) an ablation study for the hyperparameters α, a and τ, and (ii) experiment with different network architectures

4.2 Discussion

We see that both Heuristic and Gradient-based F3 perform significantly better in fairness while maintaining competitive accuracy compared to `FedAvg-DH`.

Pareto-Optimality: To quantify the improved accuracy and fairness trade-off using our approaches, we use Mahalanobis distance (MD). More concretely, we derive the distance between our approaches' performance, i.e., % Error and Δ_k, $k = \{EOpp, EO, AP\}$, from the origin in Fig. 2b. We compute the distance as $MD(\overline{x}) = \sqrt{(\overline{x} - \overline{\mu})^T S^{-1} (\overline{x} - \overline{\mu})}$. Here, $\overline{x} = (Error, \Delta_k)$ is the vector with observed % Error and Fairness Violation, $\bar{\mu}$ is a vector with mean values, and S the covariance matrix of $\bar{x}$.

An approach's lesser MD implies a better trade-off. We provide the specific distance values of our approaches and `FedAvg-DH` in the full version [8]. Our results highlight that for FairFace, FAIRGRAD provides the best accuracy and trade-off, i.e., least MD). For FFHQ and UTK, FAIRGRAD and 10-FAIRAVG provide the least distance for Δ_{EOpp}, respectively. For Δ_{EO} and Δ_{AP}, 10-FAIRACCAVG outperforms others for both datasets. These distances quantitatively show that FAIRGRAD and 10-FAIRACCAVG significantly improve the accuracy and fairness trade-off over `FedAvg-DH`.

Heuristic-Based F3 vs. Gradient-Based F3: Overall, FAIRBEST provides the least fairness violation, and the baseline `FedAvg-DH` provides the highest accuracy. Ranking the approaches using MD shows that 10-FAIRACCAVG and FAIRGRAD provide an improved accuracy and fairness trade-off than `FedAvg-DH`.

Of the two, FAIRGRAD requires comparatively higher communication overhead between the clients and the aggregator. On the other hand, finding an optimal α for 10-FAIRACCAVG that obtains a desirable trade-off may also be challenging. As α increases, the increase in accuracy also increases the fairness violation (refer [8]). As a result, a practitioner can appropriately decide between the approaches to achieve the desired accuracy and fairness trade-off.

5 Conclusion

In this paper, we focus on Fair Attribute Classification (FAC) in FL setting with data heterogeneity. We observe that existing approaches to ensure fairness in FL do not work in a heterogeneous setting due to the unavailability of demographic-specific data samples across clients. To address this, we propose F3, a novel FL framework to achieve fairness in FAC. With F3, we introduce (i) Heuristic-based F3, which includes three aggregation heuristics that ensure fairness while simultaneously maximizing the model's accuracy (ii) Gradient-based F3 to ensure clients are trained for fairness and accuracy. Experimentally, our approaches outperform the default counterpart in FL on challenging benchmark face datasets. The results suggest that F3 helps strike a practical balance between fairness and accuracy for FAC.

References

1. Agarwal, A., Beygelzimer, A., Dudik, M., Langford, J., Wallach, H.: A reductions approach to fair classification. In: ICML, pp. 60–69 (2018)
2. Augenstein, S., Hard, A., Partridge, K., Mathews, R.: Jointly learning from decentralized (federated) and centralized data to mitigate distribution shift. In: NeurIPS 2021 Workshop on Distribution Shifts: Connecting Methods and Applications (2021)
3. Chouldechova, A.: Fair prediction with disparate impact: a study of bias in recidivism prediction instruments. Big Data 5(2), 153–163 (2017)

4. Ezzeldin, Y.H., Yan, S., He, C., Ferrara, E., Avestimehr, S.: FairFed: enabling group fairness in federated learning. In: NeurIPS Workshop on New Frontiers in Federated Learning (NFFL) (2021)
5. Hardt, M., Price, E., Srebro, N.: Equality of opportunity in supervised learning. In: NeurIPS, vol. 29, pp. 3315–3323 (2016)
6. Hu, S., Wu, Z.S., Smith, V.: Provably fair federated learning via bounded group loss. In: ICLR Workshop on Socially Responsible Machine Learning (2022)
7. Jung, S., Chun, S., Moon, T.: Learning fair classifiers with partially annotated group labels. In: CVPR, pp. 10348–10357 (2022)
8. Kanaparthy, S., Padala, M., Damle, S., Sarvadevabhatla, R.K., Gujar, S.: F3: fair and federated face attribute classification with heterogeneous data. arXiv preprint arXiv:2109.02351 (2021)
9. Karkkainen, K., Joo, J.: Fairface: face attribute dataset for balanced race, gender, and age for bias measurement and mitigation. In: WACV, pp. 1548–1558 (2021)
10. Karras, T., Laine, S., Aila, T.: A style-based generator architecture for generative adversarial networks. In: CVPR, pp. 4401–4410 (2019)
11. Konečný, J., McMahan, H.B., Ramage, D., Richtárik, P.: Federated optimization: distributed machine learning for on-device intelligence. arXiv preprint arXiv:1610.02527 (2016)
12. Lokhande, V.S., Akash, A.K., Ravi, S.N., Singh, V.: FairALM: augmented Lagrangian method for training fair models with little regret. In: Vedaldi, A., Bischof, H., Brox, T., Frahm, J.-M. (eds.) ECCV 2020. LNCS, vol. 12357, pp. 365–381. Springer, Cham (2020). https://doi.org/10.1007/978-3-030-58610-2_22
13. McMahan, B., Moore, E., Ramage, D., Hampson, S., Arcas, B.A.: Communication-efficient learning of deep networks from decentralized data. In: AISTATS (2017)
14. Padala, M., Damle, S., Gujar, S.: Federated learning meets fairness and differential privacy. In: ICONIP, pp. 692–699 (2021)
15. Padala, M., Gujar, S.: FNNC: achieving fairness through neural networks. In: IJCAI, pp. 2277–2283 (2020)
16. Ruan, Y., Joe-Wong, C.: Fedsoft: soft clustered federated learning with proximal local updating. In: Proceedings of the AAAI Conference on Artificial Intelligence, vol. 36, pp. 8124–8131 (2022)
17. Terhörst, P., et al.: A comprehensive study on face recognition biases beyond demographics. IEEE Trans. Technol. Soc. 3(1), 16–30 (2021)
18. Torralba, A., Efros, A.A.: Unbiased look at dataset bias. In: CVPR, pp. 1521–1528 (2011)
19. Zafar, M.B., Valera, I., Rogriguez, M.G., Gummadi, K.P.: Fairness constraints: mechanisms for fair classification. In: AISTATS, pp. 962–970 (2017)
20. Zhang, B.H., Lemoine, B., Mitchell, M.: Mitigating unwanted biases with adversarial learning. In: AIES, pp. 335–340 (2018)
21. Zhang, D.Y., Kou, Z., Wang, D.: FairFL: a fair federated learning approach to reducing demographic bias in privacy-sensitive classification models. In: IEEE Big Data, pp. 1051–1060 (2020)
22. Zhang, J., Wu, Y., Pan, R.: Incentive mechanism for horizontal federated learning based on reputation and reverse auction. In: WWW, pp. 947–956 (2021)
23. Zhang, Z., Song, Y., Qi, H.: Age progression/regression by conditional adversarial autoencoder. In: CVPR, pp. 5810–5818 (2017)
24. Zhao, H., Gordon, G.: Inherent tradeoffs in learning fair representations. In: NeurIPS, vol. 32, pp. 15675–15685 (2019)
25. Zheng, X., Guo, Y., Huang, H., Li, Y., He, R.: A survey of deep facial attribute analysis. Int. J. Comput. Vision 128(8), 2002–2034 (2020)

Estimating the Risk of Individual Discrimination of Classifiers

Jonathan Vasquez[1,2]([✉]), Xavier Gitiaux[1], and Huzefa Rangwala[1]

[1] George Mason University, Fairfax, VA 22030, USA
{jvasqu6,xgitiaux,rangwala}@gmu.edu
[2] Universidad de Valparaiso, Valparaiso, Chile
jonathan.vasquez@uv.cl

Abstract. Data owners are increasingly liable for the potential harm caused by using their data on underprivileged communities. Stakeholders seek to identify data characteristics that lead to biased algorithms against specific demographic groups, such as race, gender, age, or religion. We focus on identifying feature subsets of datasets where the ground truth response function from features to observed outcomes differs across demographic groups. To achieve this, we propose *FORESEE*, a decision tree-based algorithm that generates a score indicating the likelihood of an individual's response varying with sensitive attributes. Our approach enables us to identify individuals most likely to be misclassified by various classifiers, including Random Forest, Logistic Regression, Support Vector Machine, Multi-Layer Perceptron, and k-Nearest Neighbors. The advantage of our approach is that it allows stakeholders to identify risky samples that may contribute to discrimination and use *FORESEE* to estimate the risk of upcoming samples.

Keywords: Algorithmic Fairness · Discrimination Risks Estimates · Bias Identification

1 Introduction

Algorithmic decision-making systems are a growing concern due to their potential to replicate or worsen existing social biases. Empirical evidence suggests that algorithmic outcomes may depend on irrelevant demographic characteristics. Examples can be found in criminal justice [24], banking and finance [12,26], education [20], and facial recognition [7]. Toolkits such as [5,6] have been developed to measure and mitigate unfairness outcomes. However, only some are suitable for removing sensitive attribute information [15] or supplementing data delivery with context-aware components to warn someone of the unfairness issues [2,22].

Unfairness issues are commonly identified in later development stages [23]. It has been argued that the root causes of unfair outcomes include social or historical biases encoded in the data [10] and heteroskedastic noise between demographic groups [8]. Therefore, assessing unfairness in the early stages can

© The Author(s), under exclusive license to Springer Nature Switzerland AG 2023
H. Kashima et al. (Eds.): PAKDD 2023, LNAI 13935, pp. 495–506, 2023.
https://doi.org/10.1007/978-3-031-33374-3_39

lead to effective and efficient mitigation of unfairness outcomes. This paper contributes to this research stream by formulating a score to evaluate the risk of unfair outcomes.

Systematic differences in the data-generating process between demographic groups predict whether future uses of the data in a machine learning (ML) pipeline will lead to outcomes that violate standard fairness measures. Our proposed risk score aims to capture those differences. We demonstrate how to obtain reasonable estimates of the proposed risk by either boosting or bagging decision trees based on Bayesian Additive Regression Trees (BART) [9] and our variant of BART (named FORESEE, a FORESt of decision trEEs algorithm), where we ensemble decision trees trained on different combinations of data features. We find that FORESEE generates fewer biased estimates than BART. Furthermore, our experiments on three benchmark datasets show that the risk correlates with standard measurements of demographic parity, equalized opportunity, and equalized odds. Specifically, we show that sub-populations with high risk are more likely to experience high unfairness by classifiers from different families trained on a sample of the data. Consequently, we argue that proper risk estimates provide useful warnings to data owners and stakeholders before a classifier ingests the data.

In a context where organizations that own data are increasingly liable for future unfair uses, this risk score could be instrumental in deciding whether data can be distributed to ML pipelines. It also allows for identifying the characteristics of subsamples that are most likely to be exposed or contribute to unfair outcomes, guiding future data collection and model development. Our contributions are as follows:

- We formalize unfairness risk assessment for potentially discriminatory future uses as the problem of measuring differences in data generation processes across demographic groups.
- We demonstrate how model averaging over diverse decision trees provides reliable estimates for the risk.
- We provide experimental evidence of how estimating the risk is a useful warning sign for the demographic disparity, inequality of odds, and opportunities of classifiers trained on a sample of the data.

2   Related Work

Few efforts have been made to estimate or warn of unfairness risks in datasets. For example, Model Cards [22], and Method Cards [1] propose adding sections related dataset in the supplementary reports to communicate key information about potential unfairness issues in the models. Similarly, the Datasheets method [13] proposes a set of questions to find these issues in the dataset. On the other hand, Lee and Singh [21] introduce a bias identification methodology and questionnaire to identify bias-related risks in the ML development pipeline, and Vetro et al. [25] propose quality-based frameworks to measure unfairness risks at the dataset level. These qualitative and quantitative methods are focused on a macro level. We complement these approaches by proposing quantitative assessment at an individual level. Finally, we highlight that our work is motivated by the definitions of Kamiran and Calders [18]. We both focus on the differences in positive class ratios and compute the definitions from a labeled dataset; however, we differentiate in that our score includes the individual fairness notion and is defined at the sample level. Also, our goal is to determine how likely a data point will lead to discrimination while they aim to correct bias in the data.

3 Problem Setting

We frame the task of finding $g : \mathcal{X} \mapsto \mathcal{Y}$ from a set of individual features and the target variable $\mathcal{X} \times \mathcal{Y}$ to support decision-making. We assume g has sufficient performance for decision-makers to use in their process. We also assume that each sampled individual has sensitive attributes $s \in \mathcal{S}$, representing protected information forbidden in classification. Furthermore, the sensitive attribute is defined by the union of I groups, i.e., $S = \cup_i S_i, i \in [I]$. The objective is to determine the likelihood that any individual (x, s, y) will lead to discrimination by a function $g : \mathcal{X} \mapsto \mathcal{Y}$ learned from a given dataset. The problem setting implies two challenges: (1) it has to be defined what *discrimination* means, and (2) the likelihood of discrimination is determined when g is still unknown. Although we address these two under a binary classification task in the following section, we argue that it is extensible to other task types.

4 Discrimination Notion and Risk Scores

Discrimination refers to the disparity between groups of individuals. Although sensitive attributes can be removed, it is not enough since the correlation with other features may still result in disparities in the outcome [11]. The Demographic Parity (*demP*) metric can measure this disparate impact by computing disparity across groups over the outcome space of g [11]. Similarly, individual fairness certifies that g outputs similar outcomes for similar individuals [17]. Using these two notions, we define a risk measure by evaluating disparity on the labeled dataset for similar individuals of a given sample. Specifically, we define the *unfairness risk score* as the conditional probability of Y being 1 (from *demP*) given $X = x$ with differences in the sensitive attribute S (from individual fairness). Formally:

Definition 1. For any individual $(x, s) \in \mathcal{X} \times \mathcal{S}$, we denote risk score $r(x)$ as

$$r(x) = |P(Y = 1 \, S = s, X = x) - P(Y = 1 \, S \neq s, X = x)|. \qquad (1)$$

Definition 1 assumes an infinite population (or at least a large enough one); however, since this is not true for most cases, we propose two estimation procedures under finite population in Sect. 5. Additionally, $r(x)$ does not depend on any classifier that will use the data, allowing it to be used in the early steps of pipelines. We also argue that Definition 1 captures the granular and aggregate fairness properties based on the *independence* and *separate* notions [20] of classifiers using the data: (i) differences in individual misclassification rates across demographic groups; (ii) differences in aggregate misclassification rates across demographic groups. Finally, we propose that $r(x)$ is directly related to the differences in misclassification rate conditional on $X = x$ for an unknown function g. We denote the latter difference as $\delta_{mis}(x)$ and define it as Eq. 2.

$$\delta_{mis}(x) = |P(g(X) \neq Y|X = x, S = s) - P(g(X) \neq Y|X = x, S \neq s)|. \quad (2)$$

We present Theorem 1,[1] which reflects the variation in misclassification rates of an unaware classifier across demographic groups. Therefore, for a given $X = x$, a high-risk score indicates that individuals in one group are more likely to be misclassified than those in the other group (i.e., disparate treatment), regardless of the classifier used.

Theorem 1. *For any unaware classifier g ($x \in \mathcal{X}$ is given but not $s \in \mathcal{S}$), we have:*

$$\delta_{mis}(x) = r(x) \left| I_{\{g(x)=0\}} - I_{\{g(x)=1\}} \right|$$

where $I_{\{g(x)=1\}}, I_{\{g(x)=0\}}$ are the characteristic functions of the sets $\{x \in \mathcal{X}|g(x) = 1\}$ and $\{x \in \mathcal{X}|g(x) = 0\}$. Moreover, for a deterministic unaware classifier, $\delta_{mis}(x) = r(x)$.

Since g is a deterministic function, then $g(x)$ must either equal 0 or 1, which implies $\delta_{mis}(x) = r(x)$. Furthermore, the risk score also encompasses the aggregate fairness properties of any classifier g. We can examine the aggregate difference in misclassification rates [8] using Eq. 3, where large values of Δ_{mis} indicate that the error rate is unequal across groups

$$\Delta_{mis}(g) = |P(g(X) \neq Y|S = s) - P(g(X) \neq Y|S \neq s)|. \quad (3)$$

Having this, we present Theorem 2, which suggests that if the distribution of features X is independent of the sensitive attribute, then the differences between the distribution of X given $S = s$ and $S \neq s$ is zero. Low aggregate risk scores also correspond to low $\Delta_{mis}(g)$. Therefore, when the distribution shift between $X|S = s$ and $X|S \neq s$ is minimal, the risk measure can help users identify data that are unlikely to result in unfair classification, as measured by differences in misclassification rates (i.e., using *separation* concepts [20]).

Theorem 2. *For any unaware classifier g,*

$$\Delta_{mis}(g) \leq \frac{1}{2} E_{x \sim P_s}[r(x)] + \frac{1}{2} E_{x \sim P_{ns}}[r(x)] + TV(P_s, P_{ns}), \quad (4)$$

where $TV(P_s, P_{ns})$ is the total variation between the distribution P_s of X conditional on $S = s$ and the distribution P_{ns} of X conditional on $S \neq s$.

---

[1] The proofs of Theorems 1, 2, and 3 can be found in the supplementary material in this github repository: jovasque156/foresee.

Sub-population level definitions of algorithmic fairness are stronger than their aggregate counterparts since they protect subgroups defined by a complex intersection of many sensitive attributes ([19]) or a structured slicing of the feature space [14]. We define sub-population differences in misclassification rates of a classifier g as follows:

$$\Delta_{sub-mis}(g,\gamma) = \max_{G:P(G)\geq\gamma} \Delta_{mis}(g|x \in G), \tag{5}$$

where $\Delta_{mis}(g|x \in G)$ is the difference in misclassification rates between demographic groups in sub-population G.

Theorem 3. *Suppose that there exists a sub-population* $G \subset \mathcal{X}$ *such that* $P(G) > \gamma$, $P_s = P_{ns}$, *and* $P(Y = 1|X = x, S = s) > P(Y = 1|X = x, S \neq s)$. *For any stochastic classifier* $g : \mathcal{X} \to [0,1]$ *such that* $inf_{x\in G}g(x) > 1/2$, *there exists* $\kappa > 0$ *such that*

$$\Delta_{sub-mis}(g,\gamma) > \kappa E[r(x)|x \in G]. \tag{6}$$

Theorem 3 suggests that if the response function $E[Y = 1|X = x, S = s]$ is greater than $E[Y = 1|X = x, S \neq s]$ for a sub-population G, a classifier that predicts $Y = 1$ for this sub-population will result in higher misclassification rates for one demographic group. The differences in misclassification rates will be bounded from below by the average risk score. Therefore, by examining sub-populations with high predicted risk scores, the user can identify sub-populations where differences in sensitive attributes S are likely to correlate with differences in misclassification rates.

5   Risk Estimation

Two approaches are tested to estimate $r(x)$: (1) A Bayesian-based method to estimate the probability $P(Y = 1|X = x, S = s)$ followed by the computation of $r(x)$ and (2) an ensemble-based approach from diverse decision trees.

5.1   Method 1: BART

This approach obtains estimates of the response functions $E[Y|X = x, S = s]$ and $E[Y|X = x, S \neq s]$, followed by computing the risk score. We use the non-parametric additive tree model BART [9], as suggested by [16], which has been successful in estimating individual treatment effects. The risk score is computed using the trees $T_1, T_2, ..., T_M$ as below. BART uses a prior to regularize the depth of each tree T_m.

$$r_{BART}(x) = \frac{1}{M}\left|\sum_{m=1}^{M} E(Y|T_m, X = x, S = s) - E(Y|T_m, X = x, S \neq s)\right|$$

BART requires estimating both response functions $E[Y|X = x, S = s]$ and $E[Y|X = x, S \neq s]$, whereas we are only interested in the difference $E[Y|X =$

$x, S = s] - E[Y|X = x, S \neq s]$. BART averages decision trees to obtain a robust Bayesian estimate of each response function. The following section presents the second method that directly averages the risk score estimates from each decision tree.

5.2   Method 2: FORESEE

Given a collection of trees $T_1, T_2, ..., T_M$, we denote $L_1(x)$, $L_2(x)$, ..., $L_M(x)$ as the leaves to which an individual belongs. For each decision tree T_m, we compute its estimate of the risk score as the difference in the estimation error rate Y_{mis} between demographic groups in the leaf $L_m(x)$, i.e., $r(x, T_m) = |E(Y_{mis}|L_m(x), S = s) - E(Y_{mis}|L_m(x), S \neq s)|$. The risk estimate is then:

$$r_{FORESEE}(x) = \frac{1}{M} \sum_{m=1}^{M} r(x, T_m).$$

Decision trees partition the feature space into regions where outcomes are assumed to be constant. Therefore, leaf-level risk scores characterize how violations of this assumption vary across demographic groups. To obtain a robust risk estimate, we average a diverse set of trees obtained as follows: (i) each tree is trained on a different random sub-sample of the instances and a random subset of the features, hence we ensemble trees that see different aspects of the data-generating process and de-correlate estimation errors; and (ii) trees with performance lower than β are filtered out, so we retain only the trees that properly and capture partial characteristics of the response function $E[Y|X = x]$. We remark that S is used for training since leaves with no members of one group would reflect a dependency between Y and S for the given $X = x$ in the leaf, hence a high risk of discrimination. However, in such a case, we face the challenge of estimating the misclassification rate for the unobserved group in the leaf. Under an optimistic scenario, the expected misclassification rate should be equal across groups, while the opposite should expect the highest error rate (i.e., equal to 1) for the unobserved group. Since we aim to flag probable discrimination, we argue that the latter is preferable. Although a better approach can be used, this assumption seems sufficient according to the results. Finally, the Algorithm 1 provides a pseudo-code for estimating the risk scores $r_{FORESEE}$. More details, like description about SUBSAMPLING, LEAVES, and PERFORMANCE function, are provided in the supplementary material.

6   Experimental Evaluation

We design our experiments[2] to answer the following four research questions: **RQ1**: Do risk estimates from *FORESEE* and *BART* average to the ground truth risk in numerical simulations? **RQ2**: Do high-risk and low-risk groups

---

[2] The code of our experiments can be found in this repository: jovasque156/foresee.

Algorithm 1. FORESEE Algorithm

Input: D_{train}, D_{est} to estimate risk
Parameters: M number of decision trees, π portion of instances for sampling, ϕ number of features for sampling, β performance threshold
Output: $r_{FORESEE}$, r estimates

1: $r_{FORESEE} \leftarrow \emptyset$
2: $r \leftarrow [\,]$
3: **for** $m \in [M]$ **do**
4:    $subD_{train} \leftarrow$ SUBSAMPLING(D_{train}, π, ϕ)
5:    $T_m \leftarrow$ train a Decision Tree on $subD_{train}$
6:    $L_m \leftarrow$ LEAVES(T_m)
7:    $r_{FORESEE} \leftarrow r_{FORESEE} \cup (T_m, L_m)$
8: **end for**
9: **for** $x \in D_{est}$ **do**
10:    $total\_mis \leftarrow 0$
11:    **for** $(T_m, L_m) \in r_{FORESEE}$ **do**
12:        $p \leftarrow$ PERFORMANCE(T_m)
13:        **if** $p \geq \beta$ **then**
14:            $total\_mis \leftarrow total\_mis+$ MV(L_m, x)
15:        **end if**
16:    **end for**
17:    $risk = ave(total\_mis)$
18:    append $risk$ to r
19: **end for**
20: **return** $r_{FORESEE}$, r

---

have differing levels of fairness? **RQ3**: Does our measure of risk allow the user to describe the sub-populations that are the most likely to be under-served by different classifiers families using the data as inputs?

Synthetic Data. First, we rely on numerical simulations to test whether FORESEE and BART are unbiased estimators of the true risk score. We uniformly draw features from $[0,1]^2$ and assign a sensitive attribute s with $P(S = 1) = 0.5$. For $S = 1$, we assign a label $Y = 1$. For $S = 0$, we assign a label $Y = 1$ with probability equal to $1 - \frac{x_1+x_2}{2}$. Therefore, for a given $X = x$, the ground truth risk is equal to $r(x) = \frac{x_1+x_2}{2}$. We draw 5000 samples from this synthetic data, run FORESEE and BART to estimate $r(x)$, and compare their values to the ground truth. We repeat the protocol 20 times with different seeds to compute the mean and standard deviation of the risk estimates.

Real World Datasets. We use two benchmark datasets and a private dataset from an educational context. **Adults** (also known as the Census Income Dataset) consists of $48,844$ individuals described by 14 features such as marital status, education, and working hours. The binary sensitive attribute is gender, and the outcome is whether the income is larger than $50K$ [3]. **Compas** contains information about 7, 214 individuals in the criminal justice system to assess their recidivism risk using features such as the number of felonies and misdemeanors. The sensitive attribute is race, and the outcome is whether an individual re-commits a crime within two years [24]. Finally, **Dropout** gathers academic records of 4, 706 undergraduate students. Features include grades and academic support program participation. The sensitive attribute is gender; the outcome reflects dropping out within the first two years. We split each dataset into training and testing sets using a 70/30 ratio.

Experiments Protocol. To explore how the risk score is informative of future unfair uses of the data independently of the type of classifier, we train four models from different families – Logistic Regression (LR), Random Forest (RF), k-Nearest Neighbors (KNN), and Support Vector Machine (SVM), and Multi-Layer Perceptron (MLP). We compute accuracy to evaluate the models' performance in Compas and F-1 score in Adults and Dropout. We also compute three standard fairness metrics: demographic disparity δ_{demP}, inequality of opportunity δ_{opp}, and inequality of odds δ_{odd} [23]. We also classify individuals into High and Low-Risk groups based on whether their risk score is larger or smaller than λ, taking value depending on the context.

7   Results and Discussions

RQ1: Figure 1 shows estimates' mean (line) and standard deviation (area) against the ground truth risk (dash-dot-black line) in the synthetic dataset. We observe that FORESEE's estimates are less biased than BART. BART estimates have a higher upward bias for low values of the true risk score $r(x)$ and a bigger downward bias for large values of $r(x)$. This observation confirms our intuition that it is preferable to directly average risk estimates from local approximations.

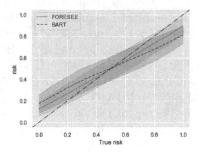

Fig. 1. FORESEE and BART risk estimates against ground truth risks.

Table 1. Performances of machine learning models on each dataset. Higher F-1 and accuracy (acc) are better, whereas lower fairness metric scores are better. Figures are in absolute values.

Model	Adult				Dropout				Compas			
	F-1	δ_{opp}	δ_{odd}	δ_{demP}	F-1	δ_{opp}	δ_{odd}	δ_{demP}	acc	δ_{opp}	δ_{odd}	δ_{demP}
LR	.61	.40	.32	.35	**.65**	.06	.03	**.03**	.65	.22	.18	.21
RF	**.70**	.20	.20	.33	.61	.06	.04	.04	**.66**	.17	.17	.20
KNN	.64	.20	.14	.19	.55	.08	.05	.04	.65	.17	.16	.19
SVM	.66	.21	.24	.38	.62	.05	.03	.04	.65	.18	.17	.19
MLP	.65	**.12**	**.09**	**.17**	.56	**.02**	**.02**	.03	.64	**.14**	**.14**	**.16**

RQ2: Table 1 shows the classification performance of the four models across the three real-world datasets. The best classifier for each dataset generates significant unfairness outcomes, which is particularly interesting whether our risk score anticipates these fairness issues. Figure 2 plots the distribution of FORESEE's estimates of risks, where we highlight that for Compas the entire population shows risks larger than 0.4. These high-risk scores confirm findings in existing studies where it is claimed that the Compas dataset is not well-suited for training automated decision systems without significantly harming a demographic group [4]. Additionally, we arbitrarily set λ equal to 0.5 for Adults and Compas (i.e., potential discrimination is higher than non-discrimination) and 0.06 for Dropout to group individuals into High and Low. The results are depicted in Table 2. From the table, we observe that the fairness metrics are higher for High group across all models. This is evidence that the risk score allows identifying individuals for whom a given classifier's aggregate performances would significantly vary across demographic groups. We highlight that the differences are not significant for Compas, flagging once again the issues of using this dataset for automated training.

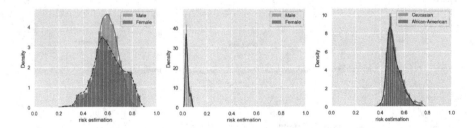

Fig. 2. Distribution of risk score along the three datasets.

Table 2. Fairness metrics in absolute value for High and Low risk groups, defined by setting the thresholds to 0.5 for Adult and Compas, and 0.06 for Dropout. Values closer to 0 are preferable. Figures are truncated to two decimal places.

Model		Adult			Dropout			Compas		
		δ_{opp}	δ_{odd}	δ_{demP}	δ_{opp}	δ_{odd}	δ_{demP}	δ_{opp}	δ_{odd}	δ_{demP}
LR	High	.49	.37	.39	.50	.25	.14	.29	.19	.21
	Low	.01	.03	.06	.09	.05	.06	.16	.19	.20
RF	High	.24	.24	.37	.33	.20	.04	.30	.20	.22
	Low	.02	.02	.04	.17	.09	.06	.07	.16	.17
KNN	High	.23	.16	.22	.22	.13	.08	.32	.20	.22
	Low	.05	.03	.01	.04	.02	.04	.05	.14	.15
SVM	High	.24	.28	.42	.14	.12	.03	.28	.19	.21
	Low	.00	.02	.04	.09	.05	.06	.09	.15	.17
MLP	High	.17	.13	.20	.23	.15	.02	.28	.18	.20
	Low	.11	.06	.01	.03	.02	.04	.02	.10	.11

RQ3: Figure 3 shows the differences between the datasets' High and Low-Risk profiles. The high and low-risk groups comprise the top and lowest 20% of samples with high and low-risk scores drawn equally from both groups in S. From Fig. 3, we can characterize high risks individuals as (i) *Adult*: aged individuals, a high number of working hours per week, and married or divorced, (ii) *Dropout*: students with low academic performances and higher numbers of academic warnings, and (iii) *Compas*: younger defendants, fewer criminal history, mostly felony as charge degree, and spending fewer days in jail. Our risk score provides stakeholders and data owners with information on the characteristics of individuals for which predictions would be affected by their sensitive attributes. A useful implication of the profiles in Fig. 3 is to use with caution any data mining algorithm that predicts the target variable if the individual's risk estimate is high.

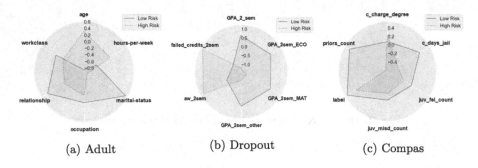

(a) Adult   (b) Dropout   (c) Compas

Fig. 3. Average feature for high and low fairness risk groups.

8   Conclusions

Fair machine learning literature offers metrics to measure the harmful effects. However, stakeholders would like to anticipate these fairness pitfalls before investing in model development or designing a data-generating process. We propose a score to measure whether future classifiers using the data will vary with sensitive attributes. Moreover, we propose *FORESEE* to estimate it from a finite sample. Our experiments demonstrate that the proposed score anticipates correctly whether a model using the data will under-serve some demographic groups. Furthermore, our approach offers useful guidance: (1) to decide whether a dataset is appropriate for training automated decision systems and designing proper data-generating procedures, (2) to anticipate potentially discriminatory outcomes and mitigate them early, and (3) to decide when to trust the outcome of the algorithm. We believe FORESEE is still computationally expensive (training and estimation time complexity are $O(dkn\log(n))$ and $O(k\log(n))$, with k as the number of decision trees and d as the dataset dimensionality); hence more estimators will be explored in the future. Furthermore, the assumption about rate variation for the absent group in a leaf is an interesting aspect to be analyzed in the forthcoming studies as well.

Acknowledgements. This study was supported by the National Agency for Research and Development (ANID - Agencia Nacional de Investigación y Desarrollo/Subdirección de Capital Humano), "Becas Chile" Doctoral Fellowship 2020 program; Grant No. 72210492 to Jonathan Patricio Vasquez Verdugo.

References

1. Adkins, D., et al.: Method cards for prescriptive machine-learning transparency. In: 2022 IEEE/ACM 1st International Conference on AI Engineering-Software Engineering for AI (CAIN), pp. 90–100. IEEE (2022)
2. Adkins, D., et al.: Prescriptive and descriptive approaches to machine-learning transparency. In: CHI Conference on Human Factors in Computing Systems Extended Abstracts, pp. 1–9 (2022)
3. Asuncion, A., Newman, D.: UCI machine learning repository (2007)
4. Bao, M., et al.: It's compaslicated: the messy relationship between rai datasets and algorithmic fairness benchmarks. arXiv preprint arXiv:2106.05498 (2021)
5. Bellamy, R.K.E., et al.: AI Fairness 360: an extensible toolkit for detecting, understanding, and mitigating unwanted algorithmic bias (2018)
6. Bird, S., et al.: Fairlearn: a toolkit for assessing and improving fairness in AI. Technical Report. MSR-TR-2020-32, Microsoft (2020)
7. Buolamwini, J., Gebru, T.: Gender shades: intersectional accuracy disparities in commercial gender classification. In: Conference on Fairness, Accountability and Transparency, pp. 77–91. PMLR (2018)
8. Chen, I., Johansson, F.D., Sontag, D.: Why is my classifier discriminatory? Adv. Neural Inf. Process. Syst. **31**, 1–12 (2018)
9. Chipman, H.A., George, E.I., McCulloch, R.E.: Bart: bayesian additive regression trees. Ann. Appl. Stat. **4**(1), 266–298 (2010)

10. Ensign, D., Friedler, S.A., Neville, S., Scheidegger, C., Venkatasubramanian, S.: Runaway feedback loops in predictive policing. In: Conference on Fairness, Accountability and Transparency, pp. 160–171. PMLR (2018)
11. Feldman, M., Friedler, S.A., Moeller, J., Scheidegger, C., Venkatasubramanian, S.: Certifying and removing disparate impact. In: Proceedings of the 21th ACM SIGKDD International Conference on Knowledge Discovery and Data Mining, pp. 259–268 (2015)
12. Fu, R., Huang, Y., Singh, P.V.: Crowds, lending, machine, and bias. Inf. Syst. Res. **32**(1), 72–92 (2021)
13. Gebru, T., et al.: Datasheets for datasets. Commun. ACM **64**(12), 86–92 (2021)
14. Gitiaux, X., Rangwala, H.: Multi-differential fairness auditor for black box classifiers. arXiv preprint arXiv:1903.07609 (2019)
15. Gitiaux, X., Rangwala, H.: Fair representations by compression. In: Proceedings of the AAAI Conference on Artificial Intelligence, vol. 35, pp. 11506–11515 (2021)
16. Hill, J.L.: Bayesian nonparametric modeling for causal inference. J. Comput. Graph. Stat. **20**(1), 217–240 (2011)
17. Ilvento, C.: Metric learning for individual fairness. arXiv preprint arXiv:1906.00250 (2019)
18. Kamiran, F., Calders, T.: Data preprocessing techniques for classification without discrimination. Knowl. Inf. Syst. **33**(1), 1–33 (2012)
19. Kearns, M., Neel, S., Roth, A., Wu, Z.S.: Preventing fairness gerrymandering: auditing and learning for subgroup fairness. In: International Conference on Machine Learning, pp. 2564–2572. PMLR (2018)
20. Kizilcec, R.F., Lee, H.: Algorithmic fairness in education. arXiv preprint arXiv:2007.05443 (2020)
21. Lee, M.S.A., Singh, J.: Risk identification questionnaire for detecting unintended bias in the machine learning development lifecycle. In: Proceedings of the 2021 AAAI/ACM Conference on AI, Ethics, and Society, pp. 704–714 (2021)
22. Mitchell, M., et al.: Model cards for model reporting. In: Proceedings of the Conference on Fairness, Accountability, and Transparency, pp. 220–229 (2019)
23. Pessach, D., Shmueli, E.: Algorithmic fairness. arXiv preprint arXiv:2001.09784 (2020)
24. ProPublica: How we analyzed the compas recidivism algorithm. ProPublica (2016)
25. Vetrò, A., Torchiano, M., Mecati, M.: A data quality approach to the identification of discrimination risk in automated decision making systems. Gov. Inf. Q. **38**(4), 101619 (2021)
26. Wen, M., Bastani, O., Topcu, U.: Fairness with dynamics. arXiv preprint arXiv:1901.08568 (2019)

Multi-fair Capacitated Students-Topics Grouping Problem

Tai Le Quy[1](✉)[iD], Gunnar Friege[2][iD], and Eirini Ntoutsi[3][iD]

[1] L3S Research Center, Leibniz University Hannover, Hannover, Germany
tai@l3s.de
[2] Institute for Didactics of Mathematics and Physics, Leibniz University Hannover,
Hannover, Germany
friege@idmp.uni-hannover.de
[3] Research Institute CODE, University of the Bundeswehr Munich,
Munich, Germany
eirini.ntoutsi@unibw.de

Abstract. Group work is a prevalent activity in educational settings, where students are often divided into topic-specific groups based on their preferences. The grouping should reflect students' aspirations as much as possible. Usually, the resulting groups should also be balanced in terms of protected attributes like gender, as studies suggest that students may learn better in mixed-gender groups. Moreover, to allow a fair workload across the groups, the cardinalities of the different groups should be balanced. In this paper, we introduce a *multi-fair capacitated* (MFC) grouping problem that fairly partitions students into non-overlapping groups while ensuring balanced group cardinalities (with a lower and an upper bound), and maximizing the diversity of members regarding the protected attribute. To obtain the MFC grouping, we propose three approaches: a greedy heuristic approach, a knapsack-based approach using vanilla maximal knapsack formulation, and an MFC knapsack approach based on group fairness knapsack formulation. Experimental results on a real dataset and a semi-synthetic dataset show that our proposed methods can satisfy students' preferences and deliver balanced and diverse groups regarding cardinality and the protected attribute, respectively.

Keywords: Fairness · Grouping · Knapsack · Educational data · Nash social welfare

1 Introduction

Teamwork plays a vital role in educational activities, as students can work together to achieve shared learning goals while learning about leadership, higher-order thinking, and conflict management [6]. A common approach to group students into teams is as follows: the instructor provides a list of topics, projects, tasks, etc. (shortly: *topics*), according to which the different non-overlapping groups of students should be formed. The grouping procedure can be performed randomly or based on students' preferences [14] typically expressed as a ranking over the provided topics. Or, the instructor just says: "Find yourself into

© The Author(s) 2023
H. Kashima et al. (Eds.): PAKDD 2023, LNAI 13935, pp. 507–519, 2023.
https://doi.org/10.1007/978-3-031-33374-3_40

groups"; in this case, a grouping is not random and does not consider students' preferences w.r.t. topics but it is triggered by social connections. The common case in educational settings is the grouping w.r.t. students' preferences.

The grouping process should consider various requirements. First, students' preferences should be taken into account (i.e., *student satisfaction*). A grouping is considered satisfactory if it can satisfy the students' preferences as much as possible. Second, the groups should be balanced in terms of their cardinalities, so all students share a similar workload (i.e., *group cardinality*) because when groups have unequal sizes, and the minority group is smaller than a critical size, the minority cohesion widens inequality [17]. Third, the instructor might be interested in fair-represented groups w.r.t. some protected attributes like gender or race [8] (i.e., *group fairness*), as studies suggest that mixed-gender grouping may have a positive effect on groups' performance [4].

These requirements have been discussed in the related work but are typically treated independently. For example, fairness w.r.t. workload distribution and students' preferences has been discussed in group assignments [6], assignment of group members to tasks [14] or students to projects [19]. Student satisfaction is typically assessed as the number of topics staffed [11] or the sum of the utilities of the topics assigned to students based on the ranking of preferences chosen by students [12]. The group cardinality can be satisfied by the heuristic method [15], or the hierarchical clustering approach [9]. However, providing a grouping solution that simultaneously satisfies all three requirements is hard [19].

To this end, we introduce *multi-fair capacitated (MFC) grouping* problem that aims to ensure fairness of the resulting groups in multiple aspects. In particular, we target fairness in terms of i) maximizing students' satisfaction, ii) ensuring fairness in group representation w.r.t. the protected attribute, and iii) balancing group cardinalities. For the satisfaction aspect, we employ the Nash social welfare notation [16]; for the fairness w.r.t protected attribute we use the balance score notion [3]. To solve the MFC problem, we propose three approaches: i) a greedy heuristic algorithm; ii) a knapsack-based approach that reformulates the assignment step as a maximal knapsack problem; iii) an MFC knapsack model based on the group fairness knapsack formulation [18].

2   Related Work

Agrawal et al. [1] proposed the problem of grouping students in a large class w.r.t. the overall gain of students. Miles et al. [14] investigated the problem of assignment of group members to tasks w.r.t. the workload distribution. Concerning a diversity of features such as skills, genders, and academic backgrounds, Krass et al. [8] investigated the problem of assigning students to multiple non-overlapping groups. However, students' preferences were not considered. To consider both efficiency and fairness, Magnanti et al. [12] solved a CPLEX integer programming formulation with two objectives: maximizing the total utility computed by the rank of student's preferences (efficiency) and minimizing the number of students assigned to the projects which they do not prefer (fairness). Besides, Rezaeinia et al. [19] introduced a lexicographic approach to prioritize the goals. The efficiency objective is computed based on the utility, similar to [12]. A related problem is the problem of assigning reviewers to papers [7]. Each reviewer can be assigned to several papers, and each paper can be assigned to several reviewers [7]. However, in the students grouping problem, we attempt to generate non-overlapping groups [8], where each student can be assigned to only one group [19].

The knapsack problem formulation has been used for finding good clustering assignments [9] without students' preference and the minimum capacity of a group (cluster) is not considered. Recently, Stahl et al. [20] introduced a fair knapsack model to balance the price given by the data provider and the suggested price by the customer. Fluschnik et al. [5] proposed three concepts of fair knapsack (individually best, diverse and fair knapsack) to solve the problem of choosing a subset of items where the total cost is not greater than a given *budget* while taking into account the preferences of the voters. Fairness of the knapsack is measured by the Nash social welfare (or Nash equilibrium) [16]. The group fairness definition for the knapsack problem was investigated recently by Patel et al. [18]. In their study, each item is characterized by a *category*, their goal is to select a subset of items such that the total value of the selected items is maximized, and the total weight does not surpass a given weight while each category is *fairly* represented.

3   Problem Definition

Let $X = \{x_1, x_2, \cdots, x_n\}$ be a set of n students, $T = \{t_1, t_2, \cdots, t_m\}$ be a set of m topics. For an integer n we use [n] to denote the set $\{1, 2, \cdots, n\}$. Each student can choose h topics as their preference ($h \ll m$). The students' preferences are stored in matrix *wishes*. Row $wishes_i$ contains the list of h topics preferred by student i. We use a matrix V to record the student's level of interest in the topics. The preference of topic t_j chosen by student x_i is represented by a number v_{ij}. The more preferred topic will have a higher value of v_{ij}. Matrix V is computed as: $V_{i,wishes_{io}} = h/o$ with $o \in [h]$, where o indicates the order of preferences. Likewise, each topic t_j can be chosen by several students. A priority matrix W consists of values computed based on the registration time, where w_{ij} represents the priority of student x_i on topic t_j. Students who register earlier will have a higher value of w_{ij}. If the topic t_j is not preferred by student x_i then $v_{ij} = 0$ and $w_{ij} = 0$.

Let $\psi : V \times W \to \mathbb{R}$ be the aggregate function of matrices V and W. For each student x_i, we define a *welfare* value w.r.t. topic t_j: $welfare_{ij} = \psi(v_{ij}, w_{ij})$. In detail, $\psi(v_{ij}, w_{ij}) = \alpha v_{ij} + \beta w_{ij}$, where α and β are the parameters indicating the weight of each component. Figure 1 illustrates a dataset with 5 students and 4 topics. The matrix $welfare$ is computed with $\alpha = 1$ and $\beta = 1$ (preferences and registration time are equally considered).

The goal of a grouping problem is to distribute n students into k disjoint groups $\mathcal{G} = \{G_1, G_2, \cdots, G_k\}$, $(k \leq m)$, that maximizes the students' preferences w.r.t. the registration time, formulated by the objective function:

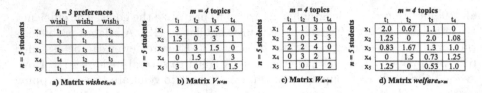

Fig. 1. A dataset with matrices *wishes*, V, W and *welfare*.

$$\mathcal{L}(X, \mathcal{G}) = \prod_{r=1}^{k} \left(1 + \sum_{i=1}^{n} welfare_{ij_r} \times y_{ij_r} \right) \tag{1}$$

In other words, the goal is to maximize the product of the total *welfare* obtained from each group G_r. In Eq. 1, a set of indexes $J = \{j_1, j_2, \cdots, j_k\}$ of k selected topics is defined as $J = \{j | x_i \in G_r, welfare_{ij} > 0\}, \forall r \in [k]$. Variable y_{ij_r} is the flag of x_i; $y_{ij_r} = 1$ if x_i is assigned to the group of topic t_{j_r}, otherwise $y_{ij_r} = 0$. Equation 1 is the representation of the Nash social welfare [16] function[1]. Therefore, we can call a grouping satisfactory if it maximizes the product in the objective function $\mathcal{L}(X, \mathcal{G})$. Furthermore, we add one to the sum $\sum_{i=1}^{n} welfare_{ij_r} \times y_{ij_r}$ to avoid the phenomenon that the sum of *welfare* in a certain group might be zero.

Fairness of Grouping w.r.t. a Protected Attribute: Assume that each student is characterized by a binary protected attribute $P = \{0, 1\}$, where 0 is the protected group (e.g., *gender* = *female*) and 1 is the non-protected group (e.g., *gender* = *male*). $\varphi : X \to P$ is the demographic category to which the student belongs. Fairness of a grouping $\mathcal{G}$ w.r.t. protected attribute [3] is computed as:

$$balance(\mathcal{G}) = \min_{\forall G_r \in \mathcal{G}} balance(G_r) \tag{2}$$

where fairness of a group G_r is the minimum ratio between two categories:

$$balance(G_r)_{\forall G_r \in \mathcal{G}} = \min \left(\frac{|\{x \in G_r \mid \varphi(x) = 0\}|}{|\{x \in G_r \mid \varphi(x) = 1\}|}, \frac{|\{x \in G_r \mid \varphi(x) = 1\}|}{|\{x \in G_r \mid \varphi(x) = 0\}|} \right) \tag{3}$$

Capacitated Grouping: Inspired by the capacitated clustering problem [15], we call a grouping *capacitated* if the cardinality of each group G_r, i.e., $|G_r|$, is between a given lower bound $C^l \geq 0$ and an upper bound $C^u \geq C^l$.

Definition 1. MFC grouping problem. We describe the MFC problem as finding a grouping $\mathcal{G} = \{G_1, G_2, \cdots, G_k\}$ that distributes a set of students X into k groups corresponding to k topics, and satisfies the following requirements:
1) The assignment is fair, i.e., maximizing students' satisfaction (Eq. 1);
2) The balance of each group G_r is maximized, i.e., the fairness constraint w.r.t. the protected attribute (Eq. 2);
3) The cardinality of each group $G_r \in \mathcal{G}$ is bounded within $[C^l, C^u]$.

---

[1] The Nash social welfare was defined as $\prod_{v_i \in V}(1 + \sum_{a \in S} u_i(a))$ [5] (the typical formula is $\prod_{v_i \in V} \sum_{a \in S} u_i(a)$, where v_i is a voter in a set of voters V, a is an item of the knapsack S, and $u_i(a)$ represents the extent to which v_i enjoys a. The knapsack S is fair if that product is maximized.

4   Methodology for the MFC Grouping Problem

To solve the MFC grouping problem, we first propose a greedy heuristic algorithm (Sect. 4.1); then we formulate the assignment phase as a vanilla maximal knapsack (Sect. 4.2) or a group fairness knapsack problem (Sect. 4.3).

4.1   A Greedy Heuristic Approach

We apply a 2-phase greedy strategy (Algorithm 1). Step 1: we maximize the students' preferences by assigning them to their most preferred topic. If a topic is preferred by many students we select the student who has the highest *welfare* value (lines 4, 5). Step 2: we adjust the assignment to satisfy the requirements by *GroupAdjustment* function (Algorithm 2). The number of students w.r.t. protected attribute $(p_0^l, p_0^u, p_1^l, p_1^u)$ are computed based on the resulting groups' cardinalities (C^l, C^u) and the balance score θ (line 2). If there exists ungrouped students, we try to assign them to the existing groups (lines 3 - 6). If all groups are full, we choose the most prevalent topic preferred by the remaining ungrouped students and assign them to such a topic (lines 7 - 11). We disband groups containing too few students and assign those ungrouped students to other groups until all groups have the desired capacity (lines 13 - 18).

Complexity: Step 1 consumes $\mathcal{O}(n \times h)$ and step 2 costs $\mathcal{O}(C^l \times n \times m)$ as the algorithm has to deal with every group having cardinality less than C^l. As $C^l \ll n$ and $C^u \ll n$, the complexity of the greedy heuristic model is $\mathcal{O}(n \times m)$.

Algorithm 1: Greedy heuristic algorithm

Input: X: a set of students; n: #students; h: #preferences; m: #topics; C^l, C^u: capacities ; matrices $wishes_{n \times h}$, $V_{n \times m}$, $W_{n \times m}$; θ: balance score

Output: A grouping with k groups

1 $groups \leftarrow \emptyset; welfare \leftarrow \psi(V, W);$ //Step 1: Assign students to groups;
2 **for** $i \leftarrow 1$ **to** n **do**
3   **for** $j \leftarrow 1$ **to** h **do**
4   **if** *(topic $wishes_{ij}$ is the most preferred topic of student i) and (welfare$_{i,wish_{ij}}$ is the highest value among students choosing topic $wishes_{ij}$) and (len(groups[$wishes_{ij}$] < C^l))* **then**
5 | $groups[wishes_{ij}]$.append(i);
6   GroupAdjustment($groups$) //Step 2: Adjustment;
7 **return** $groups$;

4.2   A Knapsack-Based Approach

The assignment of the greedy heuristic approach can be detrimental to students' satisfaction because there may be some students who have no more topics to be assigned. Therefore, we propose an approach to select the most suitable students for each topic by a *maximal knapsack* problem [13]. Let *capacity* be a cardinality array with $capacity_i = 1, \forall i \in [n]$; $welfare_{ij} = \psi(v_{ij}, w_{ij})$ and the indexes of k

---

Algorithm 2: Group adjustment algorithm

Input: *groups*: a set of groups; n: #students; h: #preferences; m: #topics;
C^l, C^u: capacities; θ: balance score

Output: An adjusted grouping

1 **Function** GroupAdjustment(*groups*):

2      $p_0^l \leftarrow \left\lceil \frac{C^l}{\frac{1+\theta}{\theta}} \right\rceil$; $p_0^u \leftarrow \left\lceil \frac{C^u}{\frac{1+\theta}{\theta}} \right\rceil$; $p_1^l \leftarrow C^l - p_0^l$; $p_1^u \leftarrow C^u - p_0^u + 1$;

3   **for** $i \leftarrow 1$ **to** n **do**

4   **for** $q \leftarrow 1$ **to** m **do**

5   **if** *(i ∉ groups[q]) and len(groups[q] < C^l) and ((n_students_0 < p_0^l) or (n_students_1 < p_1^l))* **then**

6   *groups[q]*.append(*i*);

7   **while** *len(unassigned_students)* > 0 **do**

8   *id* ← the most prevalent topic preferred by remaining students;

9   **for** $i \in$ *unassigned_students* **do**

10   **if** *len(groups[id]) < C^u and ((n_students_0 < p_0^u) or (n_students_1 < p_1^u))* **then**

11   *groups[id]*.append(*i*);

12   *n_items* ← 1;

13   **while** *(cardinalities of all groups ∉ [C^l, C^u])* **do**

14   **if** *n_items < C^l* **then**

15   Resolve the groups with cardinality *n_items*;

16   **if** *(n_students_0 < p_0^u) or (n_students_1 < p_1^u)* **then**

17   Assign ungrouped students to the remaining groups having cardinality < C^u;

18   *n_items* + +;

19 **return** *groups*;

---

topics $J = \{j_1, j_2, \cdots, j_k\}$ will be chosen for the resulting groups. For each topic $t_{j_r} \in T$, $\forall r \in [k]$, i.e., r is the index of the selected knapsack, the goal is to select a subset of students (G_r), such that:

$$\text{maximize} \sum_{i=1}^{n} welfare_{ij_r} \times y_{ij_r} \text{s.t.} \begin{cases} \sum_{i=1}^{n} capacity_i \times y_{ij_r} \leq C^u \text{ or} \\ \sum_{i=1}^{n} capacity_i \times y_{ij_r} \leq C^l \end{cases} \quad (4)$$

where $y_{ij_r} = 1$ if x_i is assigned to the group of topic t_{j_r}, otherwise $y_{ij_r} = 0$.

In other words, for each selected topic, we find a set of students that maximizes the total *welfare*, while the total *capacity*, is within the given bounds. The pseudo-code is described in Algorithm 3 with two steps. Step 1: we find the most suitable candidates among the unassigned students by the solution of a maximal knapsack problem [13] for each topic. We use dynamic programming to solve the maximal knapsack problem (Eq. 4). Step 2 is presented in Algorithm 2 which performs a fine-tuning of the assignment.

Complexity: In step 1, the complexity is $\mathcal{O}(m \times n \times C^u)$ since it costs $\mathcal{O}(n \times C^u)$ for each topic to solve the knapsack problem. The running time of step 2 is $\mathcal{O}(C^l \times n \times m)$. Therefore, the complexity is $\mathcal{O}(n \times m)$.

Algorithm 3: Knapsack-based algorithm

Input: X: a set of students; n: #students; h: #preferences; m: #topics; C^l, C^u:
capacities; matrices $wishes_{n \times h}$; $V_{n \times m}$; $W_{n \times m}$.

Output: A grouping with k groups

1 $groups \leftarrow \emptyset$ //Step 1: Assign students to groups ;
2 $welfare \leftarrow \psi(V, W)$;
3 **for** $id \leftarrow 1$ **to** m **do**
4      $capacity \leftarrow$ get_capacity($unassigned\_students$);
5      $values \leftarrow$ get_welfare($unassigned\_students, welfare$);
6      $n\_items \leftarrow$ len($unassigned\_students$);
7   **if** $n\_items > 0$ **then**
8   **if** $n \bmod C^l = 0$ **then**
9              $selected\_students \leftarrow$ knapsack($values, capacity, n, C^l$);
10   **else**
11             $selected\_students \leftarrow$ knapsack($values, capacity, n, C^u$);
12         $groups[id] \leftarrow selected\_students$;
13   GroupAdjustment($groups$) //Step 2: Adjustment;
14 **return** $groups$;

4.3 An MFC Knapsack Approach

In the knapsack-based approach, the fairness constraint w.r.t. the protected attribute is not directly considered in the knapsack formulation. Inspired by the knapsack problem with *group fairness* constraints of Patel et al. [18], we propose an *MFC knapsack* algorithm to find the group of suitable students, which satisfies the MFC problem' requirements. The goal of the MFC knapsack is to select a subset of student (G_r), such that:

$$\text{maximize} \sum_{i=1}^{n} welfare_{ij_r} \times y_{ij_r} \text{ s.t.} \quad \begin{cases} \sum_{i=1}^{n} capacity_i \times y_{ij_r} \leq C^u \text{ or} \\ \sum_{i=1}^{n} capacity_i \times y_{ij_r} \leq C^l \\ balance(G_r) \text{ is maximized} \end{cases} \quad (5)$$

where $y_{ij_r} = 1$ if x_i is assigned to the group of topic t_{j_r}, otherwise $y_{ij_r} = 0$.

We use dynamic programming to solve the MFC knapsack problem (Algorithm 4). The input parameters include a set of unassigned students $S \subseteq X$. A dynamic programming table $\mathcal{A}(p, s, w)$ is used to record the total welfare of the first s students in the set S with capacity w on group p, $\forall p \in \{0, 1\}$, e.g., $\{male, female\}$ w.r.t. protected attribute (line 3, 4). Then, we construct table $\mathcal{B}(p, w)$ to find the total welfare with capacity w w.r.t. the protected attribute. The number of students in the protected group and the non-protected group is computed based on a given balance score θ (line 6). We apply a two-phase approach to solve the MFC grouping problem. Step 1, we assign students to groups based on the MFC knapsack's solution. We replace the *knapsack* function in Algorithm 3 with the new *MFC knapsack* function (Algorithm 4). Step 2, we use the group adjustment algorithm (Algorithm 2) to fine-tune the assignment.

Complexity: The MFC knapsack takes $\mathcal{O}(n \times C^u)$ for each topic. To solve the MFC problem, step 1 consumes $\mathcal{O}(m \times n \times C^u)$, and step 2 costs $\mathcal{O}(C^l \times n \times m)$. Therefore, the complexity of the MFC knapsack approach is $\mathcal{O}(n \times m)$.

Algorithm 4: MFC knapsack algorithm

Input: $\mathcal{S} = \{x_1, x_2, \ldots, x_z\}$: a set of unassigned students; C^l, C^u: capacities; $welfare_{n \times m}$: a welfare matrix; θ: balance score

Output: An optimal total welfare value

1   $avg = \dfrac{\sum_{i=1}^{n} welfare_{ij_r}}{(C^l + C^u)/2}$;

2   Let $\mathcal{A}(p, s, w), \forall p \in \{0,1\}$, be the total welfare of the first s students in the set $\mathcal{S}$ with capacity w on group p ;

3   Initialize $\mathcal{A}(p, 0, w) \leftarrow 0$; $\mathcal{A}(p, s, 0) \leftarrow 0$;

4   $\mathcal{A}(p, s, w) \leftarrow max\{\mathcal{A}(p, s-1, w), \mathcal{A}(p, s-1, w-1) + \sum_{i=1}^{s} welfare_{ij_r}\}$;

5   Let $\mathcal{B}(p, w)$ be the total welfare of group p with capacity w;

6   $p_0^l \leftarrow \left\lceil \dfrac{C^l}{\frac{1+\theta}{\theta}} \right\rceil$; $p_0^u \leftarrow \left\lceil \dfrac{C^u}{\frac{1+\theta}{\theta}} \right\rceil$; $S_0 \leftarrow \{x \in \mathcal{S} | \varphi(x) = 0\}$; $S_1 \leftarrow \{x \in \mathcal{S} | \varphi(x) = 1\}$;

7   $\mathcal{B}(0, w) \leftarrow max\{\mathcal{A}(0, |S_0|, w) | p_0^l \leq w \leq p_0^u\}$;

8   $\mathcal{B}(1, w) \leftarrow max\{\mathcal{B}(0, w') + \mathcal{A}(1, |S_1|, w - w') | C^l - p_0^l \leq w - w' \leq C^u - p_0^u, p_0^l \leq w' \leq p_0^u$, and $\frac{w'}{w-w'} \geq \theta\}$;

9   **return** $argmax\{\mathcal{B}(1, w) | min\{\mathcal{B}(1, w) - avg\}\}$;

---

5 Evaluation

5.1 Datasets

We evaluate our proposed methods on two variations of the student performance dataset [10] and a real data science dataset collected at our institute (Table 1).

Real Data Science Dataset. This dataset is collected in a seminar on data science at our institute. Students have to register 3 desired topics out of 16 topics. The advisor assigns students into groups based on their preferences and the registration time. The data contain demographic information of students (*ID, Name, Gender*) with their preferences (*wish1, wish2, wish3*), registration time (*Time*) and priority matrix W represented by 16 attributes (*T1, ..., T16*).

Student Performance Dataset[2]. The dataset consists of demographic, including the protected attribute *gender* which is used in the evaluation, school-related attributes and grades of students in Mathematics and Portuguese subjects of two Portuguese schools in 2005 - 2006. Because there is no given information about the topics and preferences of students in the original dataset, we create a *semi-synthetic* version by generating preferences and topics. For each student, we randomly generate h different preferred topics. Then, for each topic, we list the students who select the topic and randomly generate (different) priorities and store them in m attributes (matrix W). Hence, the *semi-synthetic* version has $(h + m)$ new attributes apart from the original attributes.

---

[2] https://archive.ics.uci.edu/ml/datasets/Student+Performance.

Table 1. An overview of the datasets.

Dataset	#instances	#attributes	Protected attribute	Balance score
Real data science	24	23	Gender (F: 8, M: 16)	0.5
Student - Mathematics	395	33	Gender (F: 208, M: 187)	0.899
Student - Portuguese	649	33	Gender (F: 383; M: 266)	0.695

5.2 Experimental Setup

Parameter Selection. We set the number of wishes $h = 3$ for the student performance dataset in order to be consistent with the real data science dataset. The number of topics, $m = 200$ and $m = 325$, are set for the student performance dataset - Mathematics and Portuguese subjects, respectively, to ensure that each group has at least 2 students. Besides, we set the parameters $\alpha = 1.0$ and $\beta = 1.0$, i.e., each component has the same weight. The balance scores θ are computed based on the datasets (Table 1). Furthermore, since the real data science dataset is very small, our methods are evaluated with the lower bound C^l in the range of $(2, \ldots, 8)$. Regarding the student performance dataset, we set $C^l = (2, \ldots, 18)$, as the average number of students per group should not exceed 20 [21]. The upper bound C^u is set as $C^u = C^l + 1$ for all datasets.

Baseline. The CPLEX integer programming model which considers both efficiency and fairness [12].

Evaluation Measures. We report the results on the following measures:

- *Nash Social Welfare.* The Nash social welfare is computed by Eq. 1. However, the number of groups (k) is determined during the group assignment process, i.e., k is different for the same set C^l, C^u, for each method. Hence, we normalize the Nash social welfare of the final grouping by $Nash = log_k \mathcal{L}(X, \mathcal{G})$.

- *Balance.* The fairness in terms of the protected attribute (Eq. 2).

- *Satisfaction Level.* It is computed by the ratio of the number of satisfied students, i.e., the students are assigned to their preferred topic, out of the total number of students: $Satisfaction = \dfrac{\mid \{i | wishes_{ip} = k, i \in groups_k, p \in [h]\} \mid}{n}$.

5.3 Experimental Results

Real Data Science Dataset. In Fig. 2, we present the performance of proposed methods on various evaluation measures. The MFC knapsack method is better in terms of the Nash social welfare and satisfaction level (Fig. 2-a, c). In terms of fairness w.r.t. protected attribute, the MFC knapsack method outperforms others when a group has at least 4 people (Fig. 2-b). CPLEX fails to assign students while maintaining only a constant number of groups (Fig. 2-d).

Student Performance - Mathematics Dataset. The knapsack-based approach outperforms others regarding Nash social welfare and satisfaction level in most experiments (Fig. 3-a, c). The satisfaction level tends to decrease because students have only a limited number of preferences (3 topics). When the group's

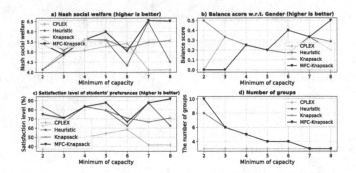

Fig. 2. Performance of methods on the real data science dataset.

cardinality increases, the desired topics become more diverse, and it is challenging to satisfy most students. In terms of fairness w.r.t. protected attribute (*gender*), the knapsack-based and MFC knapsack methods tend to achieve a higher balance score in comparison to the heuristic method (Fig. 3-b). When groups' cardinality is less than 4, the greedy heuristic and MFC knapsack methods tend to create more groups than the knapsack-based method (Fig. 3-d). The CPLEX method cannot return a solution when the groups' cardinality is less than 9 and it also fails since it is not possible to assign all students to groups.

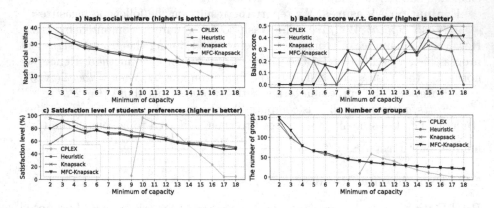

Fig. 3. Performance of methods on the student performance - Mathematics dataset.

Student Performance - Portuguese Dataset. The knapsack-based method once again demonstrates the ability to create groups with higher Nash social welfare and satisfaction level than others in many cases (Fig. 4-a and Fig. 4-c). Regarding fairness w.r.t. gender, a higher and more stable balance score is observed in the grouping generated by the MFC knapsack model (Fig. 4-b). The main reason for this phenomenon can be attributed to the model's emphasis on maximizing the balance constraint w.r.t. protected attribute. Besides, the

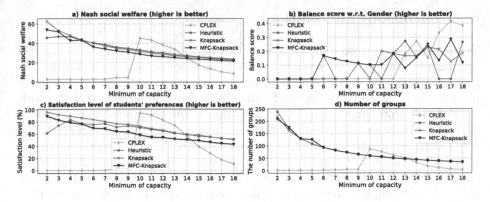

Fig. 4. Performance of methods on the student performance - Portuguese dataset.

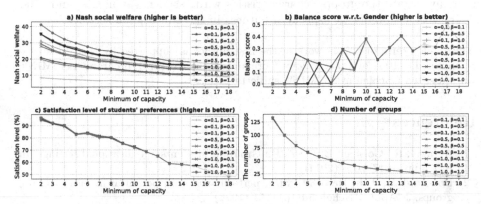

Fig. 5. Impact of α, β parameters on the knapsack-based model (student performance - Mathematics dataset).

MFC knapsack and greedy heuristic models divide students into more groups (Fig. 4-d) while the CPLEX also cannot assign all students to groups.

Impact of Parameters. The influence of α, β parameters is illustrated in Fig 5. The knapsack-based model shows the best performance with the combination of $\alpha = 1.0$ and $\beta = 1.0$.

Summary of Results. In general, the knapsack-based approach outperforms other models regarding Nash social welfare and satisfaction level. The MFC knapsack method shows its preeminence in terms of fairness w.r.t. gender in many cases, especially when the resulting groups have more members. However, in some cases, the knapsack-based approach tends to create fewer groups than the greedy heuristic method, i.e., the groups' cardinality is higher, which has both advantages and disadvantages. On the one hand, the larger groups can produce more ideas in brainstorming and discussions [2]. On the other hand, the group's performance may decline with the increase in the group's size [22].

6   Conclusions and Outlook

In this work, we introduced the MFC grouping problem that ensures fairness in multiple aspects: i) in terms of student satisfaction and ii) regarding the protected attribute and maintaining the groups' cardinality within the given bounds. We proposed three methods: the greedy heuristic approach that prioritizes the students' preferences in the assignment; the knapsack-based approach with the assignment step is formulated as a maximal knapsack problem; the MFC knapsack method considers fairness, cardinality, and students' preferences in the MFC knapsack formulation. The experiments show that our methods are effective regarding student satisfaction and fairness w.r.t. the protected attribute while maintaining cardinality within the given bounds. In the future, we plan to extend our approach to more than one protected attribute, as well as to further investigate the groups' characteristics w.r.t. students' abilities, and other definitions with different aspects of fairness in the educational settings.

Acknowledgements. The work of the first author is supported by the Ministry of Science and Culture of Lower Saxony, Germany, within the Ph.D. program "LernMINT: Data-assisted teaching in the MINT subjects".

References

1. Agrawal, R., Golshan, B., Terzi, E.: Grouping students in educational settings. In: KDD, pp. 1017–1026 (2014)
2. Bouchard Jr, T.J., Hare, M.: Size, performance, and potential in brainstorming groups. J. Appl. Psychol. **54**(1p1), 51 (1970)
3. Chierichetti, F., Kumar, R., Lattanzi, S., Vassilvitskii, S.: Fair clustering through fairlets. In: NeurIPS, pp. 5036–5044 (2017)
4. Fenwick, G.D., Neal, D.J.: Effect of gender composition on group performance. Gender Work Organization **8**(2), 205–225 (2001)
5. Fluschnik, T., Skowron, P., Triphaus, M., Wilker, K.: Fair knapsack. In: AAAI, vol. 33, pp. 1941–1948 (2019)
6. Ford, M., Morice, J.: How fair are group assignments? a survey of students and faculty and a modest proposal. J. Inf. Technol. Educ. Res. **2**(1), 367–378 (2003)
7. Hartvigsen, D., Wei, J.C., Czuchlewski, R.: The conference paper-reviewer assignment problem. Decis. Sci. **30**(3), 865–876 (1999)
8. Krass, D., Ovchinnikov, A.: The university of Toronto's rotman school of management uses management science to create MBA study groups. Interfaces **36**(2), 126–137 (2006)
9. Le Quy, T., Roy, A., Friege, G., Ntoutsi, E.: Fair-capacitated clustering. In: The 14th International Conference on Educational Data Mining, pp. 407–414 (2021)
10. Le Quy, T., Roy, A., Vasileios, I., Wenbin, Z., Ntoutsi, E.: A survey on datasets for fairness-aware machine learning. WIREs Data Min. Knowl. Discov. **12**(3) (2022)
11. Lopes, L., Aronson, M., Carstensen, G., Smith, C.: Optimization support for senior design project assignments. Interfaces **38**(6), 448–464 (2008)
12. Magnanti, T.L., Natarajan, K.: Allocating students to multidisciplinary capstone projects using discrete optimization. Interfaces **48**(3), 204–216 (2018)

13. Mathews, G.B.: On the partition of numbers. Proc. Lond. Math. Soc. **1**(1), 486–490 (1896)
14. Miles, J.A., Klein, H.J.: The fairness of assigning group members to tasks. Group Organization Manage. **23**(1), 71–96 (1998)
15. Mulvey, J.M., Beck, M.P.: Solving capacitated clustering problems. Eur. J. Oper. Res. **18**(3), 339–348 (1984)
16. Nash, J.F.: The bargaining problem. Econometrica **18**(2), 155–162 (1950)
17. Oliveira, M., Karimi, F., Zens, M., Schaible, J., Génois, M., Strohmaier, M.: Group mixing drives inequality in face-to-face gatherings. Commun. Phys. **5**(1) (2022)
18. Patel, D., Khan, A., Louis, A.: Group fairness for knapsack problems. In: AAMAS, pp. 1001–1009 (2021)
19. Rezaeinia, N., Góez, J.C., Guajardo, M.: Efficiency and fairness criteria in the assignment of students to projects. Annals of Operations Research, pp. 1–19 (2021)
20. Stahl, F., Vossen, G.: Fair knapsack pricing for data marketplaces. In: ADBIS, pp. 46–59. Springer (2016)
21. Urbina Nájera, A.B., De La Calleja, J., Medina, M.A.: Associating students and teachers for tutoring in higher education using clustering and data mining. Comput. Appl. Eng. Educ. **25**(5), 823–832 (2017)
22. Yetton, P., Bottger, P.: The relationships among group size, member ability, social decision schemes, and performance. Organ. Behav. Hum. Perform. **32**(2) (1983)

GroupMixNorm Layer for Learning Fair Models

Anubha Pandey$^{(\boxtimes)}$ ⓘ, Aditi Rai ⓘ, Maneet Singh, Deepak Bhatt ⓘ,
and Tanmoy Bhowmik

AI Garage, Mastercard, Delhi, India
{anubha.pandey,aditi.rai,maneet.singh,deepak.bhatt}@mastercard.com

Abstract. Recent research has identified discriminatory behavior of automated prediction algorithms towards groups identified on specific protected attributes (e.g., gender, ethnicity, age group, etc.). When deployed in real-world scenarios, such techniques may demonstrate biased predictions resulting in unfair outcomes. Recent literature has witnessed algorithms for mitigating such biased behavior mostly by adding convex surrogates of fairness metrics such as demographic parity or equalized odds in the loss function, which are often not easy to estimate. This research proposes a novel in-processing based *GroupMixNorm* layer for mitigating bias from deep learning models. The GroupMixNorm layer probabilistically mixes group-level feature statistics of samples across different groups based on the protected attribute. The proposed method improves upon several fairness metrics with minimal impact on overall accuracy. Analysis on benchmark tabular and image datasets demonstrates the efficacy of the proposed method in achieving state-of-the-art performance. Further, the experimental analysis also suggests the robustness of the GroupMixNorm layer against new protected attributes during inference and its utility in eliminating bias from a pre-trained network.

Keywords: Deep Learning · Ethics and fairness · Bias Mitigation

1 Introduction

Most AI algorithms process large quantities of data to identify patterns useful for accurate predictions. Such pipelines are mostly automated in nature without any human intervention, along with large data processing, high efficiency, and high accuracy. Despite the benefits of automated processing, current AI systems are marred with the challenge of biased predictions resulting in unfavourable outcomes. One of the most infamous examples of such behavior is that of an AI-based recruitment tool[1], which disfavoured applications from women because it was trained on resumes from the mostly male workforce. In order to rectify such biases and support advancement in society, we need models that generate fair results without any discrimination towards certain individuals or groups. To

---

[1] https://tinyurl.com/5apv7xeu.

H. Kashima et al. (Eds.): PAKDD 2023, LNAI 13935, pp. 520–531, 2023.
https://doi.org/10.1007/978-3-031-33374-3_41

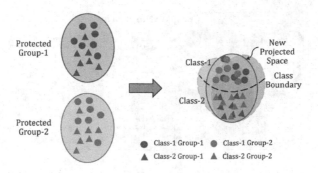

Fig. 1. The proposed GroupMixNorm layer projects the features of different classes and protected attributes onto a space which minimizes the distinction between the protected attributes, thus promoting a fairer classification model.

this effect, this research proposes a novel *GroupMixNorm* layer for learning an unbiased model for ensuring fair outcomes across different groups.

In the literature, research has focused on achieving fairness by introducing techniques at the pre-processing stage (transforming the input before feeding to the classification model) or the post-processing stage (transforming the output produced by the classification model). It is our hypothesis that these methods may not result in optimal accuracy, since they treat the classifier as a black box and focus on removing bias from the input representations or the output predictions only. Different from the above, *in-processing* techniques focus on learning bias-invariant models by incorporating additional constraints during training, thus resulting in more effective models [4]. Existing in-processing techniques mostly aim to solve a constraint optimization problem to ensure fairness [20–22] by introducing a penalty term in the loss function corresponding to the convex surrogates of the fairness objective like *demographic parity* or *equalized odds*. However, as observed in literature, it is challenging to formulate surrogates for different fairness constraints that is a reasonable estimate of the original [16].

In this research, we formulate the problem of bias mitigation as distribution alignment of several groups of the protected attribute (Fig. 1). The proposed *GroupMixNorm* layer is applied at the in-processing stage which promotes the model to learn unbiased features for classification. The formulation is motivated by the observation that Deep Learning based algorithms tend to explore the difference in the distribution among the groups of the protected attributes (e.g., male and female with similar features like age and education may have different salaries, thus resulting in different distributions) to lift the overall performance. The GroupMixNorm layer mixes the group-level feature statistics and transforms all the features in a training batch based on the interpolated group statistics. This enables the classifier to learn features invariant to the protected attribute. Further, transforming the data towards the interpolated groups regularizes the classifier and improves the generalizability at inference. Key highlights of this research are as follows:

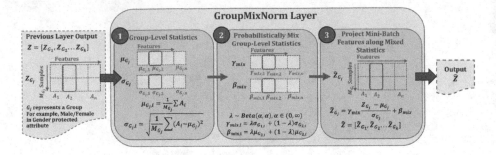

Fig. 2. The GroupMixNorm layer takes as input the previous layer's output (**Z**) along with each sample's protected attribute. Group-level statistics are computed, followed by the probabilistic mixing and projection of the mini-batch features along the mixed statistics to obtain new features (**Ẑ**).

- This research proposes a novel *GroupMixNorm* layer for learning fairer classification models. The proposed layer is applied at the architectural level and is an in-processing technique that focuses on distribution alignment of different groups during model training.
- GroupMixNorm operates at the feature level, thus making it flexible to be placed across various layers of a neural network-based model and fits well into the mini-batch gradient-based training. Experimental analysis suggests that with limited data, GroupMixNorm can be applied to mitigate the existing bias in classifiers as well, thus avoiding the need for re-training from scratch.
- The GroupMixNorm layer produces fairer results when evaluated for new groups at test time as well. We believe that the GroupMixNorm layer makes the model robust against distribution changes across sensitive groups, thus being able to generalize well for unseen groups at test time.
- The efficacy of the proposed approach has been demonstrated on different datasets (structured and unstructured), where it achieves improved performance while achieving multiple fairness constraints such as demographic parity, equal opportunity, and equalized odds simultaneously. For example, on the UCI Adult Income dataset [9], GroupMixNorm achieves an average precision of 0.77, while maintaining different fairness metrics below 0.03.

2   Related Work

Group fairness can be ensured in a machine learning system via *pre-processing*, *in-processing*, and *post-processing*. Pre-processing and post-processing methods consider the classifier as a black-box model, and try to mitigate bias from the input features or the classifier's prediction. On the other hand, *in-processing* based bias mitigation techniques solve the constraint optimization problem for different fairness objectives. To ensure independence between the predictions and sensitive attributes, Woodworth et al. [20] regularize the covariance between them. Zafar et al. [21] minimize the disparity between the sensitive groups by regularizing the decision boundary of the classifier. Game theory based approaches [1,7] provide analytical solutions and theoretical guarantees for generalizability in fair classifier but are limited by the scalability factor. Recent techniques [13,22] introduce an adversary network additional to the predictor network that predicts the sensitive label based on the classifier's output, while other algorithms [2,3,17,24] learn unbiased representations through invariant risk minimization and attention-based feature learning. Research has also focused on eliminating superficial correlations and paying more attention on task related causal features [12,14]. Recently, Cheng et al. [5] utilize contrastive learning to minimize the correlation between sentence representations and biasing words, while mixup [19,23] techniques have proved to be effective in bias mitigation. For example, Chuang et al. [6] utilize mixup as a data augmentation strategy to improve the generalizability of the model while optimizing the fairness constraints, and Du et al. [8] utilize mixup for feature neutralization to remove the correlation between the sensitive information and class labels from the encoder feature.

Instead of focusing on optimizing surrogates of the fairness metrics, this research proposes a novel GroupMixNorm layer which operates at the *architectural* level of the classifier. GroupMixNorm focuses on learning unbiased representations which results in satisfying several fairness constraints across groups.

3 Proposed GroupMixNorm Layer

As discussed before, recent research has observed that deep learning models often tend to learn group-specific characteristics, making it easier to obtain a higher performance on the underlying classification task. As an ancillary effect, the learned group-specific features often also result in discriminative behavior towards specific groups based on the protected attribute. For example, a recruitment tool may learn features based on the gender of the applicant, resulting in unintended discrimination towards applicants from the under-represented group. In order to address the above limitation, the GroupMixNorm layer focuses on eliminating the difference between the group statistics during training.

As part of the GroupMixNorm layer, we normalize each group of a protected attribute in a batch separately to collect group specific statistics (i.e. for the gender attribute, normalize all male samples and female samples in a batch separately) and further take a probabilistic convex combination between the group-level statistics and apply across all the samples in a batch. This process ensures that any protected group related diversity is removed from the internal representation of a neural network and doesn't allow the network to explore this information to lift the overall performance. The introduction of additional inductive bias in the network structure enforces it to learn invariant features pertaining to the protected attributes while training the network.

The GroupMixNorm layer is implemented as a plug-and-play module. It can be inserted between the fully connected layers of a neural network-based classifier

---

Algorithm 1. GroupMixNorm Layer

Input: Z: Learned representation of the input batch obtained from the previous layer
α, β: Hyper-parameters for the Beta distribution (default: 0.1)
Output: $\hat{Z}$: Transformed samples after the GroupMixNorm layer

1: **if** not in training mode **then**
2:     **return** Z
3: **end if**
4: Compute μ_{G_j} and σ_{G_j} for a group G_j in a protected attribute
5: Sample mixing coefficient $\lambda \sim Beta(\alpha, \alpha)$
6: Compute γ_{mix} and β_{mix} as shown in Eq. 2
7: Normalize and transform all samples in a batch to compute $\hat{Z}_{G_j}$ as shown in Eq. 3

8: $\hat{Z} = \{\hat{Z}_{G_j}\}_{j=1}^K$, where K is the number of groups identified in a protected attribute

9: **return** $\hat{Z}$

---

during training (Algorithm 1). Let X, Y, and S be the input features, class labels, and protected attribute labels in a training batch, respectively. As illustrated in Fig. 2, let Z be an n dimensional representation obtained from the previous layer and A_i represent the feature along dimension i. We identify the groups G_j in a batch based on the protected attribute labels S, and calculate their respective mean ($\mu_{G_j,i}$) and variance ($\sigma_{G_j,i}$) along each dimension (step-1 of Fig. 2). Next we calculate the weighted average of mean $\gamma_{mix,i}$ and variance $\beta_{mix,i}$ along each dimension (Eq. 1), followed by concatenation to create a single vector (Eq. 2). As we mix statistics of two groups at a time, the mixing coefficient λ is sampled from a symmetric Beta distribution $Beta(\alpha, \alpha)$, for $\alpha \in (0, \infty)$. The hyper-parameters α controls the strength of interpolation.

Finally, we normalize all the samples by applying the calculated γ_{mix} and β_{mix} to each sample as shown in Eq. 3. For the ease of notation, we have considered two groups i.e. binary protected attributes. However, the proposed solution can easily be applied to non-binary protected attributes as well.

$$\gamma_{mix,i} = \lambda\sigma_{G_1,i} + (1-\lambda)\sigma_{G_2,i}; \quad \beta_{mix,i} = \lambda\mu_{G_1,i} + (1-\lambda)\mu_{G_2,i}; \quad (1)$$

$$\gamma_{mix} = [\gamma_{mix,1}, ...\gamma_{mix,n}]; \quad \beta_{mix} = [\beta_{mix,1}, ...\beta_{mix,n}] \quad (2)$$

$$\hat{Z}_{G_j} = \gamma_{mix}\frac{(Z_{G_j} - \mu_{G_j})}{\sigma_{G_j}} + \beta_{mix} \quad (3)$$

The updated features $\hat{Z} = [\hat{Z}_{G_1}, \hat{Z}_{G_2}]$ are then provided as input to the following layer of the neural network for further processing. The process of mixing group level statistics in a GroupMixNorm layer occurs in the feature space and has no learnable parameters. The GroupMixNorm layer is easy to implement and fits perfectly into mini-batch training. Further, it is turned off during inference, thus eliminating the need for protected attributes during inference. The training procedure of GroupMixNorm layer is shown in Algorithm 1.

4 Datasets and Experimental Details

The GroupMixNorm layer has been evaluated on two datasets with different fairness evaluation metrics and compared with state-of-the-art techniques. Details regarding the dataset protocols are as follows:

- **UCI Adult Dataset** [9] contains 50,000 samples with 14 attributes to describe each data point (individual) (e.g., gender, education level, age, etc.) from the 1994 US Census. The classification task is to predict the income of an individual. It's a binary classification task, where class 1 represents salary $\geq$ 50K and class 0 represents salary $<$ 50K. We select gender as the protected attribute for the fairness evaluation. The dataset is imbalanced such that only 24% of the samples belong to class 1, with only 15.13% female samples.
- **CelebA Dataset** [15] contains 200,000 celebrity faces with 40 binary attributes associated with each image. Following the literature [6,8], we select gender as the protected attribute and wavy hair attribute for the binary classification task. The dataset has 18.36% male samples as compared to female samples in the positive class.

4.1 Fairness Evaluation Metrics

The most widely used fairness metrics [18] are: Demographic Parity, Equal Opportunity, and Equalized Odds [10]. The metrics are elaborated in detail below, where Y ($\widehat{Y}$) is actual (predicted) class label and S is protected attribute:

- **Demographic Parity Difference (DP)** suggests that the probability of favourable outcomes should be same for all the subgroups:

$$DPD = |P[\widehat{Y} = 1|S = 1] - P[\widehat{Y} = 1|S \neq 1]| \qquad (4)$$

- **Equality of Opportunity Difference (EO)** emphasises that there should be equal opportunities for all the subgroups having positive outcomes to have positive prediction i.e. true positive rates for all the groups should be same:

$$EOP = |P[\widehat{Y} = 1|S = 1, Y = 1] - P[\widehat{Y} = 1|S \neq 1, Y = 1]| \qquad (5)$$

- **Equalized Odds Difference (EOD)** focuses on equalizing false positive rates along with the same true positive rates for all the subgroups:

$$EOD = |P[\widehat{Y} = 1|S = 1, Y = 1] - P[\widehat{Y} = 1|S \neq 1, Y = 1]|+ \atop |P[\widehat{Y} = 1|S = 1, Y = 0] - P[\widehat{Y} = 1|S \neq 1, Y = 0])| \qquad (6)$$

For a fair algorithm, DP, EO and EOD values must be closer to 0.

4.2 Implementation Details

The GroupMixNorm layer has been implemented in the PyTorch framework on Ubuntu 16.04.7 OS with the Nvidia GeForce GTX 1080Ti GPU. For a fair

(a) Demographic Parity   (b) Equal Opportunity   (c) Equalized Odds

Fig. 3. Fairness-AP trade-off curves on the **Adult dataset**, where GroupMixNorm demonstrates improved performance. Results are obtained by varying the trade-off parameter as suggested in their respective publications: Adversarial Debiasing: $[0.01 \sim 1.0]$, Fair Mixup DP: $[0.1 \sim 0.7]$, Fair Mixup EO: $[0.5 \sim 5.0]$, and RNF: $[0.05, 0.015, 0.025, 0.035]$. For a fair algorithm, it is desirable to have the AP closer to 1, and the fairness metrics (DP, EO, EOD) closer to 0.

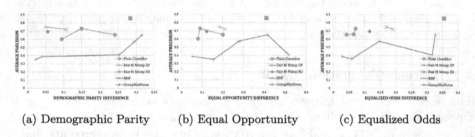

(a) Demographic Parity   (b) Equal Opportunity   (c) Equalized Odds

Fig. 4. Fairness-AP trade-off curves of GroupMixNorm layer and other comparative algorithms on the **CelebA dataset**, where the proposed approach achieves state-of-the-art performance. Results are obtained by varying the trade-off parameter as suggested in their respective publications: Fair M Mixup DP: $[25, 50]$, Fair M Mixup EO: $[1, 10, 50]$, RNF: $[0.1, 0.5, 1, 5, 10]$.

comparison with existing literature, we have followed the same dataset pre-processing and protocols as the fair mixup approach [6]. For the Adult dataset, we use four fully connected layers with hidden dimension 50. Each layer except the last output layer is followed by SiLU activation and the proposed Group-MixNorm layer. The model is trained for 10 epochs with a 1000 batch size. For each epoch, the dataset is randomly split into 60-20-20 split of train, val, and test set, respectively. We select the best-performing model on the validation set across 10 independent runs and report the mean Average Precision and fairness metrics defined above. For the CelebA dataset, we use the ResNet-18 [11] model for feature extraction followed by two fully connected layers for the classification task. We apply SiLU activation and GroupMixNorm layer between the two FC layers. We use the original split of the dataset and train the model for 100 epochs with 128 batch size. Both the models are trained with the Adam optimizer with learning rate $1e - 4$. In all experiments, mixing coefficient λ (Eqs. 1 and 2) is randomly sampled from $Beta(\alpha, \alpha)$. The value of α is empirically set to 0.1.

5   Results and Analysis

Figures 3–6 and Table 1 present the results and analysis of the GroupMixNorm layer and comparison with the state-of-the-art in-processing bias mitigation techniques. Detailed analysis is given in the following subsections:

5.1   Comparison with State-of-the-art Algorithms

Since the GroupMixNorm layer focuses on mitigating bias during the training process, comparison has been performed with algorithms that optimize fairness constraints during training: (i) Adversarial Debiasing [22], (ii) Fair Mixup: Fairness via Interpolation (Fair Mixup) [6], (iii) Fairness via Representation Neutralization (RNF) [8], and (iv) plain classifier. Fair Mixup uses two separate regularizing terms for optimizing the fairness metrics of Demographic Parity (DP) and Equal Opportunity (EO), and thus can solve for either DP or EO at a time. In this paper, we refer to these two variants of Fair Mixup as *Fair Mixup DP* and *Fair Mixup EO*. To calculate the DP, Chuang *et al.* [6] have computed the difference between the predicted probability across the protected groups. Similarly, for EO, Chuang *et al.* [6] compute class-wise difference between the predicted probability across protected groups. As part of this research, we have used the actual definitions of EO and DP for computing the fairness metrics (Eqs. 4–5). Parallely, the Representation Neutralization (RNF) technique [8] has shown the bias mitigation performance via two variants: (i) in model-1, proxy labels are generated for the protected attribute, while (ii) in model-2, ground-truth protected attribute labels are used. As part of this research, we have compared our results with their second variant (model-2), referred to as *RNF*.

For a fair comparison, we evaluate all the models under the same setting. Techniques such as Adversarial Debiasing, Fair Mixup, and RNF introduce a regularization term in the loss function to improve fairness via a hyper-parameter α that controls the trade-off between the average precision (AP) and fairness metrics (DP, EO, and EOD). We have reported the results on varying values of α as suggested in their respective papers. In our case, the GroupMixNorm layer is proposed towards architecture design and not the loss function, thus there is no such trade-off. Performance analysis on different datasets is as follows:

(a) Comparison on the UCI Adult Income Dataset: Figure 3 shows the performance comparison on the UCI Adult dataset, where the GroupMixNorm layer produces fairer results as compared to other techniques across all fairness metrics (DP, EO, EOD) with minimal impact on average precision. Since Fair Mixup solves separate constraint optimizations to achieve lower DP and EO, it minimizes either DP or EO at a time. In terms of the fairness metrics, RNF produces fair results, however the average precision is relatively lower, thus making it unsuitable for the classification task.

(b) Comparison on the CelebA Dataset (Fig. 4): Consistent with the published manuscript [6], for Fair Mixup, comparison has been performed with the combination of manifold mixup [19] (Fair M Mixup DP and Fair M Mixup EO). Similar to the previous experiments, it is observed that either Fair M

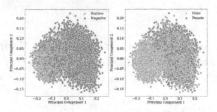
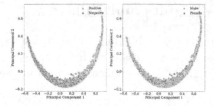

(a) Without GroupMixNorm Layer   (b) With GroupMixNorm Layer

Fig. 5. PCA visualizations of the features for the MLP classifiers trained without and with GroupMixNorm. The left plots show the class distribution, and the right plots show the gender distribution (protected attribute). The model trained with Group-MixNorm demonstrates minimal distinction on the gender attribute.

Table 1. Cosine similarity between the learned weight parameters of C_{sens} and C_{cls} linear classifiers (the former is trained for predicting the sensitive attribute, while the later is trained for class label prediction). A lower score represents less biased models since lesser similarity is observed between the weight parameters.

Method	Cosine Similarity
Plain Classifier	0.205
RNF	0.075-0.2
GroupMixNorm	0.06

Mixup DP or Fair M Mixup EO achieves optimal performance at a time. Further, the RNF model produces fair results across fairness metrics, however achieves lower average precision. GroupMixNorm achieves comparable performance to the best performing model across all the metrics, while maintaining a high average precision, thus suggesting high utility for real-world applications.

5.2   Learned Representation Analysis

Experiments have been performed for (a) feature visualization and (b) auxiliary prediction task for understanding feature quality. The key findings are as follows:
(a) **Feature Visualization:** Figure 5 presents the 2D projections obtained by using the sigmoid kernel Principal Component Analysis (PCA). Figure 5a presents the features learned by a biased MLP classifier (trained without Group-MixNorm layer), where the features appear both class and gender (protected attribute) discriminative. There is an overlap of male samples with the positive class samples, both lying majorly in the lower left side of the distribution. On the other hand, Fig. 5b shows features learned with the GroupMixNorm layer appear to be class discriminative, while not being gender discriminative. Further, both male and female samples are evenly distributed, thus preventing the model to get biased against a particular sensitive group.

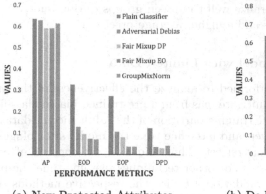
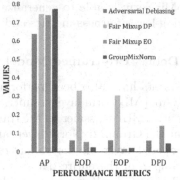

(a) New Protected Attributes   (b) De-biasing Pre-trained Classifier

Fig. 6. Average precision and fairness metrics obtained by different techniques (a) when evaluated on new protected attributes and (b) for de-biasing a pre-trained classifier with limited training data. Experiments have been performed on the Adult Income dataset with race and gender as the protected attribute, respectively. GroupMixNorm presents improved performance across metrics.

(b) Auxiliary Prediction Task: Similar to Du *et al.* [8], we use an auxiliary prediction task to analyze the quality of the learned features. The objective is to analyze how well the model can reduce the correlation between the class labels and the sensitive attributes. To this effect, we train two linear classifiers C_{sens} and C_{cls} that take the representation vector as input and predicts class labels and sensitive attributes, respectively. Next, we compare the learned weight matrix of C_{sens} and C_{cls} using cosine similarity. A higher similarity would signify similar weights and thus higher correlation between the two tasks. Table 1 shows that our model has the least cosine similarity indicating that the classifier focuses more on task relevant information than sensitive information. It is important to note that the cosine similarity for the RNF model varies from 0.2 to 0.075, based on the fairness-accuracy trade-off parameter, while the GroupMixNorm layer based model achieves a cosine similarity of 0.06 only.

5.3   Generalizability to New Protected Groups

With time, as the data evolves, new sensitive groups often get introduced. For example, gender attribute values may change from binary to non-binary. A robust classification model must remain unbiased even with the introduction of additional sensitive groups during inference. In order to simulate this setup, the proposed solution was evaluated for new groups at test time without any re-training. Experiments were performed on the Adult Income dataset where data pertaining to two races (White and Black) was used for training, while the data from White, Black, and Others (Asian-Pac-Islander, Amer-Indian-Eskimo, Other) racial groups were used for testing. Figure 6a presents the performance of the GroupMixNorm layer along with other comparative techniques, where

GroupMixNorm is able to generalize well to unseen groups during inference by obtaining lower fairness metrics and a higher average precision.

5.4 Debias Pre-trained Model with Limited Data

Experiments have also been performed to analyze the effectiveness of the proposed GroupMixNorm layer to mitigate bias from a pre-trained biased classifier. We train an MLP classifier on the training partition of the Adult Income dataset, without the GroupMixNorm layer, and later fine-tune the model after plugging the proposed layer on the validation set. The validation set consists of only 20% samples of the entire dataset. For other techniques, we fine-tune the pre-trained biased classifier on the validation set with the respective methods. We evaluate the model for fairness on the Adult dataset with gender as the protected attribute. Figure 6b presents the results obtained by the proposed Group-MixNorm layer as well as other comparative techniques. It can be observed that the proposed solution produces fairer results as compared to other algorithms across the different fairness metrics, while achieving the highest average precision. The experiment suggests that even with a small training set, the proposed GroupMixNorm can aid in eliminating bias from a pre-trained network.

6 Conclusion and Future Work

Learning bias-invariant models are the need of the hour for the research community. While existing research has focused on proposing novel solutions for learning unbiased classifiers, most of the techniques incorporate an additional term in the loss function for modeling the model fairness. We believe that it is often difficult to extrapolate the learnings of such an optimization function to the test set, especially under the challenging scenario of new protected attributes during evaluation. To this effect, this research proposes a novel *GroupMixNorm* layer, which promotes learning fairer models at the architectural level. Group-MixNorm is a distribution alignment strategy operating across the different protected groups, enabling attribute-invariant feature learning. Across multiple experiments, GroupMixNorm demonstrates improved fairness metrics while maintaining higher average precision levels, as compared to the state-of-the-art algorithms. Further analysis suggests high model generalizability to new protected attributes during evaluation, possibly due to the transformation of samples to interpolated groups resulting in model regularization during training. As an extension of this research, future research directions include studying the impact of GroupMixNorm on different convolution layers and extending the scope to evaluation on NLP datasets and tasks.

References

1. Agarwal, A., Beygelzimer, A., Dudík, M., Langford, J., Wallach, H.M.: A reductions approach to fair classification. In: ICML. vol. 80, pp. 60–69 (2018)

2. Ahuja, K., Shanmugam, K., Varshney, K.R., Dhurandhar, A.: Invariant risk mini-mization games. In: ICML. **119**, 145–155 (2020)
3. Arjovsky, M., Bottou, L., Gulrajani, I., Lopez-Paz, D.: Invariant risk minimization. CoRR abs/1907.02893 (2019)
4. Barocas, S., Hardt, M., Narayanan, A.: Fairness and Machine Learning. fairml-book.org (2019), http://www.fairmlbook.org
5. Cheng, P., Hao, W., Yuan, S., Si, S., Carin, L.: FairFil: Contrastive neural debiasing method for pretrained text encoders. In: ICLR (2021)
6. Chuang, C., Mroueh, Y.: Fair mixup: Fairness via interpolation. In: ICLR. Virtual Event, Austria, May 3–7, 2021 (2021)
7. Cotter, A. et al.: Training well-generalizing classifiers for fairness metrics and other data-dependent constraints. In: ICML. vol. 97, pp. 1397–1405 (2019)
8. Du, M., Mukherjee, S., Wang, G., Tang, R., Awadallah, A., Hu, X.: Fairness via representation neutralization. In: NeurIPS, vol. 34 (2021)
9. Dua, D., Graff, C.: UCI machine learning repository (2017), http://archive.ics.uci.edu/ml
10. Hardt, M., Price, E., Srebro, N.: Equality of opportunity in supervised learning. In: NeurIPS, pp. 3315–3323 (2016)
11. He, K., Zhang, X., Ren, S., Sun, J.: Deep residual learning for image recognition. In: IEEE CVPR, pp. 770–778 (2016)
12. Kilbertus, N. et al.: Avoiding discrimination through causal reasoning. In: NeurIPS, pp. 656–666 (2017)
13. Kim, B., Kim, H., Kim, K., Kim, S., Kim, J.: Learning not to learn: Training deep neural networks with biased data. In: IEEE CVPR, pp. 9012–9020 (2019)
14. Kusner, M.J., Loftus, J.R., Russell, C., Silva, R.: Counterfactual fairness. In: NeurIPS, pp. 4066–4076 (2017)
15. Liu, Z., Luo, P., Wang, X., Tang, X.: Large-scale celebfaces attributes (celeba) dataset. Retrieved August **15**(2018), 11 (2018)
16. Manisha, P., Gujar, S.: FNNC: Achieving fairness through neural networks. In: IJCAI, pp. 2277–2283 (2020)
17. Singh, K.K., Mahajan, D., Grauman, K., Lee, Y.J., Feiszli, M., Ghadiyaram, D.: Don't judge an object by its context: Learning to overcome contextual bias. In: IEEE/CVF CVPR, pp. 11067–11075 (2020)
18. Verma, S., Rubin, J.: Fairness definitions explained. In: International Workshop on Software Fairness, pp. 1–7 (2018)
19. Verma, V. et al.: Manifold mixup: Better representations by interpolating hidden states. In: ICML. vol. 97, pp. 6438–6447 (2019)
20. Woodworth, B.E., Gunasekar, S., Ohannessian, M.I., Srebro, N.: Learning non-discriminatory predictors. In: COLT. **65**, 1920–1953 (2017)
21. Zafar, M.B., Valera, I., Gomez-Rodriguez, M., Gummadi, K.P.: Fairness con-straints: Mechanisms for fair classification. In: AIStat. vol. 54, pp. 962–970 (2017)
22. Zhang, B.H., Lemoine, B., Mitchell, M.: Mitigating unwanted biases with adver-sarial learning. In: AAAI/ACM AIES, pp. 335–340 (2018)
23. Zhang, H., Cissé, M., Dauphin, Y.N., Lopez-Paz, D.: Mixup: Beyond empirical risk minimization. In: ICLR (2018)
24. Zunino, A. et al.: Explainable deep classification models for domain generalization. In: IEEE CVPRW, pp. 3233–3242 (2021)

Quantifying the Bias of Transformer-Based Language Models for African American English in Masked Language Modeling

Flavia Salutari[1,2(✉)], Jerome Ramos[1], Hossein A. Rahmani[1], Leonardo Linguaglossa[2], and Aldo Lipani[1]

[1] University College London, London, UK
{jerome.ramos.20,hossein.rahmani.22,aldo.lipani}@ucl.ac.uk
[2] Telecom Paris, Paris, France
{flavia.salutari,linguaglossa}@telecom.paris.fr

Abstract. In recent years, groundbreaking transformer-based language models (LMs) have made tremendous advances in natural language processing (NLP) tasks. However, the measurement of their fairness with respect to different social groups still remains unsolved. In this paper, we propose and thoroughly validate an evaluation technique to assess the quality and bias of language model predictions on transcripts of both spoken African American English (AAE) and Spoken American English (SAE). Our analysis reveals the presence of a bias towards SAE encoded by state-of-the-art LMs such as BERT and DistilBERT and a lower bias in distilled LMs. We also observe a bias towards AAE in RoBERTa and BART. Additionally, we show evidence that this disparity is present across all the LMs when we only consider the grammar and the syntax specific to AAE.

Keywords: Language Model · Transformers · Bias and Fairness · Evaluation

1 Introduction

Since their inception [8], transformers-based bidirectional encoder representations language models (LMs) have gained significant scientific interest due to their sizable improvements on a wide range of NLP tasks. The success of BERT pushed researchers to expand the state-of-the-art by introducing a plethora of model variants with differences in architecture [30], size [21,31,37] and training [22,24]. However, a growing concern in the research community has arisen: the potential societal risks coming from the pervasive adoption of these models [2]. Several studies highlight that this adoption would hinder equitable and inclusive access to NLP technologies and have real-world negative consequences in different areas, such as education, work, and politics [32]. Given the consistent emergence of new LMs trained on Web-based corpora, it is crucial to identify and measure the bias and fairness of these models.

Given the sheer size and heterogeneity of the Web, one might expect these models to be bias-free. However, even before the explosion of transformer-based LMs, a variety of biases have been identified in standard word embeddings [3]. Recently, some effort has

H. Kashima et al. (Eds.): PAKDD 2023, LNAI 13935, pp. 532–543, 2023.
https://doi.org/10.1007/978-3-031-33374-3_42

been devoted to highlighting the presence of possible biases encoded by transformer-based LMs along gender, race, ethnicity, and disability status through techniques such as sentiment analysis and named entity recognition tasks. In contrast, we focus on tasks used in conversational systems, where a word could be unheard or unrecognized by the automatic speech recognition system and would therefore need to be predicted.

In particular, we focus on spoken language, as it tends to have incomplete sentences, spontaneous self-corrections, and interruptions, and its register is more informal with respect to written language, which is typically more structured. We study the presence of potential bias towards English dialects spoken by underrepresented and historically discriminated groups, such as African American English (AAE). In linguistics, AAE and *mainstream* U.S. English, referred in this paper as Spoken American English (SAE), are regarded as two different languages because they are highly structured and possess their own phonological, syntactic and morphological rules [14]. In fact, AAE highlights the regional, societal and cultural environments in which individuals have learned to speak [13]. However, SAE speakers often believe that AAE is a version of SAE with mistakes and that AAE speakers belong to deficient cultures [28,36].

It is difficult to estimate the number of AAE speakers because some African Americans may speak a variety that aligns more with SAE, and not all AAE speakers are African Americans. Nevertheless, a 2019 census [29] estimates that approximately 13% of the U.S. population is African American, suggesting that there is a significant number of AAE speakers. Thus, the presence of potential linguistic biases may have discriminatory consequences towards a considerable group of individuals.

For these reasons, we set out to measure the robustness and quality of 7 transformer-based LMs in the prediction of *missed* words when the input is either SAE or AAE. We resort to two renowned corpora of spoken SAE and AAE and evaluate the LMs in a Masked Language Modeling (MLM) task. In particular, we formulate a *fill-in-the-blank* task, where we mask and predict a token, simulating its absence in every utterance. Next, we define two metrics, Probability Difference and Complementary Reciprocal Rank, to compare the likelihood that the model assigns the predicted token to the actual *masked* one and use that as a proxy of quality and fairness for the model itself.

We rigorously quantify the model bias and find that BERT, in both its cased and uncased variants, exposes a non-negligible bias towards SAE (up to 21% more accurate results with respect to AAE). Surprisingly we find that RoBERTa and BART models are biased towards AAE. We additionally observe distilled variants of these LMs to be fairer with respect to their teachers. Finally, our analysis reveals that the majority of bias resides in the AAE structural differences, specifically the particles, pronouns, and adpositions.

2   Related Work

Some of the major factors behind the success of transformer-based LMs include the large architectures and the training done on huge amounts of textual data. This recently raised the interest of the research community towards the potential societal risks linked to the employment of these models for either generating text tasks or as components of classification systems [2]. These works have studied the effects of transferring the stereotypical associations present in the training datasets to LMs, which cause an unintended bias towards underrepresented groups. Significant research efforts have been made to identify race and gender bias embedded in large models [1,5,20,26,33]. [18] highlights the presence of topical biases in words predicted by BERT on sentences mentioning disabilities.

In addition to works on bias measurement, researchers have proposed methods to mitigate societal biases with debiasing techniques [23,34]. In regards to research on bias towards languages, most studies have focused on offensive language and hate speech detection [7,27], whereas research on bias against dialects spoken by underrepresented groups is quite recent [10]. In contrast to the above works, which mostly focus on the negative sentiment and stereotypical associations towards specific groups in BERT [8], this work focuses on quantifying the linguistic bias towards AAE for 7 different LMs: BERT, RoBERTa [24], BART [22], DistilBERT and DistilRoBERTa [31], including both their cased and uncased versions.

Previous works have proven that the large dimension of the training datasets for state-of-the-art LMs may not lead to diversity and inclusion for underrepresented groups [2]. Therefore, our analysis is essential to provide a framework to assess, reveal, and counteract the existing biases in order to improve the performance of large language models with regard to linguistic biases.

3   Methodology

To capture and provide an accurate and comprehensive account of societal biases embedded in state-of-the-art LMs, we leverage two corpora of spoken English. These are widely used in the linguistics field because linguists consider them a fair representation of their spoken language. Although there is a 15-year gap between the collection date of these corpora, we argue that the core structure of the language remains the same and that any bias captured due to the difference in periods will be minimal. Additionally, although LMs are generally trained using text data, we argue that spoken conversational agents leverage these same LMs when communicating with users. Thus, it is still appropriate to analyze the bias of LMs using spoken corpora. We also note that while this paper is not the first to study the presence of societal biases, to the best of our knowledge, this is the first to provide a thorough characterization of it for AAE across different models tested on an MLM task. We summarize LMs' performance by means of statistical metrics, which are used to characterize both the bias and the quality of the models.

3.1   Corpora for Spoken English

For SAE, we leverage the Santa Barbara Corpus of Spoken American English (SBC-SAE) [11], which is widely adopted for different applications, such as the assessment of political risk faced by U.S. firms [16], the measure of grammatical convergence in bilingual individuals [4], and the exploration of new-topic utterances in naturally occurring dialogues [25]. It includes conversations recorded in various real everyday life situations from a wide variety of people who differ in gender, occupation, and social background. All the audio recordings are also complemented with their transcriptions.

Table 1. Corpora summary: with and without filtering utterances ($\mathcal{U}$) based on their length. With $\langle \ell_u \rangle$ we indicate the average utterance length; with L, the length of the corpus in number of words, and; with $|\mathcal{T}|$, the number of terms (unique words).

Type	Corpus	Language	$|\mathcal{U}|$	$\langle \ell_u \rangle$	L	$|\mathcal{T}|$
Original	CORAAL	AAE	90,493	6.22	563,037	17,214
	SBCSAE	SAE	40,838	7.14	291,513	12,324
Filtered	CORAAL	AAE	63,814	8.23	525,067	16,352
	SBCSAE	SAE	25,113	8.38	210,430	10,540

Since SBCSAE consists of speakers from several regional origins (except for the African American speakers that we preliminary filter out), we ensure that we do not craft the results by inducing unwanted bias when comparing AAE with a version of SAE that could be more similar to the Written American English, which is instead rather different from the spoken *mainstream* U.S. English.

For AAE, we use the Corpus of Regional African American Language (CORAAL) [19], which also provides the audio recordings along with their time-aligned orthographic transcriptions, of particular interest for this work. CORAAL includes 150 sociolinguistic interviews for over a million words. It is periodically updated and is the only publicly available corpus of AAE. As such, it has been used in the literature for a plethora of tasks, ranging from dialect-specific speech recognition [10] to cross-language transfer learning [17].

In this work, we only focus on the CORAAL:DCB portion, since it is comprised of the most recent interviews (carried out between 2015 and 2017) and contains the largest amount of data (more than 500k words). It includes conversations from 48 speakers raised in Washington, DC, a city with a long-standing African American population.

For each corpus, we define $\mathcal{U} = \{u_1, u_2, ..., u_n\}$ as the set of all the available utterances and $\mathcal{T} = \{t_1, t_2, ..., t_n\}$ as the set of all terms (unique words). Since we perform an utterance-level analysis, we first filter out noise. In particular, we discard both short utterances (composed of just one or two words) and very long ones (greater than 50 words).

In Table 1, we report a summary of the corpora statistics, both before and after having applied the filtering based on the utterance length. Even though the sizes of the two datasets are very different, not only in terms of the number of utterances $|\mathcal{U}|$, but also in terms of the total number of words L and terms $|\mathcal{T}|$, we can see that, after the filtering, the average utterance length $\langle \ell_u \rangle$ across corpora is very similar (~ 8 words per utterance).

3.2 Bias in Masked Language Modeling

In order to measure the bias in LMs we perform an MLM task. We leverage the transformer-based $\text{BERT}_{\text{base}}$ LM [8] and its recent variants, including $\text{DistilBERT}_{\text{base}}$ [31], in both their cased and uncased flavors, $\text{RoBERTa}_{\text{base}}$ [24], $\text{DistilRoBERTa}_{\text{base}}$, and $\text{BART}_{\text{base}}$ [22]. These LMs have all been pre-trained using an

MLM objective, which consists of randomly masking 15% of the tokens using a special [MASK] token. Note that these models are trained on different corpora, such as OpenWebText and BooksCorpus.

Therefore, by directly querying the underlying MLM in each LM, we simulate the typical scenario where a conversational system has to infer a *missed* word in an utterance. In particular, we encode each utterance of the two corpora with the *tokenizer* of the LM considered. We then iteratively mask each word w_{mask} and predict the masked word by feeding the model with only a context of 10 tokens surrounding w_{mask}.

The LM provides each run with a list of possible terms to *fill-in-the-blank*. In the vocabulary set $\mathcal{T}$, we select the predicted term t_p with the highest probability $P(t_p|c)$, that is, ranks first in the list $\rho(t_p|c) = 1$, where c is the context surrounding t_p and ρ is the rank of $t|c$. In this notation, a word w is a term t in a context c ($t|c$). We next retrieve the corresponding probability $P(t_m|c)$ and the rank $\rho(t_m|c)$ for the actual masked token t_m from the vocabulary of possible terms $\mathcal{T}$. The latter provides a measure of how likely the LM will choose t_m as a candidate token to replace the masked one w_{mask}. We then employ the probabilities difference $\Delta P(t|c)$ as a proxy of the quality of the prediction for a single token, defined as:

$$\Delta P(t|c) = P(t_p|c) - P(t_m|c) = \Delta P(w). \tag{1}$$

We further define for each token $t|c$ the Complementary Reciprocal Rank (CRR) as:

$$\mathrm{CRR}(t|c) = 1 - \rho(t_m|c)^{-1} = \mathrm{CRR}(w). \tag{2}$$

Note that this is the difference between the reciprocal rank (RR) of the predicted token, which is always equal to 1 ($\rho(t_p|c)^{-1} = 1$), and the RR of the masked token.

We then define the probability difference for an utterance by averaging the probability difference for each token in the utterance:

$$\Delta P(u) = \frac{1}{\ell_u} \sum_{w \in u} \Delta P(w), \tag{3}$$

with ℓ_u being the length of the utterance in terms of tokens. Similarly, we define the CRR for an utterance as:

$$\mathrm{CRR}(u) = \frac{1}{\ell_u} \sum_{w \in u} \mathrm{CRR}(w). \tag{4}$$

Note that the metrics based on the ranks $\rho(t|c)$ generated by the LMs are necessary to fully capture the bias embedded in the models, since $\Delta P(t|c)$ alone could be insufficient. This is because the $\Delta P(t|c)$ strongly depends on how the LM assigns the probability. For example, the probability distribution of $P(t|c)$ could be more unifor—m and, consequently, would lead to, on average, a smaller $\Delta P(t|c)$. Conversely, a more skewed distribution would cause larger differences $\Delta P(t|c)$. Thus, CRR is used because it is unaffected by such differences in the output probability distribution of $P(t|c)$.

Table 2. MAE and MSE of $\Delta P(u)$ and CRR(u) measured on AAE and SAE corpora: results obtained through the *fill-in-the-blank* task with different language models. † signifies that the AAE and SAE expectations are statistically significant according to Welch's two-tailed t-test (p-value < 0.05). The column d contains their effect size computed according to Cohen's d.

Model	MAE								MSE							
	$\Delta P(u)$				CRR(u)				$\Delta P(u)$				CRR(u)			
	AAE	SAE	Δ[%]	d	AAE	SAE	Δ[%]	d	AAE	SAE	Δ[%]	d	AAE	SAE	Δ[%]	d
BERT$_{cased}$	0.217	**0.171**	21†	0.417	0.497	**0.441**	11†	0.272	0.060	**0.040**	33†	0.345	0.289	**0.233**	20†	0.262
BERT$_{uncased}$	0.242	**0.198**	18†	0.352	0.494	**0.446**	10†	0.232	0.074	**0.053**	29†	0.297	0.288	**0.238**	18†	0.230
DistilBERT$_{cased}$	0.113	**0.108**	5†	0.081	0.627	**0.589**	6†	0.188	0.017	**0.016**	2†	0.015	0.436	**0.385**	12†	0.203
DistilBERT$_{uncased}$	0.126	**0.118**	6†	0.104	0.578	**0.530**	8†	0.222	0.021	**0.020**	1	0.007	0.380	**0.325**	15†	0.223
RoBERTa	**0.223**	0.261	−15†	0.368	**0.536**	0.592	−9†	0.252	**0.061**	0.079	−23†	0.311	**0.337**	0.396	-15†	0.225
DistilRoBERTa	**0.143**	0.153	−7†	0.137	**0.644**	0.668	−4†	0.117	**0.026**	0.029	-11†	0.112	**0.457**	0.487	-6†	0.115
BART	**0.156**	0.193	−20†	0.506	**0.613**	0.682	−10†	0.346	**0.030**	0.043	−31†	0.447	**0.418**	0.501	−17†	0.328

4   Results and Discussion

In this section, we first provide an accurate overview of the measured fairness of LMs and then further analyze the discovered biases from different viewpoints. We show how they vary when we take into account the syntactical, grammatical, and lexical patterns typical of AAE language.

4.1   Measuring the Bias of LMs

As described in Sect. 3, we test the fairness of transformer-based LMs by running experiments in an MLM setting. We use ΔP and CRR as metrics for measuring the quality and the fairness of the models towards the two investigated languages. We observe the expected behaviour of the LMs with respect to each utterance and consider an aggregate measure of the metrics on a per-utterance level.

Table 2 reports an overview of the results of $\Delta P(u)$ and CRR(u). Using a Welch's t-test [35], we find that the difference between the means of AAE and SAE for both $\Delta P(u)$ and CRR(u) is significant (p-value < 0.05). We then measure their effect size using Cohen's d [6] , which is reported in the last two columns of Table 2. According to Cohen's classification, there is a *small* effect for both metrics and a *medium* effect for BART on $\Delta P(u)$ (d>0.5). We summarize the quality of the prediction in the corpora using Mean Absolute Error (MAE) and Mean Squared Error (MSE) for both $\Delta P(u)$ and CRR(u).

These error measures are used to quantify the quality of the predicted terms, where an MAE and MSE closer to 0 corresponds to an utterance having more accurately predicted terms. Therefore, in Table 2, we highlight the values leading to the smallest error between AAE and SAE. Additionally, we emphasize the presence of bias by pointing out the percentage of bias change of each LM $\Delta[\%]$, which is calculated with respect to the model with the largest bias.

Three main patterns clearly emerge from Table 2. First, BERT and DistilBERT, in both their cased and uncased variants, show a bias towards SAE for all the metrics. Specifically, BERT not only presents a non-negligible bias against AAE but also

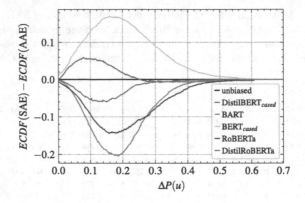

Fig. 1. The difference between the ECDFs of SAE and AAE for the $\Delta P(u)$ measure. When the values are greater than zero, the LMs are more biased towards SAE, *vice versa* otherwise.

is the LM which leads to the highest relative bias. In particular, we observe that the MAE($\Delta P(u)$) for SAE is more than 20% lower than AAE, 11% lower for the MAE(CRR(u)), 33% for the MSE($\Delta P(u)$), and 20% for the MSE(CRR(u)).

Second, DistilBERT, in both its cased and uncased flavours, and DistilRoBERTa, are the models which perform better with regards to the average probability difference $\Delta P(u)$. This is true both in terms of MAE and MSE, which are approximately half and one-third of the other LMs. On the one hand, this could seem somewhat unexpected since one could argue that DistilBERT is less accurate than BERT, achieving only 97% of its performance [31]. On the other hand, this is in line with recent work [2] reporting that such LMs sometimes exceed the performance of the original ones. However, as mentioned in Sect. 3, it is crucial to also look at the CRR(u) because better behaviour in terms of $\Delta P(u)$ could, in practice, be tied to the fact that the model generates more uniformly distributed probabilities $P(t|c)$ with respect to the others.

Finally, we observe that BART, despite having good prediction quality for AAE (MAE($\Delta P(u)$) and MSE($\Delta P(u)$) are lower than BERT), shows an opposite trend with respect to BERT and DistilBERT. This unexpected bias towards AAE is also introduced by RoBERTa and DistilRoBERTa. This is somewhat surprising and could possibly be attributed to the type of datasets they have been trained on. RoBERTa and BART are pre-trained with 1000% more data than BERT. By diving into the type of data involved, we discover multiple sources, ranging from English language encyclopedias and literary works (same as BERT) to news articles and Web content. Specifically, RoBERTa, BART, and DistilRoBERTa leverage OPENWEBTEXT [12], a corpus which includes filtered Web content obtained by scraping the social media platform Reddit, which may expose the LMs to less *standard* American English. It would be interesting to explore in future studies whether there is a significant difference between shared colloquial transcriptions in OpenWebText and AAE when compared to OpenWebText and SAE.

Since Table 2 only reports a summary of the distributions of the bias metrics computed on both datasets, we also analyze the bias measured by subtracting the empirical cumulative distribution functions (ECDFs) of $\Delta P(u)$ of AAE to that of SAE, which is shown in Fig. 1. This figure includes the bias measured for the LMs and, for the sake

Table 3. A sample of AAE utterances selected based on their syntactical features and their translations to SAE. In brackets the prevalence of the feature over utterances in the AAE corpus.

Original	Translated
Double Negative (0.7%)	
• *You don't need nothing but you.*	• *You don't need anything but you.*
• *I don't know nobody over there no more.*	• *I don't know anyone over there anymore.*
Verb *be* (2.8%)	
• *And I be okay with it.*	• *And I am okay with it.*
• *All of my friends was from like DC.*	• *All of my friends were from DC.*
Contractions (4.6%)	
• *I'm'a ask you.*	• *I'm going to ask you.*
• *something gonna happen.*	• *something is going to happen.*

Table 4. Similar to Table 2 but calculated over a sample of 50 utterances of AAE and their translated version (AAEᵀ) for each feature of AAE.

Model	MAE								MSE							
	$\Delta P(u)$				CRR(u)				$\Delta P(u)$				CRR(u)			
	AAE	AAEᵀ	Δ[%]	d	AAE	AAEᵀ	Δ[%]	d	AAE	AAEᵀ	Δ[%]	d	AAE	AAEᵀ	Δ[%]	d
Double Negative [50 utterances]																
BERT$_{cased}$	0.202	**0.159**	21†	0.591	0.391	**0.334**	15†	0.493	0.046	**0.030**	34†	0.526	0.166	**0.125**	25†	0.436
BERT$_{uncased}$	0.216	**0.187**	14	0.358	0.404	**0.340**	16†	0.503	0.053	**0.041**	23	0.319	0.179	**0.130**	27†	0.476
DistilBERT$_{cased}$	0.137	**0.106**	22†	0.548	0.506	**0.441**	13†	0.523	0.022	**0.014**	37†	0.504	0.267	**0.213**	21†	0.457
DistilBERT$_{uncased}$	0.148	**0.117**	21†	0.485	0.479	**0.394**	18†	0.701	0.025	**0.018**	27	0.293	0.240	**0.174**	28†	0.611
RoBERTa	0.202	**0.181**	10	0.227	0.434	**0.383**	12	0.328	0.048	**0.042**	14	0.180	0.208	**0.175**	16	0.243
DistilRoBERTa	0.170	**0.134**	21†	0.572	0.581	**0.498**	14†	0.628	0.034	**0.020**	41†	0.567	0.347	**0.272**	22†	0.529
BART	0.164	**0.140**	15†	0.422	0.534	**0.471**	12†	0.469	0.030	**0.023**	22	0.368	0.297	**0.245**	18	0.392
Verb *be* [50 utterances]																
BERT$_{cased}$	0.252	**0.184**	27†	0.691	0.589	**0.408**	31†	1.142	0.074	**0.043**	42†	0.622	0.373	**0.190**	49†	1.109
BERT$_{uncased}$	0.287	**0.216**	25†	0.642	0.595	**0.417**	30†	1.009	0.094	**0.059**	37†	0.520	0.383	**0.205**	46†	0.943
DistilBERT$_{cased}$	0.134	**0.119**	11	0.273	0.703	**0.540**	23†	0.910	0.021	**0.017**	16	0.198	0.519	**0.329**	37†	0.893
DistilBERT$_{uncased}$	0.138	**0.118**	14†	0.339	0.678	**0.513**	24†	0.904	0.022	**0.017**	25	0.344	0.485	**0.302**	38†	0.856
RoBERTa	0.246	**0.211**	14†	0.403	0.609	**0.458**	25†	0.800	0.069	**0.051**	26	0.380	0.405	**0.246**	39†	0.766
DistilRoBERTa	0.169	**0.142**	16†	0.425	0.723	**0.554**	23†	0.947	0.032	**0.024**	25	0.389	0.549	**0.343**	38†	0.931
BART	0.161	**0.144**	11	0.305	0.672	**0.556**	17†	0.672	0.029	**0.024**	18	0.246	0.474	**0.344**	27†	0.627
Contractions [50 utterances]																
BERT$_{cased}$	0.225	**0.181**	19†	0.507	0.470	**0.347**	26†	0.848	0.058	**0.040**	32†	0.436	0.247	**0.136**	45†	0.786
BERT$_{uncased}$	0.258	**0.205**	21†	0.605	0.482	**0.355**	26†	0.880	0.075	**0.049**	34†	0.541	0.257	**0.143**	45†	0.796
DistilBERT$_{cased}$	0.135	**0.114**	16	0.381	0.584	**0.463**	21†	0.746	0.022	**0.016**	28	0.316	0.369	**0.237**	36†	0.743
DistilBERT$_{uncased}$	0.140	**0.113**	19†	0.477	0.538	**0.410**	24†	0.799	0.023	**0.016**	33	0.374	0.318	**0.191**	39†	0.761
RoBERTa	0.215	**0.193**	10	0.264	0.500	**0.402**	20†	0.584	0.054	**0.043**	20	0.242	0.281	**0.186**	34†	0.574
DistilRoBERTa	0.154	**0.130**	16†	0.436	0.601	**0.488**	19†	0.668	0.027	**0.020**	28†	0.411	0.386	**0.268**	31†	0.635
BART	0.143	**0.136**	5	0.117	0.567	**0.475**	16†	0.562	0.023	**0.023**	1	0.015	0.346	**0.255**	26†	0.520

of simplicity, only reports the *cased* variants of BERT and DistilBERT. The solid black line at $y = 0$ shows the optimal unbiased LM and visually separates what is biased against AAE (on the positive y-axis) from what is biased against SAE (on the negative y-axis). Thus, we clearly see the behaviours of the LMs leading to the two worst biases, *i.e.*, RoBERTa and BERT$_{cased}$, which are consistently biased towards one side. They both present the maximum bias when $\Delta P(u)$ is close to 0.2 and instead mitigate for larger values. A similar behaviour is observed for the CRR(u).

4.2 Bias on AAE Features

Next, we investigate how results change when we acknowledge the lexical, syntactical, morphological, and phonological rules of AAE. Following AAE grammar [15], we
choose to focus on three major syntactical features: (i) the use of *double* negatives, (ii)
the different usage of verb *be* and, finally, (iii) the contractions of words and groups of
words.

For (i), we search for the close presence of multiple forms of grammatical negation
(which in standard English are ungrammatical) in all the utterances of the AAE corpus
and find that 0.7% of the utterances contain this feature. We then focus on feature (ii)
and select the AAE utterances that exhibit the use of the *aspectual be* verb, typically
used to denote habitual or iterative meaning (*e.g.*, *I be okay with it* in Table 3). Additionally, we filter for utterances with the verb tense in the *-ing* form where the verb is
either omitted (*e.g.*, *It depends on where you going to* in Table 3) or left at the base form
(*e.g.*, *they be getting mad* in Table 3), for a total of 2.8% of utterances. Finally, for (iii),
we include utterances containing non-standard contractions, *e.g.*, *I'm'a, ain't* or omit-
ting the auxiliary before *gonna*, *e.g.*, *something gonna happen* in Table 3. We do not
include contractions which are popular in SAE, as *wanna, won't, aren't, etc.* We obtain
4.6% of the utterances in this class. After filtering the utterances corresponding to the
specific grammar patterns, we carefully manually validate our selection by randomly
picking and inspecting 1% of them. We check that the 1% random sampled utterances
satisfy our criteria. From this manual labelling, we double-check our syntactical-rules-
based selection strategies and find that they are 99% accurate for all three cases.

Next, we randomly choose 50 utterances from each AAE case and build a ground
truth by *translating* the AAE utterances into a version compliant to SAE, which we
define as AAE^T. We keep the translation process as neutral as possible by preserving
the contractions typical of standard English and considered in dictionaries and grammar
books [9] as *short form* or *informal* and only *adjust* the selected grammar rules. Table 3
reports some examples of the utterances extracted from each AAE grammar case bucket
and the corresponding translated ones.

Finally, we repeat the MLM experiments, as described in Sect. 3, on these 150 translated utterances AAE^T and measure the bias. We report the results in Table 4. According
to Cohen's classification, there is a prevalent *medium* effect for both the metrics, with
the exception of $MSE(CRR(u))$ for the *verb be* class, where it is *large*.

At first glance, we observe that the errors for the set of the AAE utterances in the
verb *be* class are larger than the other two classes and the whole AAE corpus (reported
in Table 2). We observe that, on average, the three classes show a less accurate average
prediction with respect to the overall AAE corpus. Instead, we find that the translated
utterances AAE^T are better predicted with respect to AAE, surprisingly for all seven
LMs.

Notably, we observe that for the translated utterances in the *double negative* class,
all four metrics are always smaller (and hence a sign of better performance) than those
measured for the SAE corpus. This is somewhat unexpected since RoBERTa and BART
showed a bias towards AAE. However, we note that this may be attributed to the fact
that the SAE corpus, *SBCSAE*, is made up of conversations collected from people with
different regional origins. Consequently, despite the effort we make in trying not to

excessively standardize the utterances during the translation process, we could be generating sentences which are free from regional biases and consequently *"cleaner"* than those found in the SAE corpus.

5   Conclusion

This work proposes a methodology for evaluating the fairness of transformer-based language models. We assess and analyze the bias for two corpora, one for SAE and one for AAE. By directly querying the underlying MLM in seven LMs, we study the quality and bias of their predictions from several angles.

Results presented in this paper suggest that different models embed different biases. For example, the most popular state-of-the-art LMs, namely BERT and DistilBERT, show a non-negligible bias towards SAE, with the quality of the predictions being up to 21% more accurate than AAE. In contrast, BART, RoBERTa and DistilRoBERTa exhibit the opposite effect, with a bias leaning towards AAE. Our experiments also reveal that the distilled variants of BERT and RoBERTa are the fairest among the seven tested LMs.

Although this paper provides the first insightful snapshot of linguistic bias embedded in different LMs, it opens up a number of research questions. First, can fairer prediction outcomes be achieved with an ensemble learner of LMs embedding opposite biases, as, for instance, $BERT_{cased}$ and BART? Second, our results give insights into how the bias could be consistently mitigated with more inclusive corpora, by taking into account AAE features. Finally, special care could be put into the analysis of the distilled LMs, narrowing the gap on the causes which lead them to fairer predictions with respect to their teacher models, with a particular emphasis on the Web-based corpora used for training.

References

1. Basta, C., Costa-jussà, M.R., Casas, N.: Evaluating the underlying gender bias in contextualized word embeddings. In: Proceedings of the First Workshop on Gender Bias in Natural Language Processing, pp. 33–39. Association for Computational Linguistics (Aug 2019). https://doi.org/10.18653/v1/W19-3805,https://www.aclweb.org/anthology/W19-3805
2. Bender, E.M., Gebru, T., McMillan-Major, A., Shmitchell, S.: On the dangers of stochastic parrots: Can language models be too big? In: ACM Conference on Fairness, Accountability, and Transparency (FAccT) (2021). http://faculty.washington.edu/ebender/papers/Stochastic_Parrots.pdf
3. Bolukbasi, T., Chang, K.W., Zou, J., Saligrama, V., Kalai, A.: Man is to computer programmer as woman is to homemaker? debiasing word embeddings, p. 4356–4364. NIPS'16, Curran Associates Inc. (2016)
4. Cacoullos, R.T., Travis, C.E.: Bilingualism in the Community: Code-switching and Grammars in Contact. Cambridge University Press (2018)
5. Chada, R.: Gendered pronoun resolution using BERT and an extractive question answering formulation. In: Proceedings of the First Workshop on Gender Bias in Natural Language Processing, pp. 126–133. Association for Computational Linguistics (Aug 2019). https://doi.org/10.18653/v1/W19-3819,https://www.aclweb.org/anthology/W19-3819

6. Cohen, J.: Statistical Power Analysis for the Behavioral Sciences. Lawrence Erlbaum Associates (1988)
7. Davidson, T., Bhattacharya, D., Weber, I.: Racial bias in hate speech and abusive language detection datasets. In: Proceedings of the Third Workshop on Abusive Language Online, pp. 25–35. Association for Computational Linguistics (Aug 2019). https://doi.org/10.18653/v1/W19-3504,https://www.aclweb.org/anthology/W19-3504
8. Devlin, J., Chang, M.W., Lee, K., Toutanova, K.: BERT: Pre-training of deep bidirectional transformers for language understanding. In: Proceedings of the 2019 Conference of the North American Chapter of the Association for Computational Linguistics: Human Language Technologies, Volume 1 (Long and Short Papers), pp. 4171–4186 (Jun 2019). https://doi.org/10.18653/v1/N19-1423https://www.aclweb.org/anthology/N19-1423
9. Dictionary, C.E.: Cambridge english dictionary (2021), https://dictionary.cambridge.org/
10. Dorn, R.: Dialect-specific models for automatic speech recognition of African American Vernacular English. In: Proceedings of the Student Research Workshop Associated with RANLP 2019, pp. 16–20. INCOMA Ltd. (Sep 2019). https://doi.org/10.26615/issn.2603-2821.2019_003https://www.aclweb.org/anthology/R19-2003
11. Du Bois, J.W., Chafe, W.L., Meyer, C., Thompson, S.A., Martey, N.: Santa barbara corpus of spoken american english (2000). https://www.linguistics.ucsb.edu/research/santa-barbara-corpus/
12. Gokaslan, A., Cohen, V.: Openwebtext corpus (2019). http://web.archive.org/save/http://Skylion007.github.io/OpenWebTextCorpus
13. Gorski, P.C.: Reaching and teaching students in poverty: Strategies for erasing the opportunity gap. Teachers College Press (2017)
14. Green, L.J.: Introduction, pp. 1–11. Cambridge University Press (2002). https://doi.org/10.1017/CBO9780511800306.005
15. Green, L.J.: Syntax part 1: verbal markers in AAE, p. 34–75. Cambridge University Press (2002). https://doi.org/10.1017/CBO9780511800306.005
16. Hassan, T.A., Hollander, S., van Lent, L., Tahoun, A.: Firm-level political risk: measurement and effects. Q. J. Econ. **134**(4), 2135–2202 (2019)
17. Huang, J., et al.: Cross-language transfer learning, continuous learning, and domain adaptation for end-to-end automatic speech recognition. arXiv preprint arXiv:2005.04290 (2020)
18. Hutchinson, B., Prabhakaran, V., Denton, E., Webster, K., Zhong, Y., Denuyl, S.: Social biases in NLP models as barriers for persons with disabilities. In: Proceedings of the 58th Annual Meeting of the Association for Computational Linguistics, pp. 5491–5501. Association for Computational Linguistics (Jul 2020). https://doi.org/10.18653/v1/2020.acl-main.487https://www.aclweb.org/anthology/2020.acl-main.487
19. Kendall, T., Farrington, C.: The corpus of regional african american language (2018). http://lingtools.uoregon.edu/coraal/
20. Kurita, K., Vyas, N., Pareek, A., Black, A.W., Tsvetkov, Y.: Measuring bias in contextualized word representations. In: Proceedings of the First Workshop on Gender Bias in Natural Language Processing. Association for Computational Linguistics (2019). https://www.aclweb.org/anthology/W19-3823
21. Lan, Z., Chen, M., Goodman, S., Gimpel, K., Sharma, P., Soricut, R.: Albert: A lite bert for self-supervised learning of language representations. In: Proceedings of the 2020 International Conference on Learning Representations (2020). https://openreview.net/pdf?id=H1eA7AEtvS
22. Lewis, M., et al.: BART: Denoising sequence-to-sequence pre-training for natural language generation, translation, and comprehension. In: Proceedings of the 58th Annual Meeting of the Association for Computational Linguistics, pp. 7871–7880. Association for Computational Linguistics (Jul 2020). 10.18653/v1/2020.acl-main.703, https://www.aclweb.org/anthology/2020.acl-main.703

23. Liang, P.P., Li, I.M., Zheng, E., Lim, Y.C., Salakhutdinov, R., Morency, L.P.: Towards debiasing sentence representations. In: Proceedings of the 58th Annual Meeting of the Association for Computational Linguistics. Association for Computational Linguistics (2020). https://www.aclweb.org/anthology/2020.acl-main.488
24. Liu, Y., et al.: Roberta: A robustly optimized bert pretraining approach. arXiv preprint arXiv:1907.11692 (2019)
25. Luu, A., Malamud, S.A.: Non-topical coherence in social talk: A call for dialogue model enrichment. In: Proceedings of the 58th Annual Meeting of the Association for Computational Linguistics: Student Research Workshop, pp. 118–133 (2020)
26. May, C., Wang, A., Bordia, S., Bowman, S.R., Rudinger, R.: On measuring social biases in sentence encoders. In: Proceedings of the 2019 Conference of the North American Chapter of the Association for Computational Linguistics: Human Language Technologies, Volume 1 (Long and Short Papers), pp. 622–628. Association for Computational Linguistics (Jun 2019). https://doi.org/10.18653/v1/N19-1063https://www.aclweb.org/anthology/N19-1063
27. Mubarak, H., Rashed, A., Darwish, K., Samih, Y., Abdelali, A.: Arabic offensive language on twitter: Analysis and experiments (2020)
28. Pullum, G.K.: African american vernacular english is not standard english with mistakes. The workings of language: From prescriptions to perspectives, pp. 59–66 (1999)
29. QuickFacts, U.C.B.: United States census, QuickFacts statistics on U.S. population origin (2019). https://www.census.gov/quickfacts/fact/table/US/PST045219
30. Radford, A., Wu, J., Child, R., Luan, D., Amodei, D., Sutskever, I.: Language models are unsupervised multitask learners. OpenAI blog 1(8), 9 (2019)
31. Sanh, V., Debut, L., Chaumond, J., Wolf, T.: Distilbert, a distilled version of bert: smaller, faster, cheaper and lighter. In: NeurIPS Energy Efficient Machine Learning and Cognitive Computing Workshop (2019)
32. Shah, D.S., Schwartz, H.A., Hovy, D.: Predictive biases in natural language processing models: A conceptual framework and overview. In: Proceedings of the 58th Annual Meeting of the Association for Computational Linguistics. pp. 5248–5264. Association for Computational Linguistics (2020). https://doi.org/10.18653/v1/2020.acl-main.468https://www.aclweb.org/anthology/2020.acl-main.468
33. Sheng, E., Chang, K.W., Natarajan, P., Peng, N.: The woman worked as a babysitter: On biases in language generation. In: Proceedings of the 2019 Conference on Empirical Methods in Natural Language Processing and the 9th International Joint Conference on Natural Language Processing (EMNLP-IJCNLP), pp. 3407–3412. Association for Computational Linguistics (Nov 2019). https://doi.org/10.18653/v1/D19-1339 https://www.aclweb.org/anthology/D19-1339
34. Utama, P.A., Moosavi, N.S., Gurevych, I.: Mind the trade-off: Debiasing nlu models without degrading the in-distribution performance (2020)
35. Welch, B.L.: The generalization ofstudent's' problem when several different population variances are involved. Biometrika 34(1/2), 28–35 (1947)
36. Wheeler, R., Thomas, J.: And "still" the children suffer: The dilemma of standard english, social justice, and social access. In: JAC, pp. 363–396 (2013)
37. Xu, C., Zhou, W., Ge, T., Wei, F., Zhou, M.: BERT-of-theseus: Compressing BERT by progressive module replacing, pp. 7859–7869 (Nov 2020). https://www.aclweb.org/anthology/2020.emnlp-main.633

Fairness for Robust Learning to Rank

Omid Memarrast[1]($\boxtimes$)(iD), Ashkan Rezaei[1](iD), Rizal Fathony[2](iD),
and Brian Ziebart[1](iD)

[1] University of Illinois Chicago, Chicago, IL 60607, USA
{omemar2,arezae4,bziebart}@uic.edu
[2] Grab, Jakarta 12430, Indonesia
rizal.fathony@grab.com

Abstract. While conventional ranking systems focus solely on maximizing the utility of the ranked items to users, fairness-aware ranking systems additionally try to balance the exposure based on different protected attributes such as gender or race. To achieve this type of group fairness for ranking, we derive a new ranking system from the first principles of distributional robustness. We formulate a minimax game between a player choosing a distribution over rankings to maximize utility while satisfying fairness constraints against an adversary seeking to minimize utility while matching statistics of the training data. Rather than maximizing utility and fairness for the specific training data, this approach efficiently produces robust utility and fairness for a much broader family of distributions of rankings that include the training data. We show that our approach provides better utility for highly fair rankings than existing baseline methods.

Keywords: Learning-to-rank · Fairness · Robustness

1 Introduction

Rankings often have social implications beyond the immediate utility they provide, since higher rankings provide opportunities for individuals and groups associated with the ranked items. As a consequence, biases in ranking systems, whether intentional or not, raise ethical concerns about their long-term economic and societal harming effect. Rankings that solely maximize utility or relevance can perpetuate existing societal biases that exist in training data whilst remaining oblivious to the societal detriment they cause by amplifying such biases [21].

Conventional ranking algorithms typically produce rankings to best serve the interests of those conducting searches by ordering the items by the probability of relevance so that utility to the users will be maximized [26]. Biased outcomes drawn by these models negatively impact items in marginalized protected groups in critical decision making systems such as hiring or housing where items compete for exposure and being unfair towards one group can lead to winner-takes-all dynamics that reinforce existing disparities [27].

© The Author(s), under exclusive license to Springer Nature Switzerland AG 2023
H. Kashima et al. (Eds.): PAKDD 2023, LNAI 13935, pp. 544–556, 2023.
https://doi.org/10.1007/978-3-031-33374-3_43

Protected group definitions vary between different applications, and can include characteristics such as race, gender, religion, etc. In group fairness, algorithms divide the population into groups based on the protected attribute and guarantee the same treatment for members across groups. In ranking, this treatment can be evaluated using statistical metrics defined for measuring fairness. In this paper, we focus primarily on exposure-based group fairness measures. As a notable example, *demographic parity* (DP) in ranking is satisfied if the average exposure for both groups is equal in the top k ranks. As a motivating example, in Fig. 1 we consider two rankings based on items' true relevance and group membership. As a result of ranking 1, the highest utility is achieved, and fairness is ignored. In contrast, ranking 2 satisfies the demographic parity fairness constraint while still preserving high utility.

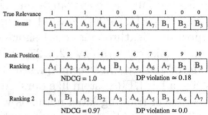

Fair ranking approaches seeking to provide group fairness properties can be categorized into *post-processing* and *in-processing* methods. *Post-processing* techniques are used to re-rank a given high utility ranking to incorporate fairness constraints while seeking to retain high utility [2,27]. These methods assume that true relevance labels are available and require other fairness-unaware learning methods (e.g., regression) to predict the true labels as a pre-processing step. Recovering from unfair regression based rankings in the re-ranking step may not be feasible in some circumstances [30].

Fig. 1. Ranking 1 ignores fairness whereas Ranking 2 satisfies the demographic parity fairness constraint while only slightly decreasing the utility.

The fair ranking problem can also be addressed as an *in-processing*, learning-to-rank (LTR) task where the algorithm learns to maximize utility subject to fairness constraints from training data [28,31]. Our algorithm falls into this category. While providing a fairness-utility trade-off, fair LTR approaches need to be robust to outliers and noisy data. For example, the label of recidivism in the COMPAS dataset is regarded to be noisy [10]. This makes prediction while incorporating fairness constraints more difficult. With improved robustness properties, a fair LTR can achieve better utility for highly fair rankings, which results in a preferable utility-fairness trade-off.

In this paper, we derive a new LTR system based on the first principles of distributional robustness to provide both fairness and robustness to label noise. We formulate a minimax game with the *ranker player* choosing a distribution over rankings constrained to satisfy fairness requirements on the training samples while maximizing utility, and an *adversary player* choosing a distribution of item relevancies that minimizes utility while being similar to training data properties. Rather than narrowly optimizing the rankings for the specific training data, this approach produces rankings that provide utility and fairness robustly for a family of distributions that includes the training data.

We show that our approach is able to trade-off between utility and fairness much better at high levels of fairness than existing baseline methods. Furthermore, the robustness properties of our approach enable it to outperform existing baselines in the presence of varying degrees of label noise in the training data. To the best of our knowledge, this is the first distributionally robust fair LTR method.

2 Related Works

Fairness in Ranking. We can broadly group existing fair ranking approaches into various categories based on their notions of fairness. Metric-based works base their fairness constraints on statistical parity for pairwise ranking across item groups [1,15,20]. Several works argue that economic opportunities (e.g., exposure, clickthroughs, etc.) should be allocated on the basis of merit, not a winner-take-all strategy [2,9,27]. While our approach falls into this category, none of the existing techniques utilizes a distributionally robust approach to derive a fair LTR system like ours. As a result their performance degrades in the presence of training label noise, as we will show in our experiments.

There have also been recent studies that focus on other aspects of fair ranking. Several works have looked at fair ranking in the presence of noisy protected attributes [19]. Another line of research aims to select individuals distributed across different groups fairly when there is implicit group bias [6,16]. Recent studies have also investigated how uncertainty about protected attributes, labels, and other features of the machine learning model affect its fairness properties [12,22]. Contrary to this line of work, [29] takes into account the presence of uncertainty when estimating merits and defining a corresponding merit-based notion of fairness.

3 Preliminary

3.1 Probabilistic Ranking

To formulate the ranking task, we consider a dataset of ranking problems $\mathcal{D} = \{\mathcal{R}^i\}_{i=1}^N$ for N different queries, where each $\mathcal{R}^i = \{d_j\}_{j=1}^M$ is a candidate item set of size M for a single query. For every item d_j in this set, we denote $rel(d_j)$ as its corresponding relevance judgment. We denote the utility of a ranking (permutation) π for a single query as $\text{Util}(\pi)$. The optimization problem can be written as: $\pi^* = \text{argmax}_{\pi \in \Pi_{\text{fair}}} \text{Util}(\pi)$. Utility measures used for rankings are based on the relevance of the individual items being ranked for a particular ranking problem, $\mathcal{R}_{|\text{query}} = \{d_j\}_{j=1}^M$. For example, the Discounted Cumulative Gain (DCG) [14], which is a common evaluation measure for ranking systems that discounts the utility for lower-ranked items,

$$DCG(\pi) = \sum_{d_j \in \mathcal{R}} \frac{2^{rel(d_j)} - 1}{\log(1 + \pi_j)} \Rightarrow \text{Util}(\pi) = \sum_{j=1}^M u_j v_{\pi_j}, \tag{1}$$

is a member of the more general family of linear utility functions $u_j = 2^{rel(d_j)} - 1$ representing the utility of a single item d_j based on its relevance $rel(.)$ and $v_k = \frac{1}{\log(1+k)}$ providing the degree of attention that item d_j receives by being placed at rank k by permutation π, i.e., $\pi_{d_j} = k$.

The space of all permutations of items is exponential in the number of items, making naïve methods that find a utility-maximizing ranking subject to fairness constraints intractable. To overcome this problem, we consider a probabilistic ranking in which instead of a single ranking, a distribution over rankings is used. We define the probability of positioning item d_j at rank k as $P_{j,k}$. Then $\mathbf{P}$ constructs a doubly stochastic matrix of size $M \times M$ where entries in each row and each column must sum up to 1. By employing the idea of probabilistic ranking, we express the ranking utility in (1) as an expected utility of a probabilistic ranking:

$$U(\mathbf{P}) = \sum_{j=1}^{M} \sum_{k=1}^{M} \mathbf{P}_{j,k} \, u_j \, v_k = \mathbf{u}^T \mathbf{P} \mathbf{v}, \qquad (2)$$

which we equivalently express in a vectorized format where $\mathbf{u}$ and $\mathbf{v}$ are both column vectors of size M. Following [27], the fair ranking optimization can be expressed as a linear programming problem:

$$\max_{\mathbf{P} \in \Delta \cap \Gamma_{\text{fair}}} \mathbf{u}^T \mathbf{P} \mathbf{v} \quad \text{where:} \quad \Delta : \mathbf{P1} = \mathbf{P}^\top \mathbf{1} = \mathbf{1}, \quad \mathbf{P}_{j,k} \geq 0, \quad \forall_{1 \leq j,k \leq M} \qquad (3)$$

and Γ_{fair} denotes any linear constraint set of the form $\mathbf{f}^\top \mathbf{P} \mathbf{g} = h$. Choosing $\mathbf{f}$ as the utility of items according to groups and $\mathbf{g}$ as the exposure of ranking position, enforces equality of exposure across protected groups. In contrast to [27], which uses this framework to re-rank the items to satisfy fairness constraints (i.e., a post-processing method), we extend this linear perspective to derive a *learning-to-rank* approach that learns to optimize utility and fairness simultaneously during training (i.e., an in-processing method).

Demographic parity of exposure, for a set of disjoint group members $G_1, \ldots,$ $G_{|S|}$, requires that: $\frac{1}{|G_s|} \sum_{d_j \in G_s} \sum_{k=1}^{M} \mathbf{P}_{j,k} v_k = \frac{1}{|G_{s'}|} \sum_{d_j \in G_{s'}} \sum_{k=1}^{M} \mathbf{P}_{j,k} v_k, \forall s,$ $s' \in S$.

In this paper, we assume binary groups and construct $\mathbf{f}_j = \frac{1_{d_j \in G_s}}{|G_s|} - \frac{1_{d_j \in G'_s}}{|G'_s|}$, which makes the constraint $\mathbf{f}^\top \mathbf{P} \mathbf{v} = 0$. For more than two groups, multiple pairwise constraints of this form can be enforced.

4   Methodology

We adopt a distributionally robust approach to the LTR problem by constructing a worst-case adversarial distribution on item utilities. We formulate the robust fair ranking construction as a minimax game between two players: a fair predictor $\mathbf{P}$ that makes a probabilistic prediction over the set of all possible rankings to maximize expected ranking utility; and an adversary $\mathbf{q}$ that approximates a probability distribution for the utility of items which minimizes the expected ranking utility. The adversary is additionally constrained to match the feature moments of the empirical training distribution. Since we solve the problem for a given query, the query-dependent terms are omitted from the formulation for simplicity.

In our notation, we represent ranking items d by their feature representation $\mathbf{X} \in \mathbb{R}^{M \times L}$ as a matrix of M items with L features. For a given item set $\mathbf{X}$, the expected ranking utility of a probabilistic ranking $\mathbf{P}$ against a utility distribution $\mathbf{q}$ can be expressed as:

$$U(\mathbf{X}, \mathbf{P}, \mathbf{q}) = \sum_{j=1}^{M} \mathbb{E}_{u_j | \mathbf{x} \sim \mathbf{q}} \left[u_j \mathbb{E}_{\pi_j | \mathbf{X} \sim \mathbf{P}} \left[v_{\pi_j} \right] \right]. \tag{4}$$

Then, the utility-maximizing optimization problem under fairness constraints can be formulated as:

Definition 1. *Given a training dataset of N ranking problems $\mathcal{D} = \{(\mathbf{X}^i, \mathbf{u}^i)\}_{i=1}^{N}$, with $\mathbf{u} \in \mathbb{R}^M$ being the true relevance and $\mathbf{X} \in \mathbb{R}^{M \times L}$ the feature representation of ranking problem of size M. The fair probabilistic ranking $\mathbf{P}(\pi) \in \mathbb{R}^{M \times M}$ in adversarial learning-to-rank learns a fair ranking that maximizes the worst-case ranking utility approximated by an adversary $\mathbf{q}(\breve{u})$, constrained to match the feature statistics of the training data.*

$$\max_{\mathbf{P}(\pi | \mathbf{X}) \in \Delta \cap \Gamma_{fair}} \min_{\mathbf{q}(\breve{u} | \mathbf{X})} \mathbb{E}_{\mathbf{X} \sim \widetilde{P}} \left[U(\mathbf{X}, \mathbf{P}, \mathbf{q}) \right] \tag{5}$$

$$s.t. \ \mathbb{E}_{\mathbf{X} \sim \widetilde{P}} \left[\sum_{j=1}^{M} \mathbb{E}_{\breve{u}_j | \mathbf{X} \sim \mathbf{q}} \left[\breve{u}_j \mathbf{X}_{j,:} \right] \right] = \mathbb{E}_{\mathbf{X}, \mathbf{u} \sim \widetilde{P}} \left[\sum_{j=1}^{M} u_j \mathbf{X}_{j,:} \right] \tag{6}$$

where $\widetilde{P}$ denotes the empirical distribution over ranking dataset $\mathcal{D} = \{(\mathbf{X}^i, \mathbf{u}^i)\}_{i=1}^{N}$, $\breve{u}$ denotes the random variable for adversary relevance, and Δ denotes the set of doubly stochastic matrices.

This general adversarial formulation plays a foundational role in constructing probability models and prediction techniques [11,13]. This approach has been utilized to provide fair and robust predictions under covariate shift [25] as well as for constructing reliable predictors for fair log loss classification [24]. Similar to this line of work, our proposed approach imposes fairness constraints on predictor $\mathbf{P}$. Our formulation in Definition 1 accepts generic utility values. In our paper, we focus on binary utility, which is one of the common applications of the ranking problem, where the utility label indicates if a particular item is relevant or not. For the binary utility problem, the expected utility can be further simplified as:

$$U(\mathbf{X}, \mathbf{P}, \mathbf{q}) = \sum_{j=1}^{M} \mathbb{E}_{u_j | \mathbf{x} \sim \mathbf{q}} \left[u_j \mathbb{E}_{\pi_j | \mathbf{X} \sim \mathbf{P}} \left[v_{\pi_j} \right] \right] = \sum_{j=1}^{M} \sum_{k=1}^{M} \mathbf{q}(u_j = 1 | \mathbf{X}) \mathbf{P}(\pi_j = k | \mathbf{X}) v_k = \mathbf{q}^\top \mathbf{P} \mathbf{v},$$

where the entries in the vector $\mathbf{q}$ contains the relevance probability of item d_j. In the following sections, we use this vector notation to simplify the optimization formulation.

5 Optimization

We solve the constrained minimax formulation in Definition 1 in Lagrangian dual form, where we optimize the dual parameters $\theta \in \mathbb{R}^{L \times 1}$ for the feature matching constraint of L features by gradient descent. Rewriting the optimization in matrix notation yields:

$$\max_{\theta} \mathbb{E}_{\mathbf{x},\mathbf{u} \sim \tilde{P}} \left[\max_{\mathbf{P} \in \Delta} \min_{0 \leq \mathbf{q} \leq 1} \ \mathbf{q}^{\top} \mathbf{P} \mathbf{v} + \left\langle \mathbf{q} - \mathbf{u}, \sum_{l} \theta_l \mathbf{X}_{:,l} \right\rangle \right] \quad \text{s.t.} \quad \mathbf{f}^{\top} \mathbf{P} \mathbf{v} = 0, \quad (7)$$

where: $\mathbf{P}(\pi) \in \mathbb{R}^{M \times M}$ is a doubly stochastic matrix, and the value of cell $\mathbf{P}_{j,k}$ represents the probability that $\pi_j = k$; $\mathbf{u} \in \mathbb{R}^{M \times 1}$ is a vector of true labels whose j^{th} values is 1 when the item j is relevant to the query, i.e., $u_j = 1$ and 0 otherwise; $\mathbf{q} \in \mathbb{R}^{M \times 1}$ is a probability vector of the adversary's estimation of each item being relevant; $\mathbf{X}_{:,l} \in \mathbb{R}^{M \times 1}$ denotes the lth feature of M samples; S is the set of protected attributes; and $\mathbf{v} \in \mathbb{R}^{M \times 1}$ is a vector containing the values of position bias function for each position. To denote the Frobenius inner product between two matrices $\langle ., . \rangle$ is used, i.e., $\langle A, B \rangle = \sum_{i,j} A_{i,j} B_{i,j}$.

For optimization purposes, using strong duality, we push the maximization over $\mathbf{q}$ to the outermost level in (7). Since the objective is non-smooth, for both $\mathbf{P}$ and $\mathbf{q}$, we add strongly convex prox-functions to make the objective smooth. Furthermore, to make our approach handle feature sampling error, we add a regularization penalty to the parameter θ. To apply (7) on training data, we replace empirical expectation with an average over all training samples. The new formulation is as follows:

$$\min_{\{0 \leq \mathbf{q}^i \leq 1\}_{i=1}^{N}} \max_{\theta} \frac{1}{N} \sum_{i=1}^{N} \max_{\mathbf{P}^i \in \Delta} \left[\mathbf{q}^{i\top} \mathbf{P}^i \mathbf{v}^i - \left\langle \mathbf{q}^i - \mathbf{u}^i, \sum_{l} \theta_l \mathbf{X}_{:,l}^i \right\rangle \right.$$
$$\left. + \lambda \mathbf{f}^{i\top} \mathbf{P}^i \mathbf{v}^i - \frac{\mu}{2} \left\| \mathbf{P}^i \right\|_F^2 + \frac{\mu}{2} \left\| \mathbf{q}^i \right\|_2^2 \right] - \frac{\gamma}{2} \|\theta\|_2^2, \quad (8)$$

where superscript i is the ith sample from N ranking problems in the training set. We denote λ, γ and μ as the fairness penalty parameter (which can be adjusted to obtain different trade-offs between fairness and utility, rather than strictly optimized), a regularization penalty parameter and a smoothing penalty parameter, respectively. The inner minimization over $\mathbf{P}$ and θ can be solved separately, given a fixed $\mathbf{q}$. The minimization over θ has a closed-form solution where the lth element of θ^* is:

$$\theta_l^* = -\frac{1}{\gamma N} \sum_{i=1}^{N} \left\langle \mathbf{q}^i - \mathbf{u}^i, \mathbf{X}_{:,l}^i \right\rangle. \quad (9)$$

Independently from θ, we can solve the inner minimization over $\mathbf{P}$ for every training sample using a projection technique. The optimal $\mathbf{P}$ for i^{th} training sample (i.e., $\mathbf{P}^{i^*}$) is:

$$\mathbf{P}^{i^*} = \underset{\mathbf{P}^i \in \Delta}{\arg\max} \ {\mathbf{q}^i}^\top \mathbf{P}^i \mathbf{v}^i + \lambda {\mathbf{f}^i}^\top \mathbf{P}^i \mathbf{v}^i - \frac{\mu}{2} \left\| \mathbf{P}^i \right\|_F^2$$

$$\mathbf{P}^{i^*} = \underset{\mathbf{P}^i \in \Delta}{\arg\min} \ \frac{\mu}{2} \left\| \mathbf{P}^i - \frac{1}{\mu}(\mathbf{q}^i + \lambda \mathbf{f}^i){\mathbf{v}^i}^\top \right\|_F^2 - \frac{1}{2\mu} \left\| \mathbf{q}^i {\mathbf{v}^i}^\top \right\|_F^2. \qquad (10)$$

As derived in (10), the minimization takes the form of $\min_{\mathbf{P} \geq 0} \|\mathbf{P} - \mathbf{R}\|_F^2$, and we can interpret this minimization as projecting matrix $\frac{1}{\mu}(\mathbf{q}^i + \lambda \mathbf{f})\mathbf{v}^{i^\top}$ into the set of doubly-stochastic matrices. The projection from an arbitrary matrix $\mathbf{R}$ to the set of doubly-stochastic matrices can be solved using the ADMM projection algorithm [3]. Since each entry in $\mathbf{q}$ represents a probability, the outer optimization over $\mathbf{q}$ is solved using the L-BFGS-B algorithm with a bounded constraint of the probability simplex [4]. The algorithm optimizes the quadratic approximation of the objective function (using limited memory Quasi-Newton) on the convex set with each iteration. In each update step, a projection to the probability simplex is needed. Based on the above optimization, the adversary's optimal relevance probability $\mathbf{q}^*$ can be obtained. Following (9) we compute the θ^* over the optimal $\mathbf{q}^*$. Algorithm 1 shows the steps for training.

5.1 Inference and Runtime Analysis

For prediction, we use θ and μ learned from training data while performing the optimization in (8). After removing the constant terms, we solve a similar optimization problem for test data. That is:

Algorithm 1: The Fair-Robust LTR

Input: Training dataset $\mathcal{D} = \{(\mathbf{X}^i, \mathbf{u}^i)\}_{i=1}^N$, fairness penalty parameter λ.
Output: $\theta^*, \mathbf{P}^*, \mathbf{q}^*$
$\mathbf{q} \leftarrow$ random initialization;
repeat
  update θ by (9) with $\mathbf{q}$.
  update $\mathbf{P}$ by (10) with $\mathbf{q}$.
  update $\mathbf{q}$ by (8) with $\{\mathbf{P}, \theta\}$.
until *convergence*;

$$\underset{\{0 \leq \mathbf{q}^i \leq 1\}_{i=1}^{N^{\text{test}}}}{\min} \frac{1}{N^{\text{test}}} \sum_{i=1}^{N^{\text{test}}} \underset{\mathbf{P}^i \in \Delta}{\max} \left[{\mathbf{q}^i}^\top \mathbf{P}^i \mathbf{v}^i - \left\langle \mathbf{q}^i, \sum_l \theta_l^* \mathbf{X}_{:,l}^i \right\rangle + \lambda \mathbf{P}^i \mathbf{v}^i - \frac{\mu}{2} \left\| \mathbf{P}^i \right\|_F^2 + \frac{\mu}{2} \left\| \mathbf{q}^i \right\|_2^2 \right],$$

where superscript i pertains to the ith ranking problem in the test set of size N^{test}. We follow the steps for solving the optimization in training. There is no gradient learning of θ as in training, and true relevance labels ($\mathbf{u}$) are not used in inference. After convergence, we use the resulting $\mathbf{P}^*$ from the optimization to predict the ranking of items in the test set. We employ the Hungarian algorithm [17] to solve the problem of matching items to positions.

Runtime Analysis. Solving optimization in (8) involves running a projected gradient descent algorithm. In each iteration, it requires the computation of the gradient and the projection to box constraints. The box constraint projection's runtime is linear in terms of the number of variables, hence costing $\mathcal{O}(NM)$. The gradient computation requires solving for θ^*, which costs $\mathcal{O}(NML)$ from the dot product computations; and solving for $\mathbf{P}^*$, which can be posed as a doubly-stochastic matrix projection. We employ an ADMM algorithm to perform the projection to doubly stochastic matrix, which has linear convergence due to the strong convexity of the objective [7]. Each step inside the ADMM consists of M projections to M-element simplex, hence costing $\mathcal{O}(M^2)$ computations in total.

6 Experiments

In order to compare our proposed framework with existing fair LTR solutions, we use simulated and real-world datasets to carry out in-depth empirical evaluations. The learning task is to determine the feature function in the training based on the items' ground truth utilities and fairness constraints. At testing time, this feature function coupled with a penalty for fairness violation is used to determine the ranking for the items in the test set with maximum utility while satisfying fairness constraints.

6.1 Fairness Benchmark Datasets

Setup. We follow steps discussed in [28] to adapt German, Adult and COMPAS datasets to a LTR task. These datasets are inherently biased, making them viable alternatives for evaluation when no real world datasets exist for a fair LTR task. First, we split each dataset randomly into a disjoint train and test set. Then from each train/test set we construct a corresponding LTR train/test set. For each query, we sample randomly with replacement a set of 10 candidates each, representative of both relevant and irrelevant items, where on average four individuals are relevant. Each individual in the candidate set is a member of a group G_s based on its protected attribute. The training data consists of 500 ranking problems. We evaluate our learned model on 100 separate ranking problems serving as the test set. We repeat this process 10 times and report the 95% confidence interval in the results. The regularization constant γ and smoothing penalty parameter μ in (8) are chosen by 3-fold cross validation. We describe datasets used in our experiments:

- UCI `Adult`, census income dataset [8]. The goal is to predict whether income is above \$50K/yr on the basis of census results.
- The `COMPAS` criminal recidivism risk assessment dataset [18] is designed to predict whether a defendant is likely to reoffend based on criminal history.

Table 1. Dataset characteristics.

Dataset	n	Features	Attribute
Adult	45,222	12	Gender
COMPAS	6,167	10	Race
German	1,000	20	Gender

– UCI German dataset [8]. Based on personal information and credit history, the goal is to classify good and bad credit.

Table 1 shows the statistics of each dataset with their protected attributes.

Baseline Methods. To evaluate the performance of our model, we compare it against three different baselines that have similarities to and differences from our model: FAIR-PGRank [28] and DELTR [31] are in-processing, LTR methods, like ours; the Post-Processing method of [27] employs the fairness constraint formulation that we build our optimization framework based on. We also add a Random baseline that ranks items in each query randomly to give context to NDCG. We discuss baseline methods in more details[1]:

– **Post Processing** (POST-PROC) [27] To make a fair comparison with LTR approaches, we first learn a linear regression model using all query-item sets in the training data and predict the relevance of an item to a query in test set. Then, these estimated relevances are used as input to the linear program optimization described in [27] with a demographic parity constraint.
– **Fair Policy Ranking** (FAIR-PGRANK) [28] An end-to-end, in-processing LTR approach that uses a policy gradient method, directly optimizing for both utility and fairness measures.
– **Reducing Disparate Exposure** (DELTR) [31] An in-processing LTR method optimizing a weighted sum of a loss function and a fairness criterion. The loss function is a cross entropy designed for ranking [5] and fairness objective is a squared hinge loss based on disparate exposure.

Evaluation Metrics. We use the *normalized discounted cumulative gain* (NDCG) [14], as the utility measure. This is defined as: $NDCG(\pi) = DCG(\pi)/Z$, where Z is the DCG for ideal ranking and is used to normalize the ranking so that a perfect ranking would give a NDCG score of 1.

For the fairness evaluation in our approach we use *demographic parity* as our fairness violation metric which is based on disparity of average exposure across two groups:

$$\hat{D}_{group}(\mathbf{P}) = |\text{Ex}(G_0|\mathbf{P}) - \text{Ex}(G_1|\mathbf{P})|. \tag{11}$$

Results. Figure 2 shows the performance of our model (FAIR-ROBUST) against baselines on the three benchmark datasets. We observe a trade-off between fairness and utility in both FAIR-PGRANK and FAIR-ROBUST, i.e., as we increase the fairness penalty parameter (λ), demographic parity difference (as a measure of fairness violation) and NDCG both drop. While DELTR and POST-PROC achieve comparable NDCG when $\lambda = 0$, they fail to satisfy demographic parity

---

[1] We use the implementation from https://github.com/ashudeep/Fair-PGRank for all baselines.

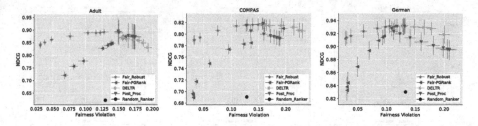

Fig. 2. Average *NDCG* versus average *difference of demographic parity* (DP) on test samples, for increasing degrees of fairness penalty λ in each method. FAIR-ROBUST: $\lambda \in [0, 20]$, FAIR-PGRANK: $\lambda \in [0, 20]$, DELTR: $\lambda \in [0, 10^6]$, POST-PROC: $\lambda \in [0, 0.2]$.

as we increase λ and are unable to provide a sufficient utility-fairness trade-off when high levels of fairness are desired.

In all three datasets, FAIR-ROBUST outperforms FAIR-PGRANK in terms of ranking utility when fairness is a priority. When comparing the utility-fairness trade-off between the two approaches, we observe that FAIR-ROBUST can retain higher NDCG in high levels of fairness and provides a preferable trade-off. One notable point is that, even in a noisy dataset like the COMPAS dataset, our approach performs better than other methods due to its robustness.

Robustness Test. One key benefit of our approach is its robustness to label noise in the learning process. This allows our method can be trained on data with noisy labels and outliers, and still perform well on the test data. To test this property, we repeat the previous experiment with noise added to the training data. After sampling rankings for the training and test sets, we randomly flip $x\%$ of the labels in each ranking problem in the training set. In our experiments, we test various amounts of noise in the training data where x can be 20%, 30%, or 40%. Figure 3 shows the results for robustness test. Similar to the previous experiment, we observe a trade-off between fairness and utility for FAIR-ROBUST. As the amount of the noise increases FAIR-PGRANK performs poorly and can't maintain its trade-off. Note that when $\lambda = 0$, FAIR-PGRANK still performs well but for other values of λ its NDCG gets close to random ranking.

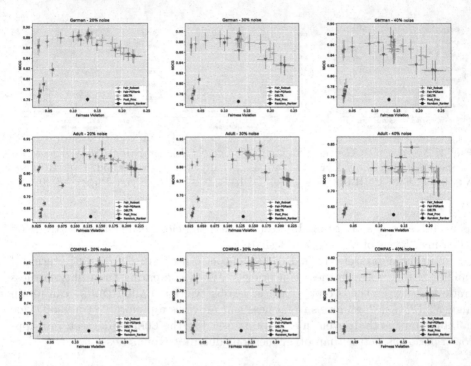

Fig. 3. Robustness test on `German`, `Adult` and `COMPAS` datasets with varying degrees of noise in the training data.

6.2 Microsoft Learning to Rank Dataset

Setup. In the previous experiments, we used datasets with inherent demographic biases but the LTR tasks were simulated and constructed from a classification task. In this experiment, we evaluate its performance on Microsoft's Learning to Rank dataset [23] which is a real world LTR dataset. We follow the steps discribed in [30] to pre-process the dataset. We compare our method to FAIR-PGRANK, as both methods are able to trade-off between fairness and utility. Additionally, we include a random baseline, which sorts each item in a query randomly, to give context to NDCG. Similar to the previous experiments, we use NDCG as the utility measure and *demo-*

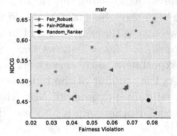

Fig. 4. *NDCG* versus *difference of demographic parity* for increasing degrees of fairness penalty λ in each method.

graphic parity as our fairness violation metric, which is based on the disparity of average exposure across two groups.

Results. Figure 4 shows the fairness and accuracy trade-off on the test set. With large fairness regularization, FAIR-PGRANK drops below a random ranking in terms of NDCG, making it inconsistant. This plot shows that FAIR-ROBUST smoothly trades-off group fairness for NDCG. FAIR-PGRANK's NDCG and group exposure, on the other hand, deteriorate for increasing regularization strength, as [30] also observed.

7 Conclusions

In this paper, we developed a new LTR system that achieves fairness of exposure for protected groups while maximizing utility to the users. We show that our method is able to trade-off between utility and fairness much better at high levels of fairness than existing baseline methods. Our work addresses the problem of providing more robust fairness given a chosen fairness criterion, but does not answer the broader question of which fairness criterion is appropriate for a particular ranking application. More extensive evaluations based on incorporating other fairness metrics, such as disparate treatment, and generalization of this approach beyond binary utility are two important future directions.

Acknowledgements. This work was supported by the National Science Foundation Program on Fairness in AI in collaboration with Amazon under award No. 1939743.

References

1. Beutel, A., et al.: Fairness in recommendation ranking through pairwise comparisons. In: ACM SIGKDD International Conference on Knowledge Discovery & Data Mining(KDD), pp. 2212–2220 (2019)
2. Biega, A.J., Gummadi, K.P., Weikum, G.: Equity of attention: Amortizing individual fairness in rankings. In: The 41st International ACM Sigir Conference on Research and Development in Information Retrieval, pp. 405–414 (2018)
3. Boyd, S., Parikh, N., Chu, E.: Distributed optimization and statistical learning via the alternating direction method of multipliers. Now Publishers Inc (2011)
4. Byrd, R.H., Lu, P., Nocedal, J., Zhu, C.: A limited memory algorithm for bound constrained optimization. SIAM J. Sci. Comput. **16**(5), 1190–1208 (1995)
5. Cao, Z., Qin, T., Liu, T.Y., Tsai, M.F., Li, H.: Learning to rank: from pairwise approach to listwise approach. In: Proceedings of the 24th International Conference on Machine Learning, pp. 129–136 (2007)
6. Celis, L.E., Mehrotra, A., Vishnoi, N.K.: Interventions for ranking in the presence of implicit bias. In: Proceedings of the 2020 Conference on Fairness, Accountability, and Transparency, pp. 369–380 (2020)
7. Deng, W., Yin, W.: On the global and linear convergence of the generalized alternating direction method of multipliers. J. Sci. Comput. **66**(3) (2016)
8. Dheeru, D., Karra Taniskidou, E.: UCI machine learning repository (2017). http://archive.ics.uci.edu/ml
9. Diaz, F., Mitra, B., Ekstrand, M.D., Biega, A.J., Carterette, B.: Evaluating stochastic rankings with expected exposure. In: Proceedings of the 29th ACM International Conference on Information & Knowledge Management, pp. 275–284 (2020)

10. Eckhouse, L.: Big data may be reinforcing racial bias in the criminal justice system. The Washington Post (2017)
11. Fathony, R., Liu, A., Asif, K., Ziebart, B.: Adversarial multiclass classification: A risk minimization perspective. In: NeurIPS (2016)
12. Ghosh, A., Dutt, R., Wilson, C.: When fair ranking meets uncertain inference. arXiv preprint arXiv:2105.02091 (2021)
13. Grünwald, P.D., Dawid, A.P.: Game theory, maximum entropy, minimum discrepancy, and robust Bayesian decision theory. Ann. Stat. **32**, 1367–1433 (2004)
14. Järvelin, K., Kekäläinen, J.: Cumulated gain-based evaluation of ir techniques. ACM Trans. Inform. Syst. (TOIS) **20**(4), 422–446 (2002)
15. Kallus, N., Zhou, A.: The fairness of risk scores beyond classification: Bipartite ranking and the xauc metric. In: Advances in Neural Information Processing Systems, pp. 3438–3448 (2019)
16. Kleinberg, J., Raghavan, M.: Selection problems in the presence of implicit bias. In: 9th Innovations in Theoretical Computer Science Conference (ITCS 2018) (2018)
17. Kuhn, H.W.: The hungarian method for the assignment problem. Naval Res. Logist. Quart. **2**(1–2), 83–97 (1955)
18. Larson, J., Mattu, S., Kirchner, L., Angwin, J.: How we analyzed the compas recidivism algorithm. ProPublica 9 (2016)
19. Mehrotra, A., Celis, L.E.: Mitigating bias in set selection with noisy protected attributes. In: Proceedings of the 2021 ACM Conference on Fairness, Accountability, and Transparency, pp. 237–248 (2021)
20. Narasimhan, H., Cotter, A., Gupta, M., Wang, S.: Pairwise fairness for ranking and regression. In: Proceedings of the AAAI Conference on Artificial Intelligence. vol. 34, pp. 5248–5255 (2020)
21. O'Neil, C.: Weapons of math destruction: How big data increases inequality and threatens democracy. Broadway Books (2016)
22. Prost, F., et al.: Measuring model fairness under noisy covariates: A theoretical perspective. arXiv preprint arXiv:2105.09985 (2021)
23. Qin, T., Liu, T.Y.: Introducing letor 4.0 datasets. arXiv preprint arXiv:1306.2597 (2013)
24. Rezaei, A., Fathony, R., Memarrast, O., Ziebart, B.: Fairness for robust log loss classification. In: Proceedings of the AAAI Conference on Artificial Intelligence. vol. 34, pp. 5511–5518 (2020)
25. Rezaei, A., Liu, A., Memarrast, O., Ziebart, B.D.: Robust fairness under covariate shift. In: Proceedings of the AAAI Conference on Artificial Intelligence. vol. 35, pp. 9419–9427 (2021)
26. Robertson, S.E.: The probability ranking principle in ir. J. Document. **33**(4), 294–304 (1977)
27. Singh, A., Joachims, T.: Fairness of exposure in rankings. In: Proceedings of the 24th ACM SIGKDD International Conference on Knowledge Discovery & Data Mining, pp. 2219–2228. ACM (2018)
28. Singh, A., Joachims, T.: Policy learning for fairness in ranking. Adv. Neural. Inf. Process. Syst. **32**, 5426–5436 (2019)
29. Singh, A., Kempe, D., Joachims, T.: Fairness in ranking under uncertainty. In: Advances in Neural Information Processing Systems, p. 34 (2021)
30. Yadav, H., Du, Z., Joachims, T.: Policy-gradient training of fair and unbiased ranking functions. In: Proceedings of the 44th International ACM SIGIR Conference on Research and Development in Information Retrieval, pp. 1044–1053 (2021)
31. Zehlike, M., Castillo, C.: Reducing disparate exposure in ranking: A learning to rank approach. In: Proceedings of The Web Conference 2020, pp. 2849–2855 (2020)

Author Index

Printed in the United States
by Baker & Taylor Publisher Services